CYCLIC-NUCLEOTIDE PHOSPHODIESTERASES IN THE CENTRAL NERVOUS SYSTEM

Wiley Series in Drug Discovery and Development

Binghe Wang, Series Editor

CYCLIC-NUCLEOTIDE PHOSPHODIESTERASES IN THE CENTRAL NERVOUS SYSTEM
From Biology to Drug Discovery

Edited by

NICHOLAS J. BRANDON
AstraZeneca Neuroscience
Cambridge, Massachusetts, USA

ANTHONY R. WEST
Rosalind Franklin University of Medicine and Science
North Chicago, Illinois, USA

Library of Congress Cataloging-in-Publication Data:
Cyclic-nucleotide phosphodiesterases in the central nervous system : from biology to drug discovery /
edited by Nicholas J. Brandon, Anthony R. West.
    p. ; cm.
Includes bibliographical references.
ISBN 978-0-470-56668-8 (cloth)
I. Brandon, Nicholas J., editor of compilation. II. West, Anthony R., 1970- editor of compilation
[DNLM: 1. 3',5'-Cyclic-AMP Phosphodiesterases–metabolism. 2. 3',5'-Cyclic-AMP
Phosphodiesterases–therapeutic use. 3. Central Nervous System–physiology. 4. Central Nervous
System Diseases–drug therapy. 5. Drug Discovery. QU 136]
    QP370
    612.8'2–dc23
                                                                        2013042742

CONTENTS

PREFACE

Cyclic-nucleotide phosphodiesterases (PDEs) are critically involved in the regulation of cellular processes at work from cell birth to death. PDEs are produced by and operate within all cells of the body, and their key role in dampening or redirecting cyclic adenosine monophosphate (cAMP) and cyclic guanosine monophosphate (cGMP) signaling cascades makes them essential for cell health. In both the brain and spinal cord, PDEs show intricate patterns of cellular localization, both regionally and at the subcellular level. Such an infrastructure undoubtedly contributes to the tremendous computational power needed for the effective execution of sensorimotor, cognitive, and affective functions. On the flipside, we are entering a period of time when diseases of the central nervous system (CNS), which disrupt these essential functions, will affect more and more of us. For reasons described in this book, it is now clear that PDEs have enormous potential as targets for new medicines. In this book we have brought together the expertise of leading researchers from both basic and applied sciences to highlight the beautiful biology of the diverse superfamily of PDEs, as well as the medical potential of targeting PDEs for the treatment of disorders of the CNS. Indeed, numerous applications for small-molecule inhibitors selective for specific PDE isoforms are being investigated for the treatment of CNS diseases, including schizophrenia, depression, Alzheimer's disease, Parkinson's disease, Huntington's disease, spinal cord injury, and others. Drug discovery for disorders of the CNS is exceptionally difficult, but undoubtedly, our understanding of PDE biology and PDE-based therapeutics will continue to evolve and hopefully lead to the development of novel medicines of value for patients suffering from these devastating disorders.

We thank all of our wonderful colleagues who have contributed chapters to this book, as well as the numerous reviewers who have provided constructive criticism of its content. We hope that this work will render important insights into PDE biology and therapeutics that will inspire a new generation of researchers interested in this field.

NICHOLAS J. BRANDON
ANTHONY R. WEST

CONTRIBUTORS

EVA P.P. BOLLEN, Department of Psychiatry and Neuropsychology, School for Mental Health and Neuroscience, Maastricht University, Maastricht, The Netherlands

NICHOLAS J. BRANDON, AstraZeneca Neuroscience iMED, Cambridge, MA, USA

ERIK I. CHARYCH, Lundbeck Research USA, Paramus, NJ, USA

MARCO CONTI, Center for Reproductive Sciences, Department of Obstetrics, Gynecology and Reproductive Sciences, School of Medicine, University of California, San Francisco, San Francisco, CA, USA

MARIE T. FILBIN, Department of Biological Sciences, Hunter College, City University of New York, New York, NY, USA

JOSEPH P. HENDRICK, Intra-Cellular Therapies Inc., New York, NY, USA

YINGCHUN HUANG, Department of Biochemistry and Biophysics and Lineberger Comprehensive Cancer Center, The University of North Carolina, Chapel Hill, NC, USA; Biomedical and Pharmaceutical Department, Biochemical Engineering College, Beijing Union University, Beijing, China

HENGMING KE, Department of Biochemistry and Biophysics and Lineberger Comprehensive Cancer Center, The University of North Carolina, Chapel Hill, NC, USA

MICHY P. KELLY, Department of Pharmacology, Physiology & Neuroscience, University of South Carolina School of Medicine, Columbia, SC, USA

FRANK S. MENNITI, Mnemosyne Pharmaceuticals, Inc., Providence, RI, USA

ELENA NIKULINA, The Feinstein Institute for Medical Research, Manhasset, NY, USA

AKINORI NISHI, Department of Pharmacology, Kurume University School of Medicine, Fukuoka, Japan

JAMES O'DONNELL, School of Pharmacy & Pharmaceutical Sciences, The State University of New York at Buffalo, Buffalo, NY, USA

NIELS PLATH, H. Lundbeck A/S, Valby, Denmark

JOS PRICKAERTS, Department of Psychiatry and Neuropsychology, School for Mental Health and Neuroscience, Maastricht University, Maastricht, The Netherlands

OLGA A.H. RENEERKENS, Department of Psychiatry and Neuropsychology, School for Mental Health and Neuroscience, Maastricht University, Maastricht, The Netherlands

WITO RICHTER, Center for Reproductive Sciences, Department of Obstetrics, Gynecology and Reproductive Sciences, School of Medicine, University of California, San Francisco, San Francisco, CA, USA

DAVID P. ROTELLA, Department of Chemistry and Biochemistry, Montclair State University, Montclair, NJ, USA

KRIS RUTTEN, Department of Psychiatry and Neuropsychology, School for Mental Health and Neuroscience, Maastricht University, Maastricht, The Netherlands

AKIRA SAWA, Department of Psychiatry, Johns Hopkins University School of Medicine, Baltimore, MD, USA

CHRISTOPHER J. SCHMIDT, Pfizer Global Research and Development, Neuroscience Research Unit, Cambridge, MA, USA

GRETCHEN L. SNYDER, Intra-Cellular Therapies Inc., New York, NY, USA

ALESSANDRA STANGHERLIN, Wellcome Trust-MRC Institute of Metabolic Science, Addenbrooke's Hospital, University of Cambridge, Cambridge, UK

HARRY W.M. STEINBUSCH, Department of Psychiatry and Neuropsychology, School for Mental Health and Neuroscience, Maastricht University, Maastricht, The Netherlands

NIELS SVENSTRUP, H. Lundbeck A/S, Valby, Denmark

SARAH THRELFELL, Department of Physiology, Anatomy and Genetics, University of Oxford, Oxford, UK

HUANCHEN WANG, Department of Biochemistry and Biophysics and Lineberger Comprehensive Cancer Center, The University of North Carolina, Chapel Hill, NC, USA

ANTHONY R. WEST, Department of Neuroscience, Rosalind Franklin University of Medicine and Science, North Chicago, IL, USA

YING XU, School of Pharmacy & Pharmaceutical Sciences, The State University of New York at Buffalo, Buffalo, NY, USA

MENGCHUN YE, Department of Biochemistry and Biophysics and Lineberger Comprehensive Cancer Center, The University of North Carolina, Chapel Hill, NC, USA

MANUELA ZACCOLO, Department of Physiology, Anatomy and Genetics, University of Oxford, Oxford, UK

HAN-TING ZHANG, Departments of Behavioral Medicine & Psychiatry and Physiology & Pharmacology, West Virginia University Health Sciences Center, Morgantown, WV, USA

SANDRA P. ZOUBOVSKY, Department of Psychiatry, Johns Hopkins University School of Medicine, Baltimore, MD, USA

CHAPTER 1

PHOSPHODIESTERASES AND CYCLIC NUCLEOTIDE SIGNALING IN THE CNS

MARCO CONTI and WITO RICHTER

Center for Reproductive Sciences, Department of Obstetrics, Gynecology and Reproductive Sciences, School of Medicine, University of California, San Francisco, San Francisco, CA, USA

INTRODUCTION

Discovery of PDEs, Historical Perspectives, and Progress in Understanding the Complexity of PDE Functions

Soon after the discovery of the second messenger cAMP by Sutherland and Rall [1], it was observed that cyclic nucleotides are unstable in tissue extracts. This observation paved the way for the identification of the enzymatic activities responsible for their destruction [1]. Sutherland and coworkers correctly attributed this activity to a Mg^{2+}-dependent, methylxanthine-inhibited enzyme that cleaves the cyclic nucleotide phosphodiester bond at the 3′-position, hence the name phosphodiesterase (PDE) (Figure 1.1). With the discovery of cGMP and the improvement of protein separation protocols [2], it also became apparent that multiple PDE isoforms with different affinities for cAMP and cGMP and sensitivity to inhibitors coexist in a cell (Figure 1.1). Only with the application of protein sequencing and molecular cloning techniques has it been realized that 21 genes code for PDEs in humans and that close to 100 proteins are derived from these genes, forming a highly heterogeneous superfamily of enzymes (Figures 1.1 and 1.2) [3].

Although PDEs were implicated early on in the control of intracellular levels of cAMP and cGMP and the termination of the neurotransmitter or hormonal signal, 30 additional years of research have been necessary to understand that PDEs are not simply housekeeping enzymes. The activity of PDEs is finely regulated by a myriad of regulatory loops and integrated in a complex fashion with the cyclic nucleotide signaling machinery and other signaling pathways. Blockade of PDE activity does not

Cyclic-Nucleotide Phosphodiesterases in the Central Nervous System: From Biology to Drug Discovery, First Edition. Edited by Nicholas J. Brandon and Anthony R. West.
© 2014 John Wiley & Sons, Inc. Published 2014 by John Wiley & Sons, Inc.

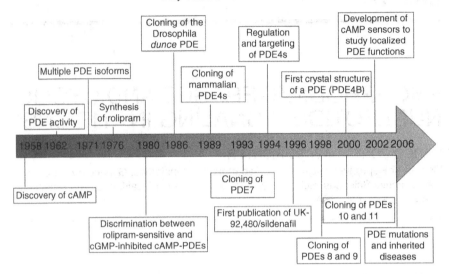

FIGURE 1.1 Timeline of the major discoveries related to the field of phosphodiesterases.

exclusively lead to an increase in cyclic nucleotides and a gain of function, as one would predict from the removal of cyclic nucleotide degradation. On the contrary, complex changes in cellular responses are associated with PDE inhibition, often causing loss of function, as documented by the phenotypes of natural mutations or engineered inactivation of the PDE genes [4–7]. These findings imply that PDEs and their regulation are indispensable to faithfully translate extracellular cues into appropriate biological responses. Indeed, in neurons as in other cells, the biological outcome of activation of a receptor is defined by the multiple dimensions of the cyclic nucleotide signal. This specificity of the response depends on the changes in concentration of the cyclic nucleotide, the time frame in which these changes occur, and the subcellular locale in which the nucleotides accumulate. Because cyclic nucleotide accumulation is dependent on the steady state of cAMP/cGMP production as well as hydrolysis, degradation by PDEs is a major determining factor of all three dimensions of the cyclic nucleotide signal.

In spite of seemingly comparable enzymatic functions, each of the several PDEs expressed within a cell appears to serve unique roles. This view is paradoxical because it implies, as fittingly summarized by L.L. Brunton, that "Not all cAMP has access to all cellular PDEs" [8]. As an extension of this concept, a PDE may play critical functions in a cell even if it represents a minor fraction of the overall hydrolytic activity, a view with considerable impact on pharmacological strategies targeting PDEs. The discovery of macromolecular complexes involving PDEs has confirmed this concept and added a new dimension to the function of these enzymes in signaling. In those complexes in which they are associated with cyclic nucleotide targets, it is likely that PDEs play an essential role in controlling or limiting the access of cyclic

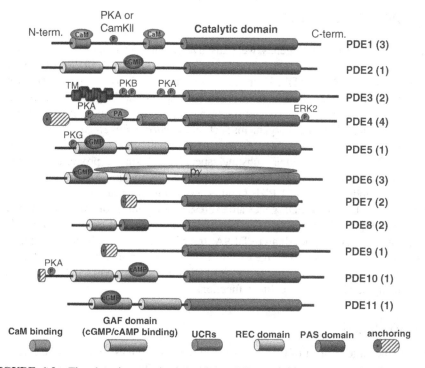

FIGURE 1.2 The domain organization of the different families of phosphodiesterases. Domains are depicted as "barrels" connected by "wires" indicating linker regions. Phosphorylation sites are shown as red circles with the respective kinase phosphorylating this site listed above. PDEs are composed of a C-terminal catalytic domain (shown in red) and distinct regulatory domains at the N-terminus. These include Ca^{2+}/calmodulin (CaM)-binding sites (PDE1), GAF domains that function as cAMP or cGMP sensors (PDE2, PDE5, PDE6, PDE10, and PDE11), the UCRs that include a phosphatidic acid (PA)-binding site in PDE4, and the PAS domain (PDE8). The inhibitory gamma subunit of PDE6 is indicated as a yellow ellipse. Domains functioning as targeting sequences by mediating membrane–protein or protein–protein interactions are indicated as red striated barrels and the transmembrane (TM) domains of PDE3 are indicated in blue. The number of PDE genes belonging to each PDE family is indicated in parentheses beside the PDE family name. (See the color version of the figure in Color Plates section.)

nucleotides to their effectors. Since protein kinase A (PKA), protein kinase G (PKG), GTP exchange protein activated by cAMP (EPAC), and cyclic nucleotide-gated (CNG) channels are tethered to specific subcellular compartments, PDEs likely contribute to the compartmentalization of cyclic nucleotide signaling and to the spatial dimension of the signal. PDEs may also have scaffolding properties within these complexes, opening the possibility that PDEs serve functions beyond their catalytic activity and that a dynamic formation and dissolution of these complexes may contribute to the allosteric regulation of PDE activities.

THE PDE SUPERFAMILY

After several more PDE genes were discovered through homology screening of nucleotide sequence databases between 1996 and 2000 (PDE8, PDE9, PDE10, and PDE11), the completion of the Human Genome Project in 2001 eventually established that there are 21 PDE genes in humans [9]. Orthologs of all 21 genes are encoded in the genomes of rats and mice and might be present in the same number in other mammals. Metazoan model organisms such as *Caenorhabditis elegans* or *Drosophila melanogaster* express orthologs of some, but usually not all of the mammalian PDEs [3]. Based upon their substrate specificities, kinetic properties, inhibitor sensitivities, and, ultimately, their sequence homology, the 21 mammalian PDE genes are subdivided into 11 PDE families, each consisting of 1 to a maximum of 4 genes (Table 1.1). Most PDE genes are expressed as a number of variants through the use of multiple promoters and alternative splicing. The PDE6 genes, with only 1 transcript per gene reported, and PDE9A, for which more than 20 putative variants have been proposed, represent the extremes in the number of variants generated from individual genes. Together, close to 100 PDE proteins are generated in mammals, each likely serving unique cellular functions.

Nomenclature

Due to the large number of PDE variants present in mammals, an initial classification based on regulatory properties and inhibitor sensitivities of newly discovered enzymes as well as their order of discovery soon became inadequate. It was subsequently replaced with a consensus nomenclature in which the first two letters indicate the species followed by the three letters "PDE", an Arabic numeral indicating the PDE family, a letter indicating the gene within the PDE family, and finally another Arabic numeral indicating the precise PDE variant. For example, HsPDE4D3 identifies the species as *Homo sapiens*, the PDE family as 4, the gene as D, and the variant as 3. This nomenclature was widely adopted in 1994 [27]. For a complete list of PDE genes and variants as well as information regarding the nomenclature used before 1994, please see http://depts.washington.edu/pde/pde.html or Ref. [28].

Overall Protein Domain Arrangement

Despite their multitude and diversity, all PDEs share several structural and functional properties. One of the most obvious is their modular structure consisting of a relatively conserved catalytic domain located in the C-terminal half of the protein and N-terminal domains that are structurally diverse, but all function to regulate enzyme activity (Figure 1.2). The C-terminal catalytic domain contains all residues required for catalysis and determines the enzyme kinetics unique to each PDE subtype. The characteristic features of the N-terminal regulatory domains are highly conserved modules such as Ca^{2+}/calmodulin-binding domains (PDE1), GAF domains (cGMP-activated PDEs, adenylyl cyclase, and Fh1A; PDE2, PDE5, PDE6, PDE10, and PDE11), UCR domains (upstream conserved regions; PDE4),

TABLE 1.1 The Properties of the Mammalian PDE Genes

Chromosome Region[a]	Gene Symbol	Name/Aliases	Predominant CNS Distribution[b]	Disease Association[c]	References
2q32.1	PDE1A	Phosphodiesterase 1A, Ca^{2+}/calmodulin-dependent	Cortex	Schizophrenia	[10]
12q13	PDE1B	Phosphodiesterase 1B, Ca^{2+}/calmodulin-dependent	Striatum, hippocampus		
7p14	PDE1C	Phosphodiesterase 1C, Ca^{2+}/calmodulin-dependent	Cerebellum		
11q13.4	PDE2A	Phosphodiesterase 2A, cGMP-stimulated	Striatum, hippocampus, cerebellum		
12p12	PDE3A	Phosphodiesterase 3B, cGMP-inhibited	Striatum, hippocampus	RR and QT interval	[11]
11p15.1	PDE3B	Phosphodiesterase 3B, cGMP-inhibited	Hippocampus		
19p13.2	PDE4A	Phosphodiesterase 4A, cAMP-specific (*Drosophila dunce* homolog, DPDE2, PDE2)	Cortex, hippocampus		
1p31	PDE4B	Phosphodiesterase 4B, cAMP-specific (*Drosophila dunce* homolog, DPD4, PDE4)	Cortex, striatum, hippocampus	Schizophrenia, bipolar disorder, depression, alcohol responses, multiple sclerosis	[12–15]
19p13.1	PDE4C	Phosphodiesterase 4C, cAMP-specific (*Drosophila dunce* homolog, DPD1, PDE1)	Low level in CNS		
5q12	PDE4D	Phosphodiesterase 4D, cAMP-specific (*Drosophila dunce* homolog, DPD3, PDE3, STRK1)	Cortex, hippocampus, brain stem	Stroke susceptibility, asthma, sleep disorders, neuroticism, schizophrenia	[10,16–21]
4q25-q27	PDE5A	Phosphodiesterase 5A, cGMP-specific	Cerebellum, hippocampus		

(continued)

TABLE 1.1 (*Continued*)

Chromosome Region[a]	Gene Symbol	Name/Aliases	Predominant CNS Distribution[b]	Disease Association[c]	References
5q31.2–q34	PDE6A	Phosphodiesterase 6A, cGMP-specific, rod, alpha	Retina	Retinitis pigmentosa	[5]
4p16.3	PDE6B	Phosphodiesterase 6B, cGMP-specific, rod, beta	Retina	Night blindness, congenital stationary type 3	[22]
10q24	PDE6C	Phosphodiesterase 6C, cGMP-specific, cone, alpha prime	Retina	Cone dystrophy	[7]
8q13–q22	PDE7A	Phosphodiesterase 7A, HCP1	Hippocampus		
6q23–q24	PDE7B	Phosphodiesterase 7B	Striatum, hippocampus		
15q25.3	PDE8A	Phosphodiesterase 8A	Cortex	Major depressive disorder	
5q13.3	PDE8B	Phosphodiesterase 8B	Striatum, hippocampus	Pigmented nodular adrenocortical disease	
21q22.3	PDE9A	Phosphodiesterase 9A	Cortex, cerebellum	Bipolar disorder	[23,24]
1q11	PDE10A	Phosphodiesterase 10A	Striatum	Autism	[25]
2q31.2	PDE11A	Phosphodiesterase 11A	Hippocampus CA1	Pigmented nodular adrenocortical disease	[26]

[a]Shown is the chromosomal localization of the PDE genes in humans.

[b]Expression data in human based on *in situ* hybridization; where not available, data from rat and mouse brain were used.

[c]The association with inherited diseases is based on published data, GWAS (http://gwas.nih.gov/) analysis, and OMIM (http://www.ncbi.nlm.nih.gov/omim) data.

REC domains (receiver; PDE8), and PAS domains (period, aryl hydrocarbon receptor nuclear translocator (ARNT), and single minded; PDE8), as well as phosphorylation sites, which mediate the regulation of enzyme activity by posttranslational modifications and/or ligand binding. As a result of this modular structure, truncated PDEs encoding only the catalytic domain not only retain enzyme activity, but also exhibit kinetic properties and substrate specificity similar to those of the holoenzyme, whereas regulation of enzyme activity is lost or is altered compared to full-length proteins [29–31]. In most cases, the inhibitor sensitivity of the full-length proteins is also retained in the catalytic domain constructs. There are some exceptions, however. The catalytic domain of PDE4, for example, exhibits an affinity for the prototypical PDE4 inhibitor rolipram that is about 100-fold lower compared to full-length proteins, whereas the sensitivity toward structurally unrelated compounds is not affected by the truncation [30,31]. Thus, catalytic domain constructs may have limited value for the development of PDE4 inhibitors.

The regulatory domains, in turn, may function in a similarly independent manner. The GAF domains, for example, are found in PDEs, adenylyl cyclases, and the *Escherichia coli* Fh1A protein. Chimeras between the GAF domains of different PDEs (PDE5, PDE10, and PDE11) and the catalytic domain of the cyanobacterial cyaB1 adenylyl cyclase were found to be fully functional in that the PDE-GAF domains sense cyclic nucleotide levels and mediate activation of the catalytic domain of the cyclase [32,33]. This finding, together with the similar domain organization of all PDEs, led to the proposal that despite their structurally distinct N-termini, the mechanism by which the N-terminal domains regulate enzyme activity is perhaps conserved among all PDE subtypes. The functional properties of the different N-terminal regulatory domains are discussed in greater detail later in this chapter.

The unique combination of catalytic domain kinetics and regulatory domain properties defines a PDE family and is conserved among its members. The variants generated from a single PDE gene, in turn, contain with few exceptions the identical catalytic domain, but often possess variant-specific N-termini. These variants are divided into those that encode the entire regulatory module characteristic of a given PDE family, such as GAF domains or UCR domains, and those that encode only a portion of the regulatory module or lack it altogether. The latter include so-called short PDE4 variants, which contain only a portion of the UCR module, and PDE11 variants that lack GAF domains. Underlining the critical role of the regulatory domains, variants that encode the entire regulatory module exhibit similar mechanisms of regulation, whereas variants that lack part or all of the regulatory domains are either insensitive to ligand binding or posttranslational modification *per se* or respond differently (Figure 1.2) [3,34]. The extreme N-termini unique to individual PDE variants are encoded by variant-specific first exons (Figure 1.3) and often mediate subcellular targeting through protein–protein or protein–lipid interactions, thus allowing the cell to specifically target PDE variants to subcellular compartments [35].

Although the mechanisms of regulation of PDE activity have been described biochemically in great detail, structural aspects of enzyme regulation had remained elusive (see Chapter 6 for a detailed discussion). Pandit et al. [36] recently provided a

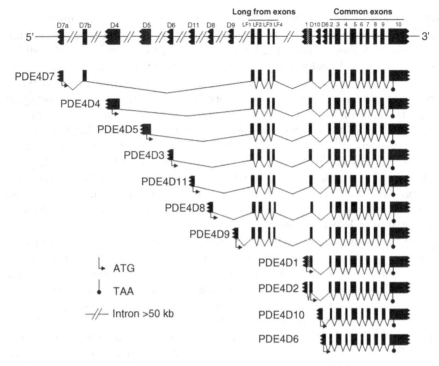

FIGURE 1.3   Structure of the PDE4D locus. Schematic representation of the structure of the mammalian PDE4D locus (top) and the encoded mRNAs. Exons are presented as filled bars, introns are drawn as solid lines, and a noncoding exon is depicted as a striated box. Indicated are variant-specific first exons, long form (LF) exons shared by all so-called long PDE4 isoforms, and common exons shared by most PDE4D variants. The scheme is not drawn to scale.

major breakthrough on the question of how modification of the N-terminal domains by posttranslational modifications and/or ligand binding exerts its effect on the conformation of the distal catalytic domain, thereby controlling PDE activity. The authors crystallized a PDE2A that, although lacking some N- and C-terminal sequence of the holoenzyme, contains the critical components of the PDE2 structure: a tandem set of GAF domains linked to the C-terminal catalytic domain. PDE2A crystallized as a linear structure that extends along a GAF-A/GAF-B/catalytic domain axis. The enzyme forms head-to-head dimers with the dimer interface spanning the entire length of the molecule with interactions between the GAF-A and GAF-A, GAF-B and GAF-B, and between the two catalytic domains of the individual monomers. When the GAF domains are unoccupied, the substrate-binding pockets of the two catalytic domains are packed against each other, essentially closing off access to the substrate. As a mechanism for the allosteric activation of PDE2, the authors propose that cGMP binding to the GAF-B domain results in a reorientation of the linker regions connecting GAF-B and the catalytic domain, which in turn leads to a disruption of the dimer interface between the two catalytic domains, thus promoting an "open" conformation of the enzyme that allows substrate access. Both the general organization of the PDE2 structure and the mechanism of PDE activation proposed by Pandit et al. [36] are in agreement with many structure–function relationships observed in other PDEs. Most PDEs have been reported to form homo- or hetero-dimers [29,34,37–41], and critical dimerization domains were identified in the N-terminal domains. The catalytic domains also retain some affinity as evidenced by the fact that several PDE catalytic domains form dimers in purified protein preparations as well as in crystal structures [42–44]. In addition, most PDEs also possess the elongated structure described for PDE2 as indicated by their high frictional ratios [39,45,46]. Electron microscopic images of PDE5 and PDE6 show an elongated structure highly similar to the atomic structure of PDE2, with points of contact between the GAF-As, the GAF-Bs, and the catalytic domains of the individual monomers [47,48]. Dimerization mediated by the N-terminal domains of PDE2 plays a critical role in stacking the substrate-binding sites at the catalytic domain against each other, thus preventing substrate access. This is in agreement with the observation that N-terminal domains in various PDEs exert an inhibitory constraint on the active site, which can be uncovered through deletion mutagenesis or proteolytic digest of full-length enzyme [49,50]. Taken together, these similarities suggest that the atomic structure of PDE2 might represent a model for a general organization of PDEs and a mechanism of enzyme activation.

However, Burgin et al. [51] recently suggested an alternative mechanism of how inhibitory constraint and regulation of enzyme activity is achieved in PDE4. In crystal structures of a truncated PDE4, substrate access to the catalytic site is prevented not by stacking of the catalytic domains against one another, as proposed for PDE2 [36], but by direct binding of a helix in the regulatory UCR2 domains to the substrate-binding pocket in the catalytic domain. PDE4 variants are divided into so-called long forms that contain the complete UCR1/2 module and short forms that lack UCR1 but still contain all or a portion of UCR2. The constructs crystallized by Burgin et al. lack UCR1 and encode only a portion of UCR2, thus encoding a PDE4 that resembles short forms. There are significant structural and functional differences between long and short forms including oligomerization, enzyme activation, and inhibitor sensitivity [40]. Thus, it remains to be determined whether the mechanism of enzyme inhibition/activation proposed by Burgin et al. [51] reflects properties of all PDE4 isoforms or whether this model describes properties inherent only to short PDE4 isoforms, whereas long isoforms are regulated differently. If the former is the case, PDE4 regulation of enzyme activity would be different from models described for PDE2, PDE5, and PDE6. This in turn would suggest that distinct modes of regulating PDE activity evolved for the different PDE families.

Catalytic Site Properties and Interaction with the Substrates

PDEs are divalent metal ion-dependent enzymes and share with other metal-dependent phosphohydrolases an HD(X2)H(X4)N motif, which defines residues forming the metal ion-binding site. Much progress has been made in understanding the structure and properties of the catalytic domains since crystal structures have become available in 2000 [43]. The catalytic domain consists of 16 α-helices folded into a compact structure. Whereas the sequence homology of the catalytic domains can be as low as 25% for members of different PDE families, the three-dimensional structure of the catalytic domain aligns all residues that are invariant or semiconserved among all PDEs to form the substrate-binding pocket. These residues include an invariant glutamine that forms hydrogen bonds with the 1- and/or 6-positions of the cyclic nucleotides, several residues that form a "hydrophobic clamp" that anchors the purine ring, and residues that form two metal ion-binding sites, termed M1 and M2, which are positioned at the bottom of the substrate-binding pocket. Based on biochemical data and X-ray diffraction, M1 is likely occupied by Zn^{2+}, whereas M2 is occupied by Mg^{2+} or Mn^{2+} in the native enzyme. The two metal ions function to activate the substrate phosphate and to coordinate a water molecule that acts as a nucleophile in the PDE reaction. Recent studies suggest that the water molecule coordinated by the metal ions may partially dissociate into a hydroxide ion [52–55], and that this hydroxide acts as the nucleophile on the phosphorus, eventually becoming part of the outgoing phosphate, whereas a nearby strictly conserved histidine assists with the protonation of the O3' group.

Defining the first atomic structure of any PDE, Xu et al. [43] reported the crystal structure of the catalytic domain of PDE4B in 2000. Since then, crystal structures for the catalytic domain of most PDE families have been reported, providing further insight into the distinct mechanisms of inhibitor binding and substrate specificity [42,43,56–64].

On the basis of their substrate specificity, the 11 PDE families can be divided into three groups. PDE4, PDE7, and PDE8 selectively hydrolyze cAMP, whereas PDE5, PDE6, and PDE9 are selective for cGMP hydrolysis. The remaining PDEs (PDE1, PDE2, PDE3, PDE10, and PDE11) bind and hydrolyze both cyclic nucleotides with varying efficiency. An invariant glutamine residue that is conserved among all PDEs and that was shown to form hydrogen bonds with either AMP or GMP in crystal structures has been proposed as the major determinant of PDE substrate specificity. As both cyclic nucleotides were thought to bind to the substrate-binding pocket in the same conformation and the hydrogen-bonding character of the 1- and 6-positions of adenine and guanine is essentially reversed, a so-called glutamine switch mechanism was previously proposed to determine PDE substrate specificity [43,63]. According to this model, the amide group of the invariant glutamine can rotate by 180° to accommodate binding of either cAMP or cGMP. It was assumed that PDEs in which additional residues limit the mobility of the invariant glutamine are selective for one of the cyclic nucleotides, whereas PDEs that allow a free rotation of the invariant glutamine could bind both. However, several recent findings suggest that this cannot be the only mechanism that determines substrate specificity among PDEs. Mutation of this invariant glutamine in PDE5A, for example, did reduce affinity of the enzyme for its physiological substrate cGMP but did not enhance binding of cAMP [65]. In addition, mutation of an aspartic acid residue conserved among the cAMP-PDEs ablates the substrate specificity of PDE4 isoenzymes, suggesting that this residue represents an additional evolutionary conserved component of substrate specificity [66,67]. Most important are recent studies by Wang et al. [61,68] that demonstrate that PDE10 binds the substrates cAMP and cGMP in *syn*-conformation, whereas the PDE reaction products AMP and GMP dock in *anti*-conformation in crystal structures. This suggests that previous reports, which relied on the analysis of cocrystals of various PDEs with AMP or GMP, might not reflect the binding of substrates in the native enzyme. In addition, Wang et al. show that rotation of the invariant glutamine is unlikely the cause for the dual-substrate specificity of PDE10, as the position of this residue is locked through additional hydrogen bonds, and that PDE4 forms only one hydrogen bond with cAMP rather than two as suggested by the "glutamine switch model" [61,68]. Thus, in addition to the conserved glutamine residue, there are likely additional, perhaps PDE family-selective determinants of substrate specificity.

Properties of the Regulatory Domains

Three critical functions have been identified for the PDE sequences N-terminal to the catalytic domain. These are the regulation of enzyme catalytic activity, oligomerization, and subcellular targeting. These properties are examined in detail in the subsequent sections of this chapter.

THE PROPERTIES OF THE GENES ENCODING PDEs

Organization of the PDE Genes

A common feature of the PDE genes already present in the ancestral *dunce* PDE locus of the *D. melanogaster* is the complex arrangement of transcription start sites and exons. The *dunce* locus, coding for a PDE involved in learning, memory, and development, spans over more than 148 kb (1 kb = 10^3 bases or base pairs) of genomic DNA and includes at least 16 exons with exceedingly long introns that often contain exons of unrelated genes. This complex arrangement is further augmented in the mammalian PDE4 genes. For instance, the human *PDE4D* locus (5q12) spans more than a million base pairs with at least 25 exons and 10 different start sites (see Figure 1.3). The exons encoding the catalytic domain are clustered together, whereas the exons coding for the amino termini and the regulatory/dimerization domains (UCRs) are distributed over a large stretch of genomic DNA. Other PDE loci have similar characteristics. For instance, the *PDE10A* gene spans more than 200 kb and contains 24 exons, whereas the *PDE11A* gene covers approximately 300 kb of genomic DNA and contains 23 exons [69]. The four PDE11A variants have different amino termini due to separate promoters and transcriptional start sites [70–72]. The *PDE1A* gene spans over 120 kb and contains 17 exons, again with multiple promoters and start sites [73]. Comparison of *PDE1B1* genomic sequence with other PDE genes has indicated that two splice junctions within the region encoding the catalytic domain are conserved in rat *PDE4B* and *PDE4D*, as well as in the *Drosophila dunce*

PDE [74,75]. This commonality in the splicing junction strongly suggests that the catalytic domains of PDEs are derived from a common ancestral gene, a view further supported by the high conservation of the amino acid sequence of the catalytic domain.

PDE loci have been associated with several neurological disorders (see Table 1.1). These associations are derived from genome-wide association studies (GWAS), meta-analyses, or candidate gene approach studies. It should be noted that numerous single-nucleotide polymorphisms (SNPs) have been identified in PDE loci and that SNPs are in linkage disequilibrium in several studies of patient populations (Table 1.1). However, nonsynonymous SNPs in the coding regions are very rare. For example, the *PDE4D* gene has 1919 SNPs in introns, but none in exons.

Multiple Transcripts

The presence of multiple transcripts derived from each PDE gene was initially revealed by Northern blot analysis. For instance, multiple RNAs ranging in size from 4.2 to 9.6 kb were detected for the *Drosophila dunce* PDE [76,77]. PDE4D cDNA probes hybridize to at least three or four different transcripts ranging between 2.3 and 5 kb depending on the tissue used [78]. PDE11A is expressed as at least three major transcripts of ~10.5, ~8.5, and ~6.0 kb, thus again suggesting the existence of multiple subtypes [72]. Further studies using PCR and exon-specific primers have provided a better understanding of the exon composition of each transcript.

Most of the heterogeneity in transcript size is due to the presence of multiple transcription start sites. In general, these start sites are under the control of promoters that often function in a tissue- and cell-specific manner as described for PDE1 [79–81], PDE4 [82–84], PDE7 [85,86], and PDE9 [87]. This property provides one of several explanations for the complexity of the PDE genes, whereby different promoters have evolved to allow developmental and cell-specific expression of a given PDE gene. However, it is also clear that PDEs encoded in different transcripts exclude or include regions with regulatory function, and show differences in subcellular localization and interaction with other proteins [35,88]. Thus, multiple transcriptional units generate PDEs with distinct functions.

Alternate Splicing

Splicing and alternate exon usage has been reported for many PDEs and most of this splicing occurs at the amino terminus. PDE4 splicing variants have been extensively characterized and are generated mostly by alternate promoters and first exon usage. Most of the splicing variants predicted by transcript analysis have been confirmed with antibodies specific for the N-termini of PDE4s [84]. Different variants show differences in interaction with other proteins and in subcellular localization. The largest number of splicing variants has been reported for PDE9. Through PCR amplification and alignment of EST sequences, 20 N-terminal mRNA variants have been identified [87,89]. However, only the proteins encoded by PDE9A1 and PDE9A5 have been expressed and characterized [90]. Unlike PDE4, these variants use the same transcriptional start site but are alternatively spliced to produce unique mRNAs that are distinct at the 5′-end. The functional consequence(s) of these amino acid sequence changes is (are) unclear as the affected regions are outside the catalytic domain or any recognized regulatory domain. PDE1C is another example of alternative splicing where proteins with identical catalytic domain and divergent amino termini are generated. Although uncommon, splicing at the carboxyl terminus is also present, as shown for PDE1A, PDE1C, PDE7A, and PDE10A [91]. Several splicing variants have been detected for PDE10. Omori and coworkers have suggested that this splicing controls the PKA phosphorylation of the enzyme and possibly subcellular localization [92]. This possibility has recently been experimentally verified by Charych et al. [93] *as* detailed in Chapter 10.

The importance of transcript splicing is underscored by analysis of RNA processing of the *PDE6B* gene in patients with autosomal recessive retinitis pigmentosa [94]. An acceptor splice site mutation in intron 2 of the *PDE6B* gene leads to the accumulation of a pre-mRNA with intermediate lariats, generating a PDE6B transcript that is 12 nucleotides shorter than wild type. In the normal PDE6B mRNA, these 12 nucleotides code for four amino acids highly conserved in the putative noncatalytic cGMP-binding domain GAF-A of PDE6B and are probably important for the correct folding and function of the protein.

Promoter Regulation

Regulation of PDE protein synthesis was observed in the 1970s when it was shown that treatment of cultured cells with cAMP analogs produced a large increase in PDE activity, which was blocked by protein synthesis inhibition [95,96]. With the cloning of different PDE4 variants, it was demonstrated that transcriptional activation of specific PDE4s accounts for some of the increased PDE activities described. Using an endocrine model and hormones that regulate cAMP synthesis, Swinnen et al. were the first to show that the accumulation of mRNA for PDE4D1 and PDE4B2 is dependent on cAMP-mediated transcriptional activation of these open reading frames (ORFs) [78]. Subsequently, an intronic promoter controlling the transcription of PDE4D1 mRNA was identified and functionally characterized [97]. In cortical neurons, D'sa et al. demonstrated that dibutyryl-cAMP (db-cAMP) induces expression of PDE4B2 and PDE4D1/PDE4D2 [98], whereas the splice variants PDE4A1, PDE4A5, PDE4A10, PDE4B3, PDE4B1, PDE4D3, and PDE4D4, although present in these cells, were not regulated at the transcriptional level by db-cAMP. Dominant negative mutants of the cAMP response element-binding (CREB) protein suppress PDE4B2 promoter activity and a constitutively active form of CREB stimulates it, confirming CRE- and CREB-dependent cAMP-mediated transcription of the short PDE4 forms. Thus, cAMP-dependent regulation of PDE4 transcription is a negative feedback loop contributing to long-term adaptation of cAMP signaling. It should be pointed out that robust PDE4B2 synthesis is also induced by activation of Toll-like receptor 4 (TLR4) and signaling through NF-κB and related transcription factors, suggesting that additional signaling pathways control the expression of PDE4 short forms [6].

During the circadian rhythm and in synchrony with cAMP synthesis, an increase in PDE activity that peaks in the middle of the night has been reported in the pineal gland. This nocturnal increase in PDE activity results from a fivefold increase in abundance of PDE4B2 mRNA [99]. The increase in PDE4B2 mRNA is followed by increases in PDE4B2 protein and PDE4 enzyme activity. These changes are dependent on the activation of adrenergic receptors and require PKA activation. The findings in this pineal gland model are an additional demonstration that PDE4B2 expression is regulated by cAMP. They also document the involvement of this regulatory loop in the circadian rhythm, providing evidence that PDE induction is a physiologically relevant mechanism of feedback regulation.

Using promoter/reporter assays, CRE elements were also identified in the promoter region of PDE4D5 [100]. This PDE4D variant accumulates in response to increased cellular cAMP, although at much lower levels than that reported for the short PDE4D1/2 and PDE4B2 forms. Site-directed mutational analysis revealed that the CRE at position −210 from the transcription start site is the principal element underlying cAMP responsiveness. The authors further determined that cAMP induced PDE4D5 expression in primary cultured human airway smooth muscle cells, leading to upregulation of phosphodiesterase activity.

CREB-mediated and cAMP-dependent regulation of PDE expression is not restricted to PDE4. For instance, PDE7B, a cAMP-specific PDE that is predominantly expressed in the striatum, is also regulated at the transcriptional level by cAMP [101]. Transcriptional activation of rat PDE7B following activation of the dopaminergic system has been demonstrated in primary striatal cultures. RT-PCR analysis revealed that dopamine D_1 agonists, forskolin, or 8-Br-cAMP stimulated PDE7B transcription in striatal neurons, whereas D_2 agonists did not. The cAMP-dependent regulation of PDE7B transcription, like that of PDE4D and PDE4B genes, was also variant-specific, because only PDE7B1 transcription was activated by a D_1 agonist. Also in this case, functional CRE elements were identified in the promoter region.

With the above exceptions, little information is available about the mechanisms of transcriptional regulation of other PDE genes in spite of the fact that numerous reports show altered PDE expression in pathological conditions or after pharmacological treatments. For instance, alterations in PDE7 and PDE8 isozyme mRNA expression were observed in Alzheimer's disease brains examined by *in situ* hybridization [102]. McLachlan et al. [103] found that PDE4D isoforms 1–9 were expressed in the hippocampus of healthy human adults as well as a patient with advanced Alzheimer's disease. However, the expression for the majority of the PDE4D isoforms was reduced in the Alzheimer's disease patient compared to normal controls, whereas PDE4D1, a short isoform under the transcriptional control of cAMP, was increased twofold. PDE4D is also regulated by treatment with antidepressants [104] and chronic haloperidol and clozapine treatment causes altered expression of PDE1B, PDE4B, and PDE10A in rat striatum [105]. The above are mostly correlative studies but imply that PDE promoters are finely regulated by a variety of extracellular signals and that pathological conditions affecting cyclic nucleotide signaling also affect PDE expression.

PATTERN OF GENE AND PROTEIN EXPRESSION IN THE CNS

General Concepts

Many of the PDE variants encoded in the human genome are expressed in a tissue- and cell-specific manner. Among the various tissues examined, the CNS is remarkable in the fact that it expresses one of the highest amounts of PDE protein and activity; essentially all PDE genes are expressed in the CNS and most cells express a multitude of variants. This complexity documents the tight control of cyclic nucleotide levels in cells whose primary function is processing and integration of information. Chapter 2 shows this clearly with some original data sets, which document PDE expression in the adult mouse brain.

The majority of studies examining PDE expression patterns in the brain have focused on PDE genes rather than individual PDE variants. This information is sufficient when exploring the therapeutic potential of PDE inhibitors, which are currently designed to inactivate all members of a PDE family or, at least, all variants generated from a single PDE gene. PDE inhibitors were often developed to inactivate the major PDE subtype expressed in the target cells or tissues as a strategy that has been successfully applied in the development of PDE5 inhibitors for erectile dysfunction, PDE4 inhibitors for inflammatory airway diseases, and PDE3 inhibitors for heart disease [91]. However, with the idea of compartmentalized cAMP signaling gaining momentum (see Chapter 3), it should be realized that the amount of a PDE subtype expressed in a cell is not proportional to its functional significance or therapeutic potential. While the predominantly expressed PDEs are certainly impor- tant, one cannot discount that PDE variants that contribute only a minor fraction of the total PDE activity in a cell may yet play critical physiological roles and may well be targeted for therapeutic intervention. In this context, the focus on where a PDE is expressed most abundantly does not necessarily predict specific functions. Further- more, the localization of PDE variants to specific subcellular compartments is thought to be necessary to stabilize cAMP microdomains of signaling (see below). As a result, displacement of a unique PDE variant from signaling complexes, rather than inacti- vation of catalytic activity, has been proposed for future drug development [106]. As subcellular localization is often unique to individual variants generated, understand- ing the nature of these complexes becomes a critical issue to develop new therapeutic strategies.

Individual PDE Gene Pattern of Expression

PDE1. The three *PDE1* genes, *PDE1A*, *PDE1B*, and *PDE1C*, are all expressed in the brain. *PDE1A* is widely distributed with high levels in cortex, hippocampus, cerebellum, olfactory bulb, and striatum. *PDE1B* message is also distributed widely, but is particularly enriched in striatum with PDE1B1 activity in mouse striatum being 3–17-fold higher than that in any other brain region. *PDE1C* is more selectively expressed in olfactory epithelium, cerebellum, and striatum [79–81,107–111]. *PDE1B* has received particular attention, as *PDE1B* knockout mice exhibit increased locomotor activity after D-amphetamine administration as well as spatial learning deficits. *PDE1B* knockout mice show increased levels of Thr34-DARPP-32 phosphorylation in response to dopamine D_1 receptor stimulation suggesting that elevated cyclic nucleotide levels resulting from *PDE1B* inactivation act synergistically to dopaminergic agonism [112]. Inactivation of PDE1 has been proposed as a possible therapeutic approach in Parkinson's disease because antiparkinson agents, such as amantadine or deprenyl, were shown to inhibit PDE1 *in vitro* [113–116]. In addition, SNPs in the *PDE1A* gene are associated with remission of antidepressant treatment response [117]. Chapter 7 explores the effect of PDE1B inhibition in memory disorders.

PDE2. Although the cGMP-stimulated *PDE2* is expressed in many tissues, the highest levels are found in the brain, where it is expressed in olfactory bulb and tubercle, cortex, amygdala, striatum, and hippocampus [108,118–122]. PDE2 inhibitors increase cyclic nucleotides in cortical and hippocampal neurons [123,124], promote enhanced long-term potentiation [123] and object recognition [125], and attenuate the learning impairment induced by acute tryptophan depletion [126]; all together suggesting that inactivation of PDE2 could boost memory functions. The potential of PDE2 inactivation in memory and cognition is presented in detail in Chapter 7.

PDE3. The two *PDE3* genes, *PDE3A* and *PDE3B*, are widely expressed throughout the brain, with PDE3A being enriched in striatal and hippocampal regions [127–129]. PDE3B has been suggested to play a role in hypothalamic leptin signaling [130].

PDE4. These enzymes are widely expressed throughout the body, but highest levels of activity and protein of any tissue are found in the brain, suggesting a critical role of PDE4 in cyclic nucleotide signaling in the CNS. Indeed, the mammalian *PDE4* genes are orthologs of the *D. melanogaster dunce* gene, whose ablation produces a phenotype of learning and memory deficits [131,132]. In addition, rolipram, the prototypical inhibitor defining the PDE4 family, was shown more than 25 years ago to produce antidepressant-like effects in rats and humans [133–135]. These studies have since been greatly expanded and PDE4 inhibitors are now investigated for potential therapeutic benefits for a plethora of CNS diseases, including Huntington's disease, Parkinson's disease, Alzheimer's disease, schizophrenia, and stroke [136].

While PDE4A, PDE4B, and PDE4D are all highly expressed in the brain, a detailed analysis reveals that their respective expression patterns are highly segregated, an observation that is further augmented if the expression patterns of individual variants, rather than the combined messages expressed from a PDE4 gene, are considered [82,98,104,137–148]. *PDE4A* is expressed at high levels in cortex, hippocampus, olfactory bulb, and brain stem. *PDE4B* displays a distinct pattern of accumulation with highest expression levels in striatum, amygdala, hypothalamus, and thalamus [12,83,149]. There is only a limited expression of *PDE4C* in the CNS, with some *PDE4C* found in cortex, some in thalamic nuclei and in the cerebellum in humans, but only in olfactory bulb in rat [148]. *PDE4D* is widely expressed with high levels in the cortex, olfactory bulb, and hippocampus [82].

The unique PDE4 expression patterns suggest distinct functions of *PDE4* genes and this assumption has been confirmed by the distinct behavioral phenotypes of mice deficient in individual *PDE4* genes. PDE4D is the primary target of the antidepressant effects of rolipram treatment [150]. However, the therapeutic use of PDE4 inhibitors has thus far been prevented by the significant side effects produced by this class of drugs, particularly emesis and nausea. PDE4D is highly expressed in the *area postrema* and nucleus of solitary tract, regions that are thought to mediate the emetic response, and PDE4DKO mice display a phenotype of shortened α_2-adrenoceptor-mediated anesthesia, a correlate of emesis in mice [138,143,148,151]. This has led to the assumption that PDE4D inactivation is the cause and is inevitably tied to the occurrence of emesis and nausea. This finding provided the rationale for an effort to design drugs that either do not penetrate the brain or are more selective for the other PDE4 subtypes, in particular PDE4B. In both cases, these efforts have forsaken the potential therapeutic use of PDE4D inactivation in CNS disorders. However, a recent report by Burgin et al. [51] suggests an alternative strategy of drug design that would overcome the obstacle of unwanted side effects. The authors describe the development of allosteric PDE4 inhibitors that target specific conformations of PDE4 rather than behaving as purely competitive inhibitors at the active site and do not completely inactivate enzyme activity even at highest concentrations. Even though these compounds are highly selective for PDE4D, they exhibit a reduced emetic activity in various animal models while retaining full cognition-enhancing properties. The authors propose that the partial inhibition of enzyme activity is the key to dissociating therapeutic benefits from side effects because spatial and temporal properties of cAMP signaling are maintained to some extent, which in turn lowers target-based toxicity. Given the absence of emetic side effects, this class of compounds can be used to target PDE4D for CNS diseases.

PDE4B has been linked to several other CNS disorders, in particular schizophrenia. *DISC1*, a well-established genetic marker associated with schizophrenia, has been shown to interact with PDE4B [13]. In addition, several studies have shown altered, generally reduced expression of PDE4B in schizophrenia patients versus healthy controls [12,152]. The changed expression levels likely result from SNPs in the *PDE4B* gene that are associated with schizophrenia in a Scottish family as well as in Japanese, Finnish, Caucasian, and African American populations [12–14,152,153]. There might be population differences or other genetic factors involved because two other studies did not detect associations of *PDE4B* SNPs with schizophrenia [154,155]. PDE4B has multiple additional CNS effects. Mice deficient in PDE4B exhibit an anxiogenic-like behavior [156] as well as decreased prepulse inhibition, decreased baseline motor activity, and an exaggerated locomotor response to amphetamine [157]. Changes in expression levels of PDE4B variants have also been associated with long-term potentiation [158–160] and levels of PDE4B and PDE4A expression were associated with autism [161].

PDE4 as a therapeutic target for CNS disorders is discussed in greater detail in several subsequent chapters.

PDE5. PDE5 exhibits a very selective expression pattern in the CNS with high levels in Purkinje cells of the cerebellum, spinal cord, pyramidal cells of the hippocampus, and little expression elsewhere [118–120,162–166]. Several selective inhibitors of PDE5 are used for the treatment of erectile dysfunction including sildenafil (Viagra®), vardenafil (Levitra®), and tadalafil (Cialis®). Availability of these drugs and the fact that they are generally well tolerated in patients have triggered a flurry of trials to probe the efficacy of these drugs in conditions where elevated cGMP is considered beneficial. These include memory and cognition deficits that are well established to be affected by cGMP levels. Indeed, inhibition of PDE5 was shown to increase cGMP levels in mouse hippocampus [167,168] and beneficial effects of PDE5 inhibition have been established in a range of animal models including attenuation of learning impairment induced by cholinergic muscarinic receptor blockade, NOS inhibition, or acute tryptophan depletion [126,169,170]. Improvement in "early" object recognition memory [167,168,171–173], synaptic function, and memory has also been reported in a mouse model of Alzheimer's disease [174]. It remains to be established whether expression of PDE5 as well as its beneficial effects on CNS functions can be translated to humans. Initial studies showed no overt effects on cognition in patients with schizophrenia [175], a number of psychological performance tests (with the exception of reduced simple choice reaction time) [176], and behavioral tests (although a positive effect on the ability to focus was noted) [177]. The function of PDE5 in the CNS is explored in more detail in Chapter 9.

PDE6. PDE6 is a key molecule in the phototransduction cascade and is selectively expressed in the retina. Expression of PDE6 has been reported in the light-sensitive pineal gland in chickens where it might play a similar role in phototransduction as in the eye [178].

PDE7 and PDE8. PDE7 and PDE8 families consist of two genes each. All are widely expressed throughout the brain and encode for cAMP-specific enzymes with a substrate affinity that is one order of magnitude higher compared to PDE4. This has led to the hypothesis that PDE7 and PDE8 function inherently differently from PDE4 and control microdomains with low cAMP levels or act under basal conditions. mRNA levels for PDE7A are highest in olfactory bulb and tubercle, hippocampus, and cerebellum, whereas PDE7B is highly expressed in striatum and hippocampus [85,86,101,102,179,180]. Activation of dopamine D_1 receptors potentiates transcription of PDE7B1 mRNA in striatal neurons through a cAMP/PKA/CREB pathway, thus implicating PDE7B in dopamine D_1 signaling [101]. While both PDE8A and PDE8B are present in the brain, PDE8B expression appears predominant and is found in most brain regions except cerebellum [102,181]. mRNA levels for PDE8B, but not PDE7A, PDE7B, or PDE8A were elevated in cortical and hippocampal regions of Alzheimer's disease patients [102] suggesting that specific PDE isoforms may be involved in this disease.

PDE9. Messenger RNAs for PDE9 are widely distributed throughout the body, including the CNS, with high levels in cerebellar Purkinje cells, olfactory system, cortex, caudate putamen, and hippocampus [87,108,118–120,182]. Development of PDE9-selective inhibitors, including BAY 73-6691 and PF-04447943, has recently been reported [184–188]. Treatment with BAY 73-6691 improved long-term memory in object and social recognition tasks and attenuated scopolamine- or MK-801-induced memory deficits in passive avoidance test and T-maze alternation task, respectively [184,185]. Similarly, PDE9A inhibition with PF-04447943 reversed memory deficits induced by ketamine or scopolamine and improved cognitive performance in several rodent models, including spatial, social, and novel object recognition [186–188]. Thus, PDE9 inhibition may represent a novel approach for the treatment of age- and/or disease-related memory deficits. Chapter 7 will explore the role of PDE9 in regulation of memory.

PDE10. The dual-substrate enzyme PDE10 exhibits a unique tissue distribution, with high levels found only in brain and testes but not in other peripheral organs and tissues. This expression pattern by itself has raised hopes that modulation of PDE10 activity may produce CNS-specific effects. Moreover, PDE10 expression in the brain is especially high in the medium spiny neurons of the striatum with lower amounts present in cortex, hippocampus, and pituitary gland [136,189–194]. As striatal dysfunction is implicated in the pathophysiology of various CNS diseases, including schizophrenia, Huntington's disease, Parkinson's disease, addiction, and obsessive-compulsive disorder, PDE10 is seen as a promising target of therapeutic intervention in diseases with striatal hypofunction. Indeed, pharmacologic or genetic inactivation of PDE10 raises striatal cAMP and cGMP levels, increases phosphorylation of established markers of neuronal activation such as CREB, extracellular signal-regulated kinase (ERK), and DARPP-32, and enhances striatal neuronal output [195–204]. PDE10 inhibitors are active in a range of models predictive of antipsychotic activity. Antipsychotic effects of PDE10 inactivation, such as a reduced conditional avoidance response and reduced locomotor activity in response to PCP or MK-801, have been shown in several animal models [197,198,205–208]. The PDE10 gene has also been implicated as a risk factor for the development of autism [25]. Chapter 10 explores the application of PDE10 inhibitors for the treatment of schizophrenia.

PDE11. Thus far, expression and functions of PDE11 in the CNS have not been extensively explored, likely because selective inhibitors are not available and knockout animals have only recently been generated. Kelly et al. [209] reported a very restricted expression pattern in mouse brain with PDE11 detected only in hippocampus CA1, subiculum, and the amygdalohippocampal area. PDE11 knockout mice display subtle behavioral phenotypes, such as open-field hyperactivity, increased sensitivity to MK-801, a male-specific stranger avoidance, and reduced social odor recognition memory. Together with anatomical and biochemical abnormalities such as an enlargement of the lateral ventricle and increased activity in ventral CA1, the phenotype of PDE11 knockout mice suggests that despite its restricted expression, PDE11 may have critical functions in the brain. Expression of PDE11 has also been reported for the pituitary gland [210] and SNPs in the *PDE11* gene have been identified as a risk factor for developing major depressive disorder (MDD) and as a determinant for antidepressant treatment response [117].

MECHANISMS OF REGULATION OF PDE ACTIVITY IN THE CNS

Phosphorylation

Most members of the PDE families undergo posttranslational phosphorylation. Undoubtedly, this is a major means of regulating PDE activity in a cell. With some exceptions (see below), a phosphorylated PDE hydrolyzes the substrate more efficiently or exhibits a higher affinity for the substrate. Phosphorylation sites have been mapped at the amino terminus of PDE1 [211–213], PDE3 [145,214,215], PDE4 [183,216–218], PDE5 [219], PDE10 [92], and PDE11 [220]. The kinases involved include PKA, PKG, protein kinase B (PKB/AKT), ERK, and other kinases.

PDE phosphorylation by cyclic nucleotide-dependent kinases defines negative feedback loops where cyclic nucleotides activate their own degradation (see Figure 1.4), a major mechanism of desensitization in cyclic nucleotide signaling as demonstrated for PDE4 [221,222]. PDE5 phosphorylation by PKA or PKG has multiple effects in addition to activation of the enzyme, including a change in conformation of the enzyme detected by chromatography, increased affinity of the allosteric cGMP-binding site for the nucleotide, and increased affinity of the catalytic domain for cGMP [223,224].

PDE4 variants are phosphorylated by PKA at an evolutionarily conserved residue present at the N-terminus of UCR1 (Ser54 in PDE4D3) [183,216,225]. The UCR1/2 module, which is conserved in all PDE4s, plays an important regulatory function, and deletions in these regions cause a loss of PKA-mediated activation, altered interaction with the prototypical PDE4 inhibitor rolipram [40], and altered oligomerization [40], suggesting that this region has profound effects on the catalytic domain. It has been proposed that phosphorylation alters the interaction of these domains with the catalytic domain, thus affecting inhibitor binding [226]. This concept has been recently consolidated by Burgin et al. through a crystal structure containing a portion of the UCR2 domain (Asn191 to Asp203 of PDE4D3) together with the catalytic domain. A helix encoded by UCR2 rests above the catalytic pocket and several UCR2 residues, notably Phe196 and Phe201, make contacts with residues in the catalytic pocket [51]. The movements of this domain may induce an open or closed conformation of the catalytic domain. Of note, a role of the UCR2 domain in regulating PDE4 catalytic function had been suggested by several other experimental approaches such as deletion mutagenesis and use of antibodies against this domain that affect PDE4 activity [49]. According to these authors, phosphorylation causes the UCR2 domain to move away from the catalytic center, causing an increased hydrolytic activity. Allosteric inhibitors making contact with this UCR2 domain have been synthesized and their properties conform to this model [51]. It should be also noted that, in an intact cell, the phosphorylation of PDE4s likely takes place within macromolecular complexes composed of scaffold proteins such as A-kinase anchoring proteins (AKAPs) that bring together PKA and the PDE [227,228,267,268] (Figure 1.4).

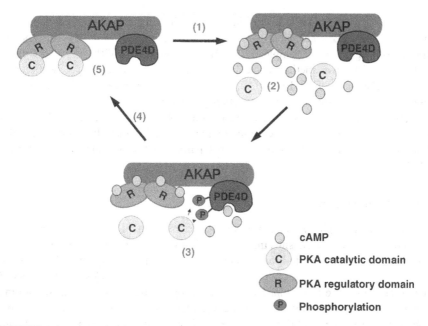

FIGURE 1.4 Localized feedback regulation mediated by a macromolecular complex tethering PKA and PDE4D. Elevation of cAMP levels (1) triggers activation of PKA through cAMP binding to the regulatory domains (R) and release (2) of the active catalytic domains (C). In addition to other downstream targets, PKA phosphorylates PDE4 isoforms (3) at a conserved residue (Ser54 in PDE4D3) resulting in activation of the enzymes (phosphorylation is indicated by red circles). The PKA-dependent PDE4 activation provides a negative feedback mechanism that results in a rapid downregulation of cAMP levels (4), thus resetting the signaling circuit (5). AKAPs promote this regulatory feedback signaling by tethering PKA and PDE4 in close proximity to each other. In PDE4D3, a second PKA phosphorylation site at Ser13 is thought to play a role in PDE4D3/mAKAP assembly [268]. (See the color version of the figure in Color Plates section.)

PDE4A5 is phosphorylated by the p38 MAPK-activated kinase MK2 (MAP-KAPK2), a modification that markedly attenuates the activation of PDE4A5 by PKA phosphorylation and alters the interaction of PDE4A5 with other proteins [229]. PDE4s are also phosphorylated by ERK kinases in a domain at the carboxyl terminus of the protein [230]. This regulatory pathway of a PDE4 has been recently implicated in brain-derived neurotrophic factor (BDNF) action and repair of spinal cord injury. Among other effects, BDNF signaling overcomes myelin inhibition of neuron regrowth by increasing cAMP. This increase is ERK-dependent and likely due to phosphorylation and inhibition of PDE4s [231].

In some instances, phosphorylation affects the interaction of PDEs with other proteins. As an example, phosphorylation of PDE4D3 at serine 13, a PKA consensus site unique to this PDE4 variant [183], enhances its association with mAKAP [232, 268] but reduces its interaction with Ndel1 [233]. DISC1, which is a scaffold protein associated with PDE4B and potentially other PDE4 proteins, binds PDE4B and represses its activity [13]. PKA phosphorylation of the anchored PDE4 uncouples the complex in some instances, an effect that is specific to both the PDE4 isoform and the DISC1 isoform within the complex [13,234]. Phosphorylation of PDE1 proteins alters the kinetics of their interaction with calmodulin [235–237]. The phosphorylation of a splicing variant of PDE10, PDE10A2, coincides with translocation of the protein to the cytosolic compartment, again suggesting that this posttranslational modification controls the interaction with anchoring proteins [92]. The molecular basis for this translocation has been recently clarified by the observation that phosphorylation of a threonine at the amino terminus prevents palmitoylation of the splicing variant PDE10A2 [93].

Small-Molecule Binding

In addition to phosphorylation, a second mechanism of PDE regulation relies on ligand binding to allosteric sites that are usually located at the amino terminus of the protein. Cyclic nucleotide-binding, cyclic nucleotide-hydrolyzing PDEs were described in the 1970s, and cooperativity between cGMP and cAMP was observed soon after [238]. The molecular basis for this property has been elucidated with the demonstration that PDEs interact with the two nucleotides *via* modular GAF domains. In PDE2, two tandem GAF domains per PDE monomer are present (GAF-A and GAF-B). However, only the GAF-B domain of PDE2 is occupied by cGMP [239]. Conversely, GAF-A is the high-affinity cGMP-binding site in PDE5 [240]. Whereas the *Anabaena* adenylyl cyclase GAF domains clearly bind cAMP, thereby activating this enzyme, the interaction of GAF domains with cAMP in any of the PDEs is still controversial [32,241]. As mentioned above, the structural basis for translating GAF domain occupancy into changes in PDE activity has been proposed with the crystal structure of a PDE2 holoenzyme [36].

Other small-molecule regulators of PDEs include phosphatidic acid, which has been shown to bind to and activate PDE4s [242]. A PAS domain has been identified in PDE8 [243]. This modular structure has the potential to interact with other molecules even though a ligand for PDE8 has not been identified. The Ca^{2+}-dependent regulation of PDE1 is mediated by calmodulin and will be described in the next section.

Protein–Protein Interactions

Interactions of PDE1 with calmodulin [244,245] and PDE6 with a regulatory gamma subunit [246] were the first demonstrations that PDEs bind other proteins, thus providing a paradigm of PDE regulation. Ca^{2+} in complex with calmodulin regulates and activates all PDE1s [245]. Calmodulin binding to PDE1 isoforms causes an increase in V_{max} of the enzymes without affecting the affinity for the substrate [244]. Although the interaction is well characterized and the regulation implicated in many processes, the exact physiological function of calmodulin activation of PDE1 has been difficult to assess, perhaps because specific inhibitors for these enzymes are not readily available and knockout mice have been generated for only one PDE1 gene,

PDE1B. A role for the Ca^{2+}/calmodulin-dependent regulation of PDE1 has been described for olfactory epithelium [247]. Together with the Ca^{2+}-dependent regulation of cyclases and phosphatases, this regulation is a mechanism for integration of Ca^{2+} and cAMP signaling pathways.

In the rod cell of the retina, a dimer of PDE6A and PDE6B is in complex with a gamma subunit that maintains the holoenzyme in an inactive state [248]. The points of contact between the gamma subunit and the catalytic subunit have been tentatively mapped to multiple domains within the catalytic domain and the amino terminus of the protein [249]. Light-activated rhodopsin promotes the GTP loading in the GTP-binding protein, transducin, which in turn promotes the dissociation of the gamma subunit from the PDE6 holoenzyme causing its activation.

Binding of the immunophilin XAP2 to the PDE4 isoform PDE4A5 also has a modulatory function [250]. Binding of this protein causes a decrease in the hydrolytic activity, modifies the interaction with rolipram, and hinders the PKA-mediated phosphorylation of the protein [250]. The significance of this regulation in a physiological context remains to be defined. PDE3s interact with the adaptor protein 14-3-3 in a phosphorylation-dependent manner [251].

A myriad of interactions with other proteins have been described in the last 5 years. Although in several cases the interaction alters the kinetic properties of the PDE, the major function for these protein–protein interactions is the subcellular anchoring, the localization of the PDE, and the generation of points of crosstalk between different pathways. These properties will be briefly discussed in the next sections.

MECHANISMS OF SUBCELLULAR LOCALIZATION OF PDEs IN THE CELLS OF THE CNS

The recovery of PDE activity in both the soluble and particulate fractions is a common finding in most cells or tissues. The anchoring to particulate structures is either direct through PDE association with the membrane or indirect through PDE interactions with membrane- or cytoskeleton-anchored proteins. PDE3 is localized by immunofluorescence to the endoplasmic reticulum (ER) and its activity is recovered in the ER fractions of heart and adipocytes [252]. With the cloning of PDE3A and PDE3B [253], a large hydrophobic domain with six predicted transmembrane helices was identified at the amino terminus of both isoforms. Subsequent mutagenesis studies have confirmed that this N-terminal domain is responsible for anchoring these proteins to the membrane [254]. Similarly, PDE2 is expressed as three splicing variants, PDE2A1, PDE2A2, and PDE2A3, with divergent N-terminal sequences. A hydrophobic domain at the N-terminus of rat PDE2A2 may be involved in its association with membranes in the brain [255]. PDE6s are inserted into the membrane through posttranslational prenylation and carboxymethylation [256].

A short variant of PDE4A, PDE4A1, is recovered in the particulate fraction of brain homogenates and in synaptosome populations, in particular those enriched in postsynaptic densities [142,257]. Deletion mutagenesis has indicated that its unique amino terminus is responsible for targeting the protein to the particulate fraction when expressed ectopically [257]. This domain also targets fused soluble proteins to the membrane. NMR analyses have shown that the first 23 amino acids of PDE4A1 encode a domain that interacts with phospholipids [258]. PDE7 and PDE9 variants are also targeted to particulate structures through their N-terminal domains even though the exact mechanisms have not been completely characterized [136].

By far the most common mechanism of anchoring PDEs to particulate structures is through protein anchors. These interactions have been extensively studied in PDE4s, which are found in complex with scaffold proteins such as myomegalin, AKAPs, β-arrestins, Rack1, and Disc1, as well as β-adrenergic receptors and other signaling molecules present in the membrane (ERK, SRC, Fyn). A comprehensive survey of PDE4 interactions with different proteins has appeared recently [35].

PDEs AND COMPARTMENTALIZATION OF SIGNALING

The diffusion of cyclic nucleotides within the intracellular space is the third dimension of the cyclic nucleotide signal under the control of PDEs. The general developing view is that multiple splice variants of any PDE provide a means for selective subcellular targeting. This localization defines microdomains in which the cyclic nucleotide signal is regulated in a manner distinct from the bulk cytosolic pool of cyclic nucleotides. Differential PDE localization was, for instance, observed in olfactory neurons where PDE4A was present in the body and PDE1 in the cilia [247]. This and many other examples of subcellular PDE compartmentalization have suggested nonuniform cyclic nucleotide distribution within the cell. More recently, the use of real-time sensors of cyclic nucleotides in live cells has further consolidated this concept of nonhomogeneous distribution of cyclic nucleotides [259–262]. Initially verified in cell lines and in cardiac myocytes [259], recent studies in the CNS are consistent with this idea. For instance, in neurons present in the mushroom bodies, a *Drosophila* brain region involved in olfactory learning, PKA activation by dopamine is evident in axons but absent in the dendrites [263]. However, in a *dunce* mutant that affects a PDE4 ortholog, the spatially restricted activation of PKA is lost, suggesting that this PDE is essential for the spatial specificity of cAMP/PKA signaling. PDEs have been proposed to function as a barrier to diffusion or sinks for cyclic nucleotides [264–266]. Although the exact role still needs to be ascertained, it is most likely that PDEs, when present in macromolecular complexes, have a major role in defining the kinetics of activation/deactivation of cyclic nucleotide targets, including cyclic nucleotide-regulated kinases, EPACs, or CNG channels. This topic is extensively reviewed in Chapter 3.

ACKNOWLEDGMENTS

We wish to thank all the investigators who have shared unpublished or in press data with us. We are indebted to Colleen Scheitrum for editorial work on the manuscript. Work done in the authors' laboratory is supported by NIH grants R01 HL092788 and

R21 HL107960, and grants from the Cystic Fibrosis Foundation and Fondation Leducq.

REFERENCES

1. Sutherland, E.W. and Rall, T.W. (1958) Fractionation and characterization of a cyclic adenine ribonucleotide formed by tissue particles. *J. Biol. Chem.*, **232**:1077–1091.

2. Thompson, W.J. and Appleman, M.M. (1971) Multiple cyclic nucleotide phosphodiesterase activities in rat brain. *Biochemistry*, **10**:311–316.

3. Conti, M. and Beavo, J. (2007) Biochemistry and physiology of cyclic nucleotide phosphodiesterases: essential components in cyclic nucleotide signaling. *Annu. Rev. Biochem.*, **76**:481–511.

4. Jin, S.L.C., Richter, W., and Conti, M. (2006) Insights into the physiological functions of PDE4 from knockout mice. In: Beavo, J., Francis, S., and Houslay, M. (Eds.), *Cyclic Nucleotide Phosphodiesterases in Health and Disease*, Boca Raton, FL: CRC Press, pp. 323–346.

5. Huang, S.H., Pittler, S.J., Huang, X., Oliveira, L., Berson, E.L., and Dryja, T.P. (1995) Autosomal recessive retinitis pigmentosa caused by mutations in the alpha subunit of rod cGMP phosphodiesterase. *Nat. Genet.*, **11**:468–471.

6. Jin, S.L. and Conti, M. (2002) Induction of the cyclic nucleotide phosphodiesterase PDE4B is essential for LPS-activated TNF-alpha responses. *Proc. Natl. Acad. Sci. USA*, **99**:7628–7633.

7. Chang, B., Grau, T., Dangel, S., Hurd, R., Jurklies, B., Sener, E.C., Andreasson, S., Dollfus, H., Baumann, B., Bolz, S., Artemyev, N., Kohl, S., Heckenlively, J., and Wissinger, B. (2009) A homologous genetic basis of the murine cpfl1 mutant and human achromatopsia linked to mutations in the PDE6C gene. *Proc. Natl. Acad. Sci. USA*, **106**:19581–19586.

8. Steinberg, S.F. and Brunton, L.L. (2001) Compartmentation of G protein-coupled signaling pathways in cardiac myocytes. *Annu. Rev. Pharmacol. Toxicol.*, **41**:751–773.

9. Venter, J.C., Adams, M.D., Myers, E.W., Li, P.W., Mural, R.J., Sutton, G.G., Smith, H.O., Yandell, M., Evans, C.A., Holt, R.A., Gocayne, J.D., Amanatides, P., Ballew, R.M., Huson, D.H., Wortman, J.R., Zhang, Q., Kodira, C.D., Zheng, X.H., Chen, L., Skupski, M., Subramanian, G., Thomas, P.D., Zhang, J., Gabor Miklos, G.L., Nelson, C., Broder, S., Clark, A.G., Nadeau, J., McKusick, V.A., Zinder, N., Levine, A.J., Roberts, R.J., Simon, M., Slayman, C., Hunkapiller, M., Bolanos, R., Delcher, A., Dew, I., Fasulo, D., Flanigan, M., Florea, L., Halpern, A., Hannenhalli, S., Kravitz, S., Levy, S., Mobarry, C., Reinert, K., Remington, K., Abu-Threideh, J., Beasley, E., Biddick, K., Bonazzi, V., Brandon, R., Cargill, M., Chandramouliswaran, I., Charlab, R., Chaturvedi, K., Deng, Z., Di Francesco, V., Dunn, P., Eilbeck, K., Evangelista, C., Gabrielian, A.E., Gan, W., Ge, W., Gong, F., Gu, Z., Guan, P., Heiman, T.J., Higgins, M.E., Ji, R.R., Ke, Z., Ketchum, K.A., Lai, Z., Lei, Y., Li, Z., Li, J., Liang, Y., Lin, X., Lu, F., Merkulov, G.V., Milshina, N., Moore, H.M., Naik, A.K., Narayan, V.A., Neelam, B., Nusskern, D., Rusch, D.B., Salzberg, S., Shao, W., Shue, B., Sun, J., Wang, Z., Wang, A., Wang, X., Wang, J., Wei, M., Wides, R., Xiao, C., Yan, C., et al. (2001) The sequence of the human genome. *Science*, **291**:1304–1351.

10. O'Donovan, M.C., Craddock, N., Norton, N., Williams, H., Peirce, T., Moskvina, V., Nikolov, I., Hamshere, M., Carroll, L., Georgieva, L., Dwyer, S., Holmans, P., Marchini,

J.L., Spencer, C.C., Howie, B., Leung, H.T., Hartmann, A.M., Moller, H.J., Morris, D.W., Shi, Y., Feng, G., Hoffmann, P., Propping, P., Vasilescu, C., Maier, W., Rietschel, M., Zammit, S., Schumacher, J., Quinn, E.M., Schulze, T.G., Williams, N.M., Giegling, I., Iwata, N., Ikeda, M., Darvasi, A., Shifman, S., He, L., Duan, J., Sanders, A.R., Levinson, D.F., Gejman, P.V., Cichon, S., Nothen, M.M., Gill, M., Corvin, A., Rujescu, D., Kirov, G., Owen, M.J., Buccola, N.G., Mowry, B.J., Freedman, R., Amin, F., Black, D.W., Silverman, J.M., Byerley, W.F., and Cloninger, C.R. (2008) Identification of loci associated with schizophrenia by genome-wide association and follow-up. *Nat. Genet.*, **40**:1053–1055.

11. Marroni, F., Pfeufer, A., Aulchenko, Y.S., Franklin, C.S., Isaacs, A., Pichler, I., Wild, S.H., Oostra, B.A., Wright, A.F., Campbell, H., Witteman, J.C., Kaab, S., Hicks, A.A., Gyllensten, U., Rudan, I., Meitinger, T., Pattaro, C., van Duijn, C.M., Wilson, J.F., and Pramstaller, P.P. (2009) A genome-wide association scan of RR and QT interval duration in 3 European genetically isolated populations: the EUROSPAN project. *Circ. Cardiovasc. Genet.*, **2**:322–328.

12. Fatemi, S.H., King, D.P., Reutiman, T.J., Folsom, T.D., Laurence, J.A., Lee, S., Fan, Y.T., Paciga, S.A., Conti, M., and Menniti, F.S. (2008) PDE4B polymorphisms and decreased PDE4B expression are associated with schizophrenia. *Schizophr. Res.*, **101**:36–49.

13. Millar, J.K., Pickard, B.S., Mackie, S., James, R., Christie, S., Buchanan, S.R., Malloy, M.P., Chubb, J.E., Huston, E., Baillie, G.S., Thomson, P.A., Hill, E.V., Brandon, N.J., Rain, J.C., Camargo, L.M., Whiting, P.J., Houslay, M.D., Blackwood, D.H., Muir, W.J., and Porteous, D.J. (2005) DISC1 and PDE4B are interacting genetic factors in schizophrenia that regulate cAMP signaling. *Science*, **310**:1187–1191.

14. Tomppo, L., Hennah, W., Lahermo, P., Loukola, A., Tuulio-Henriksson, A., Suvisaari, J., Partonen, T., Ekelund, J., Lonnqvist, J., and Peltonen, L. (2009) Association between genes of Disrupted in schizophrenia 1 (DISC1) interactors and schizophrenia supports the role of the DISC1 pathway in the etiology of major mental illnesses. *Biol. Psychiatry*, **65**:1055–1062.

15. Numata, S., Ueno, S., Iga, J., Song, H., Nakataki, M., Tayoshi, S., Sumitani, S., Tomotake, M., Itakura, M., Sano, A., and Ohmori, T. (2008) Positive association of the PDE4B (phosphodiesterase 4B) gene with schizophrenia in the Japanese population. *J. Psychiatr. Res.*, **43**:7–12.

16. Gretarsdottir, S., Thorleifsson, G., Reynisdottir, S.T., Manolescu, A., Jonsdottir, S., Jonsdottir, T., Gudmundsdottir, T., Bjarnadottir, S.M., Einarsson, O.B., Gudjonsdottir, H.M., Hawkins, M., Gudmundsson, G., Gudmundsdottir, H., Andrason, H., Gudmundsdottir, A.S., Sigurdardottir, M., Chou, T.T., Nahmias, J., Goss, S., Sveinbjornsdottir, S., Valdimarsson, E.M., Jakobsson, F., Agnarsson, U., Gudnason, V., Thorgeirsson, G., Fingerle, J., Gurney, M., Gudbjartsson, D., Frigge, M.L., Kong, A., Stefansson, K., and Gulcher, J.R. (2003) The gene encoding phosphodiesterase 4D confers risk of ischemic stroke. *Nat. Genet.*, **35**:131–138.

17. Rosand, J., Bayley, N., Rost, N., and de Bakker, P.I. (2006) Many hypotheses but no replication for the association between PDE4D and stroke. *Nat. Genet.*, **38**:1091–1092; author reply 1092–1093.

18. Gretarsdottir, S., Gulcher, J., Thorleifsson, G., Kong, A., and Stefansson, K. (2005) Comment on the phosphodiesterase 4D replication study by Bevan et al. *Stroke*, **36**:1824.

19. Shifman, S., Bhomra, A., Smiley, S., Wray, N.R., James, M.R., Martin, N.G., Hettema, J.M., An, S.S., Neale, M.C., van den Oord, E.J., Kendler, K.S., Chen, X., Boomsma, D.I.,

Middeldorp, C.M., Hottenga, J.J., Slagboom, P.E., and Flint, J. (2008) A whole genome association study of neuroticism using DNA pooling. *Mol. Psychiatry*, **13**:302–312.

20. Gottlieb, D.J., O'Connor, G.T., and Wilk, J.B. (2007) Genome-wide association of sleep and circadian phenotypes. *BMC Med. Genet.*, **8**(Suppl. 1):S9.

21. Himes, B.E., Hunninghake, G.M., Baurley, J.W., Rafaels, N.M., Sleiman, P., Strachan, D.P., Wilk, J.B., Willis-Owen, S.A., Klanderman, B., Lasky-Su, J., Lazarus, R., Murphy, A.J., Soto-Quiros, M.E., Avila, L., Beaty, T., Mathias, R.A., Ruczinski, I., Barnes, K.C., Celedon, J.C., Cookson, W.O., Gauderman, W.J., Gilliland, F.D., Hakonarson, H., Lange, C., Moffatt, M.F., O'Connor, G.T., Raby, B.A., Silverman, E.K., and Weiss, S.T. (2009) Genome-wide association analysis identifies PDE4D as an asthma-susceptibility gene. *Am. J. Hum. Genet.*, **84**:581–593.

22. Gal, A., Orth, U., Baehr, W., Schwinger, E., and Rosenberg, T. (1994) Heterozygous missense mutation in the rod cGMP phosphodiesterase beta-subunit gene in autosomal dominant stationary night blindness. *Nat. Genet.*, **7**:64–68.

23. Straub, R.E., Lehner, T., Luo, Y., Loth, J.E., Shao, W., Sharpe, L., Alexander, J.R., Das, K., Simon, R., Fieve, R.R., et al. (1994) A possible vulnerability locus for bipolar affective disorder on chromosome 21q22.3. *Nat. Genet.*, **8**:291–296.

24. Detera-Wadleigh, S.D., Badner, J.A., Goldin, L.R., Berrettini, W.H., Sanders, A.R., Rollins, D.Y., Turner, G., Moses, T., Haerian, H., Muniec, D., Nurnberger, J.I., Jr., and Gershon, E.S. (1996) Affected-sib-pair analyses reveal support of prior evidence for a susceptibility locus for bipolar disorder, on 21q. *Am. J. Hum. Genet.*, **58**: 1279–1285.

25. Talkowski, M.E., Rosenfeld, J.A., Blumenthal, I., Pillalamarri, V., Chiang, C., Heilbut, A., Ernst, C., Hanscom, C., Rossin, E., Lindgren, A.M., Pereira, S., Ruderfer, D., Kirby, A., Ripke, S., Harris, D.J., Lee, J.H., Ha, K., Kim, H.G., Solomon, B.D., Gropman, A.L., Lucente, D., Sims, K., Ohsumi, T.K., Borowsky, M.L., Loranger, S., Quade, B., Lage, K., Miles, J., Wu, B.L., Shen, Y., Neale, B., Shaffer, L.G., Daly, M.J., Morton, C.C., and Gusella, J.F. (2012) Sequencing chromosomal abnormalities reveals neurodevelopmental loci that confer risk across diagnostic boundaries. *Cell*, **149**:525–537.

26. Horvath, A., Boikos, S., Giatzakis, C., Robinson-White, A., Groussin, L., Griffin, K.J., Stein, E., Levine, E., Delimpasi, G., Hsiao, H.P., Keil, M., Heyerdahl, S., Matyakhina, L., Libe, R., Fratticci, A., Kirschner, L.S., Cramer, K., Gaillard, R.C., Bertagna, X., Carney, J.A., Bertherat, J., Bossis, I., and Stratakis, C.A. (2006) A genome-wide scan identifies mutations in the gene encoding phosphodiesterase 11A4 (PDE11A) in individuals with adrenocortical hyperplasia. *Nat. Genet.*, **38**:794–800.

27. Beavo, J.A., Conti, M., and Heaslip, R.J. (1994) Multiple cyclic nucleotide phosphodiesterases. *Mol. Pharmacol.*, **46**:399–405.

28. Bolger, G.B. (2007) Phosphodiesterase isoforms: an annotated list. In: Beavo, J.A., Francis, S.H., and Houslay, M.D. (Eds.), *Cyclic Nucleotide Phosphodiesterases in Health and Disease*, Boca Raton, FL: CRC Press, pp. 19–31.

29. Fink, T.L., Francis, S.H., Beasley, A., Grimes, K.A., and Corbin, J.D. (1999) Expression of an active, monomeric catalytic domain of the cGMP-binding cGMP-specific phosphodiesterase (PDE5). *J. Biol. Chem.*, **274**:34613–34620.

30. Jacobitz, S., McLaughlin, M.M., Livi, G.P., Burman, M., and Torphy, T.J. (1996) Mapping the functional domains of human recombinant phosphodiesterase 4A: structural requirements for catalytic activity and rolipram binding. *Mol. Pharmacol.*, **50**:891–899.

31. Richter, W., Unciuleac, L., Hermsdorf, T., Kronbach, T., and Dettmer, D. (2001) Identification of inhibitor binding sites of the cAMP-specific phosphodiesterase 4. *Cell. Signal.*, **13**:287–297.

32. Gross-Langenhoff, M., Hofbauer, K., Weber, J., Schultz, A., and Schultz, J.E. (2006) cAMP is a ligand for the tandem GAF domain of human phosphodiesterase 10 and cGMP for the tandem GAF domain of phosphodiesterase 11. *J. Biol. Chem.*, **281**: 2841–2846.

33. Hofbauer, K., Schultz, A., and Schultz, J.E. (2008) Functional chimeras of the phosphodiesterase 5 and 10 tandem GAF domains. *J. Biol. Chem.*, **283**:25164–25170.

34. Weeks, J.L., 2nd, Zoraghi, R., Francis, S.H., and Corbin, J.D. (2007) N-terminal domain of phosphodiesterase-11A4 (PDE11A4) decreases affinity of the catalytic site for substrates and tadalafil, and is involved in oligomerization. *Biochemistry*, **46**:10353–10364.

35. Houslay, M.D. (2010) Underpinning compartmentalised cAMP signalling through targeted cAMP breakdown. *Trends Biochem. Sci.*, **35**:91–100.

36. Pandit, J., Forman, M.D., Fennell, K.F., Dillman, K.S., and Menniti, F.S. (2009) Mechanism for the allosteric regulation of phosphodiesterase 2A deduced from the X-ray structure of a near full-length construct. *Proc. Natl. Acad. Sci. USA*, **106**: 18225–18230.

37. Kenan, Y., Murata, T., Shakur, Y., Degerman, E., and Manganiello, V.C. (2000) Functions of the N-terminal region of cyclic nucleotide phosphodiesterase 3 (PDE 3) isoforms. *J. Biol. Chem.*, **275**:12331–12338.

38. Muradov, K.G., Boyd, K.K., Martinez, S.E., Beavo, J.A., and Artemyev, N.O. (2003) The GAFa domains of rod cGMP-phosphodiesterase 6 determine the selectivity of the enzyme dimerization. *J. Biol. Chem.*, **278**:10594–10601.

39. Richter, W. and Conti, M. (2002) Dimerization of the type 4 cAMP-specific phosphodiesterases is mediated by the upstream conserved regions (UCRs). *J. Biol. Chem.*, **277**:40212–40221.

40. Richter, W. and Conti, M. (2004) The oligomerization state determines regulatory properties and inhibitor sensitivity of type 4 cAMP-specific phosphodiesterases. *J. Biol. Chem.*, **279**:30338–30348.

41. Martins, T.J., Mumby, M.C., and Beavo, J.A. (1982) Purification and characterization of a cyclic GMP-stimulated cyclic nucleotide phosphodiesterase from bovine tissues. *J. Biol. Chem.*, **257**:1973–1979.

42. Huai, Q., Wang, H., Zhang, W., Colman, R.W., Robinson, H., and Ke, H. (2004) Crystal structure of phosphodiesterase 9 shows orientation variation of inhibitor 3-isobutyl-1-methylxanthine binding. *Proc. Natl. Acad. Sci. USA*, **101**:9624–9629.

43. Xu, R.X., Hassell, A.M., Vanderwall, D., Lambert, M.H., Holmes, W.D., Luther, M.A., Rocque, W.J., Milburn, M.V., Zhao, Y., Ke, H., and Nolte, R.T. (2000) Atomic structure of PDE4: insights into phosphodiesterase mechanism and specificity. *Science*, **288**: 1822–1825.

44. Richter, W., Hermsdorf, T., Lilie, H., Egerland, U., Rudolph, R., Kronbach, T., and Dettmer, D. (2000) Refolding, purification, and characterization of human recombinant PDE4A constructs expressed in *Escherichia coli*. *Protein Expr. Purif.*, **19**:375–383.

45. Yamamoto, T., Manganiello, V.C., and Vaughan, M. (1983) Purification and characterization of cyclic GMP-stimulated cyclic nucleotide phosphodiesterase from calf liver. Effects of divalent cations on activity. *J. Biol. Chem.*, **258**:12526–12533.

46. Giorgi, M., Caniglia, C., Scarsella, G., and Augusti-Tocco, G. (1993) Characterization of 3′:5′ cyclic nucleotide phosphodiesterase activities of mouse neuroblastoma N18TG2 cells. *FEBS Lett.*, **324**:76–80.

47. Kajimura, N., Yamazaki, M., Morikawa, K., Yamazaki, A., and Mayanagi, K. (2002) Three-dimensional structure of non-activated cGMP phosphodiesterase 6 and comparison of its image with those of activated forms. *J. Struct. Biol.*, **139**:27–38.

48. Kameni Tcheudji, J.F., Lebeau, L., Virmaux, N., Maftei, C.G., Cote, R.H., Lugnier, C., and Schultz, P. (2001) Molecular organization of bovine rod cGMP-phosphodiesterase 6. *J. Mol. Biol.*, **310**:781–791.

49. Lim, J., Pahlke, G., and Conti, M. (1999) Activation of the cAMP-specific phosphodiesterase PDE4D3 by phosphorylation. Identification and function of an inhibitory domain. *J. Biol. Chem.*, **274**:19677–19685.

50. Sonnenburg, W.K., Seger, D., Kwak, K.S., Huang, J., Charbonneau, H., and Beavo, J.A. (1995) Identification of inhibitory and calmodulin-binding domains of the PDE1A1 and PDE1A2 calmodulin-stimulated cyclic nucleotide phosphodiesterases. *J. Biol. Chem.*, **270**:30989–31000.

51. Burgin, A.B., Magnusson, O.T., Singh, J., Witte, P., Staker, B.L., Bjornsson, J.M., Thorsteinsdottir, M., Hrafnsdottir, S., Hagen, T., Kiselyov, A.S., Stewart, L.J., and Gurney, M.E. (2010) Design of phosphodiesterase 4D (PDE4D) allosteric modulators for enhancing cognition with improved safety. *Nat. Biotechnol.*, **28**:63–70.

52. Liu, S., Mansour, M.N., Dillman, K.S., Perez, J.R., Danley, D.E., Aeed, P.A., Simons, S.P., Lemotte, P.K., and Menniti, F.S. (2008) Structural basis for the catalytic mechanism of human phosphodiesterase 9. *Proc. Natl. Acad. Sci. USA*, **105**:13309–13314.

53. Zhan, C.G. and Zheng, F. (2001) First computational evidence for a catalytic bridging hydroxide ion in a phosphodiesterase active site. *J. Am. Chem. Soc.*, **123**: 2835–2838.

54. Xiong, Y., Lu, H.T., Li, Y., Yang, G.F., and Zhan, C.G. (2006) Characterization of a catalytic ligand bridging metal ions in phosphodiesterases 4 and 5 by molecular dynamics simulations and hybrid quantum mechanical/molecular mechanical calculations. *Biophys. J.*, **91**:1858–1867.

55. Salter, E.A. and Wierzbicki, A. (2007) The mechanism of cyclic nucleotide hydrolysis in the phosphodiesterase catalytic site. *J. Phys. Chem. B*, **111**:4547–4552.

56. Barren, B., Gakhar, L., Muradov, H., Boyd, K.K., Ramaswamy, S., and Artemyev, N.O. (2009) Structural basis of phosphodiesterase 6 inhibition by the C-terminal region of the gamma-subunit. *EMBO J.*, **28**:3613–3622.

57. Huai, Q., Wang, H., Sun, Y., Kim, H.Y., Liu, Y., and Ke, H. (2003) Three-dimensional structures of PDE4D in complex with roliprams and implication on inhibitor selectivity. *Structure*, **11**:865–873.

58. Iffland, A., Kohls, D., Low, S., Luan, J., Zhang, Y., Kothe, M., Cao, Q., Kamath, A.V., Ding, Y.H., and Ellenberger, T. (2005) Structural determinants for inhibitor specificity and selectivity in PDE2A using the wheat germ *in vitro* translation system. *Biochemistry*, **44**:8312–8325.

59. Scapin, G., Patel, S.B., Chung, C., Varnerin, J.P., Edmondson, S.D., Mastracchio, A., Parmee, E.R., Singh, S.B., Becker, J.W., Van der Ploeg, L.H., and Tota, M.R. (2004) Crystal structure of human phosphodiesterase 3B: atomic basis for substrate and inhibitor specificity. *Biochemistry*, **43**:6091–6100.

60. Sung, B.J., Hwang, K.Y., Jeon, Y.H., Lee, J.I., Heo, Y.S., Kim, J.H., Moon, J., Yoon, J.M., Hyun, Y.L., Kim, E., Eum, S.J., Park, S.Y., Lee, J.O., Lee, T.G., Ro, S., and Cho, J.M. (2003) Structure of the catalytic domain of human phosphodiesterase 5 with bound drug molecules. *Nature*, **425**:98–102.

61. Wang, H., Robinson, H., and Ke, H. (2007) The molecular basis for different recognition of substrates by phosphodiesterase families 4 and 10. *J. Mol. Biol.*, **371**:302–307.

62. Wang, H., Yan, Z., Yang, S., Cai, J., Robinson, H., and Ke, H. (2008) Kinetic and structural studies of phosphodiesterase-8A and implication on the inhibitor selectivity. *Biochemistry*, **47**:12760–12768.

63. Zhang, K.Y., Card, G.L., Suzuki, Y., Artis, D.R., Fong, D., Gillette, S., Hsieh, D., Neiman, J., West, B.L., Zhang, C., Milburn, M.V., Kim, S.H., Schlessinger, J., and Bollag, G. (2004) A glutamine switch mechanism for nucleotide selectivity by phosphodiesterases. *Mol. Cell*, **15**:279–286.

64. Wang, H., Liu, Y., Chen, Y., Robinson, H., and Ke, H. (2005) Multiple elements jointly determine inhibitor selectivity of cyclic nucleotide phosphodiesterases 4 and 7. *J. Biol. Chem.*, **280**:30949–30955.

65. Zoraghi, R., Corbin, J.D., and Francis, S.H. (2006) Phosphodiesterase-5 Gln817 is critical for cGMP, vardenafil, or sildenafil affinity: its orientation impacts cGMP but not cAMP affinity. *J. Biol. Chem.*, **281**:5553–5558.

66. Herman, S.B., Juilfs, D.M., Fauman, E.B., Juneau, P., and Menetski, J.P. (2000) Analysis of a mutation in phosphodiesterase type 4 that alters both inhibitor activity and nucleotide selectivity. *Mol. Pharmacol.*, **57**:991–999.

67. Richter, W., Unciuleac, L., Hermsdorf, T., Kronbach, T., and Dettmer, D. (2001) Identification of substrate specificity determinants in human cAMP-specific phosphodiesterase 4A by single-point mutagenesis. *Cell. Signal.*, **13**:159–167.

68. Wang, H., Liu, Y., Hou, J., Zheng, M., Robinson, H., and Ke, H. (2007) Structural insight into substrate specificity of phosphodiesterase 10. *Proc. Natl. Acad. Sci. USA*, **104**:5782–5787.

69. Yuasa, K., Kanoh, Y., Okumura, K., and Omori, K. (2001) Genomic organization of the human phosphodiesterase PDE11A gene. Evolutionary relatedness with other PDEs containing GAF domains. *Eur. J. Biochem.*, **268**:168–178.

70. Yuasa, K., Kotera, J., Fujishige, K., Michibata, H., Sasaki, T., and Omori, K. (2000) Isolation and characterization of two novel phosphodiesterase PDE11A variants showing unique structure and tissue-specific expression. *J. Biol. Chem.*, **275**:31469–31479.

71. Hetman, J.M., Robas, N., Baxendale, R., Fidock, M., Phillips, S.C., Soderling, S.H., and Beavo, J.A. (2000) Cloning and characterization of two splice variants of human phosphodiesterase 11A. *Proc. Natl. Acad. Sci. USA*, **97**:12891–12895.

72. Makhlouf, A., Kshirsagar, A., and Niederberger, C. (2006) Phosphodiesterase 11: a brief review of structure, expression and function. *Int. J. Impot. Res.*, **18**:501–509.

73. Michibata, H., Yanaka, N., Kanoh, Y., Okumura, K., and Omori, K. (2001) Human Ca^{2+}/calmodulin-dependent phosphodiesterase PDE1A: novel splice variants, their specific expression, genomic organization, and chromosomal localization. *Biochim. Biophys. Acta*, **1517**:278–287.

74. Yu, J., Wolda, S.L., Frazier, A.L., Florio, V.A., Martins, T.J., Snyder, P.B., Harris, E.A., McCaw, K.N., Farrell, C.A., Steiner, B., Bentley, J.K., Beavo, J.A., Ferguson, K., and Gelinas, R. (1997) Identification and characterisation of a human calmodulin-stimulated phosphodiesterase PDE1B1. *Cell. Signal.*, **9**:519–529.

75. Reed, T.M., Browning, J.E., Blough, R.I., Vorhees, C.V., and Repaske, D.R. (1998) Genomic structure and chromosome location of the murine PDE1B phosphodiesterase gene. *Mamm. Genome*, **9**:571–576.

76. Qiu, Y. and Davis, R.L. (1993) Genetic dissection of the learning/memory gene dunce of *Drosophila melanogaster. Genes Dev.*, **7**:1447–1458.

77. Qiu, Y.H., Chen, C.N., Malone, T., Richter, L., Beckendorf, S.K., and Davis, R.L. (1991) Characterization of the memory gene dunce of *Drosophila melanogaster. J. Mol. Biol.*, **222**:553–565.

78. Swinnen, J.V., Joseph, D.R., and Conti, M. (1989) The mRNA encoding a high-affinity cAMP phosphodiesterase is regulated by hormones and cAMP. *Proc. Natl. Acad. Sci. USA*, **86**:8197–8201.

79. Polli, J.W. and Kincaid, R.L. (1994) Expression of a calmodulin-dependent phosphodi-esterase isoform (PDE1B1) correlates with brain regions having extensive dopaminergic innervation. *J. Neurosci.*, **14**:1251–1261.

80. Yan, C., Bentley, J.K., Sonnenburg, W.K., and Beavo, J.A. (1994) Differential expression of the 61 kDa and 63 kDa calmodulin-dependent phosphodiesterases in the mouse brain. *J. Neurosci.*, **14**:973–984.

81. Yan, C., Zhao, A.Z., Bentley, J.K., and Beavo, J.A. (1996) The calmodulin-dependent phosphodiesterase gene PDE1C encodes several functionally different splice variants in a tissue-specific manner. *J. Biol. Chem.*, **271**:25699–25706.

82. Miro, X., Perez-Torres, S., Puigdomenech, P., Palacios, J.M., and Mengod, G. (2002) Differential distribution of PDE4D splice variant mRNAs in rat brain suggests association with specific pathways and presynaptical localization. *Synapse*, **45**:259–269.

83. Reyes-Irisarri, E., Perez-Torres, S., Miro, X., Martinez, E., Puigdomenech, P., Palacios, J.M., and Mengod, G. (2008) Differential distribution of PDE4B splice variant mRNAs in rat brain and the effects of systemic administration of LPS in their expression. *Synapse*, **62**:74–79.

84. Richter, W., Jin, S.L., and Conti, M. (2005) Splice variants of the cyclic nucleotide phosphodiesterase PDE4D are differentially expressed and regulated in rat tissue. *Biochem. J.*, **388**:803–811.

85. Miro, X., Perez-Torres, S., Palacios, J.M., Puigdomenech, P., and Mengod, G. (2001) Differential distribution of cAMP-specific phosphodiesterase 7A mRNA in rat brain and peripheral organs. *Synapse*, **40**:201–214.

86. Sasaki, T., Kotera, J., and Omori, K. (2002) Novel alternative splice variants of rat phosphodiesterase 7B showing unique tissue-specific expression and phosphorylation. *Biochem. J.*, **361**:211–220.

87. Guipponi, M., Scott, H.S., Kudoh, J., Kawasaki, K., Shibuya, K., Shintani, A., Asakawa, S., Chen, H., Lalioti, M.D., Rossier, C., Minoshima, S., Shimizu, N., and Antonarakis, S.E. (1998) Identification and characterization of a novel cyclic nucleotide phosphodies-terase gene (PDE9A) that maps to 21q22.3: alternative splicing of mRNA transcripts, genomic structure and sequence. *Hum. Genet.*, **103**:386–392.

88. Bolger, G.B., Conti, M., and Houslay, M.D. (2007) Cellular functions of PDE4 enzymes. In: Beavo, J.A., Francis, S.H., and Houslay, M.D. (Eds.), *Cyclic Nucleotide Phospho-diesterases in Health and Disease*, Boca Raton, FL: CRC Press, pp. 99–130.

89. Rentero, C., Monfort, A., and Puigdomenech, P. (2003) Identification and distribution of different mRNA variants produced by differential splicing in the human phosphodiester-ase 9A gene. *Biochem. Biophys. Res. Commun.*, **301**:686–692.

90. Wang, P., Wu, P., Egan, R.W., and Billah, M.M. (2003) Identification and characterization of a new human type 9 cGMP-specific phosphodiesterase splice variant (PDE9A5). Differential tissue distribution and subcellular localization of PDE9A variants. *Gene*, **314**:15–27.

91. Bender, A.T. and Beavo, J.A. (2006) Cyclic nucleotide phosphodiesterases: molecular regulation to clinical use. *Pharmacol. Rev.*, **58**:488–520.

92. Kotera, J., Fujishige, K., Yuasa, K., and Omori, K. (1999) Characterization and phosphorylation of PDE10A2, a novel alternative splice variant of human phosphodiesterase that hydrolyzes cAMP and cGMP. *Biochem. Biophys. Res. Commun.*, **261**:551–557.

93. Charych, E.I., Jiang, L.X., Lo, F., Sullivan, K., and Brandon, N.J. (2010) Interplay of palmitoylation and phosphorylation in the trafficking and localization of phosphodiesterase 10A: implications for the treatment of schizophrenia. *J. Neurosci.*, **30**:9027–9037.

94. Piriev, N.I., Shih, J.M., and Farber, D.B. (1998) Defective RNA splicing resulting from a mutation in the cyclic guanosine monophosphate-phosphodiesterase beta-subunit gene. *Invest. Ophthalmol. Vis. Sci.*, **39**:463–470.

95. Pastan, I., Johnson, G.S., and Anderson, W.B. (1975) Role of cyclic nucleotides in growth control. *Annu. Rev. Biochem.*, **44**:491–522.

96. Ross, P.S., Manganiello, V.C., and Vaughan, M. (1977) Regulation of cyclic nucleotide phosphodiesterases in cultured hepatoma cells by dexamethasone and $N6,O2'$-dibutyryl adenosine $3':k'$-monophosphate. *J. Biol. Chem.*, **252**:1448–1452.

97. Vicini, E. and Conti, M. (1997) Characterization of an intronic promoter of a cyclic adenosine $3',5'$-monophosphate (cAMP)-specific phosphodiesterase gene that confers hormone and cAMP inducibility. *Mol. Endocrinol.*, **11**:839–850.

98. D'Sa, C., Tolbert, L.M., Conti, M., and Duman, R.S. (2002) Regulation of cAMP-specific phosphodiesterases type 4B and 4D (PDE4) splice variants by cAMP signaling in primary cortical neurons. *J. Neurochem.*, **81**:745–757.

99. Kim, J.S., Bailey, M.J., Ho, A.K., Moller, M., Gaildrat, P., and Klein, D.C. (2007) Daily rhythm in pineal phosphodiesterase (PDE) activity reflects adrenergic/$3',5'$-cyclic adenosine $5'$-monophosphate induction of the PDE4B2 variant. *Endocrinology*, **148**:1475–1485.

100. Le Jeune, I.R., Shepherd, M., Van Heeke, G., Houslay, M.D., and Hall, I.P. (2002) Cyclic AMP-dependent transcriptional up-regulation of phosphodiesterase 4D5 in human airway smooth muscle cells. Identification and characterization of a novel PDE4D5 promoter. *J. Biol. Chem.*, **277**:35980–35989.

101. Sasaki, T., Kotera, J., and Omori, K. (2004) Transcriptional activation of phosphodiesterase 7B1 by dopamine D_1 receptor stimulation through the cyclic AMP/cyclic AMP-dependent protein kinase/cyclic AMP-response element binding protein pathway in primary striatal neurons. *J. Neurochem.*, **89**:474–483.

102. Perez-Torres, S., Cortes, R., Tolnay, M., Probst, A., Palacios, J.M., and Mengod, G. (2003) Alterations on phosphodiesterase type 7 and 8 isozyme mRNA expression in Alzheimer's disease brains examined by *in situ* hybridization. *Exp. Neurol.*, **182**:322–334.

103. McLachlan, C.S., Chen, M.L., Lynex, C.N., Goh, D.L., Brenner, S., and Tay, S.K. (2007) Changes in PDE4D isoforms in the hippocampus of a patient with advanced Alzheimer disease. *Arch. Neurol.*, **64**:456–457.

104. Dlaboga, D., Hajjhussein, H., and O'Donnell, J.M. (2006) Regulation of phosphodiesterase-4 (PDE4) expression in mouse brain by repeated antidepressant treatment: comparison with rolipram. *Brain Res.*, **1096**:104–112.

105. Dlaboga, D., Hajjhussein, H., and O'Donnell, J.M. (2008) Chronic haloperidol and clozapine produce different patterns of effects on phosphodiesterase-1B, -4B, and -10A expression in rat striatum. *Neuropharmacology*, **54**:745–754.

106. McCahill, A., McSorley, T., Huston, E., Hill, E.V., Lynch, M.J., Gall, I., Keryer, G., Lygren, B., Tasken, K., van Heeke, G., and Houslay, M.D. (2005) In resting COS1 cells a dominant negative approach shows that specific, anchored PDE4 cAMP phosphodiesterase isoforms gate the activation, by basal cyclic AMP production, of AKAP-tethered protein kinase A type II located in the centrosomal region. *Cell. Signal.*, **17**:1158–1173.

107. Yan, C., Zhao, A.Z., Bentley, J.K., Loughney, K., Ferguson, K., and Beavo, J.A. (1995) Molecular cloning and characterization of a calmodulin-dependent phosphodiesterase enriched in olfactory sensory neurons. *Proc. Natl. Acad. Sci. USA*, **92**:9677–9681.

108. Andreeva, S.G., Dikkes, P., Epstein, P.M., and Rosenberg, P.A. (2001) Expression of cGMP-specific phosphodiesterase 9A mRNA in the rat brain. *J. Neurosci.*, **21**: 9068–9076.

109. Billingsley, M.L., Polli, J.W., Balaban, C.D., and Kincaid, R.L. (1990) Developmental expression of calmodulin-dependent cyclic nucleotide phosphodiesterase in rat brain. *Brain Res. Dev. Brain Res.*, **53**:253–263.

110. Furuyama, T., Iwahashi, Y., Tano, Y., Takagi, H., and Inagaki, S. (1994) Localization of 63-kDa calmodulin-stimulated phosphodiesterase mRNA in the rat brain by *in situ* hybridization histochemistry. *Brain Res. Mol. Brain Res.*, **26**:331–336.

111. Lal, S., Sharma, R.K., McGregor, C., and Macaulay, R.J. (1999) Immunohistochemical localization of calmodulin-dependent cyclic phosphodiesterase in the human brain. *Neurochem. Res.*, **24**:43–49.

112. Reed, T.M., Repaske, D.R., Snyder, G.L., Greengard, P., and Vorhees, C.V. (2002) Phosphodiesterase 1B knock-out mice exhibit exaggerated locomotor hyperactivity and DARPP-32 phosphorylation in response to dopamine agonists and display impaired spatial learning. *J. Neurosci.*, **22**:5188–5197.

113. Kakkar, R., Raju, R.V., and Sharma, R.K. (1999) Calmodulin-dependent cyclic nucleotide phosphodiesterase (PDE1). *Cell. Mol. Life Sci.*, **55**:1164–1186.

114. Laddha, S.S. and Bhatnagar, S.P. (2009) A new therapeutic approach in Parkinson's disease: some novel quinazoline derivatives as dual selective phosphodiesterase 1 inhibitors and anti-inflammatory agents. *Bioorg. Med. Chem.*, **17**:6796–6802.

115. Kakkar, R., Raju, R.V., Rajput, A.H., and Sharma, R.K. (1996) Inhibition of bovine brain calmodulin-dependent cyclic nucleotide phosphodiesterase isozymes by deprenyl. *Life Sci.*, **59**:PL337–PL341.

116. Kakkar, R., Raju, R.V., Rajput, A.H., and Sharma, R.K. (1997) Amantadine: an antiparkinsonian agent inhibits bovine brain 60 kDa calmodulin-dependent cyclic nucleotide phosphodiesterase isozyme. *Brain Res.*, **749**:290–294.

117. Wong, M.L., Whelan, F., Deloukas, P., Whittaker, P., Delgado, M., Cantor, R.M., McCann, S.M., and Licinio, J. (2006) Phosphodiesterase genes are associated with susceptibility to major depression and antidepressant treatment response. *Proc. Natl. Acad. Sci. USA*, **103**:15124–15129.

118. Van Staveren, W.C., Steinbusch, H.W., Markerink-Van Ittersum, M., Repaske, D.R., Goy, M.F., Kotera, J., Omori, K., Beavo, J.A., and De Vente, J. (2003) mRNA expression patterns of the cGMP-hydrolyzing phosphodiesterases types 2, 5, and 9 during development of the rat brain. *J. Comp. Neurol.*, **467**:566–580.

119. van Staveren, W.C., Steinbusch, H.W., Markerink-van Ittersum, M., Behrends, S., and de Vente, J. (2004) Species differences in the localization of cGMP-producing and NO-responsive elements in the mouse and rat hippocampus using cGMP immuno-cytochemistry. *Eur. J. Neurosci.*, **19**:2155–2168.

120. Reyes-Irisarri, E., Markerink-Van Ittersum, M., Mengod, G., and de Vente, J. (2007) Expression of the cGMP-specific phosphodiesterases 2 and 9 in normal and Alzheimer's disease human brains. *Eur. J. Neurosci.*, **25**:3332–3338.

121. Repaske, D.R., Corbin, J.G., Conti, M., and Goy, M.F. (1993) A cyclic GMP-stimulated cyclic nucleotide phosphodiesterase gene is highly expressed in the limbic system of the rat brain. *Neuroscience*, **56**:673–686.

122. Sonnenburg, W.K., Mullaney, P.J., and Beavo, J.A. (1991) Molecular cloning of a cyclic GMP-stimulated cyclic nucleotide phosphodiesterase cDNA. Identification and distribution of isozyme variants. *J. Biol. Chem.*, **266**:17655–17661.

123. Boess, F.G., Hendrix, M., van der Staay, F.J., Erb, C., Schreiber, R., van Staveren, W., de Vente, J., Prickaerts, J., Blokland, A., and Koenig, G. (2004) Inhibition of phosphodiesterase 2 increases neuronal cGMP, synaptic plasticity and memory performance. *Neuropharmacology*, **47**:1081–1092.

124. Suvarna, N.U. and O'Donnell, J.M. (2002) Hydrolysis of N-methyl-D-aspartate receptor-stimulated cAMP and cGMP by PDE4 and PDE2 phosphodiesterases in primary neuronal cultures of rat cerebral cortex and hippocampus. *J. Pharmacol. Exp. Ther.*, **302**:249–256.

125. Domek-Lopacinska, K. and Strosznajder, J.B. (2008) The effect of selective inhibition of cyclic GMP hydrolyzing phosphodiesterases 2 and 5 on learning and memory processes and nitric oxide synthase activity in brain during aging. *Brain Res.*, **1216**:68–77.

126. van Donkelaar, E.L., Rutten, K., Blokland, A., Akkerman, S., Steinbusch, H.W., and Prickaerts, J. (2008) Phosphodiesterase 2 and 5 inhibition attenuates the object memory deficit induced by acute tryptophan depletion. *Eur. J. Pharmacol.*, **600**:98–104.

127. Cho, C.H., Cho, D.H., Seo, M.R., and Juhnn, Y.S. (2000) Differential changes in the expression of cyclic nucleotide phosphodiesterase isoforms in rat brains by chronic treatment with electroconvulsive shock. *Exp. Mol. Med.*, **32**:110–114.

128. Reinhardt, R.R. and Bondy, C.A. (1996) Differential cellular pattern of gene expression for two distinct cGMP-inhibited cyclic nucleotide phosphodiesterases in developing and mature rat brain. *Neuroscience*, **72**:567–578.

129. Marte, A., Pepicelli, O., Cavallero, A., Raiteri, M., and Fedele, E. (2008) In vivo effects of phosphodiesterase inhibition on basal cyclic guanosine monophosphate levels in the prefrontal cortex, hippocampus and cerebellum of freely moving rats. *J. Neurosci. Res.*, **86**:3338–3347.

130. Zhao, A.Z., Huan, J.N., Gupta, S., Pal, R., and Sahu, A. (2002) A phosphatidylinositol 3-kinase phosphodiesterase 3B-cyclic AMP pathway in hypothalamic action of leptin on feeding. *Nat. Neurosci.*, **5**:727–728.

131. Dudai, Y., Jan, Y.N., Byers, D., Quinn, W.G., and Benzer, S. (1976) dunce, a mutant of *Drosophila* deficient in learning. *Proc. Natl. Acad. Sci. USA*, **73**:1684–1688.

132. Davis, R.L., Cherry, J., Dauwalder, B., Han, P.L., and Skoulakis, E. (1995) The cyclic AMP system and *Drosophila* learning. *Mol. Cell. Biochem.*, **149–150**:271–278.

133. Wachtel, H. (1982) Characteristic behavioural alterations in rats induced by rolipram and other selective adenosine cyclic 3′,5′-monophosphate phosphodiesterase inhibitors. *Psychopharmacology (Berl.)*, **77**:309–316.

134. Wachtel, H. (1983) Potential antidepressant activity of rolipram and other selective cyclic adenosine 3′,5′-monophosphate phosphodiesterase inhibitors. *Neuropharmacology*, **22**:267–272.

135. Zeller, E., Stief, H.J., Pflug, B., and Sastre-y-Hernandez, M. (1984) Results of a phase II study of the antidepressant effect of rolipram. *Pharmacopsychiatry*, **17**:188–190.

136. Menniti, F.S., Faraci, W.S., and Schmidt, C.J. (2006) Phosphodiesterases in the CNS: targets for drug development. *Nat. Rev. Drug Discov.*, **5**:660–670.

137. Bolger, G.B., Rodgers, L., and Riggs, M. (1994) Differential CNS expression of alternative mRNA isoforms of the mammalian genes encoding cAMP-specific phospho-diesterases. *Gene*, **149**:237–244.

138. Cherry, J.A. and Davis, R.L. (1999) Cyclic AMP phosphodiesterases are localized in regions of the mouse brain associated with reinforcement, movement, and affect. *J. Comp. Neurol.*, **407**:287–301.

139. Cheung, Y.F., Kan, Z., Garrett-Engele, P., Gall, I., Murdoch, H., Baillie, G.S., Camargo, L.M., Johnson, J.M., Houslay, M.D., and Castle, J.C. (2007) PDE4B5, a novel, super-short, brain-specific cAMP phosphodiesterase-4 variant whose isoform-specifying N-terminal region is identical to that of cAMP phosphodiesterase-4D6 (PDE4D6). *J. Pharmacol. Exp. Ther.*, **322**:600–609.

140. D'Sa, C., Eisch, A.J., Bolger, G.B., and Duman, R.S. (2005) Differential expression and regulation of the cAMP-selective phosphodiesterase type 4A splice variants in rat brain by chronic antidepressant administration. *Eur. J. Neurosci.*, **22**:1463–1475.

141. Fatemi, S.H., Reutiman, T.J., Folsom, T.D., and Lee, S. (2008) Phosphodiesterase-4A expression is reduced in cerebella of patients with bipolar disorder. *Psychiatr. Genet.*, **18**:282–288.

142. Iona, S., Cuomo, M., Bushnik, T., Naro, F., Sette, C., Hess, M., Shelton, E.R., and Conti, M. (1998) Characterization of the rolipram-sensitive, cyclic AMP-specific phosphodies-terases: identification and differential expression of immunologically distinct forms in the rat brain. *Mol. Pharmacol.*, **53**:23–32.

143. Lamontagne, S., Meadows, E., Luk, P., Normandin, D., Muise, E., Boulet, L., Pon, D.J., Robichaud, A., Robertson, G.S., Metters, K.M., and Nantel, F. (2001) Localization of phosphodiesterase-4 isoforms in the medulla and nodose ganglion of the squirrel monkey. *Brain Res.*, **920**:84–96.

144. Mackenzie, K.F., Topping, E.C., Bugaj-Gaweda, B., Deng, C., Cheung, Y.F., Olsen, A.E., Stockard, C.R., High Mitchell, L., Baillie, G.S., Grizzle, W.E., De Vivo, M., Houslay, M.D., Wang, D., and Bolger, G.B. (2008) Human PDE4A8, a novel brain-expressed PDE4 cAMP-specific phosphodiesterase that has undergone rapid evolutionary change. *Biochem. J.*, **411**:361–369.

145. Macphee, C.H., Reifsnyder, D.H., Moore, T.A., Lerea, K.M., and Beavo, J.A. (1988) Phosphorylation results in activation of a cAMP phosphodiesterase in human platelets. *J. Biol. Chem.*, **263**:10353–10358.

146. Miro, X., Perez-Torres, S., Artigas, F., Puigdomenech, P., Palacios, J.M., and Mengod, G. (2002) Regulation of cAMP phosphodiesterase mRNAs expression in rat brain by acute and chronic fluoxetine treatment. An *in situ* hybridization study. *Neuropharmacology*, **43**:1148–1157.

147. Mori, F., Perez-Torres, S., De Caro, R., Porzionato, A., Macchi, V., Beleta, J., Gavalda, A., Palacios, J.M., and Mengod, G. (2010) The human area postrema and other nuclei

related to the emetic reflex express cAMP phosphodiesterases 4B and 4D. *J. Chem. Neuroanat.*, **40**:36–42.

148. Perez-Torres, S., Miro, X., Palacios, J.M., Cortes, R., Puigdomenech, P., and Mengod, G. (2000) Phosphodiesterase type 4 isozymes expression in human brain examined by *in situ* hybridization histochemistry and [^{3}H]rolipram binding autoradiography. Comparison with monkey and rat brain. *J. Chem. Neuroanat.*, **20**:349–374.

149. Reyes-Irisarri, E., Sanchez, A.J., Garcia-Merino, J.A., and Mengod, G. (2007) Selective induction of cAMP phosphodiesterase PDE4B2 expression in experimental autoimmune encephalomyelitis. *J. Neuropathol. Exp. Neurol.*, **66**:923–931.

150. Zhang, H.T., Huang, Y., Jin, S.L., Frith, S.A., Suvarna, N., Conti, M., and O'Donnell, J.M. (2002) Antidepressant-like profile and reduced sensitivity to rolipram in mice deficient in the PDE4D phosphodiesterase enzyme. *Neuropsychopharmacology*, **27**:587–595.

151. Robichaud, A., Stamatiou, P.B., Jin, S.L., Lachance, N., MacDonald, D., Laliberte, F., Liu, S., Huang, Z., Conti, M., and Chan, C.C. (2002) Deletion of phosphodiesterase 4D in mice shortens alpha(2)-adrenoceptor-mediated anesthesia, a behavioral correlate of emesis. *J. Clin. Invest.*, **110**:1045–1052.

152. Numata, S., Iga, J., Nakataki, M., Tayoshi, S., Taniguchi, K., Sumitani, S., Tomotake, M., Tanahashi, T., Itakura, M., Kamegaya, Y., Tatsumi, M., Sano, A., Asada, T., Kunugi, H., Ueno, S., and Ohmori, T. (2009) Gene expression and association analyses of the phosphodiesterase 4B (PDE4B) gene in major depressive disorder in the Japanese population. *Am. J. Med. Genet. B Neuropsychiatr. Genet.*, **150B**:527–534.

153. Pickard, B.S., Thomson, P.A., Christoforou, A., Evans, K.L., Morris, S.W., Porteous, D.J., Blackwood, D.H., and Muir, W.J. (2007) The PDE4B gene confers sex-specific protection against schizophrenia. *Psychiatr. Genet.*, **17**:129–133.

154. Kahler, A.K., Otnaess, M.K., Wirgenes, K.V., Hansen, T., Jonsson, E.G., Agartz, I., Hall, H., Werge, T., Morken, G., Mors, O., Mellerup, E., Dam, H., Koefod, P., Melle, I., Steen, V.M., Andreassen, O.A., and Djurovic, S. (2010) Association study of PDE4B gene variants in Scandinavian schizophrenia and bipolar disorder multicenter case-control samples. *Am. J. Med. Genet. B Neuropsychiatr. Genet.*, **153B**:86–96.

155. Rastogi, A., Zai, C., Likhodi, O., Kennedy, J.L., and Wong, A.H. (2009) Genetic association and post-mortem brain mRNA analysis of DISC1 and related genes in schizophrenia. *Schizophr. Res.*, **114**:39–49.

156. Zhang, H.T., Huang, Y., Masood, A., Stolinski, L.R., Li, Y., Zhang, L., Dlaboga, D., Jin, S.L., Conti, M., and O'Donnell, J.M. (2008) Anxiogenic-like behavioral phenotype of mice deficient in phosphodiesterase 4B (PDE4B). *Neuropsychopharmacology*, **33**:1611–1623.

157. Siuciak, J.A., McCarthy, S.A., Chapin, D.S., and Martin, A.N. (2008) Behavioral and neurochemical characterization of mice deficient in the phosphodiesterase-4B (PDE4B) enzyme. *Psychopharmacology (Berl.)*, **197**:115–126.

158. Ahmed, T. and Frey, J.U. (2003) Expression of the specific type IV phosphodiesterase gene PDE4B3 during different phases of long-term potentiation in single hippocampal slices of rats *in vitro*. *Neuroscience*, **117**:627–638.

159. Ahmed, T. and Frey, J.U. (2005) Phosphodiesterase 4B (PDE4B) and cAMP-level regulation within different tissue fractions of rat hippocampal slices during long-term potentiation *in vitro*. *Brain Res.*, **1041**:212–222.

160. Ahmed, T., Frey, S., and Frey, J.U. (2004) Regulation of the phosphodiesterase PDE4B3-isotype during long-term potentiation in the area dentata *in vivo*. *Neuroscience*, **124**: 857–867.

161. Braun, N.N., Reutiman, T.J., Lee, S., Folsom, T.D., and Fatemi, S.H. (2007) Expression of phosphodiesterase 4 is altered in the brains of subjects with autism. *Neuroreport*, **18**:1841–1844.

162. Giordano, D., De Stefano, M.E., Citro, G., Modica, A., and Giorgi, M. (2001) Expression of cGMP-binding cGMP-specific phosphodiesterase (PDE5) in mouse tissues and cell lines using an antibody against the enzyme amino-terminal domain. *Biochim. Biophys. Acta*, **1539**:16–27.

163. Kotera, J., Fujishige, K., and Omori, K. (2000) Immunohistochemical localization of cGMP-binding cGMP-specific phosphodiesterase (PDE5) in rat tissues. *J. Histochem. Cytochem.*, **48**:685–693.

164. Loughney, K., Hill, T.R., Florio, V.A., Uher, L., Rosman, G.J., Wolda, S.L., Jones, B.A., Howard, M.L., McAllister-Lucas, L.M., Sonnenburg, W.K., Francis, S.H., Corbin, J.D., Beavo, J.A., and Ferguson, K. (1998) Isolation and characterization of cDNAs encoding PDE5A, a human cGMP-binding, cGMP-specific 3′,5′-cyclic nucleotide phosphodiesterase. *Gene*, **216**:139–147.

165. Menniti, F.S., Ren, J., Coskran, T.M., Liu, J., Morton, D., Sietsma, D.K., Som, A., Stephenson, D.T., Tate, B.A., and Finklestein, S.P. (2009) Phosphodiesterase 5A inhibitors improve functional recovery after stroke in rats: optimized dosing regimen with implications for mechanism. *J. Pharmacol. Exp. Ther.*, **331**:842–850.

166. Shimizu-Albergine, M., Rybalkin, S.D., Rybalkina, I.G., Feil, R., Wolfsgruber, W., Hofmann, F., and Beavo, J.A. (2003) Individual cerebellar Purkinje cells express different cGMP phosphodiesterases (PDEs): *in vivo* phosphorylation of cGMP-specific PDE (PDE5) as an indicator of cGMP-dependent protein kinase (PKG) activation. *J. Neurosci.*, **23**:6452–6459.

167. Prickaerts, J., Sik, A., van Staveren, W.C., Koopmans, G., Steinbusch, H.W., van der Staay, F.J., de Vente, J., and Blokland, A. (2004) Phosphodiesterase type 5 inhibition improves early memory consolidation of object information. *Neurochem. Int.*, **45**: 915–928.

168. Rutten, K., Vente, J.D., Sik, A., Ittersum, M.M., Prickaerts, J., and Blokland, A. (2005) The selective PDE5 inhibitor, sildenafil, improves object memory in Swiss mice and increases cGMP levels in hippocampal slices. *Behav. Brain Res.*, **164**:11–16.

169. Devan, B.D., Sierra-Mercado, D., Jr., Jimenez, M., Bowker, J.L., Duffy, K.B., Spangler, E.L., and Ingram, D.K. (2004) Phosphodiesterase inhibition by sildenafil citrate attenuates the learning impairment induced by blockade of cholinergic muscarinic receptors in rats. *Pharmacol. Biochem. Behav.*, **79**:691–699.

170. Devan, B.D., Pistell, P.J., Daffin, L.W., Jr., Nelson, C.M., Duffy, K.B., Bowker, J.L., Bharati, I.S., Sierra-Mercado, D., Spangler, E.L., and Ingram, D.K. (2007) Sildenafil citrate attenuates a complex maze impairment induced by intracerebroventricular infusion of the NOS inhibitor *N*-omega-nitro-ʟ-arginine methyl ester. *Eur. J. Pharmacol.*, **563**:134–140.

171. Prickaerts, J., Sik, A., van der Staay, F.J., de Vente, J., and Blokland, A. (2005) Dissociable effects of acetylcholinesterase inhibitors and phosphodiesterase type 5 inhibitors on object recognition memory: acquisition versus consolidation. *Psychopharmacology (Berl.)*, **177**:381–390.

172. Rutten, K., Basile, J.L., Prickaerts, J., Blokland, A., and Vivian, J.A. (2008) Selective PDE inhibitors rolipram and sildenafil improve object retrieval performance in adult cynomolgus macaques. *Psychopharmacology (Berl.)*, **196**:643–648.

173. Rutten, K., Van Donkelaar, E.L., Ferrington, L., Blokland, A., Bollen, E., Steinbusch, H.W., Kelly, P.A., and Prickaerts, J.H. (2009) Phosphodiesterase inhibitors enhance object memory independent of cerebral blood flow and glucose utilization in rats. *Neuropsychopharmacology*, **34**:1914–1925.

174. Puzzo, D., Staniszewski, A., Deng, S.X., Privitera, L., Leznik, E., Liu, S., Zhang, H., Feng, Y., Palmeri, A., Landry, D.W., and Arancio, O. (2009) Phosphodiesterase 5 inhibition improves synaptic function, memory, and amyloid-beta load in an Alzheimer's disease mouse model. *J. Neurosci.*, **29**:8075–8086.

175. Goff, D.C., Cather, C., Freudenreich, O., Henderson, D.C., Evins, A.E., Culhane, M.A., and Walsh, J.P. (2009) A placebo-controlled study of sildenafil effects on cognition in schizophrenia. *Psychopharmacology (Berl.)*, **202**:411–417.

176. Grass, H., Klotz, T., Fathian-Sabet, B., Berghaus, G., Engelmann, U., and Kaferstein, H. (2001) Sildenafil (Viagra): is there an influence on psychological performance? *Int. Urol. Nephrol.*, **32**:409–412.

177. Schultheiss, D., Muller, S.V., Nager, W., Stief, C.G., Schlote, N., Jonas, U., Asvestis, C., Johannes, S., and Munte, T.F. (2001) Central effects of sildenafil (Viagra) on auditory selective attention and verbal recognition memory in humans: a study with event-related brain potentials. *World J. Urol.*, **19**:46–50.

178. Morin, F., Lugnier, C., Kameni, J., and Voisin, P. (2001) Expression and role of phosphodiesterase 6 in the chicken pineal gland. *J. Neurochem.*, **78**:88–99.

179. Kelly, M.P., Adamowicz, W., Bove, S., Hartman, A.J., Mariga, A., Pathak, G., Reinhart, V., Romegialli, A., and Kleiman, R.J. (2014) Select 3′,5′-cyclic nucleotide phosphodiesterases exhibit altered expression in the aged rodent brain. *Cell. Signal.*, **26**(2):383–397.

180. Reyes-Irisarri, E., Perez-Torres, S., and Mengod, G. (2005) Neuronal expression of cAMP-specific phosphodiesterase 7B mRNA in the rat brain. *Neuroscience*, **132**: 1173–1185.

181. Kobayashi, T., Gamanuma, M., Sasaki, T., Yamashita, Y., Yuasa, K., Kotera, J., and Omori, K. (2003) Molecular comparison of rat cyclic nucleotide phosphodiesterase 8 family: unique expression of PDE8B in rat brain. *Gene*, **319**:21–31.

182. van Staveren, W.C., Glick, J., Markerink-van Ittersum, M., Shimizu, M., Beavo, J.A., Steinbusch, H.W., and de Vente, J. (2002) Cloning and localization of the cGMP-specific phosphodiesterase type 9 in the rat brain. *J. Neurocytol.*, **31**:729–741.

183. Sette, C. and Conti, M. (1996) Phosphorylation and activation of a cAMP-specific phosphodiesterase by the cAMP-dependent protein kinase. Involvement of serine 54 in the enzyme activation. *J. Biol. Chem.*, **271**:16526–16534.

184. van der Staay, F.J., Rutten, K., Barfacker, L., Devry, J., Erb, C., Heckroth, H., Karthaus, D., Tersteegen, A., van Kampen, M., Blokland, A., Prickaerts, J., Reymann, K.G., Schroder, U.H., and Hendrix, M. (2008) The novel selective PDE9 inhibitor BAY 73-6691 improves learning and memory in rodents. *Neuropharmacology*, **55**:908–918.

185. Wunder, F., Tersteegen, A., Rebmann, A., Erb, C., Fahrig, T., and Hendrix, M. (2005) Characterization of the first potent and selective PDE9 inhibitor using a cGMP reporter cell line. *Mol. Pharmacol.*, **68**:1775–1781.

186. Hutson, P.H., Finger, E.N., Magliaro, B.C., Smith, S.M., Converso, A., Sanderson, P.E., Mullins, D., Hyde, L.A., Eschle, B.K., Turnbull, Z., Sloan, H., Guzzi, M., Zhang, X., Wang, A., Rindgen, D., Mazzola, R., Vivian, J.A., Eddins, D., Uslaner, J.M., Bednar, R., Gambone, C., Le-Mair, W., Marino, M.J., Sachs, N., Xu, G., and Parmentier-Batteur, S. (2011) The selective phosphodiesterase 9 (PDE9) inhibitor PF-04447943 (6-[(3S,4S)-4-methyl-1-(pyrimidin-2-ylmethyl)pyrrolidin-3-yl]-1-(tetrahydro-2H-pyran-4-yl)-1,5-dihydro-4H-pyrazolo[3,4-d]pyrimidin-4-one) enhances synaptic plasticity and cognitive function in rodents. *Neuropharmacology*, **61**:665–676.

187. Kleiman, R.J., Chapin, D.S., Christoffersen, C., Freeman, J., Fonseca, K.R., Geoghegan, K.F., Grimwood, S., Guanowsky, V., Hajos, M., Harms, J.F., Helal, C.J., Hoffmann, W.E., Kocan, G.P., Majchrzak, M.J., McGinnis, D., McLean, S., Menniti, F.S., Nelson, F., Roof, R., Schmidt, A.W., Seymour, P.A., Stephenson, D.T., Tingley, F.D., Vanase-Frawley, M., Verhoest, P.R., and Schmidt, C.J. (2012) Phosphodiesterase 9A regulates central cGMP and modulates responses to cholinergic and monoaminergic perturbation *in vivo*. *J. Pharmacol. Exp. Ther.*, **341**:396–409.

188. Vardigan, J.D., Converso, A., Hutson, P.H., and Uslaner, J.M. (2011) The selective phosphodiesterase 9 (PDE9) inhibitor PF-04447943 attenuates a scopolamine-induced deficit in a novel rodent attention task. *J. Neurogenet.*, **25**:120–126.

189. Loughney, K., Snyder, P.B., Uher, L., Rosman, G.J., Ferguson, K., and Florio, V.A. (1999) Isolation and characterization of PDE10A, a novel human 3′,5′-cyclic nucleotide phosphodiesterase. *Gene*, **234**:109–117.

190. Coskran, T.M., Morton, D., Menniti, F.S., Adamowicz, W.O., Kleiman, R.J., Ryan, A. M., Strick, C.A., Schmidt, C.J., and Stephenson, D.T. (2006) Immunohistochemical localization of phosphodiesterase 10A in multiple mammalian species. *J. Histochem. Cytochem.*, **54**:1205–1213.

191. Seeger, T.F., Bartlett, B., Coskran, T.M., Culp, J.S., James, L.C., Krull, D.L., Lanfear, J., Ryan, A.M., Schmidt, C.J., Strick, C.A., Varghese, A.H., Williams, R.D., Wylie, P.G., and Menniti, F.S. (2003) Immunohistochemical localization of PDE10A in the rat brain. *Brain Res.*, **985**:113–126.

192. Xie, Z., Adamowicz, W.O., Eldred, W.D., Jakowski, A.B., Kleiman, R.J., Morton, D.G., Stephenson, D.T., Strick, C.A., Williams, R.D., and Menniti, F.S. (2006) Cellular and subcellular localization of PDE10A, a striatum-enriched phosphodiesterase. *Neuroscience*, **139**:597–607.

193. Fujishige, K., Kotera, J., and Omori, K. (1999) Striatum- and testis-specific phosphodiesterase PDE10A isolation and characterization of a rat PDE10A. *Eur. J. Biochem.*, **266**:1118–1127.

194. Soderling, S.H., Bayuga, S.J., and Beavo, J.A. (1999) Isolation and characterization of a dual-substrate phosphodiesterase gene family: PDE10A. *Proc. Natl. Acad. Sci. USA*, **96**:7071–7076.

195. Threlfell, S., Sammut, S., Menniti, F.S., Schmidt, C.J., and West, A.R. (2009) Inhibition of phosphodiesterase 10A increases the responsiveness of striatal projection neurons to cortical stimulation. *J. Pharmacol. Exp. Ther.*, **328**:785–795.

196. Giampa, C., Patassini, S., Borreca, A., Laurenti, D., Marullo, F., Bernardi, G., Menniti, F.S., and Fusco, F.R. (2009) Phosphodiesterase 10 inhibition reduces striatal excitotoxicity in the quinolinic acid model of Huntington's disease. *Neurobiol. Dis.*, **34**: 450–456.

197. Sano, H., Nagai, Y., Miyakawa, T., Shigemoto, R., and Yokoi, M. (2008) Increased social interaction in mice deficient of the striatal medium spiny neuron-specific phosphodiesterase 10A2. *J. Neurochem.*, **105**:546–556.

198. Siuciak, J.A., Chapin, D.S., Harms, J.F., Lebel, L.A., McCarthy, S.A., Chambers, L., Shrikhande, A., Wong, S., Menniti, F.S., and Schmidt, C.J. (2006) Inhibition of the striatum-enriched phosphodiesterase PDE10A: a novel approach to the treatment of psychosis. *Neuropharmacology*, **51**:386–396.

199. Nishi, A., Kuroiwa, M., Miller, D.B., O'Callaghan, J.P., Bateup, H.S., Shuto, T., Sotogaku, N., Fukuda, T., Heintz, N., Greengard, P., and Snyder, G.L. (2008) Distinct roles of PDE4 and PDE10A in the regulation of cAMP/PKA signaling in the striatum. *J. Neurosci.*, **28**:10460–10471.

200. Kelly, M.P. and Brandon, N.J. (2009) Differential function of phosphodiesterase families in the brain: gaining insights through the use of genetically modified animals. *Prog. Brain Res.*, **179**:67–73.

201. Kleiman, R.J., Kimmel, L.H., Bove, S.E., Lanz, T.A., Harms, J.F., Romegialli, A., Miller, K.S., Willis, A., des Etages, S., Kuhn, M., and Schmidt, C.J. (2011) Chronic suppression of phosphodiesterase 10A alters striatal expression of genes responsible for neurotransmitter synthesis, neurotransmission, and signaling pathways implicated in Huntington's disease. *J. Pharmacol. Exp. Ther.*, **336**:64–76.

202. Siuciak, J.A. (2008) The role of phosphodiesterases in schizophrenia: therapeutic implications. *CNS Drugs*, **22**:983–993.

203. Zhang, H.T. (2010) Phosphodiesterase targets for cognitive dysfunction and schizophrenia—a New York Academy of Sciences Meeting. *IDrugs*, **13**:166–168.

204. Kleiman, R.J., Kimmel, L.H., Bove, S.E., Lanz, T.A., Harms, J.F., Romegialli, A., Miller, K.S., Willis, A., des Etages, S., Kuhn, M., and Schmidt, C.J. (2012) Chronic suppression of phosphodiesterase 10A alters striatal expression of genes responsible for neurotransmitter synthesis, neurotransmission, and signaling pathways implicated in Huntington's disease. *J. Pharmacol. Exp. Ther.*, **336**:64–76.

205. Grauer, S.M., Pulito, V.L., Navarra, R.L., Kelly, M.P., Kelley, C., Graf, R., Langen, B., Logue, S., Brennan, J., Jiang, L., Charych, E., Egerland, U., Liu, F., Marquis, K.L., Malamas, M., Hage, T., Comery, T.A., and Brandon, N.J. (2009) Phosphodiesterase 10A inhibitor activity in preclinical models of the positive, cognitive, and negative symptoms of schizophrenia. *J. Pharmacol. Exp. Ther.*, **331**:574–590.

206. Schmidt, C.J., Chapin, D.S., Cianfrogna, J., Corman, M.L., Hajos, M., Harms, J.F., Hoffman, W.E., Lebel, L.A., McCarthy, S.A., Nelson, F.R., Proulx-LaFrance, C., Majchrzak, M.J., Ramirez, A.D., Schmidt, K., Seymour, P.A., Siuciak, J.A., Tingley, F.D., 3rd, Williams, R.D., Verhoest, P.R., and Menniti, F.S. (2008) Preclinical characterization of selective phosphodiesterase 10A inhibitors: a new therapeutic approach to the treatment of schizophrenia. *J. Pharmacol. Exp. Ther.*, **325**:681–690.

207. Siuciak, J.A., McCarthy, S.A., Chapin, D.S., Fujiwara, R.A., James, L.C., Williams, R.D., Stock, J.L., McNeish, J.D., Strick, C.A., Menniti, F.S., and Schmidt, C.J. (2006) Genetic deletion of the striatum-enriched phosphodiesterase PDE10A: evidence for altered striatal function. *Neuropharmacology*, **51**:374–385.

208. Hebb, A.L., Robertson, H.A., and Denovan-Wright, E.M. (2008) Phosphodiesterase 10A inhibition is associated with locomotor and cognitive deficits and increased anxiety in mice. *Eur. Neuropsychopharmacol.*, **18**:339–363.

209. Kelly, M.P., Logue, S.F., Brennan, J., Day, J.P., Lakkaraju, S., Jiang, L., Zhong, X., Tam, M., Sukoff Rizzo, S.J., Platt, B.J., Dwyer, J.M., Neal, S., Pulito, V.L., Agostino, M.J., Grauer, S.M., Navarra, R.L., Kelley, C., Comery, T.A., Murrills, R.J., Houslay, M.D., and Brandon, N.J. (2010) Phosphodiesterase 11A in brain is enriched in ventral hippocampus and deletion causes psychiatric disease-related phenotypes. *Proc. Natl. Acad. Sci. USA*, **107**:8457–8462.

210. Loughney, K., Taylor, J., and Florio, V.A. (2005) 3′,5′-Cyclic nucleotide phosphodiesterase 11A: localization in human tissues. *Int. J. Impot. Res.*, **17**:320–325.

211. Sharma, R.K. (1991) Phosphorylation and characterization of bovine heart calmodulin-dependent phosphodiesterase. *Biochemistry*, **30**:5963–5968.

212. Sharma, R.K. and Wang, J.H. (1986) Calmodulin and Ca^{2+}-dependent phosphorylation and dephosphorylation of 63-kDa subunit-containing bovine brain calmodulin-stimulated cyclic nucleotide phosphodiesterase isozyme. *J. Biol. Chem.*, **261**:1322–1328.

213. Florio, V., Sonnenburg, W., Johnson, R., Kwak, K., Jensen, G., Walsh, K., and Beavo, J. (1994) Phosphorylation of the 61-kDa calmodulin-stimulated cyclic nucleotide phosphodiesterase at serine 120 reduces its affinity for calmodulin. *FASEB J.*, abstract 465: A80.

214. Degerman, E., Smith, C.J., Tornqvist, H., Vasta, V., Belfrage, P., and Manganiello, V.C. (1990) Evidence that insulin and isoprenaline activate the cGMP-inhibited low-K_m cAMP phosphodiesterase in rat fat cells by phosphorylation. *Proc. Natl. Acad. Sci. USA*, **87**:533–537.

215. Rascon, A., Degerman, E., Taira, M., Meacci, E., Smith, C.J., Manganiello, V., Belfrage, P., and Tornqvist, H. (1994) Identification of the phosphorylation site *in vitro* for cAMP-dependent protein kinase on the rat adipocyte cGMP-inhibited cAMP phosphodiesterase. *J. Biol. Chem.*, **269**:11962–11966.

216. Sette, C., Iona, S., and Conti, M. (1994) The short-term activation of a rolipram-sensitive, cAMP-specific phosphodiesterase by thyroid-stimulating hormone in thyroid FRTL-5 cells is mediated by a cAMP-dependent phosphorylation. *J. Biol. Chem.*, **269**:9245–9252.

217. Hoffmann, R., Wilkinson, I.R., McCallum, J.F., Engels, P., and Houslay, M.D. (1998) cAMP-specific phosphodiesterase HSPDE4D3 mutants which mimic activation and changes in rolipram inhibition triggered by protein kinase A phosphorylation of Ser-54: generation of a molecular model. *Biochem. J.*, **333**:139–149.

218. Liu, H. and Maurice, D.H. (1999) Phosphorylation-mediated activation and translocation of the cyclic AMP-specific phosphodiesterase PDE4D3 by cyclic AMP-dependent protein kinase and mitogen-activated protein kinases. A potential mechanism allowing for the coordinated regulation of PDE4D activity and targeting. *J. Biol. Chem.*, **274**:10557–10565.

219. Wyatt, T.A., Francis, S.H., McAllister-Lucas, L.M., Naftilan, A.J., and Corbin, J.D. (1994) Phosphorylation of cGMP-binding cGMP-specific phosphodiesterase in vascular smooth muscle cells. *FASEB J.*, abstract 2150: A372.

220. Yuasa, K., Ohgaru, T., Asahina, M., and Omori, K. (2001) Identification of rat cyclic nucleotide phosphodiesterase 11A (PDE11A): comparison of rat and human PDE11A splicing variants. *Eur. J. Biochem.*, **268**:4440–4448.

221. Bruss, M.D., Richter, W., Horner, K., Jin, S.L., and Conti, M. (2008) Critical role of PDE4D in beta2-adrenoceptor-dependent cAMP signaling in mouse embryonic fibroblasts. *J. Biol. Chem.*, **283**:22430–22442.

222. Oki, N., Takahashi, S.I., Hidaka, H., and Conti, M. (2000) Short term feedback regulation of cAMP in FRTL-5 thyroid cells. Role of PDE4D3 phosphodiesterase activation. *J. Biol. Chem.*, **275**:10831–10837.

223. Bessay, E.P., Blount, M.A., Zoraghi, R., Beasley, A., Grimes, K.A., Francis, S.H., and Corbin, J.D. (2008) Phosphorylation increases affinity of the phosphodiesterase-5 catalytic site for tadalafil. *J. Pharmacol. Exp. Ther.*, **325**:62–68.

224. Corbin, J.D. and Francis, S.H. (1999) Cyclic GMP phosphodiesterase-5: target of sildenafil. *J. Biol. Chem.*, **274**:13729–13732.

225. MacKenzie, S.J., Baillie, G.S., McPhee, I., MacKenzie, C., Seamons, R., McSorley, T., Millen, J., Beard, M.B., van Heeke, G., and Houslay, M.D. (2002) Long PDE4 cAMP specific phosphodiesterases are activated by protein kinase A-mediated phosphorylation of a single serine residue in Upstream Conserved Region 1 (UCR1). *Br. J. Pharmacol.*, **136**:421–433.

226. Alvarez, R., Sette, C., Yang, D., Eglen, R., Wilhelm, R., Shelton, E.R., and Conti, M. (1995) Activation and selective inhibition of a cyclic AMP-specific phosphodiesterase, PDE4D3. *Mol. Pharmacol.*, **48**:616–622.

227. Dodge, K.L., Khouangsathiene, S., Kapiloff, M.S., Mouton, R., Hill, E.V., Houslay, M. D., Langeberg, L.K., and Scott, J.D. (2001) mAKAP assembles a protein kinase A/PDE4 phosphodiesterase cAMP signaling module. *EMBO J.*, **20**:1921–1930.

228. Tasken, K.A., Collas, P., Kemmner, W.A., Witczak, O., Conti, M., and Tasken, K. (2001) Phosphodiesterase 4D and protein kinase A type II constitute a signaling unit in the centrosomal area. *J. Biol. Chem.*, **276**:21999–22002.

229. Mackenzie, K.F., Wallace, D.A., Hill, E., Anthony, D.F., Henderson, D.J., Houslay, D.M., Arthur, J.S., Baillie, G.S., and Houslay, M.D. (2011) Phosphorylation of cyclic AMP phosphodiesterase-4A5 (PDE4A5) by MK2 (MAPKAPK2) attenuates its activation through PKA phosphorylation. *Biochem. J.*, **435**:755–759.

230. Hoffmann, R., Baillie, G.S., MacKenzie, S.J., Yarwood, S.J., and Houslay, M.D. (1999) The MAP kinase ERK2 inhibits the cyclic AMP-specific phosphodiesterase HSPDE4D3 by phosphorylating it at Ser579. *EMBO J.*, **18**:893–903.

231. Gao, Y., Nikulina, E., Mellado, W., and Filbin, M.T. (2003) Neurotrophins elevate cAMP to reach a threshold required to overcome inhibition by MAG through extracellular signal-regulated kinase-dependent inhibition of phosphodiesterase. *J. Neurosci.*, **23**:11770–11777.

232. Dodge-Kafka, K.L., Soughayer, J., Pare, G.C., Carlisle Michel, J.J., Langeberg, L.K., Kapiloff, M.S., and Scott, J.D. (2005) The protein kinase A anchoring protein mAKAP co-ordinates two integrated cAMP effector pathways. *Nature*, **437**:574–578.

233. Collins, D.M., Murdoch, H., Dunlop, A.J., Charych, E., Baillie, G.S., Wang, Q., Herberg, F.W., Brandon, N., Prinz, A., and Houslay, M.D. (2008) Ndel1 alters its conformation by sequestering cAMP-specific phosphodiesterase-4D3 (PDE4D3) in a manner that is dynamically regulated through Protein Kinase A (PKA). *Cell. Signal.*, **20**:2356–2369.

234. Murdoch, H., Mackie, S., Collins, D.M., Hill, E.V., Bolger, G.B., Klussmann, E., Porteous, D.J., Millar, J.K., and Houslay, M.D. (2007) Isoform-selective susceptibility of DISC1/phosphodiesterase-4 complexes to dissociation by elevated intracellular cAMP levels. *J. Neurosci.*, **27**:9513–9524.

235. Ang, K.L. and Antoni, F.A. (2002) Reciprocal regulation of calcium dependent and calcium independent cyclic AMP hydrolysis by protein phosphorylation. *J. Neurochem.*, **81**:422–433.

236. Sharma, R.K. and Wang, J.H. (1985) Differential regulation of bovine brain calmodulin-dependent cyclic nucleotide phosphodiesterase isoenzymes by cyclic AMP-dependent protein kinase and calmodulin-dependent phosphatase. *Proc. Natl. Acad. Sci. USA*, **82**:2603–2607.

237. Sharma, R.K. and Wang, J.H. (1986) Regulation of cAMP concentration by calmodulin-dependent cyclic nucleotide phosphodiesterase. *Biochem. Cell Biol.*, **64**:1072–1080.

238. Charbonneau, H., Prusti, R.K., LeTrong, H., Sonnenburg, W.K., Mullaney, P.J., Walsh, K.A., and Beavo, J.A. (1990) Identification of a noncatalytic cGMP-binding domain conserved in both the cGMP-stimulated and photoreceptor cyclic nucleotide phosphodiesterases. *Proc. Natl. Acad. Sci. USA*, **87**:288–292.

239. Martinez, S.E., Wu, A.Y., Glavas, N.A., Tang, X.B., Turley, S., Hol, W.G., and Beavo, J.A. (2002) The two GAF domains in phosphodiesterase 2A have distinct roles in dimerization and in cGMP binding. *Proc. Natl. Acad. Sci. USA*, **99**:13260–13265.

240. Rybalkin, S.D., Rybalkina, I.G., Shimizu-Albergine, M., Tang, X.B., and Beavo, J.A. (2003) PDE5 is converted to an activated state upon cGMP binding to the GAF A domain. *EMBO J.*, **22**:469–478.

241. Matthiesen, K. and Nielsen, J. (2009) Binding of cyclic nucleotides to phosphodiesterase 10A and 11A GAF domains does not stimulate catalytic activity. *Biochem. J.*, **423**:401–409.

242. Nemoz, G., Sette, C., and Conti, M. (1997) Selective activation of rolipram-sensitive, cAMP-specific phosphodiesterase isoforms by phosphatidic acid. *Mol. Pharmacol.*, **51**:242–249.

243. Soderling, S.H., Bayuga, S.J., and Beavo, J.A. (1998) Cloning and characterization of a cAMP-specific cyclic nucleotide phosphodiesterase. *Proc. Natl. Acad. Sci. USA*, **95**:8991–8996.

244. Sharma, R.K., Adachi, A.M., Adachi, K., and Wang, J.H. (1984) Demonstration of bovine brain calmodulin-dependent cyclic nucleotide phosphodiesterase isozymes by monoclonal antibodies. *J. Biol. Chem.*, **259**:9248–9254.

245. Hansen, R.S. and Beavo, J.A. (1986) Differential recognition of calmodulin-enzyme complexes by a conformation-specific anti-calmodulin monoclonal antibody. *J. Biol. Chem.*, **261**:14636–14645.

246. Deterre, P., Bigay, J., Forquet, F., Robert, M., and Chabre, M. (1988) cGMP phosphodiesterase of retinal rods is regulated by two inhibitory subunits. *Proc. Natl. Acad. Sci. USA*, **85**:2424–2428.

247. Juilfs, D.M., Fulle, H.J., Zhao, A.Z., Houslay, M.D., Garbers, D.L., and Beavo, J.A. (1997) A subset of olfactory neurons that selectively express cGMP-stimulated phosphodiesterase (PDE2) and guanylyl cyclase-D define a unique olfactory signal transduction pathway. *Proc. Natl. Acad. Sci. USA*, **94**:3388–3395.

248. Cote, R.H. (2006) Photoreceptor phosphodiesterase (PDE6): a G-protein-activated PDE regulating visual excitation in rod and cone photoreceptor cells. In: Beavo, J., Francis, S. H., and Houslay, M.D. (Eds.), *Cyclic Nucleotide Phosphodiesterases in Health and Disease*, Boca Raton, FL: CRC Press, pp. 165–193.

249. Guo, L.W., Muradov, H., Hajipour, A.R., Sievert, M.K., Artemyev, N.O., and Ruoho, A.E. (2006) The inhibitory gamma subunit of the rod cGMP phosphodiesterase binds the catalytic subunits in an extended linear structure. *J. Biol. Chem.*, **281**:15412–15422.

250. Bolger, G.B., Peden, A.H., Steele, M.R., MacKenzie, C., McEwan, D.G., Wallace, D.A., Huston, E., Baillie, G.S., and Houslay, M.D. (2003) Attenuation of the activity of the

cAMP-specific phosphodiesterase PDE4A5 by interaction with the immunophilin XAP2. *J. Biol. Chem.*, **278**:33351–33363.

251. Palmer, D., Jimmo, S.L., Raymond, D.R., Wilson, L.S., Carter, R.L., and Maurice, D.H. (2007) Protein kinase A phosphorylation of human phosphodiesterase 3B promotes 14-3-3 protein binding and inhibits phosphatase-catalyzed inactivation. *J. Biol. Chem.*, **282**:9411–9419.

252. Kono, T., Robinson, F.W., and Sarver, J.A. (1975) Insulin-sensitive phosphodiesterase. Its localization, hormonal stimulation, and oxidative stabilization. *J. Biol. Chem.*, **250**:7826–7835.

253. Taira, M., Hockman, S.C., Calvo, J.C., Belfrage, P., and Manganiello, V.C. (1993) Molecular cloning of the rat adipocyte hormone-sensitive cyclic GMP-inhibited cyclic nucleotide phosphodiesterase. *J. Biol. Chem.*, **268**:18573–18579.

254. Shakur, Y., Takeda, K., Kenan, Y., Yu, Z.X., Rena, G., Brandt, D., Houslay, M.D., Degerman, E., Ferrans, V.J., and Manganiello, V.C. (2000) Membrane localization of cyclic nucleotide phosphodiesterase 3 (PDE3). Two N-terminal domains are required for the efficient targeting to, and association of, PDE3 with endoplasmic reticulum. *J. Biol. Chem.*, **275**:38749–38761.

255. Yang, Q., Paskind, M., Bolger, G., Thompson, W.J., Repaske, D.R., Cutler, L.S., and Epstein, P.M. (1994) A novel cyclic GMP stimulated phosphodiesterase from rat brain. *Biochem. Biophys. Res. Commun.*, **205**:1850–1858.

256. Anant, J.S., Ong, O.C., Xie, H.Y., Clarke, S., O'Brien, P.J., and Fung, B.K. (1992) *In vivo* differential prenylation of retinal cyclic GMP phosphodiesterase catalytic subunits. *J. Biol. Chem.*, **267**:687–690.

257. Shakur, Y., Wilson, M., Pooley, L., Lobban, M., Griffiths, S.L., Campbell, A.M., Beattie, J., Daly, C., and Houslay, M.D. (1995) Identification and characterization of the type-IVA cyclic AMP-specific phosphodiesterase RD1 as a membrane-bound protein expressed in cerebellum. *Biochem. J.*, **306**(Part 3):801–809.

258. Baillie, G.S., Huston, E., Scotland, G., Hodgkin, M., Gall, I., Peden, A.H., MacKenzie, C., Houslay, E.S., Currie, R., Pettitt, T.R., Walmsley, A.R., Wakelam, M.J., Warwicker, J., and Houslay, M.D. (2002) TAPAS-1, a novel microdomain within the unique N-terminal region of the PDE4A1 cAMP-specific phosphodiesterase that allows rapid, Ca^{2+}-triggered membrane association with selectivity for interaction with phosphatidic acid. *J. Biol. Chem.*, **277**:28298–28309.

259. Zaccolo, M. and Pozzan, T. (2002) Discrete microdomains with high concentration of cAMP in stimulated rat neonatal cardiac myocytes. *Science*, **295**:1711–1715.

260. Nikolaev, V.O., Bunemann, M., Schmitteckert, E., Lohse, M.J., and Engelhardt, S. (2006) Cyclic AMP imaging in adult cardiac myocytes reveals far-reaching β_1-adrenergic but locally confined β_2-adrenergic receptor-mediated signaling. *Circ. Res.*, **99**:1084–1091.

261. Rich, T.C., Fagan, K.A., Tse, T.E., Schaack, J., Cooper, D.M., and Karpen, J.W. (2001) A uniform extracellular stimulus triggers distinct cAMP signals in different compartments of a simple cell. *Proc. Natl. Acad. Sci. USA*, **98**:13049–13054.

262. Rich, T.C., Tse, T.E., Rohan, J.G., Schaack, J., and Karpen, J.W. (2001) *In vivo* assessment of local phosphodiesterase activity using tailored cyclic nucleotide-gated channels as cAMP sensors. *J. Gen. Physiol.*, **118**:63–78.

263. Gervasi, N., Tchenio, P., and Preat, T. (2010) PKA dynamics in a *Drosophila* learning center: coincidence detection by rutabaga adenylyl cyclase and spatial regulation by dunce phosphodiesterase. *Neuron*, **65**:516–529.

264. Jurevicius, J. and Fischmeister, R. (1996) cAMP compartmentation is responsible for a local activation of cardiac Ca^{2+} channels by beta-adrenergic agonists. *Proc. Natl. Acad. Sci. USA*, **93**:295–299.

265. Barnes, A.P., Livera, G., Huang, P., Sun, C., O'Neal, W.K., Conti, M., Stutts, M.J., and Milgram, S.L. (2005) Phosphodiesterase 4D forms a cAMP diffusion barrier at the apical membrane of the airway epithelium. *J. Biol. Chem.*, **280**:7997–8003.

266. Terrin, A., Di Benedetto, G., Pertegato, V., Cheung, Y.-F., Baillie, G., Lynch, M.J., Elvassore, N., Prinz, A., Herberg, F.W., Houslay, M.D., and Zaccolo, M. (2006) PGE1 stimulation of HEK293 cells generates multiple contiguous domains with different [cAMP]: role of compartmentalized phosphodiesterases. *J. Cell Biol.*, **175**:441–451.

267. Asirvatham, A.L., Galligan, S.G., Schillace, R.V., Davey, M.P., Vasta, V., Beavo, J.A., and Carr, D.W. (2004) A-kinase anchoring proteins interact with phosphodiesterases in T lymphocyte cell lines. *J. Immunol.*, **173**:4806–4814.

268. Carlisle Michel, J.J., Dodge, K.L., Wong, W., Mayer, N.C., Langeberg, L.K., and Scott, J.D. (2004) PKA-phosphorylation of PDE4D3 facilitates recruitment of the mAKAP signalling complex. *Biochem. J.*, **381**:587–592.

CHAPTER 2

PUTTING TOGETHER THE PIECES OF PHOSPHODIESTERASE DISTRIBUTION PATTERNS IN THE BRAIN: A JIGSAW PUZZLE OF CYCLIC NUCLEOTIDE REGULATION

MICHY P. KELLY
Department of Pharmacology, Physiology & Neuroscience, University of South Carolina School of Medicine, Columbia, SC, USA

INTRODUCTION

Phosphodiesterases (PDEs) are the only enzymes known to degrade the cyclic nucleotides cAMP and cGMP [1,2]. There are 11 families of PDEs identified to date, and each family is comprised of one or more isoforms (see below) that are each encoded by a unique gene. As we will see in this chapter (Figures 2.1–2.3), most PDE families are expressed in the brain (PDE6 being the exception), and each brain region expresses more than one PDE isoform. For example, the cortex expresses 11, the striatum expresses 9, CA1 of hippocampus expresses 11, and the cerebellum scores the most points with 14 of the 16 brain-expressed PDE isoforms. Although there is much overlap in terms of regional expression patterns, no two PDE isoforms show the same expression pattern across brain regions. This results in an exquisite jigsaw puzzle of cyclic nucleotide regulation that enables precise control over cAMP and cGMP signaling. Great attention has been paid in recent years to the potential of specific PDE isoforms as therapeutic targets for treatment of psychiatric and neurological dysfunction. As such, it is important to gain a full understanding of exactly where in the brain each PDE isoform is expressed in order to better understand how each may influence brain function. By performing autoradiographic *in situ* hybridization in mouse brain, we here illustrate the expression pattern for each PDE isoform using unique oligonucleotide probes (Table 2.1).

Cyclic-Nucleotide Phosphodiesterases in the Central Nervous System: From Biology to Drug Discovery, First Edition. Edited by Nicholas J. Brandon and Anthony R. West.
© 2014 John Wiley & Sons, Inc. Published 2014 by John Wiley & Sons, Inc.

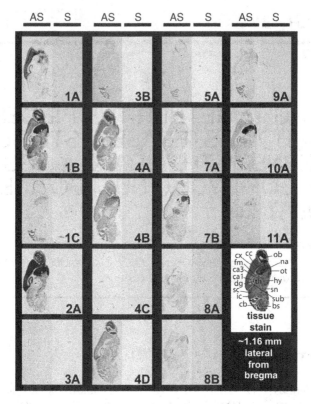

FIGURE 2.1 Autoradiographic *in situ* hybridization for all brain-expressed phosphodiesterase families using ^{35}S-labeled oligonucleotide antisense (left of each subpanel) and sense probes (right of each subpanel) 1.16 ± 0.2 mm lateral from bregma. Tissue was processed as per Ref. [3]. Slides were washed between 50 and 55 °C and apposed to film for 14 days. Shown here are results produced with one of two probe sets (see Table 2.1) that were used to label 20 μm sagittal sections taken from brains of adult male CF-1 mice. A thionin-stained tissue section is shown in the lower right of the figure panel to illustrate which brain regions are present at this level of bregma. AS, antisense; S, sense; cc, corpus callosum; cx, cortex; fm, fimbria; ca3, CA3 subfield of hippocampus; ca1, CA1 subfield of hippocampus; dg, dentate gyrus of hippocampus; sc, superior colliculus; ic, inferior colliculus; cb, cerebellum; ob, olfactory bulb; na, nucleus accumbens; ot, olfactory tubercle; hy, hypothalamus; sn, substantia nigra; sub, subiculum; bs, brain stem; cp, caudate putamen; th, thalamus.

PDE1

The dual-specificity (meaning degrades both cAMP and cGMP) PDE1 family is comprised of three gene products, PDE1A, PDE1B, and PDE1C. Consistent with previous reports [4], we see that PDE1A shows the greatest expression in CA1, CA2, and CA3 of hippocampus and subiculum, the amygdalohippocampal area, the basolateral amygdala, the nucleus accumbens core, and throughout the deep layers of cortex. Despite substantial expression of PDE1A in cortex, minimal expression is

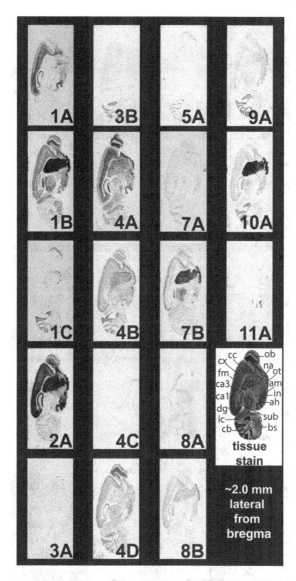

FIGURE 2.2 Autoradiographic *in situ* hybridization for all brain-expressed phosphodiester-ase families using ^{35}S-labeled oligonucleotide antisense probes 2.0 ± 0.2 mm lateral from bregma. Tissue was processed as per Ref. [3]. Slides were washed between 50 and 55 °C and apposed to film for 14 days. Shown here are results produced with one of two probe sets (see Table 2.1) that were used to label 20 μm sagittal sections taken from brains of adult male CF-1 mice. A thionin-stained tissue section is shown in the lower right of the figure panel to illustrate which brain regions are present at this level of bregma. cc, corpus callosum; cx, cortex; fm, fimbria; ca3, CA3 subfield of hippocampus; ca1, CA1 subfield of hippocampus; dg, dentate gyrus of hippocampus; ic, inferior colliculus; cb, cerebellum; ob, olfactory bulb; na, nucleus accumbens; ot, olfactory tubercle; am, amygdala; in, internal capsule; ah, amygdalohippo-campal area; sub, subiculum; bs, brain stem; cp, caudate putamen; th, thalamus.

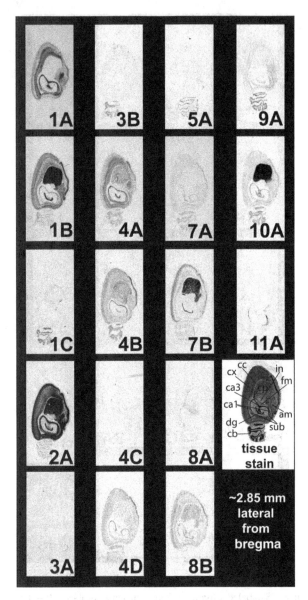

FIGURE 2.3   Autoradiographic *in situ* hybridization for all brain-expressed phosphodiesterase families using ^{35}S-labeled oligonucleotide antisense probes 2.85 ± 0.2 mm lateral from bregma. Tissue was processed as per Ref. [3]. Slides were washed between 50 and 55 °C and apposed to film for 14 days. Shown here are results produced with one of two probe sets (see Table 2.1) that were used to label 20 μm sagittal sections taken from brains of adult male CF-1 mice. A thionin-stained tissue section is shown in the lower right of the figure panel to illustrate which brain regions are present at this level of bregma. cc, corpus callosum; cx, cortex; fm, fimbria; ca3, CA3 subfield of hippocampus; ca1, CA1 subfield of hippocampus; dg, dentate gyrus of hippocampus; cb, cerebellum; in, internal capsule; am, amygdala; sub, subiculum; cp, caudate putamen.

TABLE 2.1   List of Oligonucleotide Probes Used to Assess mRNA Expression of PDE Isoforms in Brains of Adult Male CF-1 Mice

Gene	Accession No.	Nucleotides	Antisense Sequence
PDE1A	NM_001009978.1	557–518	acctgtatgaagcattatataatgcacagtttgagtgacg
PDE1A	NM_001009978.1	606–567	ctgcaaagaccatggctaaaatttccagttcagtgagcca
PDE1B	NM_008800	1153–1114	tccaactgctgcagtgctgtcttcatagtcttcacttgct
PDE1B	NM_008800	1212–1173	ggtggctgatgtcagcagcatgaagcagaagagatagggc
PDE1C	NM_001025568.1	1286–1247	tcttataaaggaggtaatgcacagtctgtgtgacgtcagc
PDE1C	NM_001025568.1	1346–1306	tggcagccgaaaagattattgcaaagatctccagctctgtc
PDE2A	NM_001143848	1660–1621	gggatgttcaggatctggcccgtagtcgccacgtggcccg
PDE2A	NM_001143848	1820–1781	aaacttgctgaaccatggcccattgatcttgttcactagc
PDE3A	NM_018779.1	2877–2838	actagtggcaaccaggaaggcatttgtcctccctgggtga
PDE3A	NM_018779.1	2937–2898	tgcagcggcgtgatggttctccagaacagaacggtcgttg
PDE3B	NM_011055.1	2530–2491	tttcatggttattatggatctgaggtaagccaggaattgg
PDE3B	NM_011055.1	2589–2550	agaaatgtaagcaatctgtccagaaccaagtctaccatct
PDE4A	NM_019798.2	1019–980	ggcgtggccagcagcacatgtgtggactgtagcacatcgg
PDE4A	NM_019798.2	1079–1040	atggcagcagcgaagagagcagcaagaatctccaggtctg
PDE4B	NM_001177980.1	1737–1698	aacctgtatccggtcagtatagttgtccaggaggagaaca
PDE4B	NM_001177980.1	1774–1738	ggttgctcaggtctgcacagtgtaccatgttgcgaag
PDE4C	NM_201607.1	403–367	ctgagccaaggtttctctgtgtcccttgagttccaat
PDE4C	NM_201607.1	463–424	gagcgtctaggagacccttgccgcccggaagcccattttc
PDE4D	NM_011056.1	1178–1139	aggacatcttcctgctccgttttaaccccaaacctgggga
PDE4D	NM_011056.1	1237–1198	gctctgctattcggaaaacgtgggaggccccacttgttcac
PDE5A	NM_153422.1	2510–2471	cttgcatacttggggattttattttcttctctctgttcat
PDE5A	NM_153422.1	2569–2531	ttcggatacgtgggtcagggcctcatacagctgcaagca
PDE7A	NM_008802.1	924–885	caatggcaagtgtgagaacaaaccagattctcttaacaag
PDE7A	NM_008802.1	984–945	actgatatccgtggctaatatcaaagcacctatctgagcc
PDE7B	NM_013875.2	1163–1124	ggcaagtgagccaggagccgtgattctcgaagcatgccaa
PDE7B	NM_013875.2	1223–1184	atatccgtggccaagatgagggagcccagctgctgttcga
PDE8A	NM_008803.2	1120–1081	tctgaacacactcagttgtggtctcaaccttattgttgcc
PDE8A	NM_008803.2	2020–1981	ggtcattttcaagcgtcagctggaaggccagtgccacgtg
PDE8B	NM_172263.1	1240–1201	acagtctcctcaagccatcagtcatcaggcccccaacaag
PDE8B	NM_172263.1	1600–1561	tcactctttcctttccaaggaagaaagccgtggcatgcag
PDE9A	NM_001163748.1	760–721	tgcggagcgtgattgggttgatgctgaagtccctgaccag
PDE9A	NM_001163748.1	1300–1261	cccacggttctgccacctccatgggacggacttcattgga
PDE10A	NM_011866	2799–2763	gaaaacaaaaagaacgaagtcacagaagcaggttgag
PDE10A	NM_011866	6125–6087	cagcaccacacaaagcaaatgcagggtaggaatatacag
PDE11A	NM_001081033	152–116	ccaccagttcctgttttccttttcgcatcaagtaatc
PDE11A	NM_001081033	1168–1129	ctgagcattagatatggcgattccacagaacggaagatac

Autoradiographic methods as per Ref. [3].

seen in the olfactory bulb. PDE1A expression appears to be notably absent from dentate gyrus of hippocampus, cerebellum, and caudate putamen. In contrast, we see that PDE1B is particularly enriched in the striatum (caudate putamen and nucleus accumbens), dentate gyrus of hippocampus, the olfactory tubercle, and the olfactory bulb. PDE1B is also highly expressed throughout most layers of cortex, the thalamus, the hypothalamus, the amygdala, CA1 of hippocampus, the substantia nigra, and the inferior and superior colliculi. Some low level of PDE1B expression can also be measured in the cerebellum and CA3 of hippocampus. Interestingly, PDE1B appears absent from subiculum, even though it is expressed before and after the subiculum in the cortex and CA1. Our observation that PDE1B is one of the most highly, widely, and variably expressed PDE isoforms in the brain is consistent with previous reports [4,5]. PDE1C shows the most restricted pattern of expression of the PDE1 family, with enriched expression in the granule cell layer of the cerebellum and low levels in the striatum and outer layers of the olfactory bulb, consistent with previous reports [6]. Expression of the PDE1 family may be dynamically regulated in CNS disease states, as increased PDE1 cAMP phosphodiesterase activity was previously observed in a mouse model of schizophrenia [3].

PDE2

The dual-specificity PDE2 family is comprised of a single gene, PDE2A. PDE2A is expressed throughout multiple brain regions, with particular enrichment in CA1, CA3, and dentate gyrus of hippocampus, the striatum (caudate putamen and nucleus accumbens), the olfactory tubercle, the amygdalohippocampal area, the basolateral and basomedial nuclei of the amygdala, and the most superficial layer of cortex (particularly entorhinal cortex). Additional high levels of PDE2A expression are seen throughout all remaining layers of cortex, the subiculum, the anteroventral (ventro-lateral and dorsal medial partitions) and reticular nuclei of the thalamus, the olfactory bulb, and other nuclei of the amygdala. Low levels of PDE2A expression can also be observed in the hypothalamus, substantia nigra, cerebellum, and the brain stem, including the inferior and superior colliculi.

PDE3

PDE3 is a dual-specificity PDE family comprised of the PDE3A and PDE3B genes. Evidence of PDE3 expression in brain is scant [1,2]. Here, we see no basal expression of PDE3A in brain of adult male CF-1 mice. It is possible that PDE3A may only express in the brain upon strong stimulation as one report suggests that PDE3A is upregulated in striatum (but not hippocampus) following repeated electroconvulsive shock treatment [7]. In the case of PDE3B, we do observe a very low level of basal expression throughout the brain, with highest expression in the cerebellum. The low expression level measured for PDE3B (or any other low-expressing isoform) should not be confused with a lack of function for this PDE isoform, because a functional role for PDE3B has been described in the hypothalamus [8].

PDE4

The PDE4 family is, arguably, the most intensely investigated PDE family, particu-larly in the context of psychiatric and neurological disorders [9]. This is discussed in detail in Chapters 7 and 8. The PDE4 family is cAMP-specific and comprised of four genes, PDE4A, PDE4B, PDE4C, and PDE4D. Consistent with previous reports [10–13], we see that PDE4A, PDE4B, and PDE4D are expressed relatively ubiquitously throughout the rodent brain in cortex, striatum, hippocampus, amygdala, thalamus, hypothalamus, substantia nigra, brain stem, cerebellum, and olfactory bulb. There are, however, key differences in the distribution of each isoform across brain regions that make the expression pattern of each PDE4 family member unique.

A key factor differentiating the expression pattern of PDE4A from other PDE4 family members is the fact that PDE4A is enriched in the dentate gyrus of hippocampus. In contrast, PDE4D and PDE4B appear to be expressed at low levels in the dentate. Another point of differentiation is that PDE4A and PDE4D appear enriched in olfactory bulb, while PDE4B is only moderately expressed there. With regard to cortex, PDE4A appears more highly expressed in the deeper layers relative to the superficial layers of cortex, whereas PDE4B is more highly expressed in the superficial layers and PDE4D is equally distributed among cortical layers. One particularly interesting point of differentiation is the fact that PDE4B appears to express well within white matter tracks (except those within hippocampus). This makes PDE4B one of only three PDE isoforms that express in white matter tracks (PDE1A and PDE8A being the other two), suggesting localization within axons and/ or oligodendrocytes. Finally, PDE4D is expressed well throughout thalamic nuclei, whereas PDE4A and PDE4B expression is strikingly lower in the medial thalamic nuclei relative to surrounding regions. We do not see measurable expression of PDE4C in the brain. It is important to note that expression of PDE4 family members is dynamically regulated in a brain region-specific manner following the administration of psychoactive agents (e.g., antidepressants, antipsychotics, etc.) [12,14,15].

PDE5

The PDE5 family catalyzes cGMP only and is comprised of a single gene, PDE5A. Expression of PDE5A in brain may differ between species. Although no PDE5A expression was observed in human brain [16], PDE5A expression was previously noted in rodent cerebellum (high), hippocampus (moderate), and cortex (low) [17–19]. This distribution pattern of PDE5A in rodent brain was confirmed in studies showing that administration of PDE5A inhibitors effectively increased cGMP levels in prefrontal cortex, hippocampus, and cerebellum of rat [20]. Here, we see that PDE5A is expressed in a unique punctuate pattern within the Purkinje cell layer of the cerebellum, as well as the most rostral layers of the olfactory bulb. In contrast to previous reports, we did not observe any reliable signal in hippocampus nor cortex of our adult CF-1 mice using either probe tested (results with one probe shown in figures).

PDE6

The cGMP-hydrolyzing PDE6 family is localized to the retina and is unique in that PDE6 is not a single molecule, but rather is comprised of two catalytic subunits

(PDE6A and PDE6B in rods and PDE6C in cones) and two regulatory subunits (gamma, sometimes modified with a delta subunit) [1,2]. Naturally occurring mutations have been identified in the PDE6 genes that lead to various forms of blindness in mice, similar to that observed in humans [1,2]. It is well accepted that expression of the PDE6 family is specific to retina, so probes were not tested here.

PDE7

PDE7 is a cAMP-specific PDE family for which two genes have been identified to date, PDE7A and PDE7B. Here we see that PDE7A is expressed at very low levels throughout much of the mouse brain, including cerebellum, cortex, CA1, CA3, and dentate gyrus of hippocampus, olfactory bulb, and brain stem. Expression of PDE7B is strikingly different from that of PDE7A. PDE7B is enriched in the dentate gyrus of hippocampus, the caudate putamen, and nucleus accumbens, in addition to the olfactory tubercle. It is also quite highly expressed in the superficial layers of cortex, thalamic nuclei, and the inferior colliculus. Areas with low levels of PDE7B expression include the olfactory bulb, deep layers of cortex, the hypothalamus, the amygdala, brain stem, and cerebellum, where the expression pattern is quite different from that of other PDE isoforms (expressed in the white matter with distinct edges along the outer edge of the granular layer). For the most part, our observations are consistent with previous reports in rats [21,22], except that we find PDE7B distinctly absent from CA1 and CA3 of hippocampus. Expression of PDE7A and PDE7B has also been described in human cortex, basal ganglia, hippocampus, and cerebellum [23].

PDE8

The PDE8 family, which encompasses PDE8A and PDE8B, is also cAMP-specific. The expression pattern of PDE8A is one of the most unique of all the PDE isoforms. Expression is very low, but PDE8A can be distinctly observed in the corpus callosum, the internal capsule, the fimbria, and the white matter of the cerebellum. This expression pattern at the mRNA level suggests that PDE8A may be preferentially expressed in oligodendrocytes, although the possibility of axonal mRNA trafficking cannot be ruled out. Low levels of PDE8A expression are also seen throughout the hypothalamus and brain stem. In contrast, PDE8B is expressed at moderate levels throughout the rodent brain in a pattern that is very reminiscent for that of activity-regulated cytoskeleton-associated protein (Arc) [3]. PDE8B is most highly expressed in the olfactory tubercle, yet appears all but absent from the olfactory bulb. PDE8B is also expressed in amygdala, cortex (with higher expression in superficial layers), and CA1, CA3, and dentate gyrus of hippocampus. We see only minimal PDE8B expression in brain stem (mostly in the colliculi) and cerebellum. Our observations are largely consistent with previous reports [24]. PDE8B mRNA expression has also been described in human cortex, hippocampus, and basal ganglia, with increased expression noted in cortex and hippocampus of patients with Alzheimer's disease [23].

PDE9

The cGMP-hydrolyzing PDE9 family has only one gene identified to date, PDE9A. Previous reports suggest that PDE9A is expressed at low levels relatively ubiquitously throughout the rodent and human brain [16,18]. Here we see that PDE9A is enriched in the cerebellum (particularly the Purkinje cell layer) and olfactory tubercle, and is expressed at very low levels throughout the rest of the brain.

PDE10

The dual-specificity PDE10 family is encoded by one gene, PDE10A. We and others [25,26] observed that PDE10A is particularly enriched in the caudate putamen, nucleus accumbens, and olfactory tubercle. High levels of PDE10A expression are seen in cerebellum and moderate expression of PDE10A can be observed in cortex, amygdala, olfactory bulb, thalamus, hypothalamus, and subiculum, CA1, CA3, and dentate gyrus of hippocampus. Finally, a very low level of PDE10A expression can be seen in brain stem.

PDE11

PDE11 is the most recently identified PDE family. It is comprised of a single gene, PDE11A, whose product hydrolyzes both cAMP and cGMP. PDE11A exhibits the most restricted expression pattern in brain of all the PDE isoforms, being expressed at appreciable levels only in CA1 and subiculum of hippocampus and the amygdalo-hippocampal area [27]. Interestingly, PDE11A is expressed at threefold higher levels in ventral versus dorsal hippocampus. Although PDE11A is expressed only within a small compartment of the brain, it clearly plays an important role there, because PDE11A$^{-/-}$ mice exhibit a number of behavioral, anatomical, and biochemical phenotypes relevant to psychiatric disease [27].

CONCLUSIONS

When one observes how distinctly each PDE isoform is expressed across brain regions, it is easy to understand how each contributes uniquely to the function of the nervous system. Here we have scratched only the surface in describing differences in overall regional expression patterns. Within each brain region expressing multiple isoforms, there are likely additional degrees of freedom with regard to localization in terms of cell population expression profiles and localization patterns for each PDE isoform within each cell type. Future studies using biochemical, pharmacological, and genetic tools with improved specificity will continue to foster our understanding of how each PDE isoform differentially contributes to brain function. Such findings will have important implications for differentiating the numerous PDE isoforms as therapeutic targets.

REFERENCES

1. Bender, A.T. and Beavo, J.A. (2006) Cyclic nucleotide phosphodiesterases: molecular regulation to clinical use. *Pharmacol. Rev.*, **58**(3):488–520.

2. Lugnier, C. (2006) Cyclic nucleotide phosphodiesterase (PDE) superfamily: a new target for the development of specific therapeutic agents. *Pharmacol. Ther.*, **109**(3):366–398.

3. Kelly, M.P., et al. (2007) Constitutive activation of Galphas within forebrain neurons causes deficits in sensorimotor gating because of PKA-dependent decreases in cAMP. *Neuropsychopharmacology*, **32**(3):577–588.

4. Yan, C., Bentley, J.K., Sonnenburg, W.K., and Beavo, J.A. (1994) Differential expression of the 61 kDa and 63 kDa calmodulin-dependent phosphodiesterases in the mouse brain. *J. Neurosci.*, **14**(3 Part 1):973–984.

5. Polli, J.W. and Kincaid, R.L. (1994) Expression of a calmodulin-dependent phosphodiesterase isoform (PDE1B1) correlates with brain regions having extensive dopaminergic innervation. *J. Neurosci.*, **14**(3 Part 1):1251–1261.

6. Yan, C., et al. (1995) Molecular cloning and characterization of a calmodulin-dependent phosphodiesterase enriched in olfactory sensory neurons. *Proc. Natl. Acad. Sci. USA*, **92**(21):9677–9681.

7. Cho, C.H., Cho, D.H., Seo, M.R., and Juhnn, Y.S. (2000) Differential changes in the expression of cyclic nucleotide phosphodiesterase isoforms in rat brains by chronic treatment with electroconvulsive shock. *Exp. Mol. Med.*, **32**(3):110–114.

8. Sahu, A. and Metlakunta, A.S. (2005) Hypothalamic phosphatidylinositol 3-kinase–phosphodiesterase 3B–cyclic AMP pathway of leptin signalling is impaired following chronic central leptin infusion. *J. Neuroendocrinol.*, **17**(11):720–726.

9. Houslay, M.D., Schafer, P., and Zhang, K.Y. (2005) Keynote review: phosphodiesterase-4 as a therapeutic target. *Drug Discov. Today*, **10**(22):1503–1519.

10. Engels, P., Abdel'Al, S., Hulley, P., and Lubbert, H. (1995) Brain distribution of four rat homologues of the *Drosophila* dunce cAMP phosphodiesterase. *J. Neurosci. Res.*, **41**(2):169–178.

11. Iona, S., et al. (1998) Characterization of the rolipram-sensitive, cyclic AMP-specific phosphodiesterases: identification and differential expression of immunologically distinct forms in the rat brain. *Mol. Pharmacol.*, **53**(1):23–32.

12. Takahashi, M., et al. (1999) Chronic antidepressant administration increases the expression of cAMP-specific phosphodiesterase 4A and 4B isoforms. *J. Neurosci.*, **19**(2):610–618.

13. Perez-Torres, S., et al. (2000) Phosphodiesterase type 4 isozymes expression in human brain examined by *in situ* hybridization histochemistry and [^{3}H]rolipram binding autoradiography. Comparison with monkey and rat brain. *J. Chem. Neuroanat.*, **20**(3–4): 349–374.

14. Dlaboga, D., Hajjhussein, H., and O'Donnell, J.M. (2006) Regulation of phosphodiesterase-4 (PDE4) expression in mouse brain by repeated antidepressant treatment: comparison with rolipram. *Brain Res.*, **1096**(1):104–112.

15. Dlaboga, D., Hajjhussein, H., and O'Donnell, J.M. (2008) Chronic haloperidol and clozapine produce different patterns of effects on phosphodiesterase-1B, -4B, and -10A expression in rat striatum. *Neuropharmacology*, **54**(4):745–754.

16. Reyes-Irisarri, E., Markerink-Van Ittersum, M., Mengod, G., and de Vente, J. (2007) Expression of the cGMP-specific phosphodiesterases 2 and 9 in normal and Alzheimer's disease human brains. *Eur. J. Neurosci.*, **25**(11):3332–3338.

17. Giordano, D., De Stefano. M.E., Citro, G., Modica, A., and Giorgi, M. (2001) Expression of cGMP-binding cGMP-specific phosphodiesterase (PDE5) in mouse tissues and cell lines using an antibody against the enzyme amino-terminal domain. *Biochim. Biophys. Acta*, **1539**(1–2):16–27.

18. Van Staveren, W.C., et al. (2003) mRNA expression patterns of the cGMP-hydrolyzing phosphodiesterases types 2, 5, and 9 during development of the rat brain. *J. Comp. Neurol.*, **467**(4):566–580.

19. Shimizu-Albergine, M., et al. (2003) Individual cerebellar Purkinje cells express different cGMP phosphodiesterases (PDEs): *in vivo* phosphorylation of cGMP-specific PDE (PDE5) as an indicator of cGMP-dependent protein kinase (PKG) activation. *J. Neurosci.*, **23**(16):6452–6459.

20. Marte, A., Pepicelli, O., Cavallero, A., Raiteri, M., and Fedele, E. (2008) *In vivo* effects of phosphodiesterase inhibition on basal cyclic guanosine monophosphate levels in the prefrontal cortex, hippocampus and cerebellum of freely moving rats. *J. Neurosci. Res.*, **86**(15):3338–3347.

21. Miro, X., Perez-Torres, S., Palacios, J.M., Puigdomenech, P., and Mengod, G. (2001) Differential distribution of cAMP-specific phosphodiesterase 7A mRNA in rat brain and peripheral organs. *Synapse*, **40**(3):201–214.

22. Reyes-Irisarri, E., Perez-Torres, S., and Mengod, G. (2005) Neuronal expression of cAMP-specific phosphodiesterase 7B mRNA in the rat brain. *Neuroscience*, **132**(4):1173–1185.

23. Perez-Torres, S., et al. (2003) Alterations on phosphodiesterase type 7 and 8 isozyme mRNA expression in Alzheimer's disease brains examined by *in situ* hybridization. *Exp. Neurol.*, **182**(2):322–334.

24. Kobayashi, T., et al. (2003) Molecular comparison of rat cyclic nucleotide phosphodiesterase 8 family: unique expression of PDE8B in rat brain. *Gene*, **319**:21–31.

25. Seeger, T.F., et al. (2003) Immunohistochemical localization of PDE10A in the rat brain. *Brain Res.*, **985**(2):113–126.

26. Coskran, T.M., et al. (2006) Immunohistochemical localization of phosphodiesterase 10A in multiple mammalian species. *J. Histochem. Cytochem.*, **54**(11):1205–1213.

27. Kelly, M.P., et al. (2010) Phosphodiesterase 11A in brain is enriched in ventral hippocampus and deletion causes psychiatric disease-related phenotypes. *Proc. Natl. Acad. Sci. USA*, **107**(18):8457–8462.

CHAPTER 3

COMPARTMENTALIZATION AND REGULATION OF CYCLIC NUCLEOTIDE SIGNALING IN THE CNS

MANUELA ZACCOLO
Department of Physiology, Anatomy and Genetics, University of Oxford, Oxford, UK

ALESSANDRA STANGHERLIN
Wellcome Trust-MRC Institute of Metabolic Science, Addenbrooke's Hospital, University of Cambridge, Cambridge, UK

INTRODUCTION

The cyclic nucleotides $3'$-$5'$-cyclic adenosine monophosphate (cAMP) and $3'$-$5'$-cyclic guanosine monophosphate (cGMP) are diffusible intracellular second messengers that act as critical modulators of neuronal function. cAMP is generated in response to binding of a number of neurotransmitters to G-protein coupled receptors (GPCRs) and subsequent activation of adenylyl cyclases (ACs). Serotonin, adrenergic, dopaminergic, adenosine, vasoactive intestinal peptide, muscarinic, γ-aminobutyric acid (GABA), and opioid receptors, among others, signal through the cAMP cascade via specific heterotrimeric G proteins. cAMP synthesis can also result from activation of Ca^{2+}-sensitive ACs as a consequence of Ca^{2+} influx through the glutamate receptors AMPA and NMDA. cAMP activates three major targets: protein kinase A (PKA), the exchange proteins activated by cAMP (EPACs), and cyclic nucleotide-gated ion channels (CNGCs). Through these effectors cAMP controls a bewildering number of neuronal functions, ranging from regulation of ion channel activity and, consequently, neuronal excitability, to cell volume control and axon guidance; from metabolism and transcription to neurotransmitter release and learning and memory formation [1]. cAMP has also been associated with the pathogenesis of a number of neurological disorders including Alzheimer's disease [2], schizophrenia [3], and depression [4].

Cyclic-Nucleotide Phosphodiesterases in the Central Nervous System: From Biology to Drug Discovery, First Edition. Edited by Nicholas J. Brandon and Anthony R. West.
© 2014 John Wiley & Sons, Inc. Published 2014 by John Wiley & Sons, Inc.

cGMP is generated in response to natriuretic peptides and nitric oxide (NO) from membrane-bound particulate guanylyl cyclase (pGC), and cytosolic, Ca^{2+} regulated, soluble guanylyl cyclases (sGCs), respectively. cGMP exerts its function via activation of protein kinase G (PKG), CNGCs, and cGMP-regulated phosphodiesterases (PDEs) [5].

Both cyclic nucleotides are degraded by the large superfamily of PDEs. This superfamily includes 11 families (PDE1–PDE11) that encompass multiple genes and a number of splice variants accounting for more than 70 different isoenzymes. Some families degrade only cAMP, others only cGMP, and some others can degrade both cyclic nucleotides [6,7]. Individual PDE enzymes exert specific functional roles as a consequence of their unique combination of regulatory mechanisms, enzyme kinetics, and intracellular localization [6,7]. PDE localization to different compartments occurs through different mechanisms involving direct binding to membrane lipids or protein–protein interactions [6,8].

Although the role of cGMP in the nervous system has been less well studied in comparison to cAMP, recent biochemical and functional data show that cGMP plays an important role in regulating critical functions such as axon guidance [9], neurodegeneration [10,11], synaptic plasticity [12], learning, and other complex behaviors, including addiction, anxiety, and the pathogenesis of schizophrenia and depression [13].

The brain contains a large number of different GPCRs. Each individual neuron can express at its plasma membrane a number of these receptors, each of which will generate a cAMP signal upon binding to its ligand. In addition, glutamatergic stimulation can generate both cAMP and cGMP signals in response to Ca^{2+} influx. Cyclic nucleotides are therefore a very common currency in brain communication. Indeed, the plethora of extracellular chemical stimuli to which a neuron is subjected to at any given time raises the question of how cAMP and cGMP messages are decoded, preserving specificity and avoiding inappropriate activation of downstream targets in response to a given input. In recent years, the view has emerged that spatial restriction of signal propagation plays a critical role in ensuring that the required specificity in signaling is achieved.

NEW TOOLS TO STUDY CYCLIC NUCLEOTIDE SIGNALING: FRET-BASED BIOSENSORS

An exciting and somewhat surprising finding has been that cyclic nucleotides do not freely diffuse in the cell as they would be expected to based on their small size and high hydrophilicity. On the contrary, they form gradients, or local pools, in which their concentration is higher than in the rest of the cell. Such spatial confinement of cAMP and cGMP signals appears to be required to achieve specificity of response [14].

The concept of compartmentalized cAMP signaling was formulated almost three decades ago when studies on isolated perfused hearts showed that isoproterenol and prostaglandin E_1 (PGE_1), although elevating intracellular cAMP to comparable levels, had very different effects on PKA substrates. Specifically, isoproterenol caused phosphorylation of phosphorylase kinase [15] and troponin I [16] and increased contractility of myocytes, whereas no increase in contractility and in the phosphorylation of these substrates was observed upon PGE_1 stimulation [17]. Over the years, further evidence of a functional compartmentalization of the cAMP signal accumulated, but it was only recently that direct documentation of restricted diffusion of cAMP was possible using real-time imaging in intact living cardiac myocytes [18]. The approach used in these studies takes advantage of genetically encoded sensors for cAMP [19] and of the phenomenon of fluorescence resonance energy transfer (FRET) [20].

Commonly, cellular cAMP levels have been measured in cell lysates by radioimmunoassays. Such an approach estimates total cAMP rather than free cAMP and offers very poor temporal resolution and no spatial resolution, therefore proving inadequate for studying the fine details of cAMP signaling [21]. FRET-based biosensors that are genetically encoded overcome such problems. FRET is a physicochemical phenomenon whereby the excited state energy of a donor fluorophore can be transferred to an acceptor fluorophore, which then emits its own characteristic fluorescence [22]. For FRET to occur, the acceptor must absorb at roughly the same wavelengths as the donor emits. FRET depends on the antiparallel alignment of the electric dipoles of donor and acceptor fluorophores and is highly sensitive to the donor-acceptor distance (between 1 and 10 nm). In fact, FRET efficiency decreases with the sixth power of the distance between donor and acceptor; consequently, a minimal perturbation of the spatial relationship between the two fluorophores can drastically alter the efficiency of energy transfer, making FRET a very sensitive tool to measure protein–protein interactions or protein conformational changes. This has been exploited to generate FRET-based sensors in which one or two cAMP binding sites are linked to spectral variants of the green fluorescent protein, typically CFP and YFP, which act as FRET donor and acceptor, respectively. When cAMP binds to the cAMP binding sites, a conformational change or protein interaction change occurs that alters the distance between the two fluorophores, thus affecting FRET. A number of these sensors have been developed and applied to define cAMP dynamics in intact living cells in real-time [23]. Similar sensors are also available to detect cGMP [24,25].

As a result of being genetically encoded, the FRET-based sensors can be easily targeted to specific cellular compartments, such as the plasma membrane, nucleus, mitochondria [26], or individual macromolecular complexes [5,27], for real-time imaging of cyclic nucleotides at different subcellular locations.

A different approach to selectively measure cAMP and cGMP changes in the subplasma membrane compartment exploits CNGCs, which are directly opened by cyclic nucleotides [28–30]. The α subunit of a rat olfactory CNGC is expressed in cells via adenoviral infection and electrophysiological measurements of Ca^{2+} or other ionic currents are used as a read out of changes in cAMP and cGMP levels.

SPECIFICITY BY COMPARTMENTALIZATION

Application of these new approaches has significantly advanced our understanding of spatial control of cyclic nucleotide signaling. Using FRET sensors it was possible to demonstrate that β-adrenergic stimulation of neonatal cardiac myocytes generates multiple and restricted microdomains with increased concentration of cAMP. Such pools of cAMP specifically activate selected subset of PKA enzymes [18], leading to the phosphorylation of selected downstream targets. Activation of PGE_1 receptors generates an increase in cAMP in different compartments and does not lead to the phosphorylation of the same targets [27]. These findings provide a molecular mechanism that explains how different hormones can generate completely different functional outcomes although they generate similar amounts of cAMP. Compartmentalization of the cAMP signal in the heart appears to be very tight, and even different β-AR subtypes seem to generate quite distinct cAMP signals [31,32], accounting for the distinct physiological and pathophysiological responses known to be elicited by $β_1$- and $β_2$-adrenoceptors [33]. A recent study, elegantly combining nanoscale live-cell scanning ion conductance and FRET microscopy, showed that spatially confined $β_2$-AR-induced cAMP signals are localized exclusively to deep invaginations of the plasma membrane known as transverse T tubules, whereas $β_1$-AR stimulation generates cAMP signals that are distributed across the entire cell. Selective stimulation of $β_1$- and $β_2$-AR generates spatially distinct pools of cAMP and the compartmentalization of the cAMP signals seems to correlate with the distinct membrane distribution of the receptor subtypes. Interestingly, in cardiomyocytes derived from a rat model of chronic heart failure, the adrenoceptor subtypes distribution appears to be altered, leading to diffuse cAMP signaling and raising the possibility that altered compartmentalization of cAMP signals might contribute to the failing myocardial phenotype [34].

Compartmentalization does not seem to apply to cAMP only. Several studies support the notion that sGC and pGC play different roles in a number of cell types [35,36], raising the possibility that cGMP is also spatially restricted and evidence is starting to emerge indicating that indeed this may be the case. For example, in a study on adult cardiac myocytes the cGMP pool generated by the pGC was shown to effectively modulate the cGMP-gated current associated with olfactory CNGCs, whereas the pool of cGMP generated upon activation of the sGC had very little effect [30], indicating that cGMP is not free to diffuse in the cell, and that, depending on the source, it has different downstream effects.

Although the paradigm of compartmentalized cyclic nucleotide signaling emerges mainly from work on cardiac myocytes, a number of recent studies show that it is a generalized phenomenon likely occurring in all cell types, including neurons. For example, in olfactory neurons the odorant receptor (OR), a G_{olf}-coupled receptor that signals via generation of cAMP, is known to be expressed both at the level of the cilia in the nasal epithelium, where it is responsible for transducing the signal from odorant molecules, and at the level of the growth cone [37], where genetic evidence supports its role in regulating axonal convergence into the olfactory bulb [38]. The mechanisms by which the OR signals at the growth cone have remained quite mysterious until recently when, using real-time imaging of cAMP *in situ* and FRET biosensors, it was shown that the OR at the growth cone signals through the generation of cAMP exactly as the OR expressed on the cilia. However, the pools of cAMP generated at the two opposite ends of these cells are independent and can thus act locally and exert their distinct downstream functions, namely, transmission of odorant stimuli at the cilia and regulation of glomerular convergence at the growth cone [39]. Evidence of spatial restriction of cAMP signals has also been reported in hippocampal neurons, as shown in both embryonic dissociated cells [40] and brain slices [41]. These studies show that in pyramidal neurons the response to β-AR activation generates gradients of cAMP with a higher concentration in the dendrites as compared to the soma. Numerical simulations indicate that cell shape may play a critical role in such spatial control of signal propagation [40], as the high surface-area-to-volume ratio in small-diameter compartments would favor the accumulation of cAMP, whereas the filling of the larger soma compartment would take longer and be less efficient. Interestingly, cell geometry does not seem to be the only relevant parameter and negative regulators of the signal, such as PDEs, appear to play a critical role in determining the size of microdomains by creating boundaries to signaling molecules [40]. Local cAMP and cGMP signals have also been implicated in the differentiation of one axon and multiple dendrites in dissociated embryonic hippocampal neurons in culture and in neuronal polarization in cortical neurons *in vivo*. In a study using FRET-based indicators and local application of stimuli, compartmentalized activities of cAMP and cGMP have been shown to be reciprocally regulated and to exert opposing effects on dendrite formation and to promote and suppress axon formation, respectively. The molecular mechanisms responsible for such a reciprocal regulation of cyclic nucleotides remain to be determined; however, there are indications that they may involve the activity of specific PDEs and protein kinases [42].

MECHANISMS RESPONSIBLE FOR CYCLIC NUCLEOTIDE COMPARTMENTALIZATION

The mechanisms responsible for compartmentalization of cAMP and cGMP have not been fully elucidated. One hypothesis put forward is that physical diffusional barriers, possibly formed by elements of the endoplasmic reticulum and localized underneath the plasma membrane, may be involved [28]. Such a hypothesis was formulated to explain the limited diffusion of cAMP from the plasma membrane to the deep cytosol in HEK293 cells upon prostaglandin stimulation [43]. Although some cell types, such as cardiac myocytes, are rich in physical submembrane microdomains, it is not clear how these may restrict diffusion of cAMP and yet allow diffusion of Ca^{2+} from the same microdomains, a phenomenon that occurs in the millisecond timescale in these cells [44]. Whatever the nature of the barrier may be, a reduced diffusion coefficient for cAMP appears to be particularly relevant for sub-plasma-membrane compartmentalization of cAMP and much less so for generation of cAMP gradients in the inner cell [45]. Another mechanism that has been suggested as contributing to cAMP compartmentalization is PKA-mediated buffering [45]. In this case, binding of cAMP to the R subunits of PKA may reduce diffusion of cAMP because of the low diffusivity of R subunits. This hypothesis is supported by the observation that a significant proportion of the total basal cAMP may be bound to PKA, at least in some cell types [46,47].

Without a doubt, however, the best-established mechanism contributing to cAMP compartmentalization involves PDEs (Figure 3.1) [48]. Again, cardiac myocytes have served as the paradigm to demonstrate that inhibition of PDE activity has a profound effect on intracellular cAMP compartmentalization [31]. In an early biochemical study in adult canine ventricular myocytes, 45% of the cAMP generated upon β-AR stimulation was recovered in the particulate fraction but the proportion of total cAMP residing in the particulate fraction declined to less than 20% in the presence of PDE inhibitors [49], indicating that PDE activity contributes to the compartmentalization of cAMP. Subsequent analysis in intact living myocytes using both electrophysiological [50,51] and imaging [18,52,53] approaches confirmed that PDEs indeed have a key role in shaping the intracellular gradients of cAMP. The mechanism by which PDEs control intracellular diffusion of cAMP appears to involve localization of PDEs to specific subcellular compartments [52]. For example, the striking difference observed between the effect of PDE3 and PDE4 inhibition in the control of cAMP levels on β-AR stimulation in the heart was shown to correlate with distinct localization of PDE3 and PDE4 enzymes within the myocytes, with PDE3 being mainly localized on internal membranes and PDE4 showing a strong sarcomeric localization. Further support for the involvement of PDEs in the spatial control of cAMP diffusion comes from a study in HEK293 cells demonstrating that the specific spatial arrangement of different PDEs generates a pattern of local drains that reduce cAMP concentration in defined locales, thus resulting in the generation of multiple gradients of cAMP [54]. Overexpression of mutant PDEs that are catalytically inactive and exert a dominant-negative effect, by displacing the cognate endogenous active PDEs from their functionally relevant anchor sites [55], was shown to be sufficient to disrupt the cAMP gradients generated in response to PGE$_1$ in these cells [54]. The efficacy of these PDE mutants in disrupting intracellular pools of cAMP confirms that the tethered PDEs are responsible for shaping the cAMP gradients.

Virtually all PDE families are expressed in the central nervous system, and specific regions of the brain express unique subsets of PDEs. Thus, in the cortex, for example, high levels of PDE1A, PDE2A, PDE4A, PDE4B, PDE4D, and PDE8A are found, whereas the hippocampus expresses PDE1B, PDE2A, PDE3A, PDE3B, PDE4A, PDE4B, PDE4D, PDE7A, PDE7B, PDE8A, PDE8B [56], and PDE11A [57]. Multiple PDEs can also be expressed within the same neuronal cell. A striking example of specific subcellular topography of individual PDEs is the localization of PDE2A and PDE10A in hippocampal pyramidal neurons where PDE10A is largely confined to the perinuclear/nuclear region, whereas PDE2A is largely excluded from the soma but abundant in the neurites (Figure 3.1) [56,58]. As another example, the medium spiny neurons in the striatum express high levels of PDE1B, PDE2A, PDE4B, PDE7B, and PDE10A (see Chapter 2). Some of these enzymes (e.g., PDE1B, PDE2A, and PDE10A) can degrade both cAMP and cGMP, but their distinct

FIGURE 3.1 Compartmentalization of cyclic nucleotides in neurons. Compartmentalization of cAMP and cGMP signaling in neuronal cells. cAMP and cGMP are generated by AC and GC in specific compartments and their diffusion is restricted by the activity of PDEs. PDE isoforms possess specific subcellular topography that contributes to the local control of individual pools of cAMP and cGMP. Anchoring of PKA to AKAPs and the generation of signaling domains bring specific receptors and the effector kinase in proximity of downstream targets and provide tight spatial control on the propagation of the signal. AC, adenylyl cyclase; AKAP, A kinase anchoring protein; cAMP, cyclic adenosine monophosphate; CaN, calcineurin; DARPP-32, dopamine- and cAMP-regulated phosphoprotein of molecular weight 32 kDa; GC, guanylyl cyclase; PDE, phosphodiesterase; pGC, particulate guanylyl cyclase; PKA, protein kinase A; PSD-95, postsynaptic density protein 95; sGC, soluble guanylyl cyclase; TH, tyrosine hydroxylase. (See the color version of the figure in Color Plates section.)

localization is likely to result in different functional roles. PDE10A is predominantly membrane bound [59], PDE1B is cytosolic and PDE2A is enriched in lipid rafts in association with high concentrations of AC V/VI and PKA (Figure 3.1) [60]. In the mouse striatum, PDE10A was found to preferentially modulate cAMP rather than cGMP-mediated signaling [61]. However, selective inhibition of PDE10A showed a strikingly different effect when compared to selective inhibition of PDE4, another cAMP-selective PDE abundantly expressed in the striatum. Selective inhibition of PDE4 with rolipram induced a large increase in the phosphorylation of tyrosine hydroxylase (TH), a presynaptic target of PKA, whereas inhibition of PDE10A with papaverine had no effect on TH phosphorylation but instead robustly increased the phosphorylation of the dopamine- and cAMP-regulated phosphoprotein of molecular weight 32 kDa (DARPP-32), a postsynaptic target of PKA (Figure 3.1). In further support of the notion that discrete subcellular localization of PDE4 and PDE10A is responsible for their distinct roles in the control of striatal dopaminergic transmission, these enzymes were shown to have distinct subcellular localizations in immuno-histochemical studies [61]. This example well illustrates how different PDEs may locally regulate the level of second messengers, thus selectively affecting downstream effectors and ultimately determining the specific functional outcome of the cyclic nucleotide signal.

Another recent example of the role of PDEs in the spatial control of cAMP signal propagation in neurons comes from studies in the fruit fly *Drosophila melanogaster*. The study of olfactory learning in this organism has provided key insights into the molecular mechanisms of memory formation. Using FRET-based reporters of PKA activity, it was possible to directly show that the mushroom bodies in *Drosophila*, a critical center for olfactory memory formation, have spatially restricted PKA activation in response to specific neuromodulators. Also, the fly PDE *dunce* is necessary for localizing the effect of dopamine signaling on PKA enzymes that are selectively localized in the *a lobe* neurites, whereas neurites in the *b lobe* of mushroom bodies remain unaffected [62].

COMPARTMENTALIZATION OF CYCLIC NUCLEOTIDE EFFECTORS: THE ROLE OF ANCHORING PROTEINS

Although compartmentalization of cyclic nucleotide signals may well contribute to a specific response, spatial control of cAMP and cGMP would not serve any purpose if their downstream effectors could diffuse freely in the cytosol. In fact, a critical contribution to specificity is provided by anchoring of the kinases that are activated by cyclic nucleotides (Figure 3.1). While this has been best documented for the main cAMP effector, PKA, there are indications that anchoring to specific subcellular compartments via interaction with scaffolding proteins may apply also to PKG [63–65].

Targeting of PKA to subcellular compartments occurs via binding to A kinase anchoring proteins (AKAPs). AKAPs are a large (>50 members) family of structurally unrelated proteins that have in common the ability to tether PKA [66] via interaction of a conserved amphipathic α-helix region of 14–18 amino acids on the AKAP [67] with a hydrophobic groove in the dimerization/docking (DD) domain located at the N-terminus of the R subunit of PKA [68,69]. AKAPs also have unique protein–lipid or protein–protein targeting domains that tether the AKAP–PKA complex to distinct subcellular locations [70,71]. Most importantly, AKAPs can anchor PKA in proximity to its targets, thereby leading to the preferential phosphorylation of a local pool of PKA substrates [72]. AKAP79, for example, interacts with β1- and β2-AR, L-type Ca^{2+} channels, M-type K+ channels, and the capsaicin receptor [73], and is localized at excitatory synapses through interactions with the membrane-associated guanylate kinases (MAGUKs) PSD-95 and SAP97 [74], scaffolding proteins that bind to NMDA and AMPA glutamate receptors, respectively. A number of studies using different manipulations to disrupt this AKAP signaling complex have demonstrated that anchoring of PKA is critical for selective PKA-mediated phosphorylation of nearby targets and have highlighted the important role of AKAPs in cAMP signaling at excitatory synapses. For example, disruption of PKA interaction with AKAPs using Ht31, a peptide that competes for RII binding to AKAPs, reduces the number of AMPA receptors on the postsynaptic membrane [75] and decreases the strength of the AMPA receptor-mediated currents [76]. In addition, PKA-dependent forms of long-term depression (LTD) appear to be disrupted in AKAP150 (rat ortholog of AKAP79) mutant mice, as are increases of forskolin-induced GluR1 phosphorylation [77], indicating that anchoring of AKAP/150 has a critical role at excitatory synapses.

A key feature of AKAPs is their ability to coordinate multiple signaling enzymes, such as kinases, phosphatases, PDEs, guanosine triphosphatases (GTPases), and other regulatory proteins into multivalent transduction complexes, thereby ensuring integration and processing of multiple signals within discrete locales [78]. As an example, AKAP79 contains binding sites for protein kinase C (PKC) [79], and the phosphatase calcineurin (CaN, PP2B) [80]. In a study using FRET microscopy, PKA-RII subunits and CaN were simultaneously imaged in a complex with AKAP79 within approximately 5 nm of each other, indicating that the AKAP recruits simultaneously both the kinase and the phosphatase to the AMPA receptor [81], thereby providing the machinery to rapidly switch on and off target activation and achieve a very tight local control (Figure 3.1).

Another example of local signal integration is provided by the AKAP Yotiao. Yotiao is the smallest splice variant of AKAP450 and it is found in the brain, heart, and near neuromuscular junctions [82]. Yotiao can anchor PKA, the NR1 subunit of the NMDA receptor [83], the IP_3 receptor [84], and the K^+ channel subunit KCNQ1 [85], which, in the heart, is responsible for I_{Ks} currents. Interestingly, mutations in either Yotiao or KCNQ1 that disrupt binding to each other are associated with long QT syndrome, a cardiac condition characterized by arrhythmia and sudden death, again emphasizing the requirement for PKA anchoring for appropriate cardiac function and the relevance of compartmentalized cyclic nucleotide signaling in pathology [86]. Like AKAP79, Yotiao brings together opposing regulators by anchoring both PKA and the protein phosphatase PP1, which activate and reduce the NMDA channel activity, respectively [87,88]. Yotiao, like other AKAPs, can also bind PDEs [89]. In cardiac myocytes, PDE4D3 has been shown to anchor to Yotiao and to modulate the PKA-mediated I_{Ks} response to cAMP [90], thereby providing another example of how AKAPs can nucleate signaling domains where the signal is tightly regulated both in time and in space.

COMPARTMENTALIZATION OF THE SIGNALING MACHINERY AT THE PLASMA MEMBRANE

If the specific response to a given hormone or neurotransmitter relies on compartmentalization of cAMP and its effector PKA being close to specific targets, one would expect to find that the signaling machinery upstream of cAMP is also spatially confined. If the localization of GPCRs and ACs in the plasma membrane was not regulated, cAMP would be made available ubiquitously, leading to unselective activation of PKA subsets, irrespective of what targets they are coupled to. In fact, a number of GPCRs, including β-ARs, serotonin receptors, adenosine receptors, to mention only a few, have been shown to localize to specific membrane microdomains (see Figure 3.1) [91]. Lipid rafts, specialized regions of the plasma membrane enriched in cholesterol and other lipids, and caveolae, a subset of lipid rafts that form flask-shaped invaginations of the plasma membrane enriched in particular proteins (such as caveolins), appear to be the sites at which these GPCRs concentrate [91].

Not only the receptors but ACs also appear to be confined to defined membrane compartments. ACs make up a family of several isoforms (nine membrane bound and one soluble) that show different regulatory mechanisms and interaction with other signaling pathways [92]. Different AC isoforms have been shown to localize to distinct compartments.

All Ca^{2+} sensitive isoforms of AC (AC1, AC3, AC5, AC6, and AC8) have been found to be localized in lipid rafts whereas the Ca^{2+} insensitive isoforms (AC2, AC4, and AC7) are excluded from these membrane compartments [92]. Destruction of lipid rafts by extraction of cholesterol disrupts regulation of AC6 and AC8 by Ca^{2+} [93], suggesting that these structures are required for at least some regulation. There is also evidence for AC complexes containing GPCRs. For example, ACs have been found in a complex with the β-AR, G proteins, PKA, phosphatases, and L-type Ca^{2+} channels [94]. Although the molecular basis for the preferential coupling of different AC isoforms with specific receptors remains largely to be elucidated, the multiplicity of AC isoforms and their unique regulatory mechanisms appear to be perfectly poised to contribute to specificity of response. Furthermore, recent studies have shed new light on how such coupling may be achieved, at least for some receptors. Evidence is emerging that indicates that membrane-bound ACs may be coupled to AKAP complexes to potentially generate local pools of cAMP. Biochemical studies on rat brain extracts indicate that AKAP79 forms a complex with AC5/AC6 and PKA [95].

As AKAP79 also interacts with the AMPA and NMDA receptors, it provides a platform where external stimuli can generate a local cAMP response that can activate a restricted subset of anchored PKA. The study of the macromolecular complex organized by AKAP79 has revealed a further mechanism of local control. PKA anchoring to AKAP79 facilitates preferential phosphorylation of AC5, which leads to inhibition of cAMP synthesis [95]. Thus, anchoring of AC5/AC6 to AKAP79 not only potentially links receptor activation with local generation of cAMP but also provides an additional feedback loop mechanism to control second messenger synthesis. Another AKAP that associates with ACs is Yotiao. Immuno-precipitation of Yotiao from brain and heart identified significant association with AC1, AC2, AC3 and AC9, but not with AC4, AC5, and AC6. Expression of Yotiao inhibited the activity of AC2 and AC3, but not AC1 and AC9, for which the AKAP seems to serve purely as a scaffold [96].

The functional relevance of scaffolding ACs to AKAPs is revealed by a number of observations. For example, several phenotypes of AKAP150 deletion have similarities with knockouts of AC5 and AC6. AC5 and AKAP150 are highly expressed in the striatum and both exhibit defects of motor coordination when deleted [97,98]. AC6 deletion results in reduced Ca^{2+} transients and other defects associated with Ca^{2+} handling in cardiac myocytes [99] and AKAP150Δ36 mice show a deficit of persistent Ca^{2+} sparklets in cardiac myocytes and have lower intracellular Ca^{2+} as a result of defective PKA-mediated control of the L-type Ca^{2+} channel. In the case of Yotiao, disruption of its interaction with AC2 in the brain gives rise to 40% increased AC activity upon stimulation [96], indicating that AC activity is clearly regulated by this interaction.

Although less well documented, there is evidence that the source of cGMP may also be compartmentalized. For example, it has been reported that the pGC that serves as the receptor for atrial natriuretic peptide is localized to caveolae in atrial myocytes [100] and that the α_2 subunit of the sGC specifically interacts with the adaptor protein PSD-95 at postsynaptic membranes [101], thereby providing means for the generation of local pools of cGMP (Figure 3.1).

CONCLUSIONS

Spatial confinement of signal propagation is emerging as a critical feature for appropriate function of cyclic nucleotide signaling. The application of novel tools that allow analysis of intracellular signaling with high spatial resolution will certainly contribute to generate a detailed description of the spatial organization of individual cyclic nucleotide microdomains and of their specific regulation, ultimately providing a map for deciphering the intricacy of cAMP and cGMP signaling. The definition of a topographical map of cyclic nucleotide pools and their downstream targets may provide the basis for the identification of novel therapeutic strategies targeting only the relevant signaling cascades while leaving the remaining signaling network unaffected, thus improving treatment efficacy while reducing side-effects.

ACKNOWLEDGMENTS

This work was supported by the joint NSF-NIH CRCNS program (E2015651), the Fondation Leducq (O6 CVD 02) and the British Heart Foundation (PG/07/091/23698).

REFERENCES

1. Neves, S.R., Ram, P.T., and Iyengar, R. (2002) G protein pathways. *Science*, **296** (5573):1636–1639.

2. Gong, B., et al. (2004) Persistent improvement in synaptic and cognitive functions in an Alzheimer mouse model after rolipram treatment. *J. Clin. Invest.*, **114**(11):1624–1634.

3. Siuciak, J.A. (2008) The role of phosphodiesterases in schizophrenia: therapeutic implications. *CNS Drugs*, **22**(12):983–993.

4. Zhang, H.T. (2009) Cyclic AMP-specific phosphodiesterase-4 as a target for the development of antidepressant drugs. *Curr. Pharm. Des.*, **15**(14):1688–1698.

5. Stangherlin, A., et al. (2011) cGMP signals modulate cAMP levels in a compartment-specific manner to regulate catecholamine-dependent signaling in cardiac myocytes. *Circ. Res.*, **108**:929–939.

6. Conti, M. and Beavo, J. (2007) Biochemistry and physiology of cyclic nucleotide phosphodiesterases: essential components in cyclic nucleotide signaling. *Annu. Rev. Biochem.*, **76**:481–511.

7. Francis, S.H., Blount, M.A., and Corbin, J.D. (2011) Mammalian cyclic nucleotide phosphodiesterases: molecular mechanisms and physiological functions. *Physiol. Rev.*, **91**(2):651–690.

8. Lynch, M.J., Hill, E.V., and Houslay, M.D. (2006) Intracellular targeting of phosphodiesterase-4 underpins compartmentalized cAMP signaling. *Curr. Top. Dev. Biol.*, **75**: 225–259.

9. Tojima, T., Itofusa, R., and Kamiguchi, H. (2009) The nitric oxide-cGMP pathway controls the directional polarity of growth cone guidance via modulating cytosolic Ca^{2+} signals. *J. Neurosci.*, **29**(24):7886–7897.

10. Maddison, L.A., et al. (2009) A gain-of-function screen in zebrafish identifies a guanylate cyclase with a role in neuronal degeneration. *Mol. Genet. Genomics*, **281**(5):551–563.

11. Pifarré, P., et al. (2010) Cyclic GMP phosphodiesterase inhibition alters the glial inflammatory response, reduces oxidative stress and cell death and increases angiogenesis following focal brain injury. *J. Neurochem.*, **112**(3):807–817.

12. Calabrese, V., et al. (2007) Nitric oxide in the central nervous system: neuroprotection versus neurotoxicity. *Nat. Rev. Neurosci.*, **8**(10):766–775.

13. Kleppisch, T. and Feil, R. (2009) cGMP signaling in the mammalian brain: role in synaptic plasticity and behaviour. *Handb. Exp. Pharmacol.*, (191):549–579.

14. Tasken, K. and Aandahl, E.M. (2004) Localized effects of cAMP mediated by distinct routes of protein kinase A. *Physiol. Rev.*, **84**(1):137–167.

15. Keely, S.L. (1977) Activation of cAMP-dependent protein kinase without a corresponding increase in phosphorylase activity. *Res. Commun. Chem. Pathol. Pharmacol.*, **18**(2): 283–290.

16. Brunton, L.L., Hayes, J.S., and Mayer, S.E. (1979) Hormonally specific phosphorylation of cardiac troponin I and activation of glycogen phosphorylase. *Nature*, **280**(5717):78–80.

17. Hayes, J.S., Brunton, L.L., Brown, J.H., Reese, J.B., and Mayer, S.E. (1979) Hormonally specific expression of cardiac protein kinase activity. *Proc. Natl. Acad. Sci. USA*, **76** (4):1570–1574.

18. Zaccolo, M. and Pozzan, T. (2002) Discrete microdomains with high concentration of cAMP in stimulated rat neonatal cardiac myocytes. *Science*, **295**(5560):1711–1715.

19. Zaccolo, M., et al. (2000) A genetically encoded, fluorescent indicator for cyclic AMP in living cells. *Nat. Cell. Biol.*, **2**(1):25–29.

20. Förster, T. (1948) Intermolecular energy migration and fluorescence. *Ann. Phys.*, **2**:55–57.

21. Zaccolo, M. (2004) Use of chimeric fluorescent proteins and fluorescence resonance energy transfer to monitor cellular responses. *Circ. Res.*, **94**(7):866–873.

22. Lakowicz, J.R. (1999) Energy transfer. *Principles of Fluorescence Spectroscopy*, Kluwer Academic/Plenum, pp. 368–391.

23. Berrera, M., et al. (2008) A toolkit for real-time detection of cAMP: insights into compartmentalized signaling. *Handb. Exp. Pharmacol.*, (186):285–298.

24. Honda, A., et al. (2001) Spatiotemporal dynamics of guanosine $3',5'$-cyclic monophosphate revealed by a genetically encoded, fluorescent indicator. *Proc. Natl. Acad. Sci. USA*, **98**(5):2437–2442.

25. Nikolaev, V.O., Gambaryan, S., and Lohse, M.J. (2006) Fluorescent sensors for rapid monitoring of intracellular cGMP. *Nat. Methods*, **3**(1):23–25.

26. DiPilato, L.M., Cheng, X., and Zhang, J. (2004) Fluorescent indicators of cAMP and Epac activation reveal differential dynamics of cAMP signalling within discrete subcellular compartments. *Proc. Natl. Acad. Sci. USA*, **101**:16513–16518.

27. Di Benedetto, G., et al. (2008) Protein kinase A type I and type II define distinct intracellular signaling compartments. *Circ. Res.*, **103**(8):836–844.

28. Rich, T.C., et al. (2000) Cyclic nucleotide-gated channels colocalize with adenylyl cyclase in regions of restricted cAMP diffusion. *J. Gen. Physiol.*, **116**(2):147–161.

29. Rich, T.C., Tse, T.E., Rohan, J.G., Schaack, J., and Karpen, J.W. (2001) In vivo assessment of local phosphodiesterase activity using tailored cyclic nucleotide-gated channels as cAMP sensors. *J. Gen. Physiol.*, **118**(1):63–78.

30. Castro, L.R., Verde, I., Cooper, D.M., and Fischmeister, R. (2006) Cyclic guanosine monophosphate compartmentation in rat cardiac myocytes. *Circulation*, **113**(18):2221–2228.

31. Fischmeister, R., et al. (2006) Compartmentation of cyclic nucleotide signaling in the heart: the role of cyclic nucleotide phosphodiesterases. *Circ. Res.*, **99**(8):816–828.

32. Nikolaev, V.O., Bunemann, M., Schmitteckert, E., Lohse, M.J., and Engelhardt, S. (2006) Cyclic AMP imaging in adult cardiac myocytes reveals far-reaching beta1-adrenergic but locally confined beta2-adrenergic receptor-mediated signaling. *Circ. Res.*, **99**(10):1084–1091.

33. Xiao, R.P. (2001) Beta-adrenergic signaling in the heart: dual coupling of the beta2-adrenergic receptor to G(s) and G(i) proteins. *Sci. STKE*, **2001**(104):re15.

34. Nikolaev, V.O., et al. (2010) Beta2-adrenergic receptor redistribution in heart failure changes cAMP compartmentation. *Science*, **327**(5973):1653–1657.

35. Rho, E.H., Perkins, W.J., Lorenz, R.R., Warner, D.O., and Jones, K.A. (2002) Differential effects of soluble and particulate guanylyl cyclase on Ca(2+) sensitivity in airway smooth muscle. *J. Appl. Physiol.*, **92**(1):257–263.

36. Rivero-Vilches, F.J., de Frutos, S., Saura, M., Rodriguez-Puyol, D., and Rodriguez-Puyol, M. (2003) Differential relaxing responses to particulate or soluble guanylyl cyclase

activation on endothelial cells: a mechanism dependent on PKG-I alpha activation by NO/cGMP. *Am. J. Physiol. Cell Physiol.*, **285**(4):C891–898.

37. Barnea, G., et al. (2004) Odorant receptors on axon termini in the brain. *Science*, **304** (5676):1468.

38. Reed, R.R. (2003) The contribution of signaling pathways to olfactory organization and development. *Curr. Opin. Neurobiol.*, **13**(4):482–486.

39. Maritan, M., et al. (2009) Odorant receptors at the growth cone are coupled to localized cAMP and Ca^{2+} increases. *Proc. Natl. Acad. Sci. USA*, **106**(9):3537–3542.

40. Neves, S.R., et al. (2008) Cell shape and negative links in regulatory motifs together control spatial information flow in signaling networks. *Cell*, **133**(4):666–680.

41. Castro, L.R., et al. (2010) Type 4 phosphodiesterase plays different integrating roles in different cellular domains in pyramidal cortical neurons. *J. Neurosci.*, **30**(17):6143–6151.

42. Shelly, M., et al. (2010) Local and long-range reciprocal regulation of cAMP and cGMP in axon/dendrite formation. *Science*, **327**(5965):547–552.

43. Rich, T.C., et al. (2001) A uniform extracellular stimulus triggers distinct cAMP signals in different compartments of a simple cell. *Proc. Natl. Acad. Sci. USA*, **98**(23):13049–13054.

44. Bers, D.M. (2008) Calcium cycling and signaling in cardiac myocytes. *Annu. Rev. Physiol.*, **70**:23–49.

45. Saucerman, J.J., et al. (2006) Systems analysis of PKA-mediated phosphorylation gradients in live cardiac myocytes. *Proc. Natl. Acad. Sci. USA*, **103**(34):12923–12928.

46. Corbin, J.D., Sugden, P.H., Lincoln, T.M., and Keely, S.L. (1977) Compartmentalization of adenosine 3′:5′-monophosphate and adenosine 3′:5′-monophosphate-dependent protein kinase in heart tissue. *J. Biol. Chem.*, **252**(11):3854–3861.

47. Khac, L.D., Harbon, S., and Clauser, H.J. (1973) Intracellular titration of cyclic AMP bound to receptor proteins and correlation with cyclic-AMP levels in the surviving rat diaphragm. *Eur. J. Biochem.*, **40**(1):177–185.

48. Houslay, M.D. (2010) Underpinning compartmentalised cAMP signaling through targeted cAMP breakdown. *Trends Biochem. Sci.*, **35**(2):91–100.

49. Hohl, C.M. and Li, Q.A. (1991) Compartmentation of cAMP in adult canine ventricular myocytes. Relation to single-cell free Ca^{2+} transients. *Circ. Res.*, **69**(5):1369–1379.

50. Jurevicius, J. and Fischmeister, R. (1996) cAMP compartmentation is responsible for a local activation of cardiac Ca^{2+} channels by beta-adrenergic agonists. *Proc. Natl. Acad. Sci. USA*, **93**(1):295–299.

51. Rochais, F., et al. (2006) A specific pattern of phosphodiesterases controls the cAMP signals generated by different Gs-coupled receptors in adult rat ventricular myocytes. *Circ. Res.*, **98**(8):1081–1088.

52. Mongillo, M., et al. (2004) Fluorescence resonance energy transfer-based analysis of cAMP dynamics in live neonatal rat cardiac myocytes reveals distinct functions of compartmentalized phosphodiesterases. *Circ. Res.*, **95**(1):67–75.

53. Mongillo, M., et al. (2006) Compartmentalized phosphodiesterase-2 activity blunts beta-adrenergic cardiac inotropy via an NO/cGMP-dependent pathway. *Circ. Res.*, **98**(2): 226–234.

54. Terrin, A., et al. (2006) PGE(1) stimulation of HEK293 cells generates multiple contiguous domains with different [cAMP]: role of compartmentalized phosphodiesterases. *J. Cell Biol.*, **175**(3):441–451.

55. Baillie, G.S., et al. (2003) beta-Arrestin-mediated PDE4 cAMP phosphodiesterase recruitment regulates beta-adrenoceptor switching from Gs to Gi. *Proc. Natl. Acad. Sci. USA*, **100**(3):940–945.

56. Menniti, F.S., Faraci, W.S., and Schmidt, C.J. (2006) Phosphodiesterases in the CNS: targets for drug development. *Nat. Rev. Drug Discov.*, **5**(8):660–670.

57. Kelly, M.P., et al. (2010) Phosphodiesterase 11A in brain is enriched in ventral hippocampus and deletion causes psychiatric disease-related phenotypes. *Proc. Natl. Acad. Sci. USA*, **107**(18):8457–8462.

58. Coskran, T.M., et al. (2006) Immunohistochemical localization of phosphodiesterase 10A in multiple mammalian species. *J. Histochem. Cytochem.*, **54**(11):1205–1213.

59. Kotera, J., et al. (2004) Subcellular localization of cyclic nucleotide phosphodiesterase type 10A variants, and alteration of the localization by cAMP-dependent protein kinase-dependent phosphorylation. *J. Biol. Chem.*, **279**(6):4366–4375.

60. Noyama, K. and Maekawa, S. (2003) Localization of cyclic nucleotide phosphodiesterase 2 in the brain-derived Triton-insoluble low-density fraction (raft). *Neurosci. Res.*, **45**(2):141–148.

61. Nishi, A., et al. (2008) Distinct roles of PDE4 and PDE10A in the regulation of cAMP/PKA signaling in the striatum. *J. Neurosci.*, **28**(42):10460–10471.

62. Gervasi, N., Tchenio, P., and Preat, T. (2010) PKA dynamics in a Drosophila learning center: coincidence detection by rutabaga adenylyl cyclase and spatial regulation by dunce phosphodiesterase. *Neuron*, **65**(4):516–529.

63. Vo, N.K., Gettemy, J.M., and Coghlan, V.M. (1998) Identification of cGMP-dependent protein kinase anchoring proteins (GKAPs). *Biochem. Biophys. Res. Commun.*, **246**(3):831–835.

64. Airhart, N., Yang, Y.F., Roberts, C.T., Jr., and Silberbach, M. (2003) Atrial natriuretic peptide induces natriuretic peptide receptor-cGMP-dependent protein kinase interaction. *J. Biol. Chem.*, **278**(40):38693–38698.

65. Yuasa, K., Michibata, H., Omori, K., and Yanaka, N. (1999) A novel interaction of cGMP-dependent protein kinase I with troponin T. *J. Biol. Chem.*, **274**(52):37429–37434.

66. Wong, W. and Scott, J.D. (2004) AKAP signalling complexes: focal points in space and time. *Nat. Rev. Mol. Cell. Biol.*, **5**(12):959–970.

67. Newlon, M.G., et al. (1999) The molecular basis for protein kinase A anchoring revealed by solution NMR. *Nat. Struct. Biol.*, **6**(3):222–227.

68. Gold, M.G., et al. (2006) Molecular basis of AKAP specificity for PKA regulatory subunits. *Mol. Cell*, **24**(3):383–395.

69. Kinderman, F.S., et al. (2006) A dynamic mechanism for AKAP binding to RII isoforms of cAMP-dependent protein kinase. *Mol. Cell*, **24**(3):397–408.

70. Dell'Acqua, M.L., Faux, M.C., Thorburn, J., Thorburn, A., and Scott, J.D. (1998) Membrane-targeting sequences on AKAP79 bind phosphatidylinositol-4, 5-bisphosphate. *EMBO J.*, **17**(8):2246–2260.

71. Trotter, K.W., et al. (1999) Alternative splicing regulates the subcellular localization of A-kinase anchoring protein 18 isoforms. *J. Cell Biol.*, **147**(7):1481–1492.

72. Zhang, J., Ma, Y., Taylor, S.S., and Tsien, R.Y. (2001) Genetically encoded reporters of protein kinase A activity reveal impact of substrate tethering. *Proc. Natl. Acad. Sci. USA*, **98**(26):14997–15002.

73. Smith, F.D. and Scott, J.D. (2006) Anchored cAMP signaling: onward and upward-a short history of compartmentalized cAMP signal transduction. *Eur. J. Cell. Biol.*, **85**(7):585–592.

74. Colledge, M., et al. (2000) Targeting of PKA to glutamate receptors through a MAGUK-AKAP complex. *Neuron*, **27**(1):107–119.

75. Snyder, E.M., et al. (2005) Role for A kinase-anchoring proteins (AKAPS) in glutamate receptor trafficking and long term synaptic depression. *J. Biol. Chem.*, **280**(17):16962–16968.

76. Tavalin, S.J., et al. (2002) Regulation of GluR1 by the A-kinase anchoring protein 79 (AKAP79) signaling complex shares properties with long-term depression. *J. Neurosci.*, **22**(8):3044–3051.

77. Lu, Y., et al. (2007) Age-dependent requirement of AKAP150-anchored PKA and GluR2-lacking AMPA receptors in LTP. *EMBO J.*, **26**(23):4879–4890.

78. Beene, D.L. and Scott, J.D. (2007) A-kinase anchoring proteins take shape. *Curr. Opin. Cell Biol.*, **19**(2):192–198.

79. Klauck, T.M., et al. (1996) Coordination of three signaling enzymes by AKAP79, a mammalian scaffold protein. *Science*, **271**(5255):1589–1592.

80. Coghlan, V.M., et al. (1995) Association of protein kinase A and protein phosphatase 2B with a common anchoring protein. *Science*, **267**(5194):108–111.

81. Oliveria, S.F., Gomez, L.L., and Dell'Acqua, M.L. (2003) Imaging kinase–AKAP79–phosphatase scaffold complexes at the plasma membrane in living cells using FRET microscopy. *J. Cell Biol.*, **160**(1):101–112.

82. Schmidt, P.H., et al. (1999) AKAP350, a multiply spliced protein kinase A-anchoring protein associated with centrosomes. *J. Biol. Chem.*, **274**(5):3055–3066.

83. Lin, J.W., et al. (1998) Yotiao, a novel protein of neuromuscular junction and brain that interacts with specific splice variants of NMDA receptor subunit NR1. *J. Neurosci.*, **18**(6):2017–2027.

84. Tu, H., Tang, T.S., Wang, Z., and Bezprozvanny, I. (2004) Association of type 1 inositol 1,4,5-trisphosphate receptor with AKAP9 (Yotiao) and protein kinase A. *J. Biol. Chem.*, **279**(18):19375–19382.

85. Marx, S.O., et al. (2002) Requirement of a macromolecular signaling complex for beta adrenergic receptor modulation of the KCNQ1-KCNE1 potassium channel. *Science*, **295**(5554):496–499.

86. Chen, L. and Kass, R.S. (2006) Dual roles of the A kinase-anchoring protein Yotiao in the modulation of a cardiac potassium channel: a passive adaptor versus an active regulator. *Eur. J. Cell. Biol.*, **85**(7):623–626.

87. Westphal, R.S., et al. (1999) Regulation of NMDA receptors by an associated phosphatase-kinase signaling complex. *Science*, **285**(5424):93–96.

88. Tingley, W.G., et al. (1997) Characterization of protein kinase A and protein kinase C phosphorylation of the *N*-methyl-D-aspartate receptor NR1 subunit using phosphorylation site-specific antibodies. *J. Biol. Chem.*, **272**(8):5157–5166.

89. Tasken, K.A., et al. (2001) Phosphodiesterase 4D and protein kinase a type II constitute a signaling unit in the centrosomal area. *J. Biol. Chem.*, **276**(25):21999–22002.

90. Terrenoire, C., Houslay, M.D., Baillie, G.S., and Kass, R.S. (2009) The cardiac IKs potassium channel macromolecular complex includes the phosphodiesterase PDE4D3. *J. Biol. Chem.*, **284**(14):9140–9146.

91. Patel, H.H., Murray, F., and Insel, P.A. (2008) G-protein-coupled receptor-signaling components in membrane raft and caveolae microdomains. *Handb. Exp. Pharmacol.*, (186):167–184.

92. Willoughby, D. and Cooper, D.M. (2007) Organization and Ca^{2+} regulation of adenylyl cyclases in cAMP microdomains. *Physiol. Rev*, **87**(3):965–1010.

93. Fagan, K.A., Smith, K.E., and Cooper, D.M. (2000) Regulation of the Ca^{2+}-inhibitable adenylyl cyclase type VI by capacitative Ca^{2+} entry requires localization in cholesterol-rich domains. *J. Biol. Chem.*, **275**(34):26530–26537.

94. Davare, M.A., et al. (2001) A beta2 adrenergic receptor signaling complex assembled with the Ca^{2+} channel Cav1.2. *Science*, **293**(5527):98–101.

95. Bauman, A.L., et al. (2006) Dynamic regulation of cAMP synthesis through anchored PKA-adenylyl cyclase V/VI complexes. *Mol. Cell*, **23**(6):925–931.

96. Piggott, L.A., Bauman, A.L., Scott, J.D., and Dessauer, C.W. (2008) The A-kinase anchoring protein Yotiao binds and regulates adenylyl cyclase in brain. *Proc. Natl. Acad. Sci. USA*, **105**(37):13835–13840.

97. Iwamoto, T., et al. (2003) Motor dysfunction in type 5 adenylyl cyclase-null mice. *J. Biol. Chem.*, **278**(19):16936–16940.

98. Tunquist, B.J., et al. (2008) Loss of AKAP150 perturbs distinct neuronal processes in mice. *Proc. Natl. Acad. Sci. USA*, **105**(34):12557–12562.

99. Tang, T., et al. (2008) Adenylyl cyclase type 6 deletion decreases left ventricular function via impaired calcium handling. *Circulation*, **117**(1):61–69.

100. Doyle, D.D., et al. (1997) Type B atrial natriuretic peptide receptor in cardiac myocyte caveolae. *Circ. Res.*, **81**(1):86–91.

101. Russwurm, M., Wittau, N., and Koesling, D. (2001) Guanylyl cyclase/PSD-95 interaction: targeting of the nitric oxide-sensitive alpha2beta1 guanylyl cyclase to synaptic membranes. *J. Biol. Chem.*, **276**(48):44647–44652.

CHAPTER 4

PHARMACOLOGICAL MANIPULATION OF CYCLIC NUCLEOTIDE PHOSPHODIESTERASE SIGNALING FOR THE TREATMENT OF NEUROLOGICAL AND PSYCHIATRIC DISORDERS IN THE BRAIN

FRANK S. MENNITI
Mnemosyne Pharmaceuticals, Inc., Providence, RI, USA

NIELS PLATH and NIELS SVENSTRUP
H. Lundbeck A/S, Valby, Denmark

CHRISTOPHER J. SCHMIDT
Pfizer Global Research and Development, Neuroscience Research Unit, Cambridge, MA, USA

INTRODUCTION

Essentially the entire PDE gene family is expressed in the mammalian central nervous system (CNS) [1]. Furthermore, it is reasonable to assume that the vast majority of neurons express one or more PDE isoforms. On the other hand, PDE isoform expression is specific to different brain regions, neuronal subtypes, and neuronal subcellular compartments (see Chapter 3). Thus, the pharmacological targeting of different PDE isoforms is predicted to have diverse, isoform-specific effects on CNS function. As presented in detail in other chapters of this book, there is tremendous interest in manipulating PDE activity as potential treatment for a variety of psychiatric and neurological disorders. Given these therapeutic potentials and the fact that the substrate binding pocket of the PDEs is an attractive target to medicinal chemists, there has been significant progress over the last decade in developing pharmacological

Cyclic-Nucleotide Phosphodiesterases in the Central Nervous System: From Biology to Drug Discovery,
First Edition. Edited by Nicholas J. Brandon and Anthony R. West.
© 2014 John Wiley & Sons, Inc. Published 2014 by John Wiley & Sons, Inc.

tools to study the functions of the PDEs in the CNS. Indeed, some of these compounds have advanced into clinical trials for various CNS indications. The continued progress is critically dependent on the availability of new chemical tools that specifically interact with the broader range of PDE isoforms and have pharmacological properties amenable to different modes of delivery in experimental systems.

This chapter reviews the current state of PDE drug development efforts with specific focus on the CNS. The chapter starts with a short commentary on the utility of PDE expression data in initially directing these drug development efforts and as a reference point for evaluation of preclinical functional data at later stages. To date, the majority of the drug discovery efforts have been directed at PDE inhibitors interacting at the catalytic site. Thus, this chapter presents a short discussion of the molecular properties that facilitate the ability of such inhibitors to cross the blood–brain barrier. It then reviews the state-of-the-art on PDE inhibitor pharmacology for PDE 2, PDE5, PDE7, PDE8, PDE9, and PDE10, as these are the focus of most current efforts with regard to the CNS. The chapter closes with a short discussion of the potential for the development of inhibitors selective for subtypes within a PDE family and for allosteric modulators of the PDEs, using as a takeoff point recent data regarding novel PDE4 inhibitors. Details related to the underlying biology behind these drug discovery efforts are elaborated elsewhere in this book.

PDE LOCALIZATION ANALYSIS IN THE DISCOVERY PROCESS

A key component in the development of PDE inhibitors for the treatment of any disease is analysis of PDE target expression patterns. In the CNS, this analysis ideally includes a determination of the neuronal circuits in which the different PDEs are expressed, as well as the subcellular distribution of the enzyme within those neurons. At the earliest stage in the discovery process, the circuit level expression pattern of a particular PDE may aid in generating hypotheses as to the potential effects of pharmacological manipulation of that PDE, based on extrapolation from the known role(s) of those circuits in physiological and behavioral processes (see Chapter 2). Expression in vulnerable neuronal and/or glial populations may also be a component in generating hypotheses around the potential for targeting particular PDEs for neuroprotection or cytoprotection. However, the analysis of circuit-level expressions is complicated by the fact that individual neurons within those circuits express multiple PDEs. A well-documented example is the striatal medium spiny neurons, all of which express high levels of mRNA and protein for PDE1B [2], PDE2A [3], and PDE10A [4], and a subpopulation of which express mRNA and protein for PDE4B [5]. These neurons also express high levels of mRNA for PDE7B [6], and there is evidence for mRNA expression and functional activity for PDE9A [7,8]. It is well accepted that PDE-regulated cyclic nucleotide signaling is highly compartmentalized within cells to regulate distinct cellular processes. Thus, it is reasonable to assume that these different PDEs within the medium spiny neurons are regulating distinct, nonoverlapping aspects of the information processing by these neurons. This, in turn, implies distinct functional effects of inhibiting these different PDEs on striatal circuitry and, therefore, distinct therapeutic utilities. A similar scenario in which multiple PDEs are expressed within individual neuronal types is highly likely to hold for most, if not all, of the different neuronal populations in the brain.

Compartmental analysis may shed light on the distinct roles of different PDEs expressed in different neuronal types. Neuronal information processing is extremely highly compartmentalized. Thus, localization of a PDE to synaptic elements suggests a role for that enzyme in synaptic transmission and/or plasticity. For example, PDE2A appears to be excluded from the cell bodies and primary dendrites of principal neurons in cortex and concentrated in axons and presynaptic terminals [3]. This suggests a role for this enzyme in regulating glutamate release from these neurons and plasticity that involves the presynaptic element. Another example where subcellular localization may play a key role in differential function is the difference in the subcellular distributions of PDE1B and PDE10A in the striatum. Both of these are dual substrate enzymes highly expressed in striatal tissue. The former is restricted to the soluble compartment, whereas the latter is primarily membrane bound [9]. PDE1B and PDE10A knockout mice both have a locomotor phenotype, consistent with a role in regulating striatal function. Significantly, these phenotypes are opposites: the PDE1B knockout mice exhibit enhanced locomotor response [10], whereas the PDE10A knockouts are hypolocomotor [11].

From a drug discovery perspective, targeting PDEs in distinct subcellular compartments is challenging. The vast majority of compounds targeting the PDEs to date are inhibitors that bind in the catalytic site. Such compounds interact with their PDE targets regardless of subcellular localization. Thus, to target PDEs in discreet subcellular compartments, new classes of compounds must be discovered that reach beyond the catalytic site. One broad approach is to develop compounds that block the recruitment of a particular PDE to a signaling complex. It is now well established that a key step in many intracellular signaling pathways is the recruitment of mediators into physical complexes to facilitate interaction and localize signaling events. For example, A kinase anchoring protein (AKAP)-mediated clustering of adenylyl cyclase, protein kinase A (PKA), and PKA substrates into a complex yields efficient transduction of G protein-coupled receptor (GPCR) activation into protein phosphorylation [12]. Also recruited into such AKAP clusters are PDE4 to terminate the cyclic adenosine monophosphate (cAMP) signal. In this scenario, inhibiting PDE4 recruitment to AKAP may be as effective in potentiating the signaling as inhibiting PDE4 enzymatic activity. Blocking recruitment of PDE4 to a particular AKAP/signaling complex yields specificity to that signaling process and avoids the consequence of PDE4 inhibition across the many different signaling cascades that utilize this enzyme. It must be cautioned that, while this approach is theoretically appealing, it involves the discovery of compounds that disrupt interactions amongst large proteins. Such interactions involve multiple points of contact, and blocking such interactions has been notoriously difficult with traditional small molecules. Perhaps this approach will be become more feasible with the advance of newer technologies, such as the development of aptamer-based strategies [13]. Alternatively, it may be feasible to discover small-molecule "hybrids," at least for PDE4 [14]. Structural studies by Burgin et al. [15] has shed light on the mechanisms by which N-terminal upstream conserved region (UCR) and C-terminal extracellular signal-regulated kinase (ERK) phosphorylation domains regulate PDE4 enzymatic activity. Significantly, PDE4 inhibitors were discovered that interact with the catalytic site to block enzymatic activity but also "reach out" of the catalytic domain to specifically interact with these regulatory domains at the mouth of the catalytic site. Such interactions can stabilize different conformations of the enzymes. In so far as enzyme conformation is also impacted by the interaction of these same regulatory domains with signaling complex partners, hybrid inhibitors may be specific for inhibition of specific enzyme conformations in distinct signaling complexes. Thus, such inhibitors take advantage of binding interactions within the catalytic domain as anchor, with additional functionality reaching out to interact with other sites accessible only in certain signaling complexes. These latter interactions may confer signaling compartment specificity to such catalytic site inhibitors. The challenge here is extending the elegant discoveries of Burgin et al. to other PDE families and then capitalizing on these discoveries for new drug development.

At later stages of the discovery process, the expression pattern of the targeted PDE may serve as a "reality check" on the therapeutic rationale developed from functional data. Simply put, how does the localization of the target PDE reconcile with the actions of the PDE inhibitor? For example, there is a growing body of data indicating effects of PDE5A inhibitors on CNS function in preclinical studies (see Chapter 9). These include procognitive effects [16], effects of amyloid precursor protein metabolism and $A\beta$ formation [17], and enhancement of recovery of function after stroke [18]. However, the expression of PDE5A in the CNS that might account for these diverse effects is controversial, with the most detailed studies indicating no appreciable expression of mRNA or protein in plausibly relevant neuronal populations [19,20]. PDE9A is another interesting example. In this case, there is replicable evidence that PDE9A mRNA is broadly expressed throughout the brain [7,21], which is congruent with a growing body of data indicating that PDE9A inhibition has widespread effects on brain cyclic guanosine monophosphate (cGMP) metabolism, and also effects synaptic plasticity and learning and memory [8,22–24]. However, to date, PDE9A protein has not been detected in brain, either because suitable antibodies have not been developed and/or expression is very low. It will be of considerable interest to determine the synaptic compartment in which PDE9A is localized that accounts for the effects of PDE9A inhibitors on synaptic plasticity. This may provide significant new insights into the synaptic events that presumably mediate the systemic effects of PDE9A inhibitors in learning/memory and other behavioral paradigms that are driving the drug development efforts. On the other hand, how will the rationale for PDE9A as a therapeutic target for cognitive dysfunction be effected should it be found that PDE9A is not localized to the synapse? Finally, reconciling PDE expression patterns with functional effects of inhibitors is an issue even for PDEs that are highly expressed in the CNS or for which multiple family members are expressed. Examples of the latter are the PDE4 isozymes PDE4A, PDE4B, and PDE4D, which are differentially expressed, in different isoform splice variants, throughout the CNS [25].

In summary, analysis of the expression patterns of different PDEs in the CNS can be an essential guide to the initial consideration of a PDE as a therapeutic target, as well as a reference point for understanding the functional effects of modulating the activity of that target. Ultimately, however, it is the functional data that drives the development of a particular PDE target for a particular therapeutic indication. Genetic manipulation of PDE activity has been an extremely valuable tool in generating such functional data. This includes the use of "traditional" gene knockout approaches in mice [26], and, more recently, site-directed knockdown using interfering RNA technology [27]. These approaches are complementary to the use of pharmacological tools to probe PDE function. It is these pharmacological tools that are then turned into potential small-molecule therapeutics. Pharmacological tools for the study of CNS indications ideally possess the pharmaceutical properties advantageous to the use of any compound (high potency and specificity for the target, solubility, systemic absorption) and one unique characteristic, that is, the ability to readily cross the blood–brain barrier. This is discussed further in the following sections.

MOLECULAR PHARMACOLOGY OF PDEs

The development of highly potent and specific PDE inhibitors is being greatly aided by the availability of high-resolution crystal structures of a number of PDEs [28] (see Chapter 6). However, for the treatment of CNS disorders, therapeutically useful compounds must possess an additional characteristic, the ability to readily cross the blood–brain barrier (BBB) to gain access to the CNS. The molecular properties of the enzyme and the characteristics of inhibitors relevant to the development of PDE inhibitors for use in treating neuropsychiatric disease are reviewed in the following sections.

Structure of Phosphodiesterases

The PDEs are modular enzymes in which the catalytic domain in the C-terminal portion of the protein is coupled to regulatory elements that reside in the N-terminal region. The 11 PDE families differ most significantly from one another within the unique N-terminal regulatory domains. On the other hand, the C-terminal catalytic domains are highly conserved with respect to specific invariant amino acids, three-dimensional structure, and catalytic mechanism [29]. Nonetheless, subtle differences within the catalytic core impart important family-specific characteristics [30]. To date, essentially all of the pharmaceutical development around the phosphodiesterases has been toward the discovery of catalytic site inhibitors; targeting the regulatory domains is discussed above but is as yet unproven. Structural information from single-crystal X-ray crystallography has played an important role in elucidating the important functional differentiating features within the catalytic domains of the 11 gene families that allow for the development of family-specific inhibitors. Indeed, current lead optimization projects without the use of some form of structure-based drug design are becoming practically unthinkable. This area of knowledge is summarized in the following.

Structures of the catalytic domains of all but two phosphodiesterase families (PDE6 and PDE11) have been solved. Since the field was last reviewed in 2007 [28], two new PDE families have been added to the list of solved structures, namely, PDE8 in its unliganded form and in complex with IBMX [31], and PDE10A with various ligands [32,33]. Characteristic of all PDE structures solved so far are the following features, which are also important for the design of new inhibitors.

The active site contains a glutamine-residue that contributes to the binding of the natural substrate cAMP or cGMP through a dual hydrogen bond. The "glutamine switch" mechanism [34] suggests that hydrogen-bonding residues surrounding the glutamine serve to either lock it in a fixed conformation (cAMP- or cGMP-selective phosphodiesterases) or allow it to change conformation (PDE1, PDE2, PDE3, PDE10, and PDE11). Although very elegant in its simplicity, the glutamine switch hypothesis remains somewhat controversial [35,36]. The glutamine is also nearly invariably involved in hydrogen bonding to PDE inhibitors, although not necessarily through two hydrogen bonds (see Ref. [37] and the references cited therein).

A phenylalanine, situated just below the plane of the bound substrate–inhibitor, participates in the substrate binding by π–π interactions. This hydrophobic region, usually referred to as the "clamp" region, explains why many PDEs appear to have a preference for flat and π-electron rich inhibitors of the sildenafil type. The only exception is PDE11A, which has a tryptophan in this position.

The metal ions in the active site may also be targeted for inhibitor binding; however, this approach is not usually addressed by design elements in PDE inhibitors intended for CNS indications. Specifically, a good ligand for the metal ion is by its very nature rather polar, thereby adding to the overall polar surface area of the inhibitor to such a degree that transport across the BBB becomes exceedingly difficult.

Structural information obtained from the X-ray crystal structures of inhibitor complexes have already demonstrated immense value during the optimization of potency on the phosphodiesterase in question, and in identifying regions of the lead molecule that do not directly contribute to binding to the PDE active site, thereby lowering the ligand efficiency.

Another use of crystallographic structure information is in the optimization of the selectivity among the various phosphodiesterases. Knowledge of the structure of an inhibitor in two different PDEs does help to identify structural features that differ between two PDEs, thereby acting as inspiration for how chemical changes in the lead structure would make it fit better in one PDE than in the other PDE. There are always several different ways of achieving selectivity, however. For example, Pfizer has published about a novel class of PDE9A inhibitors that specifically target Tyr424. This residue is unique to PDE8 and PDE9A, and the interaction of compounds with this residue contributes significantly to the good subtype selectivity found in the Pfizer PDE9A inhibitors [8]. BAY 736691, a PDE9A-selective inhibitor from the laboratories of Bayer Healthcare, achieves its selectivity in a much more subtle fashion, apparently without directly interacting with Tyr424. In another example, a unique "pocket" near the substrate-binding domain of PDE10A was discovered and exploited in a structure-based drug design for highly selective and potent PDE10A inhibitors [32].

Pharmaceutical Properties: Crossing the Blood–Brain Barrier

The development of highly potent and specific PDE inhibitors is being greatly aided by the availability of high-resolution crystal structures of a number of PDEs. For the treatment of CNS disorders, therapeutically useful compounds must possess an additional characteristic, the ability to readily cross the BBB to gain access to the CNS. This barrier is made up of two components [38]. The BBB is traditionally defined as the tight junctions between the endothelial cells that line the capillary beds of the CNS vasculature that occlude paracellular diffusion of small molecules into the brain parenchyma. It is becoming increasingly appreciated that these endothelial cells also express carrier proteins, such as the P-glycoprotein (P-gp) that actively pumps small molecules out of the brain [39]. There is relatively little published data on the CNS penetrability of PDE inhibitors used clinically or as pharmacological tools. It is reasonable to speculate that the physicochemical properties needed to achieve high brain exposure of a PDE inhibitor (an intracellular enzyme target) with high potency will not necessarily overlap completely with those historically associated with high brain exposure for a monoamine receptor (a GPCR with an extracellular binding domain). Nonetheless, there are a number of physiochemical parameters established from work on GPCR ligands that may be considered to increase the likelihood of a PDE inhibitor having desirable CNS access. These properties can be combined with those required for high affinity interaction with the PDEs to guide design of inhibitors favorable for CNS indications.

Three molecular characteristics are recognized to be significant determinants of BBB penetrability: lipophilicity, polar surface area (PSA), and size or molecular weight [40,41]. High lipophilicity (clogP in the range of 2–4) and low polar surface area (<80–$90\,\text{Å}^2$) [40–42] appear to be optimum for transcellular diffusion through the lipophilic membranes of the BBB. A smaller molecule with a low molecular weight ($\sim<400\,\text{Da}$) is also advantageous. A computational model for the prediction of BBB penetration based on such parameters has been proposed [43,44]. This method, dubbed multiparameter optimization (MPO), calculates a single parameter for a compound based on a combination of calculated logP, PSA, calculated logD, molecular weight, pK_a value, and the number of hydrogen bond donors. The resulting MPO parameter can serve as a guide to optimizing chances of identifying compounds with suitable metabolic properties, sufficient permeation, and favorable safety profile. An interesting implication is that because each parameter contributes to the MPO score in an additive way, a nonideal PSA can be somewhat compensated for by having a good score in the other categories. Key molecular characteristics that promote high affinity interaction with the active sites of the PDEs have been proposed [45]. As noted above, a major point of interaction is the conserved hydrophobic pocket or P-clamp [45]. Hydrophobicity facilitating a P-clamp interaction would, thus, also be compatible with BBB permeability. However, two other features proposed to be important for inhibitor binding—interactions to the conserved glutamine and metal site—may be deleterious to CNS penetration, as they may increase PSA of inhibitors. Specifically, molecules with more than one hydrogen bond donor rarely display significant CNS exposure. Because a dual hydrogen bond donor/acceptor binding to the glutamine appears to be the key pharmacophore, any additional hydrogen bond donors in the structure are usually not allowed. The molecular characteristics that minimize the likelihood that a compound will be a substrate for transport out of the brain are less well understood. Seelig [46] has proposed a general pattern for substrate recognized by P-gp that includes a spacing of $2.3 \pm 0.3 \, \text{Å}$ between two electron donors or a spacing of $4.6 \pm 0.6 \, \text{Å}$ between two or three electron donors. An alternative P-gp pharmacophore model proposed by Pajeva [47] includes two hydrophobic points, three hydrogen bond acceptor points, and one hydrogen bond donor point. The knowledge of the structure of the PDE active site with bound inhibitor should allow for moving away from such structures while introducing physiochemical properties favorable for high BBB permeability.

CURRENT STATUS

Interest continues to grow in the PDEs as molecular targets for developing new therapies to treat CNS diseases. This interest has fueled the continued search for PDE inhibitors with the pharmaceutical characteristics appropriate for the study of these enzymes in the brain. The most active areas of research are in identifying competitive inhibitors for PDE2, PDE5, PDE7, PDE8, PDE9, and PDE10. This is reviewed in the following sections.

PDE2A

PDE2A belongs to the dual-substrate PDEs hydrolyzing both cAMP and cGMP [48]. PDE2A is a single-gene family with three known splice variants (PDE2A1, PDE2A2, and PDE2A3) that differ with respect to their N-terminus [49–51]. It is unclear whether all splice variants are shared across species. PDE2A2 and PDE2A3 are the predominant splice variants expressed in the brain, where they are associated with membranes [52]. This localization is partially a result of N-terminal palmitoylation [53], but for PDE2A3 it is mainly mediated through N-terminal acetylation [52]. PDE2A1 is found in soluble fractions and lacks most of the N-terminal region present in PDE2A2/PDE2A3.

PDE2A is one of the PDE superfamily classes that it is allosterically activated by physiological concentrations of cGMP binding to one of the N-terminal GAF domains. In fact, the X-ray crystal structure of a PDE2A construct that contains both the GAF domains and the catalytic domain has been solved [54]. This structure strongly supported a mechanism for the allosteric regulation of PDE activity in which binding of cGMP to the GAF domains causes a conformational change in the protein, which enhances access to the active site. With regard to the physiology, in peripheral tissues, binding of cGMP to the PDE2A GAF B domain activates the degradation of cAMP [50]. This allosteric activation constitutes a mechanism of crosstalk between distinct cAMP and cGMP regulated signaling pathways [55]. However, in primary cultures of forebrain neurons, PDE2A preferentially metabolized cGMP [56], suggesting that in the CNS this enzyme may also serve as an inhibitory feedback regulator of cGMP signaling. It also indicates that, whereas PDE2A may be silent under baseline conditions, it is selectively activated upon neuronal stimulation that causes an increase in cGMP.

As for all PDEs, it is of central interest to reveal the cellular and subcellular localization of PDE2A in order to understand the impact of this selective activation on CNS function. Initial studies showed PDE2A expression reaches highest levels in the CNS, particularly within the limbic system [3,49,57]. A recent detailed immuno-histochemical analysis substantiates and extends these earlier findings [3]. PDE2A is densely expressed in forebrain parenchyma that contains the neuronal dendrites and axons, suggesting compartmentalization of the enzyme directly at the input and/or output region of neurons [3]. Interestingly, this fine punctuate pattern in neurites is pronounced in areas known to be involved in learning and memory formation, such as the hippocampus, striatum and cortex. Here, neuropil localization is accompanied by a lack of PDE2A immunoreactivity in cell bodies. In further studies, PDE2A was detected in membrane rafts [53] and synaptosomal membranes [52], substantiating evidence for a localization at the immediate site of synaptic contacts and thus in a suitable position to hydrolyze the second messengers cAMP and cGMP immediately at the synapse. Inhibition of PDE2A therefore appears attractive as it might offer a selective prolongation of cAMP and cGMP levels directly related to synaptic activation. In fact, one of the highest levels of PDE2A expression in brain appears to be the mossy fibers emanating from the hippocampal dentate granule cells and receiving input from the entorhinal cortex, one of the first brain regions showing morphological signs of pathology in Alzheimer's disease (AD) [3]. This raises the intriguing possibility that PDE2A is involved in regulating presynaptic forms of synaptic plasticity. Perhaps PDE2A is one of the mediators of retrograde nitric oxide signaling in the presynaptic terminal, either through regulating cGMP directly or by regulating cAMP levels in response to cGMP binding to the GAF domain. Interestingly, in a few brain regions, including the medial habenula, and in neuronal subsets in the cortex, substantia nigra pars compacta, or raphe nuclei, neurons show somatic staining. This heterogeneous localization pattern within different neuronal populations indicates divergent roles of PDE2A in different cell populations. The CNS expression pattern of PDE2A is preserved in mammals, including humans and remains unaltered in postmortem brain of AD patients [3,58].

A highly potent and selective PDE2A inhibitor, BAY 60-7550, with an IC_{50} for human recombinant PDE2A of 4.7 nM has been shown to enhance long-term potentiation (LTP) at the CA3/CA1 synapse in hippocampal slices [59]. Systemic administration of BAY 60-7550 to rodents has been shown to attenuate natural forgetting in young rats and improve age-related impairment on old rats in behavioral tasks addressing episodic short- and long-term memory in rats [59–61]. The compound also reverses working memory deficits in mice induced by a time decay or acute treatment with the N-methyl-D-aspartate (NMDA) receptor antagonist MK-801 [59]. The various temporal stages of memory consolidation, reaching from working to short-term and long-term memory, have been suggested to be differentially regulated by cAMP and cGMP in either presynaptic or postsynaptic terminals [62]. It was, therefore, speculated that interference with a dual-substrate

PDE that is localized both at the presynaptic and postsynaptic site should have a broad impact on different temporal stages of memory processing. The promnemonic effects achieved with BAY 60-7550 are in line with this hypothesis [62]. It should be noted that BAY 60-7550 penetrates into the CNS very poorly; thus, generalization regarding the effect of PDE2A inhibition on cognitive function await confirmatory studies with other compounds. Moreover, a PDE2A constitutive knockout mouse line cannot be utilized for behavioral studies as these were reported to be embryonically lethal [3].

Based on the CNS expression pattern, positive cooperative kinetics between cAMP and cGMP and synaptic association of PDE2A, the enzyme is believed to be a very attractive target to support signaling pathways involved in synaptic plasticity and learning and memory. However, to date, a clear link from PDE2A to AD or other cognitive disorder is missing. It should also be noted that PDE2A is widely expressed in peripheral tissues, including heart, liver, lung, and kidney, where PDE2A inhibitors have various functional effects [63]. With the identification of more brain-penetrating PDE2A inhibitors, it will be important to identify pharmacological windows between centrally mediated effects on cognition and those in the other organs. Toward this end, PDE2A inhibitors have been pursued by a number of research groups for various indications. So far, discovery of potent and selective PDE2A inhibitors with good CNS exposure has proven to be a real challenge; progress is reviewed in the following.

The main tool compounds available for mechanistic research at present are EHNA (submicromolar inhibitor of PDE2A, the first selective inhibitor described in the literature) [64], BAY 60-7550 (depending on the construct, of a nanomolar to subnanomolar inhibitor of PDE2A, structurally related to EHNA) [59], and the chemically distinct oxindole (double-digit nanomolar inhibitor of PDE2A with good selectivity) [65], as shown in Figure 4.1. All three compounds are unlikely to advance beyond the tool compound level; EHNA is not potent enough to qualify as a development candidate and is also a potent inhibitor of adenosine deaminase, whereas BAY 60-7550 has rather poor pharmacokinetic properties, and oxindole has negligible CNS penetration. Still, all three have been immensely useful as mechanistic probes of the PDE2A enzyme and for studying non-CNS pharmacology models.

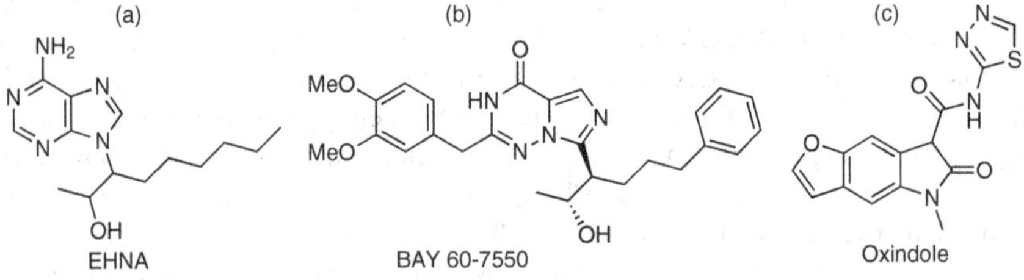

FIGURE 4.1 PDE2A inhibitor tool compounds. (a) EHNA (submicromolar inhibitor of PDE2A, the first selective inhibitor of PDE2A described in the literature). (b) BAY 60-7550 (nanomolar to subnanomolar inhibitor of PDE2A, structurally related to EHNA). (c) The chemically distinct oxindole.

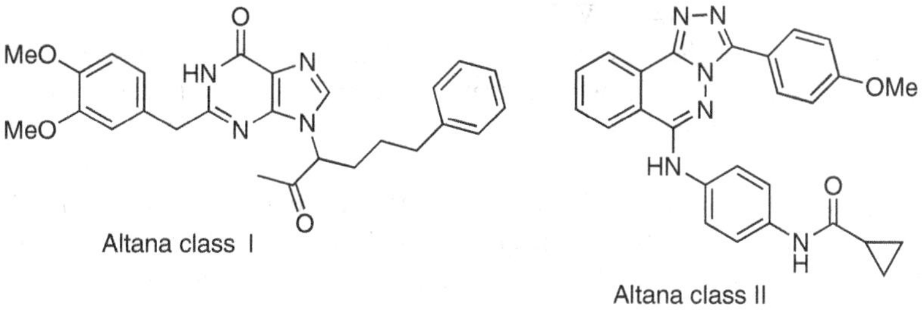

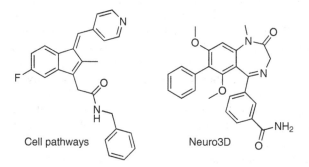

FIGURE 4.2   PDE2A inhibitors from Altana. Class I is a derivative of the Bayer series, whereas class II derives from a triazolophthalazine scaffold.

New PDE2A inhibitors, with improved pharmaceutical properties, are being explored by several pharmaceutical companies. Bayer has been pursuing the structural class around BAY 60-7550 as documented by various patent applications for CNS and cardiovascular indications (WO2008043461, WO02068423, WO00250078, WO00209713, WO00012504, WO09840384), although apparently without identifying a clinical candidate. Likewise, Pfizer has pursued the oxindole class and a class of azaquinazolines (WO2005061497), but the current development stage of this project is uncertain. Altana Pharma (now a part of Nycomed) has been addressing PDE2A through two distinct chemical classes: a BAY 60-7550–like class (WO2005021037, WO2004089953) employing the EHNA-scaffold and another class of triazolophthalazines (WO2006024640, WO2006072612, WO2006072615) (Figure 4.2). No data have been disclosed for individual compounds, but some are reported to inhibit PDE2A in the low nanomolar region. It appears that chronic obstructive pulmonary disease (COPD) and inflammation have been the relevant indications for these compounds rather than CNS indications, although the general physicochemical profile might also be compatible with CNS exposure. It is unclear if Nycomed is actively developing either of these classes of PDE2A inhibitors. The PDE2A inhibitor research program at Cell Pathways has been largely based on substituted indenes of the type shown in Figure 4.3

FIGURE 4.3   PDE2A inhibitors from cell pathways and Neuro3D. PDE2A inhibitor from cell pathways is a substituted indene (0.68 μM PDE2A), meant for non-CNS indications such as inflammatory bowel disease. Neuro3D (acquired by A Evotec AG) has developed a class of benzodiazepinones that are selective, although not especially potent, inhibitors of PDE2A.

(EP01749824, US06465494, WO02067936). Information about pharmacological properties are scarce (the best example reported in the patent literature is a 0.68 μM PDE2A inhibitor), but it seems clear that these compounds are meant for non-CNS indications, such as inflammatory bowel disease based on the wording of the patent application claims.

Finally, Neuro3D (acquired by Evotec AG) have been involved in PDE2A research with a class of benzodiazepinones (EP01548011, WO2004041258) that are reported to be selective, although not especially potent, inhibitors of PDE2A.

PDE5A

The PDE5A inhibitors Viagra (sildenafil), Levitra (vardenafil), and Cialis (tadalafil) as treatments for male erectile dysfunction (ED) are the first commercial successes targeting the PDE superfamily. These successes have generated tremendous interest in PDE5A as a therapeutic target for other disorders [66], and have provided excellent pharmacological tools that greatly facilitate such investigation. Chapter 9 looks in detail at the possibilities for PDE5A inhibition beyond ED. PDE5A specifically regulates cGMP signaling. Given the extensive literature on the role of nitric oxide (NO)–cGMP signaling in synaptic plasticity, there has been considerable interest in PDE5A inhibitors to treat cognitive disorders [62,67,68]. Indeed, there is a growing body of literature indicating that PDE5A inhibitors may improve cognitive function in preclinical species. PDE5A inhibitors robustly improve novel object recognition in rodents [16,61] (and unpublished observation) and attenuate spatial learning impairment in the 14-unit T-maze induced by cholinergic blockade, inhibition of nitric oxide synthase (NOS), or because of age in rats [69]. Rutten et al. recently reported that the PDE5A inhibitor sildenafil improves object retrieval performance in nonhuman primates [70]. However, to date, preliminary clinical trials using sildenafil have not provided evidence of procognitive effects in healthy human volunteers [71,72] or in patients suffering from schizophrenia [73].

There is also evidence to suggest a therapeutic utility of PDE5A inhibitors in the treatment of neurological and neurodegenerative conditions [68]. PDE5A inhibitors robustly facilitate functional recovery of sensorimotor function after stroke in the rat [18,20,74,75]. In these studies, PDE5A inhibitors were administered days after the stroke and had no effect on the infarct volume. Thus, it is argued that the effect of the compounds on sensorimotor recovery is through facilitating the ability of the brain to reorganize after damage; that is, through an effect on plasticity. Recently, Puzzo et al. reported effects of PDE5A inhibition that may suggest therapeutic utility in AD [17]. This group found dramatic improvements caused by the PDE5A inhibitor sildenafil on hippocampal LTP measured *in vitro* in slices and on performance in cognitive tasks in a mouse model of AD, the APP/PS1 mice. These effects were accompanied by an upregulation of cAMP response element binding (CREB) phosphorylation and a reduction in levels of Aβ.

The data reviewed above, from various laboratories and in various model systems, indicate potentially significant beneficial effects of PDE5A inhibition on brain function. However, an enigma arises from the fact that the expression of PDE5A in forebrain neuronal populations relevant to such effects is very limited. In rat forebrain, PDE5A mRNA was found only in isolated, phenotypically unidentified neurons in one report [21], and was found not at all in another report [19]. Additionally, PDE5A protein was not detected [19] or only rarely detected [20] in rat forebrain in studies in which two different antibodies were used. PDE5A mRNA was also not detected in postmortem samples of forebrain from patients suffering AD [58]. In contrast to forebrain neurons, PDE5A message and protein are robustly expressed in cerebellar Purkinje neurons, some brain stem neurons, spinal cord [19,20,76], and the cerebrovascular structure [20]. It is difficult to reconcile this distribution of the enzyme with the various effects of the PDE5A inhibitors on brain function that have been reported. It remains to be determined which biological systems in which tissues are impacted by PDE5A inhibitors to account for these effects in cognition and neurodegenerative disease models.

One pharmacological approach to gaining insight into the source of this PDE5A target is to compare the effects of PDE5A inhibitors that differ in crossing the BBB. Simply put, effects of compounds that do not get into brain would indicate that the PDE5A target is outside the brain; alternatively, lack of effects of such compounds would indicate that the PDE5A target is, in fact, in the brain. Puzzo et al. reported that sildenafil, which gains access to brain, ameliorated contextual fear conditioning and spatial working memory deficits in the APP/PS1 mouse model of AD [17]. In contrast, tadalafil, which is cited as not accessing brain, was ineffective in these assays. On the other hand, it has recently been reported that PDE5A inhibitors that do *not* cross the BBB improve performance in an object recognition task [77] and improve functional recovery after stroke [78]. Clearly, further study is needed to address the two questions raised by these results: Is it the locus of PDE5A within the brain that accounts for the procognitive effects? A role for cerebellar involvement is suggested, given that there is a consistent finding that PDE5A is expressed in cerebellar Purkinje neurons. However, it certainly is far from ruled out that PDE5A is expressed at levels below the limits of detection, or is restricted to small, but critical, neuronal populations and that these limited pools of PDE5A nonetheless account for significant effects of cognitive performance. By what mechanisms do PDE5A inhibition outside the brain impacts cognitive performance? An obvious locus of interaction is at the brain–vascular interface. Initial studies failed to find evidence for an effect of PDE5A inhibitors mediated by changes in cerebral blood flow [79]. The investigation of paracrine signaling interactions may be fruitful. These further studies are clearly warranted, given that numerous PDE5A inhibitors are immediately available for clinical use and have proven safe and well tolerated for both episodic and chronic administration [68].

PDE7 and PDE8

Extensive literature indicates that modulation of cAMP signaling has potential therapeutic benefit for a broad range of neuropsychiatric disorders. A cornerstone of this body of research is the beneficial effect of inhibiting the major cAMP-specific PDE4, in particular using rolipram [80]. However, the therapeutic potential of PDE4 inhibition has yet to be realized because inhibition of this PDE is associated with dose-limiting side effects. Discussed below are current strategies to discover PDE4 inhibitors with improved therapeutic index. However, a developing area of investigation is the possibility of modulating cAMP signaling by targeting the two other cAMP-specific phosphodiesterases PDE7 and PDE8. Progress to date is reviewed in the following.

The PDE7 family is composed of two members, PDE7A and PDE7B [6]. Unlike most other PDEs, these enzymes do not contain recognized N-terminal regulatory domains. However, a consensus site for PKA phosphorylation is present and may represent a mechanism for allosteric regulation of activity. Moreover, in humans there are three known splice variants for PDE7A that contain unique N- and C-terminal mRNA modifications that likely influence both intracellular localization and interactions with other proteins. Promoter variants have also been reported for PDE7A, offering additional subtleties with respect to cAMP responsiveness.

mRNA for both of the PDE7 subtypes is widely expressed in the mammalian CNS [81–83]. This underwrites targeting these enzymes for new treatments for neuropsychiatric diseases, although the highest level of interest is as a target to treat inflammatory disease [84,85]. In rat brain, PDE7A mRNA is expressed in high levels in the glutamatergic neurons of the hippocampus, including the dentate granule cells and pyramidal cells within the CA regions [81]. In cortex, message levels were found to be moderate and differentially distributed in different layers in different regions. For example, higher levels of expression are observed in layer 4, which is enriched for sensory input. PDE7A mRNA is only moderately expressed in the striatum and in the substantia nigra. High levels of expression are also observed in the olfactory bulb and tubercle and in some midbrain and hindbrain nuclei.

PDE7B mRNA has a distribution that is distinct from that of PDE7A [83]. Most notable is the very high expression of PDE7B mRNA in the striatum. Expression levels are comparable to other PDEs that are enriched in this brain region, including PDE1B, PDE2A, and PDE10A. Interestingly, PDE7B mRNA expression in the striatum appears to be regulated by dopamine D_1 receptor signaling [86]. Thus, it appears that PDE7B may be particularly involved in striatal information processing. PDE7B, like PDE7A, is also expressed at high levels in the dentate granule cells. However, in contrast to PDE7A, PDE7B mRNA is expressed only at low levels in the hippocampal pyramidal cell layer. Moderate expression of PDE7B is observed in cortex in layers 2 and 3 and in the deep layers. It is interesting that this cortical expression pattern of PDE7B is complementary to that of PDE7A.

There is a growing patent and medicinal chemistry literature developing around PDE7 inhibitors [85]. This includes development of screening methodology [87,88] and molecular modeling techniques [89] to identify novel inhibitors. These efforts are starting to yield compounds to serve as tools to explore the physiology and therapeutic potential of targeting PDE7 in the brain. The first such report is from Omeros Corporation, which has disclosed in a patent application that in the MPTP mouse model of Parkinson disease (PD), PDE7 inhibitors restore stride length to prelesioned levels when administered alone and also potentiate the activity of L-DOPA [90]. The model proposed for this effect is augmentation of cAMP signaling in striatum downstream of dopamine D_1 receptor activation. Striatal D_1 signaling deficits are widely considered to be one of the cardinal causes of reduced motor activity in PD that results from the loss of midbrain dopamine input into striatum. However, given that PDE7 isoforms are expressed throughout the brain, an effect of PDE7 inhibition mediated by other brain circuitry is also plausible. It would be of interest to investigate the effects of the PDE7 inhibitors in the MPTP mouse model of PD in which the gene for PDE7A or PDE7B is selectively deleted. In this type of paradigm, the knockout mice serve as a "null background" for investigating the particular isozyme of PDE involved in mediating a behavioral response. The results of such experiments may support the development of isozyme-specific PDE7 inhibitors for PD (see further discussion below). More recently, Gil and colleagues have reported that a novel PDE7 inhibitor is effective in an experimental autoimmune encephalomyelitis model [91]. The use of these PDE7 inhibitors to investigate the broader role(s) of PDE7 in regulating brain function is of significant interest.

The PDE8 family is also encoded by two genes: PDE8A and PDE8B [92–94]. Both of the PDE8 mRNAs code for putative N-terminal regulatory elements within their protein structure, although their exact function is still unknown. Each of the putative regulatory domains found in PDE8 (the "REC" domain and "PAS" domain) are unique to this particular PDE family and share homology with highly conserved regulatory domains found in bacteria and several mammalian proteins (see Chapter 1). In lower organisms, the REC domain has been characterized as a sequence responsible for receiving signals from a particular sensor protein. As yet, it is unclear whether or not the REC domain in PDE8 plays a similar role in mammals. The PAS domain has been identified in several proteins involved with regulation of circadian rhythms, and has been shown to be a potential site for ligand binding that may influence protein interactions. PDE8B mRNA expression is highest in brain, thyroid (human, but not rat), and testes [94,95]. In fact, the PDE8B1 variant appears to be expressed only in the brain. In contrast, PDE8A mRNA is not appreciably expressed in brain. Recently, a highly potent, specific, and brain-penetrable PDE8 inhibitor was disclosed [96], offering a new tool to investigate the physiological function of this enzyme in the CNS. Insights into the structural elements of the enzyme will certainly aid in the identification of additional new compounds in the near future [97].

PDE9A

PDE9A is one of the newly emerging PDE targets for which neuropsychiatric disorders are the primary therapeutic interests. PDE9A is a high affinity, cGMP-specific enzyme encoded by a single gene that is subject to a complex pattern of regulation that yields a total of 20 human splice variants [98,99]. All splice variants utilize the same transcriptional start, but generate unique changes in the 5′-region of the mRNA, possibly allowing tissue-specific expression patterns [99,100]. Functional changes mediated by these variations remain unclear, as both the C-terminal catalytic domain and the main part of the N-terminal domain remain unaltered. The primary structure of PDE9A does not contain recognized regulatory domains, such as GAF domains, and the C-terminal homology compared to other PDEs is low, resulting in insensitivity of the enzyme to most known PDE inhibitors [101,102]. Nonetheless, PDE9A is thought to be key player of regulating cGMP levels as it has the lowest K_m among the PDEs for this nucleotide [101,102].

Significantly, information is available on the expression pattern of PDE9A mRNA. Transcripts have been detected in many tissues, reaching peak levels in kidney, spleen, gastrointestinal tissues, and prostate, as well as in brain [98,99,101]. In the brain, PDE9A mRNA is widely, but very moderately, expressed [7,21]. It reaches peak levels in cerebellar Purkinje cells, and is furthermore easily detectable in the olfactory bulb, hippocampus, and cortical layer V [21]. Expression is considered primarily neuronal, but signals have been detected in astrocytes and Schwann cells [21,103]. In the postmortem of brain tissue of healthy elderly people and AD patients, PDE9A mRNA was detected in the cortex, hippocampus, and cerebellum in a pattern comparable to the rodent [58]. No differences in expression were observed in the patients with AD. However, knowledge is limited regarding PDE9A protein expression and localization, either because of the low level of expression or the lack of suitable antibodies or both. Two variants have been examined to date: PDE9A1, which was found in the nucleus, and PDE9A5, which is located in the cytoplasm [100]. A recent immunohistochemical analysis of PDE9A in the trigeminal ganglion indicated neuronal localization of the protein with a cytoplasmic subcellular distribution [103].

Sequence analysis and X-ray crystallographic evidence reveal a number of fundamental differences between PDE9A and other PDEs, but from a chemogenomic perspective, the low affinity of PDE9A to IBMX is a clear indicator that PDE9A inhibitors must fulfill other structural requirements than inhibitors of most other PDE isoforms [104]. Full-length PDE9A is inhibited by IBMX with an IC_{50} value of around $230\,\mu M$, which is significantly lower than for all other PDEs except PDE8 [105]. Nevertheless, a crystal structure of IBMX bound to the PDE9A2 catalytic domain has been obtained by crystallizing the protein with a large excess of IBMX. The X-ray crystal structure reveals a single hydrogen bond between the xanthine N-7 of IBMX and the Glu453 of PDE9A, rather than the double hydrogen bond usually observed in complexes of IBMX with other PDEs. A subsequent study of PDE9A crystallized with its natural ligand at low temperatures has provided important information about the catalytic mechanism [35].

Although initially elusive, the search for selective PDE9A inhibitors has yielded a number of interesting compounds from various classes. It appears that PDE9A has a very pronounced preference for compounds displaying variations of the purinone scaffold, that is, flat, aromatic heterobicyclic compounds capable of forming the characteristic double hydrogen bond to the active site glutamine, as observed in structures of many other PDE inhibitors, such as sildenafil and vardenafil [106]. These structural characteristics are also recognizable in the chemical classes that have resulted from the four major discovery efforts disclosed so far; these chemical classes are discussed in the following.

Bayer was the first company to publish detailed pharmacological data for a selective PDE9A inhibitor [107] and has also been involved in pursuing several compound classes and indications. BAY 736691 (WO2004099211, WO

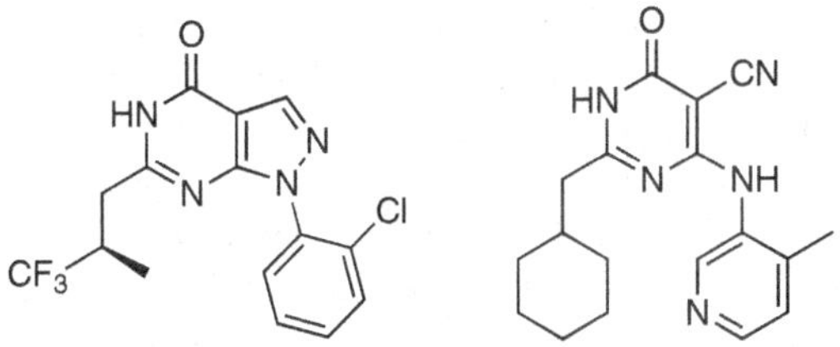

FIGURE 4.4   Bayer PDE9A inhibitors, including the prototype BAY 736691.

2004099210, WO 2004018474) belongs to the class of pyrazolopyrimidinones (Figure 4.4). This compound selectively inhibits human PDE9A with an *in vitro* IC_{50} of 55 nM and a minimum 25-fold window to other PDEs. The X-ray crystal structure of BAY 736691 and of its enantiomer bound to the active site of PDE9A was recently published [108]. This analysis shows that BAY 736691 binds to the conserved Glu453 by a dual hydrogen bond and by a direct interaction with Phe456, in analogy with the PDE9A structures of GMP and IBMX, which is also the manner that sildenafil binds to PDE5. Several published patent applications also describe a second chemical class, the cyanopyrimidinones (WO2004113306, WO2005068436, WO2006125554) exemplified by the compound in Figure 4.4. Bayer has never disclosed the structure of the clinical candidate from this series, but it appears that there is a high degree of similarity between the structure–activity relationship (SAR) in the two series.

Pfizer appears to have been involved in at least two distinct discovery programs centered on PDE9A pharmacology, namely, programs in the indications of diabetes and cardiovascular disease, as well as neurology. While the peripheral and central indications may have differing requirements of the inhibitor in terms of selectivity profile, pharmacokinetics, and organ distribution, it is interesting to see both programs in comparison. The starting point for Pfizer's first published PDE9A projects (US20040220186) was identified by screening a library of PDE inhibitors from previous projects on other PDE isoforms [109]. The initial lead, notably similar to sildenafil and depicted on the left in Figure 4.5, was a potent inhibitor of PDE9A ($IC_{50} = 10$ nM), but essentially nonselective with activity on PDE1A, PDE1B,

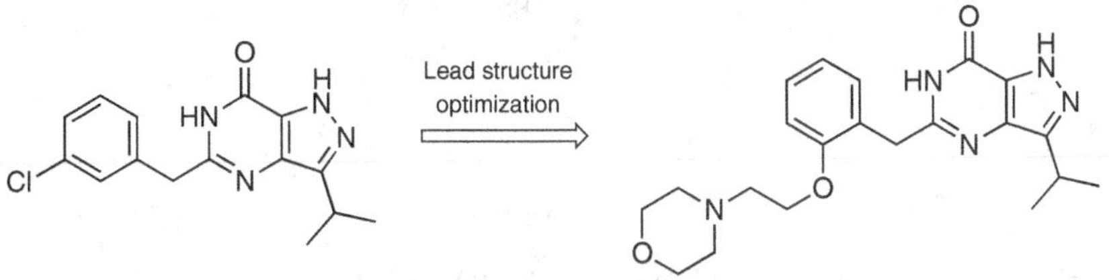

FIGURE 4.5   Structural optimization of PDE9A inhibitors by Pfizer yielding a 41 nM PDE9A inhibitor with a selectivity factor of 30 or better toward PDE1A–PDE1C and PDE5A over the starting material.

PDE1C, and PDE5A. Structural optimization over several iterations led to selective compounds such as that depicted on the right in Figure 4.5, a potent PDE9A inhibitor ($IC_{50} = 41$ nM) with a selectivity factor of 30 or better over PDE1A, PDE1B, PDE1C, and PDE5A. The later compound is active *in vivo* in mice (glucose lowering) after oral dosing of 100 mg/kg and above. Other compounds with promising *in vitro* profiles had no effect *in vivo*, probably because of poor absorption as a result of relatively high PSAs; such compounds almost certainly have very low CNS exposure.

Pfizer's second and currently most advanced PDE9A program aimed at identifying a PDE9A inhibitor for the treatment of cognitive deficits in AD and other neuro-psychiatric disorders [23,24]. The *in vitro* and *in vivo* profile of a development compound, PF-4447943, has recently been disclosed (Figure 4.6, together with two other representative examples from [WO2008139293]) [110,111]. PF-4447943 and other of the potent compounds are characterized by a high polar surface and, interestingly, the calculated properties of the presumed clinical candidate are divergent from those considered "ideal" for a clinical candidate for a CNS indication in some important aspects (see "Structure of Phosphodiesterases" section). First, the number of heteroatoms is unusually high, and the calculated log D (using our admittedly less-refined calculation tools) is below −1.5. Still, the compound appears to penetrate the CNS is several animal species as well as man, causing an elevation in the levels of cGMP in the cerebrospinal fluid (CSF) *in vivo* [24,111]. The compounds from this series bind to the active site of PDE9A in much the same way as does sildenafil, with which these compounds share the pyrazolopyrimidinone scaffold, namely, through a double hydrogen bond to the centrally located Glu453 [8]. Tyr424 seems to be engaged in a water-mediated contact to the pyrrolidine ring nitrogen, which is particularly interesting in the context of selectivity: All other PDEs apart from PDE8 carry a phenylalanine in this position, so interacting with Tyr424 may be a good strategy to ensure high selectivity.

ASKA Pharmaceutical Co. Ltd of Tokyo has been involved in PDE9A discovery projects for years; the first patent application was filed in 2006, and to this date, a total of four patent applications on PDE9A inhibitors have been made public (WO2006135080, WO2008018306, WO2008072778, and WO2008072779). Two distinct compound classes have been disclosed so far: a class of quinazolinone

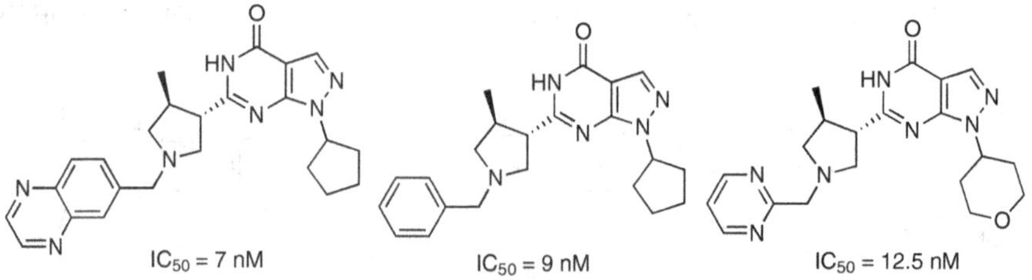

IC50 = 7 nM   IC50 = 9 nM   IC50 = 12.5 nM

FIGURE 4.6   2-Pyrimidylmethyl-derivative PDE9A inhibitors from the Pfizer series that contains the clinical candidate PF-4447943. These compounds are characterized by a high polar surface and other calculated properties that are diverging from those considered "ideal" for a clinical candidate for a CNS indication.

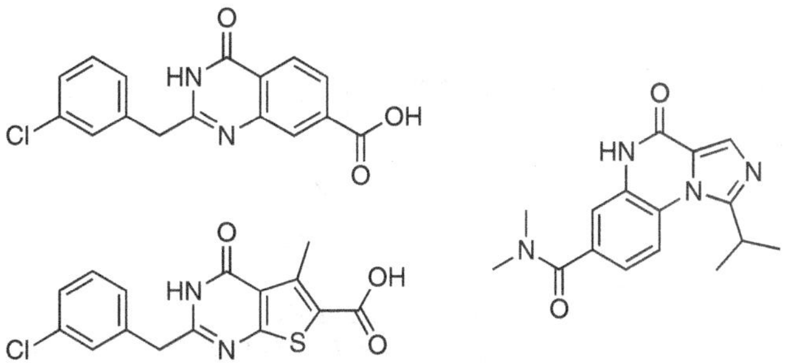

FIGURE 4.7 Two distinct classes of PDE9A inhibitors from ASKA include quinazolinone carboxylic acids on the left and a representative heterotricyclic compound on the right.

carboxylic acids and congeners (depicted on the left, Figure 4.7), and a class of heterotricyclic compounds (depicted on the right, Figure 4.7). Although various indications have been claimed for both classes (including CNS indications such as AD and general neuropathy), the carboxylic acid group contained in the compounds shown on the left makes any high CNS exposure rather unlikely, and it would appear that these compounds are targeted for peripheral indications, such as prostate disease, incontinence, or pulmonary hypertension, although no *in vivo* data have been published to support these claims. The tricyclic systems, on the other hand, seem more promising in that respect, but the general SAR appears to overlap with that of the Bayer and Pfizer programs, so there is reason to believe that the binding mode of this compound is essentially the same (as is the case for the other ASKA compounds no). However, no structural data have been published to this point. Currently, it is unclear if ASKA is still actively involved in PDE9A-related research and development; there is no mention of the project on the company homepage, although the most recent patent application was published in 2008.

The most recent player to enter the increasingly competitive field of PDE9A research is Boehringer Ingelheim, with a series of patent applications detailing inhibitors of the pyrazolopyrimidinone type (WO2009068617 and WO2009121919 have been published so far). Although no detailed biological data have been disclosed, this focused compound class seems to be quite selective versus PDE1, and generally exhibits rather potent inhibitory effects on PDE9A (Figure 4.8). The similarity to BAY 736691 and the Pfizer series is noticeable, and, interestingly, one of the original Bayer inventors appears on the Boehringer Ingelheim patent application, which seems to indicate that the Boehringer Ingelheim program is based on intellectual property acquired from Bayer. No *in vivo* data have been reported, but the overall physicochemical profile of compounds in this series seems to be compatible with penetration into brain tissue.

Considerable interest in PDE9A as a target for CNS disorders was engendered following the first characterization of BAY 736691 [22]. In a broad pharmacological assessment, BAY 736691 enhanced early LTP after weak tetanic stimulation in hippocampal slices prepared from young adult Wistar rats and old, but not young, Fischer 344 X Brown Norway (FBNF1) rats [112]. Significantly, BAY 736691 enhanced acquisition, consolidation, and retention of long-term memory in a number

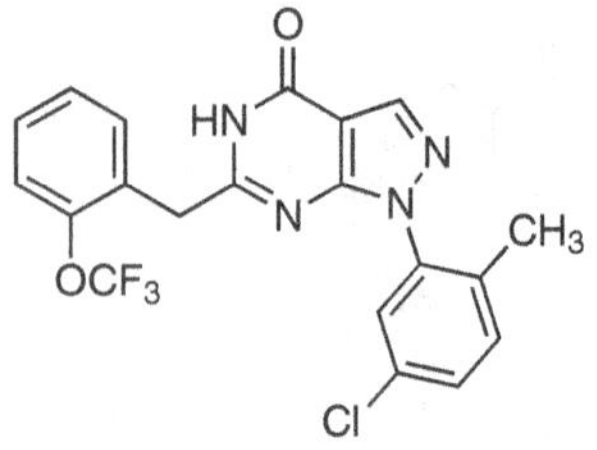

IC$_{50}$ between 10 and 500 nM
Selectivity versus PDE1:271-fold

FIGURE 4.8 Boehringer Ingelheim PDE9A inhibitors of the pyrazolopyrimidinone type.

of preclinical behavioral paradigms, including a social recognition task, a scopolamine-disrupted passive avoidance task, and a MK-801-induced short-term memory deficit in a T-maze alternation task [112]. Subsequently, it was reported that LTP is enhanced in hippocampal slices prepared from PDE9A knockout mice and that this effect is mimicked by a PDE9A inhibitor in slices prepared from the rat hippocampus [23]. There have now been two in-depth characterizations of PF-4447943 [111,113] that replicated and extended the effects of BAY 736991 on CNS functions. Taken together, these data suggested that PDE9A inhibition may provide AD patients with some therapeutic benefit and Pfizer Inc. has advanced the PDE9A inhibitor PF-4447943 into clinical development for AD. In phase I, PF-4447943 induced a measurable increase in cGMP in CSF of healthy volunteers, providing proof-of-mechanism to the concept of PDE9A inhibition in humans [114]. In a subsequent phase II trial, PF-4447943 administered as a single fixed dose once daily for 4 weeks failed to improve scores on cognitive function measured using the Alzheimer Disease Assessment Scale, cognitive subscale (ADAS-Cog). The implications of this result are discussed later in this chapter.

PDE10A

There has been significant interest in PDE10A as a target for neuropsychiatric disease since 1998, when the gene for this enzyme was cloned [115–117]. PDE10A is a dual-substrate PDE that contains tandem GAF domains in the N-terminal region (molecular characteristics are reviewed in Ref. [118]). However, unlike PDE2A and PDE5, the affinity of the PDE10A GAF domains for cGMP is low, and the enzyme instead may be regulated through GAF binding of cAMP [119].

The striking characteristic of PDE10A is that mRNA is expressed at high levels only in testes and brain [115–117], and significant protein expression is detected only in brain [120]. Furthermore, in brain, high levels of protein expression are restricted to one neuronal population, the striatal medium spiny neuron [4,9,120,121]. The striatal medium spiny neurons are a key node for information processing through the basal ganglia circuit. These neurons receive a massive, diverse glutamatergic input from the cortex and thalamus, together with an equally massive input from the midbrain dopamine system. These inputs are integrated to mediate the learning and execution of cognitive and behavioral action plans, while at the same time suppressing response to stimuli that lack salience to the organism [122]. Dysfunction in the basal ganglia circuitry is implicated in a host of neuropsychiatric conditions [123]. It is the loss of the dopaminergic input to the striatal medium spiny neurons that results in the symptoms of PD [124]. It is also the blockade of dopamine D_2 receptors on these neurons that accounts for the efficacy of drugs currently used to dampen psychotic symptoms in patients with schizophrenia [125]. Degeneration of the striatal medium spiny neurons is the hallmark pathology of Huntington's disease [126]. There is also compelling evidence to suggest basal ganglia dysfunction in obsessive compulsive disorder and Tourette's syndrome [127]. Thus, the discovery of the localization of PDE10A to striatal medium spiny neurons immediately engendered interest in the enzyme as a drug target because it seemed highly likely that inhibitors would have some therapeutic utility. Recent advances in the knowledge of the enzyme structure, X-ray cocrystal structures, and pharmacophore models that are guiding these drug design efforts have been extensively reviewed [128–131]. These are highlighted briefly in the following.

The natural product papaverine (Figure 4.9) was the first reported PDE10A inhibitor [132]. This report was followed shortly thereafter by a Pfizer patent describing the first cocrystal structure of the PDE10A catalytic domain and an inhibitor, compound 489 (Figure 4.9, US2007/0155779 A1). These early inhibitor structures proved useful in guiding further medicinal chemistry efforts as illustrated by a previous review [133] and a 2007 account from Pfizer that detailed the development of a dimethoxyqinazolylpyrrolidine series, exemplified by compound PQ-10 (3–9) [134]. Information garnered from X-ray crystal structures of PDE10A complexed with these "dialkoxyaryl" compounds enabled advancement of a pharmacophore model to explain the binding interactions of this class of inhibitors [128,134]. The patent literature now contains several examples of such compounds that ostensibly behave according to the proposed pharmacophore model. These include compounds from Lundbeck (PCT/DK2007/000351) and Memory/Amgen (PCT/US2006/032000). A patent from Pfizer (PCT/IB2007/002382) suggests that the PDE10A enzyme can also accommodate a long ether chain at position 6 of the quinazoline ring, possibly providing new avenues for the development of SAR in this series.

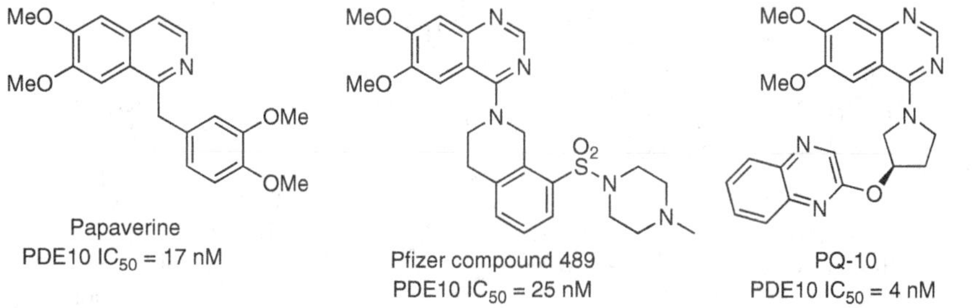

Papaverine
PDE10 IC_{50} = 17 nM

Pfizer compound 489
PDE10 IC_{50} = 25 nM

PQ-10
PDE10 IC_{50} = 4 nM

FIGURE 4.9   Early PDE10A inhibitors from Pfizer, including the natural product papaverine.

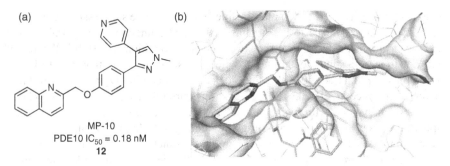

(a)

MP-10
PDE10 IC$_{50}$ = 0.18 nM
12

(b)

FIGURE 4.10 (a) The Pfizer PDE10A inhibitor MP-10 (also referred to as PF-2545920) that is in phase II for the treatment of schizophrenia. (b) MP-10 docked into a model of the catalytic pocket of PDE10A.

More recently, a number of inhibitor classes structurally distinct from the dialkoxyaryl inhibitors have been disclosed. Notable of these is a Pfizer series exemplified by MP-10 and TP-10 (Figure 4.10) [135]. MP-10 is PF-2545920, the phase II compound referred to above [32]. X-ray analysis of the binding complex formed between MP-10 and PDE10A revealed a new binding mode, which does not involve interaction with the invariant glutamine residue (3–10). Recent patents describing new heterocyclic replacements for the quinoline ether and a series of compounds that are rigidified through incorporation of a three-atom tether between the pyrazole and phenyl rings are complementary to this approach.

New entries in the field of PDE10A inhibitors include structures for which binding modes are not yet described (Figure 4.11). These include the Biotie (Elbion) pyridopyrazine (PCT/EP2007/004747), a unique series of very potent PDE10A inhibitors from Pfizer (PCT/IB2007/002000), and a series of acyclic hydrazones exhibiting submicromolar PDE10A inhibition from Omeros (PCT/20080300240). Lundbeck recently patented the use of the benzodiazepine tofisopam as a PDE10A inhibitor, citing IC$_{50}$ values of 273 and 315 nM for the resolved (*R*)- and (*S*)-enantiomers, respectively (Figure 4.12). The (*R*)-enantiomer will likely be of most interest because of modest selectivity (~5×) over PDE4. Compound NN113 is

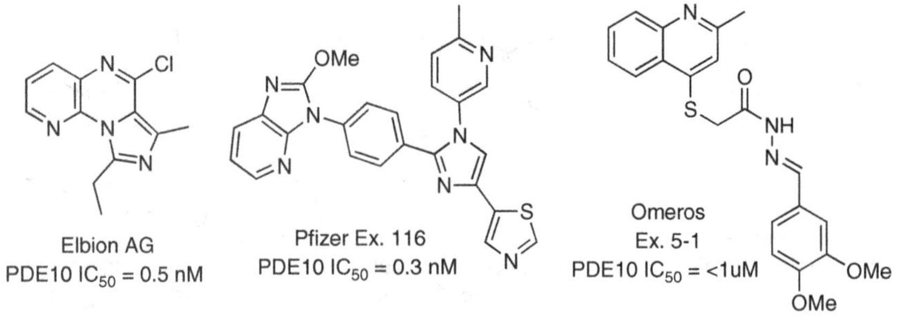

Elbion AG
PDE10 IC$_{50}$ = 0.5 nM

Pfizer Ex. 116
PDE10 IC$_{50}$ = 0.3 nM

Omeros
Ex. 5-1
PDE10 IC$_{50}$ = <1uM

FIGURE 4.11 PDE10A inhibitors from Elbion/Biotie Pfizer and Omeros.

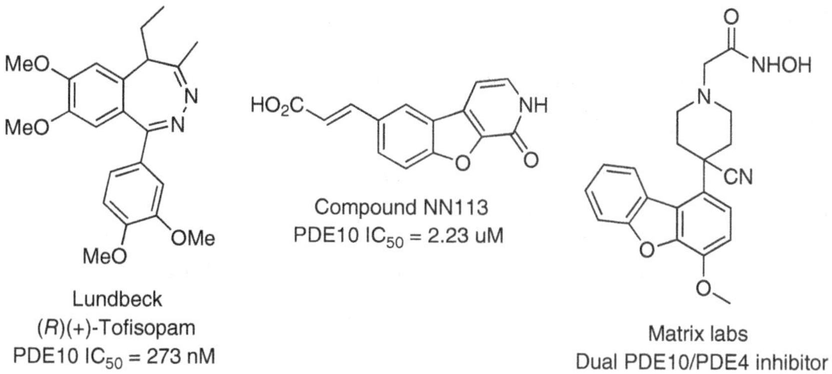

FIGURE 4.12   PDE10A inhibitors from Pfizer and Matrix.

described in a recent patent as being a 2.23 μM inhibitor of PDE10A. Finally, Matrix describes the preparation of dual PDE10A/PDE4 inhibitors (PCT/IB2007/002596).

As mentioned above, the localization of PDE10A to the medium spiny neurons of the striatum suggests a number of therapeutic possibilities for inhibitors. The most compelling rationale is for the use of PDE10A inhibitors to treat psychosis in schizophrenia, based on preclinical data from several pharmaceutical companies [131,135–138] (and see Chapter 10). Pfizer has recently reported on the completion of a large phase II study of PF-2545920 [32] in patients with acute exacerbation of psychosis in schizophrenia [139]. The compound was administered at two doses for 4 weeks; effects were compared with risperidone. There was no effect of PF-2545920 on psychotic symptoms compared to placebo, although risperidone was effective at reducing symptoms. There are also a number of active programs targeting schizophrenia at other companies.

An emerging area of interest is in the use of PDE10A inhibitors in the treatment of Huntington disease. This interest stems from the hypothesis that Huntington disease pathology results from transcriptional dysregulation, particularly caused by a loss of CREB-mediated signaling in striatal medium spiny neurons [140–142]. PDE10A inhibition is very effective at upregulating cAMP and CREB signaling in these neurons [132,135,143]. It was recently reported that the PDE10A inhibitor TP-10 reduces the destruction of striatal medium spiny neurons caused by the neurotoxin quinolinic acid in rats [144]. This compound also ameliorates neurological deficits and neuropathology in the R6/2 mouse model of Huntington disease [145]. These studies should fuel interest in the further study of PDE10A inhibitors as a new and accessible approach to the treatment of Huntington's disease [146]. Given the early and severe psychiatric symptoms suffered by Huntington disease patients, and the preclinical data suggesting utility of PDE10A inhibitors in the treatment of psychosis, PDE10A inhibitors may exert a dual impact in Huntington disease, providing early symptomatic relief and long-term slowing of disease progression.

FUTURE DIRECTIONS

To date, the vast majority of the drug discovery efforts have been directed at PDE inhibitors that are competitive with cyclic nucleotide binding to the catalytic site. Although this effort continues to generate powerful research tools and promising therapeutic candidates, this approach only begins to mine the possibilities of PDE pharmacology. Two areas of research poised for growth are the development of compounds that are selective for isozymes within a PDE family and the development of allosteric modulators. A perspective on these two areas follows.

Subtype Selectivity

Several of the PDE families of active interest to drug discovery for CNS indications are encoded by more than one gene. These include the cAMP-specific families PDE4A–PDE4D, PDE7A and PDE7B, and PDE8A and PDE8B. Also of significant interest is the PDE1A–PDE1C family, of which the PDE1A and PDE1B isoforms are expressed in the CNS (but for which there has not been recently disclosed drug discovery progress). There is considerable interest in developing compounds that are specific for the individual isozymes within these PDE families, based on differences in localization in the CNS and, therefore, putative therapeutic value. For example, PDE7A is broadly expressed throughout the CNS, whereas the expression of PDE7B is notably more restricted to the striatal medium spiny neurons and a few other regions. A similar scenario exists for PDE1, where the 1A isozyme is broadly expressed and the 1B isozyme is restricted to medium spiny neurons and dentate granule neurons. The therapeutic potential of modulating information processing by the medium spiny neurons has been discussed above with respect to PDE10A. It is possible that different aspects of striatal function may be modulated by PDE1B or PDE7B inhibitors and, thus, such compounds may have therapeutic effects that are complementary to those of PDE10A inhibitors. Alternatively, PDE1A or PDE7A inhibitors would be predicted to impact brain circuitry more broadly, portending unique therapeutic possibilities. For PDE8, only the 8B isozyme is expressed in the CNS. Thus, a PDE8B-selective inhibitor may be advantageous in capturing potential CNS therapeutic utilities while avoiding side effects from inhibition of PDE8A in the periphery.

The difficulty in identifying PDE isozyme-selective inhibitors is the high homology within families in the catalytic domain. The homology for the catalytic domains of different isozymes within a PDE family averages $\sim$75%. However, homology tends to be much higher for the key amino acids lining the substrate binding pocket that is the target for the competitive inhibitors. For example, the homology between the substrate binding pockets of PDE4B and PDE4D is 98% in this key region. Nonetheless, subtle differences in catalytic activities and substrate binding affinities are common between different isozyme within a family, indicating differences in the catalytic sites that may be exploited pharmacologically. The most extreme of these differences are within the PDE1 family, where affinities for cAMP differ by as much as two orders of magnitude [63]. Outside the catalytic sites homologies drop off significantly between isozymes within a family. These regions of dissimilarity putatively offer handles for synthetic compound interactions that may yield significant isozyme selectivity.

The most significant progress in identifying isozyme-selective PDE inhibitors is for PDE4. Compounds that are up to 100-fold selective for PDE4D over the other PDE4 isozymes have been available since 2006 [65], although the molecular basis for the selectivity was not known at that time. Significantly, Burgin et al. [147] have recently described a new series of compounds that are highly selective for PDE4D and identified a molecular mechanism that accounts for this selectivity. The deCODE group solved the crystal structures for PDE4 constructs that contained a significant portion of the UCR2 regulatory domain that is N-terminal to the catalytic domain. They report that UCR2 folds over and occludes the opening to the catalytic site. This interaction of UCR2 with the catalytic domain is proposed to be part of the regulatory mechanism for PDE4 [148]. deCODE identified and developed a series of compounds that derive a significant portion of their binding energy through an interaction with the UCR2 domain. This UCR2 interaction brings the compounds into position to occlude the active site. The critical residue that the deCODE compounds interact with to achieve high levels of PDE4D selectivity is Phe184 in PDE4D within the UCR2 domain. The analogous residue in PDE4A–PDE4C is a tyrosine, with the tyrosine hydroxyl occluding optimal compound binding. It is interesting to note that the Phe/Tyr difference is specific for the human PDE4 isoforms, whereas in other species both residues are Phe. Nonetheless, the deCODE compounds retain substantial PDE4D selectivity in analyses using mouse PDE4D and PDE4B constructs. This suggests additional interactions are contributing to isozyme selectivity, possibly offering additional opportunities to develop PDE4 isozyme-selective compounds.

The deCODE studies offer elegant proof of principles for developing isozyme selectivity by taking advantage of binding interactions outside the highly conserved catalytic domains. However, it is unclear to what extent these findings with PDE4 may be applicable to other PDEs. The strategy generalized from the deCODE studies is one of developing compounds that occlude substrate access to the catalytic site but that gain binding affinity to the enzyme through interaction with an N-terminal (or C-terminal) element that is in close proximity to the catalytic site. However, it is not known at present to what extent the close apposition of N-terminal regulatory elements and the catalytic domains occur in full-length PDEs, or whether the occlusion of the catalytic domain by an N-terminal element is a regulatory mechanism unique to PDE4. This question arises in light of analyses of X-ray crystal structures of PDE2A constructs containing the catalytic domain and the N-terminal GAF domains [54]. These PDE2A structures revealed an extended configuration for the enzyme without extensive overlap of N-terminal elements and the catalytic domain. Instead, analysis of these structures suggests a mechanism by which the GAF domains allosterically regulate catalytic domain dimerization and, thereby, access of substrate to the catalytic site. Clearly, further study is required to sort out these possibilities.

Allosteric Modulators

The discussion above segues into the topic of allosteric modulators of the PDEs. The N-terminal domains of the different PDE families contain family-unique regulatory elements (see Refs. [149,150] for review). The best studied of these are calmodulin binding and PDE1, the UCR1/UCR2 sites of PDE4, and the GAF domains of PDE2A and PDE5. PDE6, PDE10A, and PDE11A also contain GAF domains, and PDE8 contains a PAS domain, whereas PDE3, PDE7, and PDE9A do not contain such canonical regulatory elements (see Chapter 1). Nonetheless, all of the PDEs are thought to be subject to N-terminal regulation through a host of additional mechanisms, including dimerization, phosphorylation and other covalent modifications, and protein and/or membrane interactions. The C-terminal domains of the PDEs are also likely to be involved in regulation. The best example in this case is ERK phosphorylation of the C-terminal domains of PDE4.

Theoretically, the regulatory elements of the PDEs are targets for pharmacological interactions. Targeting these domains may yield compounds with unique pharmacological actions and therapeutic possibilities. For example, it is possible that PDE activities may be activated instead of inhibited through a regulatory mechanism. Targeting the more variable N-terminal domains offer another possibility for isozyme specificity. Furthermore, such an approach may yield compounds that target particular N-terminal splice variants. An example is compound that might regulate PDE4 activity through an interaction with UCR1. This regulatory element is present across the four PDE4 families but only in the long isoforms. Thus, compounds that interact with this region may be devised to only effect activity in the long isoforms while sparing the short and super short isoforms.

Perhaps the most immediately accessible route to discovery of allosteric PDE modulators is through targeting the GAF domains of PDE2A or PDE5A [151]. GAFA of PDE5A and GAFB of PDE2A are known to bind cGMP with high affinity to activate enzyme activity [48,152]. Furthermore, GAF domain elements have been used to construct sensors for cGMP [153–155]. Although these constructs were designed as monitors of intracellular cGMP levels, they appear to be well suited to the development of high-throughput screens to detect synthetic compounds that bind to the GAF domains and/or displace the binding of cGMP. Discovery of such synthetic compounds could then be rapidly characterized as activators or inhibitors of PDE2A or PDE5A using full-length enzyme constructs. These could be new tools to explore the physiology and regulation of PDE2A and PDE5, perhaps leading to therapeutic applications. Such compounds would nicely complement the insights gained from the recent structural work on PDE2A [54], and may open a new avenue for the development of PDE pharmacology.

CONCLUDING REMARKS

In the late 1990s, the elucidation of the human genome and the subsequent identification of the entire PDE gene family, coupled with the approval and rapid commercial success of sildenafil (Viagra), opened the doors to an explosion of drug discovery efforts targeting the PDEs for a broad range of therapeutic uses. This effort has been particularly robust with regard to the CNS, with PDE10A inhibitors for schizophrenia, PDE9A inhibitors for AD, and PDE5A inhibitors for stroke recovery all in early clinical development by multiple pharmaceutical companies. As noted above, initial clinical experiments with Pfizer's PDE10A inhibitor in schizophrenia and Pfizer's PDE9A inhibitor in AD did not demonstrate efficacy. In the case of PDE10A inhibitors, the primary locus of action is basal ganglia circuitry, for which there is a tremendous depth of knowledge regarding a role in CNS function and disease [156]. There is also a significant and growing body of basic research that has begun to elucidate the physiological function of PDE10A within this circuitry. This knowledge, together with the initial phase I and phase II results with PF-2545920, serves as an essential guide to the design and implementation of new clinical studies to probe the therapeutic utility of PDE10A inhibitors. The path forward for PDE9A inhibitors is comparatively more daunting, for several reasons. Two central biological questions about PDE9A still remain to be answered: (a) the nature and signaling function of the PDE9A-regulated cGMP pool and (b) the localization of PDE9A within the neurons that express the protein. These data would be extremely helpful in placing the enzyme in a disease context. Unfortunately, the negative result in the PF-4447943 clinical trial in AD provides little new insight, given our scant understanding of the underlying pathology of the disease. Fortunately, for both PDE10A and PDE9A, excellent tool compounds are available to further explore the basic physiology and therapeutic utilities, both preclinically and in the clinic. Such tool compounds are also becoming available for PDE1, PDE2A, PDE7, and PDE8 to continue to drive research and development efforts for these targets. Finally, the groundbreaking work of the deCODE group portends the opening of new avenues of research and development around PDE4 and the PDE superfamily. Thus, we predict that the next decade will yield a growing crop of PDE modulators for clinical evaluation and provide the research tools to continue to illuminate the physiology and therapeutic potential of this class of signaling enzymes.

REFERENCES

1. Menniti, F.S., Faraci, W.S., and Schmidt, C.J. (2006) Phosphodiesterases in the CNS: targets for drug development. *Nat. Rev. Drug Discov.*, **5**:660–670.
2. Repaske, D.R., Swinnen, J.V., Jin, S.L., Van Wyk, J.J., and Conti, M. (1992) A polymerase chain reaction strategy to identify and clone cyclic nucleotide phosphodiesterase cDNAs. Molecular cloning of the cDNA encoding the 63-kDa calmodulin-dependent phosphodiesterase. *J. Biol. Chem.*, **267**:18683–18688.
3. Stephenson, D., Coskran, T., Wilhelms, M., Adamowicz, W., O'Donnell, M., Muravnick, K., Menniti, F., Kleiman, R., and Morton, D. (2009) Immunohistochemical localization of PDE2A in multiple mammalian species. *J. Histochem. Cytochem.*, **57**:933–949.
4. Seeger, T.F., Bartlett, B., Coskran, T.M., Culp, J.S., James, L.C., Krull, D.L., Lanfear, J., Ryan, A.M., Schmidt, C.J., Strick, C.A., Varghese, A.H., Williams, R.D., Wylie, P.G., and Menniti, F.S. (2003) Immunohistochemical localization of PDE10A in the rat brain. *Brain Res.*, **985**:113–126.

5. Nishi, A., Kuroiwa, M., Miller, D.B., O'Callaghan, J.P., Bateup, H.S., Shuto, T., Sotogaku, N., Fukuda, T., Heintz, N., Greengard, P., and Snyder, G.L. (2008) Distinct roles of PDE4 and PDE10A in the regulation of cAMP/PKA signaling in the striatum. *J. Neurosci.*, **28**:10460–10471.

6. Michaeli, T. (2006) PDE7. In: Beavo, J., Francis, S., and Houslay, M. (Eds.), *Cyclic Nucleotide Phosphodiesterases in Health and Disease*, Boca Raton, FL: CRC Press, pp. 195–204.

7. Andreeva, S., Dikkes, P., Epstein, P., and Rosenberg, P. (2001) Expression of cGMP-specific phosphodiesterase 9A mRNA in the rat brain. *J. Neurosci.*, **21**:9068–9076.

8. Verhoest, P.R., Proulx-Lafrance, C., Corman, M., Chenard, L., Helal, C.J., Hou, X., Kleiman, R., Liu, S., Marr, E., Menniti, F.S., Schmidt, C.J., Vanase-Frawley, M., Schmidt, A.W., Williams, R.D., Nelson, F.R., Fonseca, K.R., and Liras, S. (2009) Identification of a brain penetrant PDE9A inhibitor utilizing prospective design and chemical enablement as a rapid lead optimization strategy. *J. Med. Chem.*, **52**:7946–7949.

9. Xie, Z., Adamowicz, W.O., Eldred, W.D., Jakowski, A.B., Kleiman, R.J., Morton, D.G., Stephenson, D.T., Strick, C.A., Williams, R.D., and Menniti, F.S. (2006) Cellular and subcellular localization of PDE10A, a striatum-enriched phosphodiesterase. *Neuroscience*, **139**:597–607.

10. Siuciak, J.A., McCarthy, S.A., Chapin, D.S., Reed, T.M., Vorhees, C.V., and Repaske, D.R. (2007) Behavioral and neurochemical characterization of mice deficient in the phosphodiesterase-1B (PDE1B) enzyme. *Neuropharmacology*, **53**:113–124.

11. Siuciak, J.A., McCarthy, S.A., Chapin, D.S., Fujiwara, R.A., James, L.C., Williams, R.D., Stock, J.L., McNeish, J.D., Strick, C.A., Menniti, F.S., and Schmidt, C.J. (2006) Genetic deletion of the striatum-enriched phosphodiesterase PDE10A: evidence for altered striatal function. *Neuropharmacology*, **51**:374–385.

12. Baillie, G., Scott, J., and Houslay, M. (2005) Compartmentalisation of phosphodiesterases and protein kinase A: opposites attract. *FEBS Lett.*, **579**:3264–3270.

13. Lee, J., Stovall, G., and Ellington, A. (2006) Aptamer therapeutics advance. *Curr. Opin. Chem. Biol.*, **10**:282–289.

14. Houslay, M.D. and Adams, D.R. (2010) Putting the lid on phosphodiesterase 4. *Nat. Biotechnol.*, **28**:38–40.

15. Burgin, A.B., Magnusson, O.T., Singh, J., Witte, P., Staker, B.L., Bjornsson, J.M., Thorsteinsdottir, M., Hrafnsdottir, S., Hagen, T., Kiselyov, A.S., Stewart, L.J., and Gurney, M.E. (2009) Design of phosphodiesterase 4D (PDE4D) allosteric modulators for enhancing cognition with improved safety. *Nat. Biotechnol.*, **28**:63–70.

16. Prickaerts, J., Sik, A., van Staveren, W., Koopmans, G., Steinbusch, H., van der Staay, F., de Vente, J., and Blokland, A. (2004) Phosphodiesterase type 5 inhibition improves early memory consolidation of object information. *Neurochem. Int.*, **45**:915–928.

17. Puzzo, D., Staniszewski, A., Deng, S.X., Privitera, L., Leznik, E., Liu, S., Zhang, H., Feng, Y., Palmeri, A., Landry, D.W., and Arancio, O. (2009) Phosphodiesterase 5 inhibition improves synaptic function, memory, and amyloid-beta load in an Alzheimer's disease mouse model. *J. Neurosci.*, **29**:8075–8086.

18. Zhang, R., Wang, Y., Zhang, L., Zhang, Z., Tsang, W., Lu, M., Zhang, L., and Chopp, M. (2002) Sildenafil (Viagra) induces neurogenesis and promotes functional recovery after stroke in rats. *Stroke*, **33**:2675–2680.

19. Kotera, J., Fujishige, K., and Omori, K. (2000) Immunohistochemical localization of cGMP-binding cGMP-specific phosphodiesterase (PDE5) in rat tissues. *J. Histochem. Cytochem.*, **48**:685–694.

20. Menniti, F., Ren, J., Coskran, T., Liu, J., Morton, D., Sietsma, D., Som, A., Stephenson, D., Tate, B., and Finklestein, S. (2009) PDE5A inhibitors improve functional recovery after stroke in rats: optimized dosing regimen and implications for mechanism. *J. Pharmacol. Exp. Ther.*, **331**:1–9.

21. Van Staveren, W., Steinbusch, H., Markerink-Van Ittersum, M., Repaske, D., Goy, M., Kotera, J., Omori, K., Beavo, J., and de Vente, J. (2003) mRNA expression patterns of the cGMP-hydrolyzing phosphodiesterases types 2, 5, and 9 during development of the rat brain. *J. Comp. Neurol.*, **467**:566–580.

22. Wunder, F., Tersteegen, A., Rebmann, A., Erb, C., Fahrig, T., and Hendrix, M. (2005) Characterization of the first potent and selective PDE9 inhibitor using a cGMP reporter cell line. *Mol. Pharmacol.*, **68**:1775–1781.

23. Menniti, F.S., Kleiman, R., and Schmidt, C. (2008) PDE9A-mediated regulation of cGMP: impact on synaptic plasticity. *Schizophr. Res.*, **102**:38–39.

24. Schmidt, C.J., Harms, J.F., Tingley, F.D., Schmidt, K., Adamowicz, W.O., Romegialli, A., Kleiman, R.J., Barry, C.J., Coskran, T.M., O'Neill, S.M., Stephenson, D.T., and Menniti, F.S. (2009) PDE9A-mediated regulation of cGMP: developing a biomarker for a novel therapy for Alzheimer's disease. *Alzheimers Dement.*, **5**:P331.

25. Cherry, J.A. and Davis, R.L. (1999) Cyclic-AMP phosphodiesterases are localized in regions of the mouse brain associated with reinforcement, movement, and affect. *The J. Comp. Neurol.*, **407**:287–301.

26. Kelly, M.P. and Brandon, N.J. (2009) Differential function of phosphodiesterase families in the brain: gaining insights through the use of genetically modified animals. *Prog. Brain Res.*, **179**:67–73.

27. Masood, A., Huang, Y., Hajjhussein, H., Xiao, L., Li, H., Wang, W., Hamza, A., Zhan, C.G., and O'Donnell, J.M. (2009) Anxiolytic effects of phosphodiesterase-2 inhibitors associated with increased cGMP signaling. *J. Pharmacol. Exp. Ther.*, **331**:690–699.

28. Ke, H. and Wang, H. (2007) Crystal structures of phosphodiesterases and implications on substrate specificity and inhibitor selectivity. *Curr. Top. Med. Chem.*, **7**:391–403.

29. Ke, H. and Wang, H. (2006) Structure, catalytic mechanism, and inhibitor selectivity of cyclic nucleotide phosphodiesterases. In: Beavo, J., Francis, S., and Houslay, M. (Eds.), *Cyclic Nucleotide Phosphodiesterases in Health and Disease*, Boca Raton, FL: CRC Press.

30. Scapin, G., Patel, S.B., Chung, C., Varnerin, J.P., Edmondson, S.D., Mastracchio, A., Parmee, E.R., Singh, S.B., Becker, J.W., Van der Ploeg, L.H.T., and Tota, M.R. (2004) Crystal structure of human phosphodiesterase 3B: atomic basis for substrate and inhibitor specificity. *Biochemistry*, **43**:6091–6100.

31. Wang, H., Yan, Z., Yang, S., Cai, J., Robinson, H., and Ke, H. (2008) Kinetic and structural studies of phosphodiesterase-8A and implication on the inhibitor selectivity. *Biochemistry*, **47**:12760–12768.

32. Verhoest, P.R., Chapin, D.S., Corman, M., Fonseca, K., Harms, J.F., Hou, X., Marr, E.S., Menniti, F.S., Nelson, F., O'Connor, R., Pandit, J., Proulx-Lafrance, C., Schmidt, A.W., Schmidt, C.J., Suiciak, J.A., and Liras, S. (2009) Discovery of a novel class of phosphodiesterase 10A inhibitors and identification of clinical candidate 2-[4-(1-methyl-4-pyridin-4-yl-1*H*-pyrazol-3-yl)-phenoxymethyl]-quinoline (PF-2545920) for the treatment of schizophrenia. *J. Med. Chem.*, **52**:7946–7949.

33. Wang, H., Liu, Y., Hou, J., Zheng, M., Robinson, H., and Ke, H. (2007) Structural insight into substrate specificity of phosphodiesterase 10. *Proc. Natl. Acad. Sci. USA*, **104**:5782–5787.

34. Zhang, K.Y.J., Card, G.L., Suzuki, Y., Artis, D.R., Fong, D., Gillette, S., Hsieh, D., Neiman, J., West, B.L., Zhang, C., Milburn, M.V., Kim, S.-H., Schlessinger, J., and Bollag, G. (2004) A glutamine switch mechanism for nucleotide selectivity by phosphodiesterases. *Mol. Cell*, **15**:279–286.

35. Liu, S., Mansour, M.N., Dillman, K.S., Perez, J.R., Danley, D.E., Aeed, P.A., Simons, S.P., LeMotte, P.K., and Menniti, F.S. (2008) Structural basis for the catalytic mechanism of human phosphodiesterase 9. *Proc. Natl. Acad. Sci. USA*, **105**:13309–13314.

36. Zoraghi, R., Corbin, J.D., and Francis, S.H. (2006) Phosphodiesterase-5 gln817 is critical for cGMP, vardenafil, or sildenafil affinity: its orientation impacts cGMP but not cAMP affinity. *J. Biol. Chem.*, **281**:5553–5558.

37. Sandner, P., Svenstrup, N., Tinel, H., Haning, H., and Bischoff, E. (2008) Phosphodiesterase 5 inhibitors and erectile dysfunction. *Expert Opin. Ther. Pat.*, **18**:21–33.

38. Ecker, G.F. and Noe, C.R. (2004) *In silico* prediction models for blood–brain barrier permeation. *Curr. Med. Chem.*, **11**:1617–1628.

39. Fromm, M.F. (2004) Importance of P-glycoprotein at blood–tissue barriers. *Trends Pharmacol. Sci.*, **25**:423–429.

40. Rishton, G.M., LaBonte, K., Williams, A.J., Kassam, K., and Kolovanov, E. (2006) Computational approaches to the prediction of blood–brain barrier permeability: a comparative analysis of central nervous system drugs versus secretase inhibitors for Alzheimer's disease. *Curr. Opin. Drug Discov. Devel.*, **9**:303–313.

41. Norinder, U. and Haeberlein, M. (2002) Computational approaches to the prediction of the blood–brain distribution. *Adv. Drug Deliv. Rev.*, **54**:291–313.

42. Clark, D.E. (1999) Rapid calculation of polar molecular surface area and its application to the prediction of transport phenomena. 2. Prediction of blood–brain barrier penetration. *J. Pharm. Sci.*, **88**:815–821.

43. Verhoest, P.R., Hou, X., Villalobos, A., and Wager, T. (2009) Development of a CNS multiparameter optimization design tool increasing the probability of a compound survival by aligning metabolism, permeability, and safety properties in one molecule. *237th ACS National Meeting*, pp. MEDI-165.

44. Wager, T., Chandrasekaran, R., Hou X, Troutman, M., Verhoest, P., Villalobos, A., and Will, Y. (2010) Defining desirable central nervous system drug space through the alignment of molecular properties, *in vitro* ADME, and safety attributes. *ACS Chem. Neurosci.*, **1**:420–434.

45. Card, G.L., England, B.P., Suzuki, Y., Fong, D., Powell, B., Lee, B., Luu, C., Tabrizizad, M., Gillette, S., Ibrahim, P.N., Artis, D.R., Bollag, G., Milburn, M.V., Kim, S.H., Schlessinger, J., and Zhang, K.Y. (2004) Structural basis for the activity of drugs that inhibit phosphodiesterases. *Structure*, **12**:2233–2247.

46. Seelig, A. (1998) A general pattern for substrate recognition by P-glycoprotein. *Eur. J. Biochem.*, **251**:252–261.

47. Pajeva, I.K. and Wiese, M. (2002) Pharmacophore model of drugs involved in P-glycoprotein multidrug resistance: explanation of structural variety (hypothesis). *J. Med. Chem.*, **45**:5671–5686.

48. Stroop, S. and Beavo, J. (1991) Structure and function studies of the cGMP-stimulated phosphodiesterase. *J. Biol. Chem.*, **266**:23802–23809.

49. Rosman, G.J., Martins, T.J., Sonnenburg, W.K., Beavo, J.A., Ferguson, K., and Loughney, K. (1997) Isolation and characterization of human cDNAs encoding a cGMP-stimulated 3′,5′-cyclic nucleotide phosphodiesterase. *Gene*, **191**:89–95.

50. Sonnenburg, W.K., Mullaney, P.J., and Beavo, J.A. (1991) Molecular cloning of a cyclic GMP-stimulated cyclic nucleotide phosphodiesterase cDNA: identification and distribution of isozyme variants. *J. Biol. Chem.*, **266**:17655–17661.

51. Yang, Q., Paskind, M., Bolger, G., Thompson, W.J., Repaske, D.R., Cutler, L.S., and Epstein, P.M. (1994) A novel cyclic GMP stimulated phosphodiesterase from rat brain. *Biochem. Biophys. Res. Commun.*, **205**:1850–1858.

52. Russwurm, C., Zoidl, G., Koesling, D., and Russwurm, M. (2009) Dual acylation of PDE2A splice variant 3: targeting to synaptic membranes. *J. Biol. Chem.*, **284**:25782–25790.

53. Noyama, K. and Maekawa, S. (2003) Localization of cyclic nucleotide phosphodiesterase 2 in the brain-derived triton-insoluble low-density fraction (raft). *Neurosci. Res.*, **45**: 141–148.

54. Pandit, J., Forman, M.D., Fennell, K.F., Dillman, K.S., and Menniti, F.S. (2009) Mechanism for the allosteric regulation of phosphodiesterase 2A deduced from the X-ray structure of a near full-length construct. *Proc. Natl. Acad. Sci. USA*, **106**:18225–18230.

55. Stangherlin, A., Gesellchen, F., Zoccarato, A., Terrin, A., Fields, L.A., Berrera, M., Surdo, N.C., Craig, M.A., Smith, G., Hamilton, G., and Zaccolo, M. (2011) cGMP signals modulate cAMP levels in a compartment-specific manner to regulate catecholamine-dependent signaling in cardiac myocytes. *Circ. Res.*, **108**:929–939.

56. Suvarna, N.U. and O'Donnell, J.M. (2002) Hydrolysis of N-methyl-D-aspartate receptor-stimulated cAMP and cGMP by PDE4 and PDE2 phosphodiesterases in primary neuronal cultures of rat cerebral cortex and hippocampus. *J. Pharmacol. Exp. Ther.*, **302**:249–256.

57. Sadhu, K., Hensley, K., Florio, V.A., and Wolda, S.L. (1999) Differential expression of the cyclic GMP-stimulated phosphodiesterase PDE2A in human venous and capillary endothelial cells. *J. Histochem. Cytochem.*, **47**:895–906.

58. Reyes-Irisarri, E., Markerink-Van Ittersum, M., Mengod, G., and de Vente, J. (2007) Expression of the cGMP-specific phosphodiesterases 2 and 9 in normal and Alzheimer's disease human brains. *Eur. J. Neurosci.*, **25**:3332–3338.

59. Boess, F., Hendrix, M., van der Staay, F., Erb, C., Schreiber, R., van Staveren, W., de Vente, J., Prickaerts, J., Blokland, A., and Koenig, G. (2004) Inhibition of phosphodiesterase 2 increases neuronal cGMP, synaptic plasticity and memory performance. *Neuropharmacology*, **47**:1081–1092.

60. Domek-Lopacinska, K. and Strosznajder, J.B. (2008) The effect of selective inhibition of cyclic GMP hydrolyzing phosphodiesterases 2 and 5 on learning and memory processes and nitric oxide synthase activity in brain during aging. *Brain Res.*, **1216**:68–77.

61. Rutten, K., Prickaerts, J., Hendrix, M., van der Staay, F., Sik, A., and Blokland, A. (2007) Time-dependent involvement of cAMP and cGMP in consolidation of object memory: studies using selective phosphodiesterase type 2, 4 and 5 inhibitors. *Eur. J. Pharmacol.*, **558**:107–112.

62. Reneerkens, O., Rutten, K., Steinbusch, H., Blokland, A., and Prickaerts, J. (2009) Selective phosphodiesterase inhibitors: a promising target for cognition enhancement. *Psychopharmacology*, **202**:419–443.

63. Bender, A. (2006) Calmodulin-stimulated cyclic nucleotide phosphodiesterase. In: Beavo, J.A., Francis, S.H., and Houslay, M.D. (Eds.), *Cyclic Nucleotide Phosphodiesterases in Health and Disease*, Boca Raton, FL: CRC Press, pp. 35–54.

64. Podzuweit, T., Nennstiel, P., and Muller, A. (1995) Isozyme selective inhibition of cGMP-stimulated cyclic nucleotide phosphodiesterases by erythro-9-(2-hydroxy-3-nonyl) adenine. *Cell. Signal.*, **7**:733–738.

65. Chambers, R., Abrams, K., Castleberry, T., Cheng, J., Fisher, D., Kamath, A., Marfat, A., Nettleton, D., Pillar, J., Salter, E., Sheils, A., Shirley, J., Turner, C., Umland, J., and Lam, K. (2006) A new chemical tool for exploring the role of the PDE4D isozyme in leukocyte function. *Bioorg. Med. Chem. Lett.*, **16**:718–721.

66. Sandner, P., Hutter, J., Tinel, H., Ziegelbauer, K., and Bischoff, E. (2007) PDE5 inhibitors beyond erectile dysfunction. *Int. J. Impot. Res.*, **19**:533–543.

67. Puzzo, D., Sapienza, S., Arancio, O., and Palmeri, A. (2008) Role of phosphodiesterase 5 in synaptic plasticity and memory. *Neuropsychiatr. Dis. Treat.*, **4**:371–387.

68. Farooq, M.U., Naravetla, B., Moore, P.W., Majid, A., Gupta, R., and Kassab, M.Y. (2008) Role of sildenafil in neurological disorders. *Clin. Neuropharmacol.*, **31**:353–362.

69. Devan, B., Duffy, K., Bowker, J., Bharati, I., Nelson, C., Daffin, L.J., Spangler, E., and Ingram, D. (2005) Phosphodiesterase type 5 (PDE5) inhibition and cognitive enhancement. *Drug Future*, **30**:725.

70. Rutten, K., Basile, J.L., Prickaerts, J., Blokland, A., and Vivian, J.A. (2008) Selective PDE inhibitors rolipram and sildenafil improve object retrieval performance in adult cynomolgus macaques. *Psychopharmacology (Berl.)*, **196**:643–648.

71. Schultheiss, D., Muller, S.V., Nager, W., Stief, C.G., Schlote, N., Jonas, U., Asvestis, C., Johannes, S., and Munte, T.F. (2001) Central effects of sildenafil (Viagra) on auditory selective attention and verbal recognition memory in humans: a study with event-related brain potentials. *World J. Urol.*, **19**:46–50.

72. Reneerkens, O.A.H., Sambeth, A., Rutten, K., Menniti, F.S., Ramaekers, J., Blokland, A., Steinbusch, H.W.M., and Prickaerts, J. (2009) The effects of phosphodiesterase 5 inhibition on recognition memory in rats and humans. *Society for Neuroscience*, p 886.821.

73. Goff, D., Cather, C., Freudenreich, O., Henderson, D., Evins, A., Culhane, M., and Walsh, J. (2009) A placebo-controlled study of sildenafil effects on cognition in schizophrenia. *Psychopharmacology*, **202**:411–417.

74. Zhang, L., Zhang, R.L., Wang, Y., Zhang, C., Zhang, Z.G., Meng, H., and Chopp, M. (2005) Functional recovery in aged and young rats after embolic stroke: treatment with a phosphodiesterase type 5 inhibitor. *Stroke*, **36**:847–852.

75. Zhang, L., Zhang, Z., Zhang, R.L., Cui, Y., LaPointe, M.C., Silver, B., and Chopp, M. (2006) Tadalafil, a long-acting type 5 phosphodiesterase isoenzyme inhibitor, improves neurological functional recovery in a rat model of embolic stroke. *Brain Res.*, **1118**: 192–198.

76. Kruse, L.S., Sandholdt, N.T.H., Gammeltoft, S., Olesen, J., and Kruuse, C. (2006) Phosphodiesterase 3 and 5 and cyclic nucleotide-gated ion channel expression in rat trigeminovascular system. *Neurosci. Lett.*, **404**:202–207.

77. Reneerkens, O.A., Rutten, K., Akkerman, S., Blokland, A., Shaffer, C.L., Menniti, F.S., Steinbusch, H.W., and Prickaerts, J. (2012) Phosphodiesterase type 5 (PDE5) inhibition improves object recognition memory: indications for central and peripheral mechanisms. *Neurobiol. Learn. Mem.*, **97**:370–379.

78. Menniti, F.S., Ren, J., Sietsma, D.K., Som, A., Nelson, F.R., Stephenson, D.T., Tate, B.A., and Finklestein, S.P. (2012) A non-brain penetrant PDE5a inhibitor improves functional recovery after stroke in rats. *Restor. Neurol. Neurosci.*, **30**:283–289.

79. Rutten, K., Van Donkelaar, E.L., Ferrington, L., Blokland, A., Bollen, E., Steinbusch, H.W., Kelly, P.A., and Prickaerts, J.H. (2009) Phosphodiesterase inhibitors enhance object memory independent of cerebral blood flow and glucose utilization in rats. *Neuropsychopharmacology*, **34**:1914–1925.

80. Barad, M., Bourtchouladze, R., Winder, D.G., Golan, H., and Kandel, E. (1998) Rolipram, a type IV-specific phosphodiesterase inhibitor, facilitates the establishment of long-lasting long-term potentiation and improves memory. *Proc. Natl. Acad. Sci. USA*, **95**:15020–15025.

81. Miró, X., Pérez-Torres, S., Palacios, J., Puigdomènech, P., and Mengod, G. (2001) Differential distribution of cAMP-specific phosphodiesterase 7A mRNA in rat brain and peripheral organs. *Synapse*, **40**:201–214.

82. Pérez-Torres, S., Cortés, R., Tolnay, M., Probst, A., Palacios, J., and Mengod, G. (2003) Alteration of phosphodiesterase type 7 and 8 isozyme mRNA expression in Alzheimer's disease brains examined by *in situ* hybridization. *Exp. Neurol.*, **182**:322–334.

83. Reyes-Irisarri, E., Pérez-Torres, S., and Mengod, G. (2005) Neuronal expression of cAMP-specific phosphodiesterase 7B mRNA in the rat brain. *Neuroscience*, **132**:1173–1185.

84. Giembycz, M.A. and Smith, S.J. (2006) Phosphodiesterase 7 (PDE7) as a therapeutic target. *Drug Future*, **31**:207–229.

85. Gil, C., Campillo, N.E., Perez, D.I., and Martinez, A. (2008) PDE7 inhibitors as new drugs for neurological and inflammatory disorders. *Expert Opin. Ther. Pat.*, **18**:1127–1139.

86. Sasaki, T., Kotera, J., and Omori, K. (2004) Transcriptional activation of phosphodiesterase 7B1 by dopamine D_1 receptor stimulation through the cyclic AMP/cyclic AMP-dependent protein kinase/cyclic AMP-response element binding protein pathway in primary striatal neurons. *J. Neurochem.*, **89**:474–483.

87. Malik, R., Bora, R.S., Gupta, D., Sharma, P., Arya, R., Chaudhary, S., and Saini, K.S. (2008) Cloning, stable expression of human phosphodiesterase 7A and development of an assay for screening of PDE7 selective inhibitors. *Appl. Microbiol. Biotechnol.*, **77**:1167–1173.

88. Alaamery, M.A., Wyman, A.R., Ivey, F.D., Allain, C., Demirbas, D., Wang, L., Ceyhan, O., and Hoffman, C.S. (2010) New classes of PDE7 inhibitors identified by a fission yeast-based HTS. *J. Biomol. Screen*, **15**:359–367.

89. Daga, P.R. and Doerksen, R.J. (2008) Stereoelectronic properties of spiroquinazolinones in differential PDE7 inhibitory activity. *J. Comput. Chem.*, **29**:1945–1954.

90. Bergmann, J.E., Cutshall, N.S., Demopulos, G.A., Florio, V.A., Gaitanaris, G., Gray, P., Hohmann, J., Onrust, R., and Zeng, H. (2008) US Patent application No. 20080260643, Published 10/23/2008.

91. Redondo, M., Brea, J., Perez, D.I., Soteras, I., Val, C., Perez, C., Morales-Garcia, J.A., Alonso-Gil, S., Paul-Fernandez, N., Martin-Alvarez, R., Cadavid, M.I., Loza, M.I., Perez-Castillo, A., Mengod, G., Campillo, N.E., Martinez, A., and Gil, C. (2012) Effect of phosphodiesterase 7 (PDE7) inhibitors in experimental autoimmune encephalomyelitis mice: discovery of a new chemically diverse family of compounds. *J. Med. Chem.*, **55**:3274–3284.

92. Vasta, V. (2006) cAMP-phosphodiesterase 8 family. In: Beavo, J., Francis, S.H., and Houslay, M. (Eds.), *Cyclic Nucleotide Phosphodiesterases in Health and Disease*, Boca Raton, FL: CRC Press.

93. Fisher, D.A., Smith, J.F., Pillar, J.S., St Denis, S.H., and Cheng, J.B. (1998) Isolation and characterization of PDE8A, a novel human cAMP-specific phosphodiesterase. *Biochem. Biophys. Res. Commun.*, **246**:570–577.

94. Kobayashi, T., Gamanuma, M., Sasaki, T., Yamashita, Y., Yuasa, K., Kotera, J., and Omori, K. (2003) Molecular comparison of rat cyclic nucleotide phosphodiesterase 8 family: unique expression of PDE8B in rat brain. *Gene*, **319**:221–231.

95. Perez-Torres, S., Cortes, R., Tolnay, M., Probst, A., Palacios, J.M., and Mengod, G. (2003) Alterations on phosphodiesterase type 7 and 8 isozyme mRNA expression in Alzheimer's disease brains examined by *in situ* hybridization. *Exp. Neurol.*, **182**:322–334.

96. Vang, A.G., Ben-Sasson, S.Z., Dong, H., Kream, B., DeNinno, M.P., Claffey, M.M., Housley, W., Clark, R.B., Epstein, P.M., and Brocke, S. (2010) PDE8 regulates rapid Teff cell adhesion and proliferation independent of ICER. *PLoS One*, **5**:e12011.

97. Wang, H., Yan, Z., Yang, S., Cai, J., Robinson, H., and Ke, H. (2008) Kinetic and structural studies of phosphodiesterase-8A and implication on the inhibitor selectivity. *Biochemistry*, **47**:12760–12768.

98. Guipponi, M., Scott, H.S., Kudoh, J., Kawasaki, K., Shibuya, K., Shintani, A., Asakawa, S., Chen, H., Lalioti, M.D., Rossier, C., Minoshima, S., Shimizu, N., and Antonarakis, S.E. (1998) Identification and characterization of a novel cyclic nucleotide phosphodiesterase gene (PDE9A) that maps to 21q22.3: alternative splicing of mRNA transcripts, genomic structure and sequence. *Hum. Genet.*, **103**:386–392.

99. Rentero, C., Monfort, A., and Puigdomenech, P. (2003) Identification and distribution of different mRNA variants produced by differential splicing in the human phosphodiesterase 9A gene. *Biochem. Biophys. Res. Commun.*, **301**:686–692.

100. Wang, P., Wu, P., Egan, R.W., and Billah, M.M. (2003) Identification and characterization of a new human type 9 cGMP-specific phosphodiesterase splice variant (PDE9A5): differential tissue distribution and subcellular localization of PDE9A variants. *Gene*, **314**:15–27.

101. Fisher, D.A., Smith, J.F., Pillar, J.S., St Denis, S.H., and Cheng, J.B. (1998) Isolation and characterization of PDE9A, a novel human cGMP-specific phosphodiesterase. *J. Biol. Chem.*, **273**:15559–15564.

102. Soderling, S.H., Bayuga, S.J., and Beavo, J.A. (1998) Identification and characterization of a novel family of cyclic nucleotide phosphodiesterases. *J. Biol. Chem.*, **273**:15553–15558.

103. Kruse, L.S., Moller, M., Tibaek, M., Gammeltoft, S., Olesen, J., and Kruuse, C. (2009) PDE9A, PDE10A, and PDE11A expression in rat trigeminovascular pain signalling system. *Brain Res.*, **1281**:25–34.

104. Kubinyi, H. and Müller, G. (Eds.) (2004) *Chemogenomics in Drug Discovery: A Medicinal Chemistry Perspective*, Weinheim: Wiley-VCH Verlag GmbH.

105. Huai, Q., Wang, H., Zhang, W., Colman, R.W., Robinson, H., and Ke, H. (2004) Crystal structure of phosphodiesterase 9 shows orientation variation of inhibitor 3-isobutyl-1-methylxanthine binding. *Proc. Natl. Acad. Sci. USA*, **101**:9624–9629.

106. Wang, H., Ye, M., Robinson, H., Francis, S.H., and Ke, H. (2008) Conformational variations of both phosphodiesterase-5 and inhibitors provide the structural basis for the physiological effects of vardenafil and sildenafil. *Mol. Pharmacol.*, **73**:104–110.

107. Wunder, F. (2005) Characterization of the first potent and selective PDE9 inhibitor using a cGMP reporter cell line. *Mol. Pharmacol.*, **68**:1775–1781.

108. Wang, H., Luo, X., Ye, M., Hou, J., Robinson, H., and Ke, H. (2010) Insight into binding of phosphodiesterase-9A selective inhibitors by crystal structures and mutagenesis. *J. Med. Chem.*, **53**:1726–1731.

109. Deninno, M.P., Andrews, M., Bell, A.S., Chen, Y., Eller-Zarbo, C., Eshelby, N., Etienne, J.B., Moore, D.E., Palmer, M.J., Visser, M.S., Yu, L.J., Zavadoski, W.J., and Michael Gibbs, E. (2009) The discovery of potent, selective, and orally bioavailable PDE9 inhibitors as potential hypoglycemic agents. *Bioorg. Med. Chem. Lett.*, **19**:2537–2541.

110. Verhoest, P.R., Fonseca, K.R., Hou, X., Proulx-Lafrance, C., Corman, M., Helal, C.J., Claffey, M.M., Tuttle, J.B., Coffman, K.J., Liu, S., Nelson, F., Kleiman, R.J., Menniti, F.S., Schmidt, C.J., Vanase-Frawley, M., and Liras, S. (2012) Design and discovery of 6-[(3*S*,4*S*)-4-methyl-1-(pyrimidin-2-ylmethyl)pyrrolidin-3-yl]-1-(tetrahydro-2*H*-pyr an-4-yl)-1,5-dihydro-4*H*-pyrazolo[3,4-*d*]pyrimidin-4-one (PF-04447943), a selective brain penetrant PDE9A inhibitor for the treatment of cognitive disorders. *J. Med. Chem.*, **55**:9045–9054.

111. Kleiman, R.J., Chapin, D.S., Christoffersen, C., Freeman, J., Fonseca, K.R., Geoghegan, K.F., Grimwood, S., Guanowsky, V., Hajos, M., Harms, J.F., Helal, C.J., Hoffmann, W.E., Kocan, G.P., Majchrzak, M.J., McGinnis, D., McLean, S., Menniti, F.S., Nelson, F., Roof, R., Schmidt, A.W., Seymour, P.A., Stephenson, D.T., Tingley, F.D., Vanase-Frawley, M., Verhoest, P.R., and Schmidt, C.J. (2012) Phosphodiesterase 9a regulates central cGMP and modulates responses to cholinergic and monoaminergic perturbation *in vivo. J. Pharmacol. Exp. Ther.*, **341**:396–409.

112. van der Staay, F.J., Rutten, K., Bärfacker, L., Devry, J., Erb, C., Heckroth, H., Karthaus, D., Tersteegen, A., van Kampen, M., Blokland, A., Prickaerts, J., Reymann, K.G., Schröder, U.H., and Hendrix, M. (2008) The novel selective PDE9 inhibitor BAY 73-6691 improves learning and memory in rodents. *Neuropharmacology*, **55**:908–918.

113. Hutson, P.H., Finger, E.N., Magliaro, B.C., Smith, S.M., Converso, A., Sanderson, P.E., Mullins, D., Hyde, L.A., Eschle, B.K., Turnbull, Z., Sloan, H., Guzzi, M., Zhang, X., Wang, A., Rindgen, D., Mazzola, R., Vivian, J.A., Eddins, D., Uslaner, J.M., Bednar, R., Gambone, C., Le-Mair, W., Marino, M.J., Sachs, N., Xu, G., and Parmentier-Batteur, S. (2011) The selective phosphodiesterase 9 (PDE9) inhibitor PF-04447943 (6-[(3*S*,4*S*)-4-methyl-1-(pyrimidin-2-ylmethyl)pyrrolidin-3-yl]-1-(tetrahydro-2*H*-pyran-4-yl)-1,5-dihydro-4*H*-pyrazolo[3,4-*d*]pyrimidin-4-one) enhances synaptic plasticity and cognitive function in rodents. *Neuropharmacology*, **61**:665–676.

114. Nicholas, T., Evans, R., Styren, S., Qiu, R., Wang, E.Q., Nelson, F., Le, V., Grimwood, S., Christoffersen, C., Banerjee, S., Corrigan, B., Kocan, G., Geoghegan, K., Carrieri, C., Raha, N., Verhoest, P., and Soares, H. (2009) PF-04447943, a novel PDE9A inhibitor, increases cGMP levels in cerebrospinal fluid: translation from non-clinical species to healthy human volunteers. *Alzheimers Dement.*, **5**:P330–P331.

115. Fujishige, K., Kotera, J., Michibata, H., Yuasa, K., Takebayashi, S., Okumura, K., and Omori, K. (1999) Cloning and characterization of a novel human phosphodiesterase that hydrolyzes both cAMP and cGMP (PDE10A). *J. Biol. Chem.*, **274**:18438–18445.

116. Loughney, K., Snyder, P.B., Uher, L., Rosman, G.J., Ferguson, K., and Florio, V.A. (1999) Isolation and characterization of PDE10A, a novel human 3′,5′-cyclic nucleotide phosphodiesterase. *Gene*, **234**:109–117.

117. Soderling, S.H., Bayuga, S.J., and Beavo, J.A. (1999) Isolation and characterization of a dual-substrate phosphodiesterase gene family: PDE10A. *Proc. Natl. Acad. Sci. USA*, **96**:7071–7076.

118. Strick, C.A., Schmidt, C.J., and Menniti, F.S. (2006) PDE10A: a striatum enriched, dual-substrate phosphodiesterase. In: Francis, S., Houslay, M., and Beavo, J. (Eds.), *Phosphodiesterases in Health and Disease*, Boca Raton, FL: CRC Press.

119. Gross-Langenhoff, M., Hofbauer, K., Weber, J., Schultz, A., and Schultz, J.E. (2006) cAMP is a ligand for the tandem GAF domain of human phosphodiesterase 10 and cGMP for the tandem GAF domain of phosphodiesterase 11. *J. Biol. Chem.*, **281**:2841–2846.

120. Coskran, T.M., Morton, D., Menniti, F.S., Adamowicz, W.O., Kleiman, R.J., Ryan, A.M., Strick, C.A., Schmidt, C.J., and Stephenson, D.T. (2006) Immunohistochemical localization of phosphodiesterase 10A in multiple mammalian species. *J. Histochem. Cytochem.*, **54**:1205–1213.

121. Sano, H., Nagai, Y., Miyakawa, T., Shigemoto, R., and Yokoi, M. (2008) Increased social interaction in mice deficient of the striatal medium spiny neuron-specific phosphodiesterase 10A2. *J. Neurochem.*, **105**:546–556.

122. Graybiel, A.M. (2008) Habits, rituals, and the evaluative brain. *Annu. Rev. Neurosci.*, **31**:359–387.

123. DeLong, M.R. and Wichmann, T. (2007) Circuits and circuit disorders of the basal ganglia. *Arch. Neurol.*, **64**:20–24.

124. Schapira, A.H. (2009) Etiology and pathogenesis of Parkinson disease. *Neurol. Clin.*, **27**:583–603.

125. Seeman, P. (1987) Dopamine receptors and the dopamine hypothesis of schizophrenia. *Synapse*, **1**:133–152.

126. Martin, J.B. and Gusella, J.F. (1986) Huntington's disease. Pathogenesis and management. *N. Engl. J. Med.*, **315**:1267–1276.

127. Rapoport, J.L. (1990) Obsessive compulsive disorder and basal ganglia dysfunction. *Psychol. Med.*, **20**:465–469.

128. Chappie, T.A., Humphrey, J.M., Menniti, F.S., and Schmidt, C.J. (2009) PDE10A inhibitors: an assessment of the current CNS drug discovery landscape. *Curr. Opin. Investig. Drugs*, **12**:458–467.

129. Kehler, J. and Kilburn, J.P. (2009) Patented PDE10A inhibitors: novel compounds since 2007. *Expert Opin. Ther. Pat.*, **19**:1715–1725.

130. Kehler, J., Ritzén, A., and Greve, D.R. (2007) The potential therapeutic use of phosphodiesterase 10 inhibitors. *Expert Opin. Ther. Pat.*, **17**:147–158.

131. Menniti, F.S., Chappie, T.A., Humphrey, J.M., and Schmidt, C.J. (2007) Phosphodiesterase 10A inhibitors: a novel approach to the treatment of the symptoms of schizophrenia. *Curr. Opin. Investig. Drugs*, **8**:54–59.

132. Siuciak, J.A., Chapin, D.S., Harms, J.F., Lebel, L.A., McCarthy, S.A., Chambers, L., Shrikhande, A., Wong, S., Menniti, F.S., and Schmidt, C.J. (2006) Inhibition of the striatum-enriched phosphodiesterase PDE10A: a novel approach to the treatment of psychosis. *Neuropharmacology*, **51**:386–396.

133. Kehler, J., Ritzen, A., and Greve, D.R. (2007) The potential therapeutic use of phosphodiesterase 10 inhibitors. *Expert Opin. Ther. Pat.*, **17**:147–158.

134. Chappie, T.A., Humphrey, J.M., Allen, M.P., Estep, K.G., Fox, C.B., Lebel, L.A., Liras, S., Marr, E.S., Menniti, F.S., Pandit, J., Schmidt, C.J., Tu, M., Williams, R.D., and Yang, F.V. (2007) Discovery of a series of 6,7-dimethoxy-4-pyrrolidylquinazoline PDE10A inhibitors. *J. Med. Chem.*, **50**:182–185.

135. Schmidt, C.J., Chapin, D.S., Cianfrogna, J., Corman, M.L., Hajos, M., Harms, J.F., Hoffman, W.E., Lebel, L.A., McCarthy, S.A., Nelson, F.R., Proulx-LaFrance, C., Majchrzak, M.J., Ramirez, A.D., Schmidt, K., Seymour, P.A., Siuciak, J.A., Tingley, F.D., III, Williams, R.D., Verhoest, P.R., and Menniti, F.S. (2008) Preclinical

characterization of selective phosphodiesterase 10A inhibitors: a new therapeutic approach to the treatment of schizophrenia. *J. Pharmacol. Exp. Ther.*, **325**:681–690.

136. Grauer, S.M., Pulito, V.L., Navarra, R.L., Kelly, M.P., Kelley, C., Graf, R., Langen, B., Logue, S., Brennan, J., Jiang, L., Charych, E., Egerland, U., Liu, F., Marquis, K.L., Malamas, M., Hage, T., Comery, T.A., and Brandon, N.J. (2009) Phosphodiesterase 10A inhibitor activity in preclinical models of the positive, cognitive, and negative symptoms of schizophrenia. *J. Pharmacol. Exp. Ther.*, **331**:574–590.

137. Smith, S.M., Uslaner, J.M., Cox, C.D., Huszar, S.L., Cannon, C.E., Vardigan, J.D., Eddins, D., Toolan, D.M., Kandebo, M., Yao, L., Raheem, I.T., Schreier, J.D., Breslin, M.J., Coleman, P.J., and Renger, J.J. (2013) The novel phosphodiesterase 10A inhibitor THPP-1 has antipsychotic-like effects in rat and improves cognition in rat and rhesus monkey. *Neuropharmacology*, **64**:215–223.

138. Cutshall, N.S., Onrust, R., Rohde, A., Gragerov, S., Hamilton, L., Harbol, K., Shen, H.R., McKee, S., Zuta, C., Gragerova, G., Florio, V., Wheeler, T.N., and Gage, J.L. (2012) Novel 2-methoxyacylhydrazones as potent, selective PDE10A inhibitors with activity in animal models of schizophrenia. *Bioorg. Med. Chem. Lett.*, **22**:5595–5599.

139. DeMartinis, N.A. (2012) Results of a phase 2A proof-of-concept trial with a PDE10A inhibitor in the treatment of acute exacerbation of schizophrenia. In *Society for Biological Psychiatry 2012 Annual Meeting* (Biological Psychiatry, Philadelphia, PA), p. 17S.

140. Gines, S., Seong, I.S., Fossale, E., Ivanova, E., Trettel, F., Gusella, J.F., Wheeler, V.C., Persichetti, F., and MacDonald, M.E. (2003) Specific progressive cAMP reduction implicates energy deficit in presymptomatic Huntington's disease knock-in mice. *Hum. Mol. Genet.*, **12**:497–508.

141. Sugars, K.L. and Rubinsztein, D.C. (2003) Transcriptional abnormalities in Huntington disease. *Trends Genet.*, **19**:233–238.

142. Sugars, K.L., Brown, R., Cook, L.J., Swartz, J., and Rubinsztein, D.C. (2004) Decreased cAMP response element-mediated transcription, an early event in exon 1 and full-length cell models of Huntington's disease that contributes to polyglutamine pathogenesis. *J. Biol. Chem.*, **279**:4988–4999.

143. Kleiman, R.J., Kimmel, L.H., Bove, S.E., Lanz, T.A., Harms, J.F., Romegialli, A., Miller, K.S., Willis, A., des Etages, S., Kuhn, M., and Schmidt, C.J. (2010) Chronic suppression of phosphodiesterase 10A alters striatal expression of genes responsible for neuro-transmitter synthesis, neurotransmission, and signaling pathways implicated in Hunting-ton's disease. *J. Pharmacol. Exp. Ther.*, **336**:64–76.

144. Giampà, C., Patassini, S., Borreca, A., Laurenti, D., Marullo, F., Bernardi, G., Menniti, F.S., and Fusco, F.R. (2009) Phosphodiesterase 10 inhibition reduces striatal excitotox-icity in the quinolinic acid model of Huntington's disease. *Neurobiol. Dis.*, **34**:450–456.

145. Giampà, C., Laurenti, D., Anizilotti, S., Bernardi, G., Menniti, F.S., and Fusco, F.R. (2010) Inhibition of the striatal specific phosphodiesterase PDE10A ameliorates striatal and cortical pathology in the r6/2 mouse model of Huntington's disease. *PLoS One*, **5**: e13417.

146. Munoz-Sanjuan, I. and Bates, G.P. (2011) The importance of integrating basic and clinical research toward the development of new therapies for Huntington disease. *J. Clin. Invest.*, **121**:476–483.

147. Burgin, A.B., Magnusson, O.T., Singh, J., Witte, P., Staker, B.L., Bjornsson, J.M., Thorsteinsdottir, M., Hrafnsdottir, S., Hagen, T., Kiselyov, A.S., Stewart, L.J., and Gurney,

M.E. (2010) Design of phosphodiesterase 4D (PDE4D) allosteric modulators for enhancing cognition with improved safety. *Nat. Biotechnol.*, **28**:63–70.

148. Houslay, M.D. and Adams, D.R. (2010) Putting the lid on phosphodiesterase 4. *Nat. Biotechnol.*, **28**:38–40.

149. Beavo, J.A. (1995) Cyclic nucleotide phosphodiesterases: functional implications of multiple isoforms. *Physiol. Rev.*, **75**:725–748.

150. Conti, M. and Beavo, J. (2007) Biochemistry and physiology of cyclic nucleotide phosphodiesterases: essential components in cyclic nucleotide signaling. *Annu. Rev. Biochem.*, **76**:481–511.

151. Juilfs, D., Soderling, S., Burns, F., and Beavo, J. (1999) Cyclic GMP as substrate and regulator of cyclic nucleotide phosphodiesterases (PDEs). *Rev. Physiol. Biochem. Pharmacol.*, **135**:67–104.

152. Gopal, V.K., Francis, S.H., and Corbin, J.D. (2001) Allosteric sites of phosphodiesterase-5 (PDE5): a potential role in negative feedback regulation of cGMP signaling in corpus cavernosum. *Eur. J. Biochem.*, **268**:3304–3312.

153. Nikolaev, V.O., Gambaryan, S., and Lohse, M.J. (2006) Fluorescent sensors for rapid monitoring of intracellular cGMP. *Nat. Methods*, **3**:23–25.

154. Russwurm, M., Mullershausen, F., Friebe, A., Jager, R., Russwurm, C., and Koesling, D. (2007) Design of fluorescence resonance energy transfer (fret)-based cGMP indicators: a systematic approach. *Biochem. J.*, **407**:69–77.

155. Biswas, K.H., Sopory, S., and Visweswariah, S.S. (2008) The GAF domain of the cGMP-binding, cGMP-specific phosphodiesterase (PDE5) is a sensor and a sink for cGMP. *Biochemistry*, **47**:3534–3543.

156. DeLong, M.R. and Wichmann, T. (2007) Circuits and circuit disorders of the basal ganglia. *Arch. Neurol.*, **64**:20–24.

CHAPTER 5

RECENT RESULTS IN PHOSPHODIESTERASE INHIBITOR DEVELOPMENT AND CNS APPLICATIONS

DAVID P. ROTELLA

Department of Chemistry and Biochemistry, Montclair State University, Montclair, NJ, USA

INTRODUCTION

In the last few years, there has been increased interest in phosphodiesterase (PDE) biology and inhibitor discovery for central nervous system (CNS) applications [1]. This is largely because selected PDE enzymes are established targets for treatment of erectile dysfunction (PDE5) and for various pulmonary and cardiovascular disorders (PDE3, PDE4) (see Table 5.1 and Figure 5.1). Beyond approved applications, Wong et al. investigated the association between major depressive disorder (MDD) and the PDE gene family [2]. This analysis revealed that PDE9A and PDE11A polymorphisms correlated with the occurrence of MDD. In addition, polymorphisms in PDE1A and PDE11A were associated with response to anti-depressant therapy. A variety of reports suggest that PDE4 and PDE5 inhibitors have procognitive effects in animal models [3,4]. The reports also suggest that PDE10 inhibitors demonstrate antipsychotic activity in preclinical models [5] and PDE5 inhibitors increase cerebral blood flow and improve functional outcomes in stroke models [6].

This promising range of activity in CNS disease models, coupled with an improved understanding of the role that second messengers play in key signaling pathways related to these diseases, creates a unique set of circumstances where PDE inhibitors can be investigated as potential therapeutic approaches. This chapter focuses on

Cyclic-Nucleotide Phosphodiesterases in the Central Nervous System: From Biology to Drug Discovery, First Edition. Edited by Nicholas J. Brandon and Anthony R. West.
© 2014 John Wiley & Sons, Inc. Published 2014 by John Wiley & Sons, Inc.

TABLE 5.1 Phosphodiesterase Inhibitors Approved for Clinical Use

PDE Inhibitor	PDE Target(s)	Approved Indication(s)
Sildenafil (1)	PDE5	Erectile dysfunction, pulmonary hypertension
Tadalafil (2)	PDE5	Erectile dysfunction, pulmonary hypertension
Vardenafil (3)	PDE5	Erectile dysfunction
Olprinone (4)	PDE3	Acute cardiac insufficiency
Milrinone (5)	PDE3	Acute cardiac failure
Amrinone (6)	PDE3	Acute cardiac failure, hypertension
Cilostazol (7)	PDE3	Intermittent claudication
Dipyridamole (8)	PDE1, PDE5	Platelet aggregation
Theophylline (9)	PDE3, PDE4	Asthma, bronchitis, emphysema
Aminophylline (10)	PDE3, PDE4	Asthma, bronchitis, emphysema

reports in the last few years that highlight recent advances in cell-based assays for PDE inhibition and new examples of lead discovery and optimization, including structure-based drug design. In addition, new PDE inhibitor chemotypes are highlighted, along with examples that illustrate the application of PDE inhibitors for

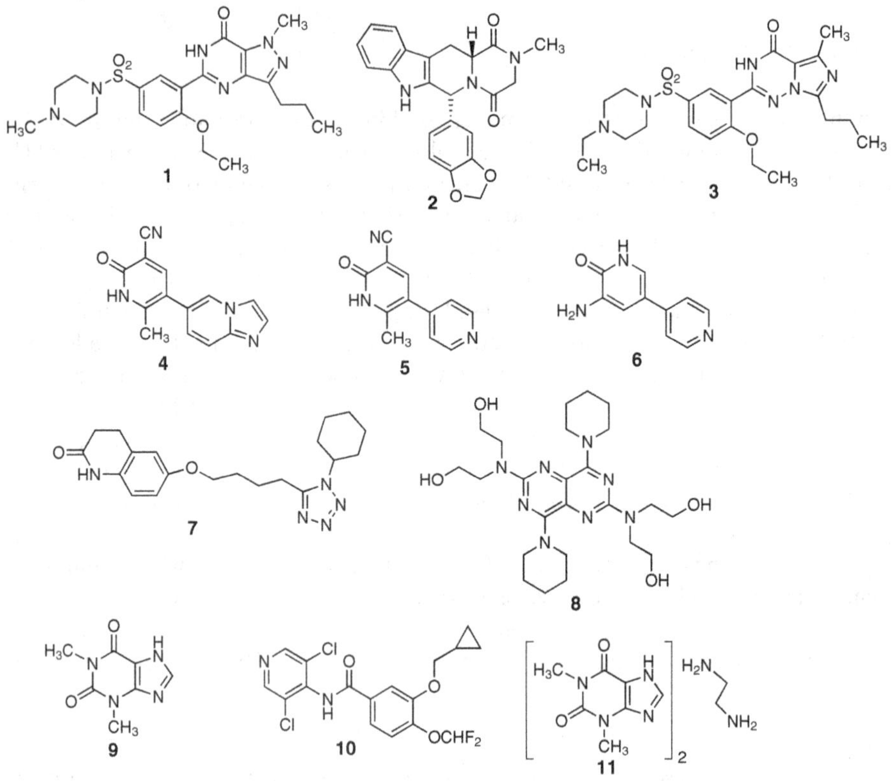

FIGURE 5.1 Approved phosphodiesterase inhibitors.

treatment of CNS disorders. Recent reviews provide summaries of medicinal chemistry advances in the field [7–11].

LEAD DISCOVERY APPROACHES

PDE inhibitor discovery has made use of several contemporary techniques to identify leads, including high-throughput screening (HTS), virtual screening, structure-based drug design, natural product screening, and optimization of existing lead structures.

Castro et al. reported the results of a ligand-based virtual screen to identify novel PDE7 chemotypes using CODES, a computational method that identifies and scores molecules from a topological perspective [12]. This program attempts to relate molecular topology to the nature of the atoms and chemical bonds in a compound, along with the relative connectivity of those atoms and bonds to arrive at a score that is used to rank examples in a virtual library. A database of 173 compounds was assembled that included 31 known PDE7 inhibitors and 11 PDE4 inhibitors. The remaining compounds in the training set were selected based on structural diversity and known biological activity in a privileged scaffold. This analysis suggested that thiadiazine and quinazoline templates might furnish interesting lead structures. The quinazolines **12–14** shown in Figure 5.2 demonstrated 40–66% inhibition of PDE7A at 10 µM, and showed some evidence of selectivity versus PDE3A and PDE4B ($\leq$20% inhibition at 10 µM). Approximately 10 thiadiazine analogs were synthesized and found to be weak PDE7A inhibitors ($\leq$25% inhibition at 10 µM).

A previous report identified a series of flavonols, isolated from *Sophora flavescens* as potent PDE5 inhibitors, with IC_{50} values as low as 13 nM, for example, compound **15** [13]. Structure activity studies reported in this paper suggested that substituents in the phenyl ring of **15** exerted a significant effect on PDE5 potency. By substituting the polyoxygenated phenyl ring for the pyrazole ring in sildenafil, followed by structure–activity relationship (SAR) studies on the phenyl ring, PDE5 potency comparable to **1** could be achieved (IC_{50} <10 nM), accompanied by a significant improvement in selectivity versus PDE6 (>1000-fold), as in compound **16** (Figure 5.3) [14].

The catalytic domains of many of PDEs have been crystallized, some with inhibitors present [15]. This offers researchers the opportunity to use these high-

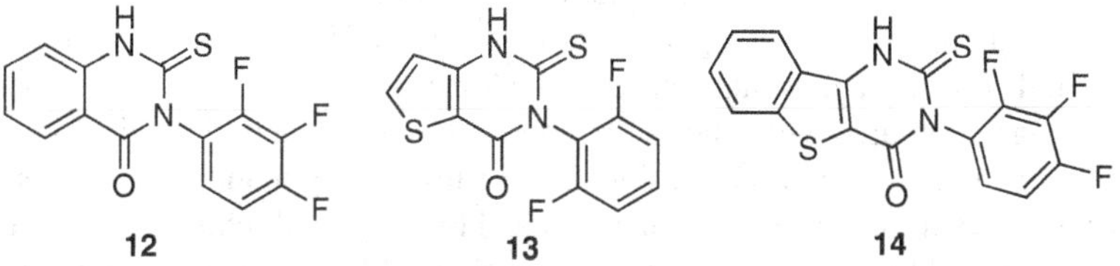

| 12 | 13 | 14 |

FIGURE 5.2   Novel PDE7A chemotypes derived from lead structures identified by CODES virtual screening.

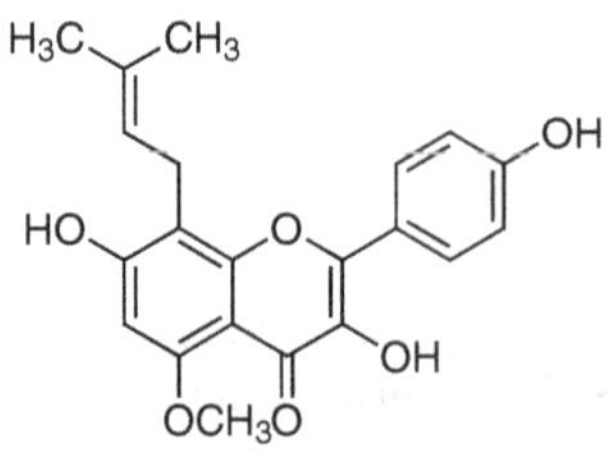

FIGURE 5.3 Hybridization of natural product lead with a known PDE inhibitor.

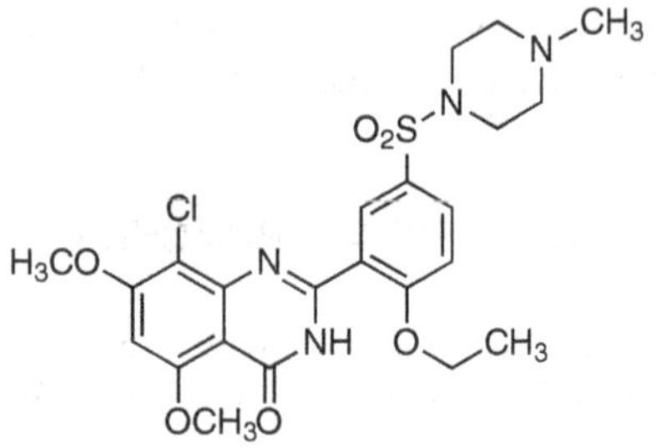

resolution structures to develop hypotheses that could improve potency and PDE selectivity of the inhibitor, and also may allow for virtual screening in the search for new chemotypes. At this writing, no reports of virtual screening to identify PDE inhibitors have been published. Ke and Wang compared the available catalytic domains of PDE1B, PDE2A, PDE3B, PDE4B, PDE5A, PDE7A, and PDE9A in 2007 and noted that overall the structures are similar [15]. In these enzymes, 16 α-helices form three smaller domains within the catalytic portion of the enzyme. Zinc and/or magnesium ions form key recognition elements within the substrate binding domain. Highly conserved aspartate and histidine residues coordinate with the zinc ion, which is present in all PDE structures. All PDE active sites contain an invariant glutamine that hydrogen bonds with the cyclic nucleotide substrate. Of the 17 amino acids within the substrate binding region, 12 are absolutely conserved across all PDE gene family members [16]. Nearly all PDE inhibitors share similar binding modes within the active site by interacting with the invariant glutamine residue and a hydrophobic interaction between an aromatic ring in the inhibitor and aromatic residues on the enzyme (so-called hydrophobic clamp) [17]. In spite of the similarity in the active site region where many inhibitors bind, it is possible to develop inhibitors with excellent (>100-fold) selectivity for some members of the PDE family. Examples of highly selective compounds are shown in the inhibitor medicinal chemistry sections that follow.

Wang et al. studied the crystal structures of PDE4D2 and PDE10A2 in an attempt to better understand features within the cyclic nucleotide binding domain that might influence selectivity [18]. Using a catalytically inactive mutant of PDE4D2, the substrate cyclic adenosine monophosphate (cAMP) binds in the substrate pocket of the enzyme and forms two hydrogen bonds between the N1 of adenine and the NH of Gln369, and between a phosphate oxygen of cAMP and an NH on His160. When this structure was compared with those obtained with two other PDE4 isoforms, the differences were considered minor. Because the invariant glutamine residue is a key amino acid needed for substrate recognition and binding of all phosphodiesterases, it had been assumed that the side chain of this amino acid would shift its conformation to allow two hydrogen bonds to form with cyclic guanosine monophosphate (cGMP), another substrate for the dual-specificity PDE10. However, careful analysis of the

PDE10A1 crystal structure with cGMP bound revealed that a nearby tyrosine residue prevented movement of the key glutamine to form two hydrogen bonds with cGMP. Additionally, when cAMP binds to PDE10A2, the adenine ring is ~2 Å displaced from the position in PDE4D2, and the ring shows nearly 180° difference in orientation between the two structures (Figure 5.4). Furthermore, the substrate binds PDE10A2 in a *syn*-orientation and in an *anti*-orientation to PDE4D2. Thus, it is possible that inhibitors that compete with cAMP can be identified that take advantage of these differences.

In a related study, the substrate specificity of two mutant and catalytically inactive PDE10A2 isoforms was investigated by crystal structure analysis using both cAMP and cGMP to evaluate the recognition and binding elements associated with each substrate (Figure 5.5) [19]. Two aspartic acid residues (D674 and D564) were converted to alanine and asparagine, respectively. In this case, both substrates bound to the enzyme in a similar *syn*-orientation, but formed different interactions with the enzyme and had distinct orientations within the substrate binding pocket. This evaluation, coupled to sequence alignment with other PDEs revealed a region of the substrate binding pocket (S-pocket) where substantial differences in amino acid structure could contribute to substrate selectivity. By extension, interactions of potential inhibitors in this pocket could also help influence inhibitor selectivity.

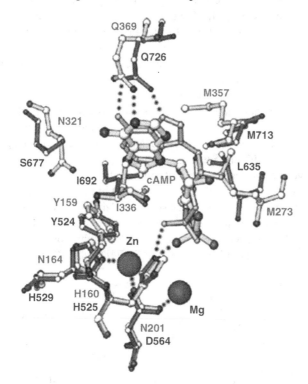

FIGURE 5.4 PDE4D2 and PDE10A1 active site comparison with cAMP bound. (See the color version of the figure in Color Plates section.)

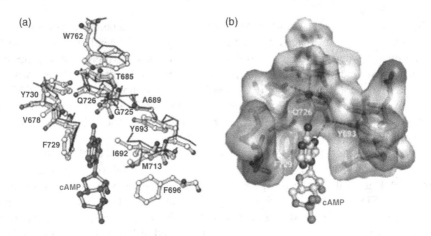

FIGURE 5.5 Hypothetical S-pocket in PDE10A. (See the color version of the figure in Color Plates section.)

Comparison of the crystal structures of PDE5A1 bound with two related inhibitors, sildenafil (**1**, $IC_{50} = 3.5$ nM) and thiophenyl sulfonamide analog **17** ($IC_{50} = 110$ nM) (Figure 5.6) revealed that the pyrazolopyrimidine portion of both molecules were essentially superimposable [20]. However, the enzyme underwent a more substantial conformational change, up to 1.5 Å, to accommodate the sulfonamide region of **18**. The authors suggest that a small hydrophobic loop on the enzyme must move to allow **17** to bind and that this movement contributes to inhibitor affinity, in spite of the observation that the SO_2NH_2 moiety does not appear to interact directly with the protein backbone. The authors also suggest that the alkoxy moiety on the aromatic ring (ethoxyl in sildenafil and **17**) fit into a small hydrophobic pocket that can also influence potency. As evidence, they cite the unexpectedly potent activity of a sildenafil analog, **18**, that lacks the sulfonamide, but has a propoxy ether moiety on the benzene ring (Figure 5.6).

The discovery of isoform-selective PDE4 inhibitors may be aided by the recent work of Wang et al. [21]. The clinical development of PDE4 inhibitors for a variety of applications, including CNS disorders such as Alzheimer's disease, has been hampered by adverse events in the clinic and in preclinical models, which may be related to insufficient PDE4 isozyme selectivity. The PDE4 inhibitor NVP, **19** (Figure 5.7), was crystallized with PDE4A, PDE4B, and PDE4D, while PDE4C was crystallized

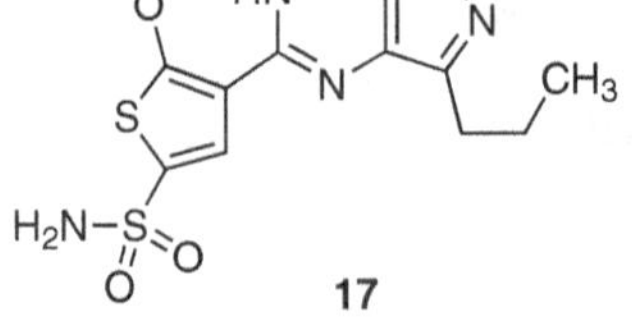

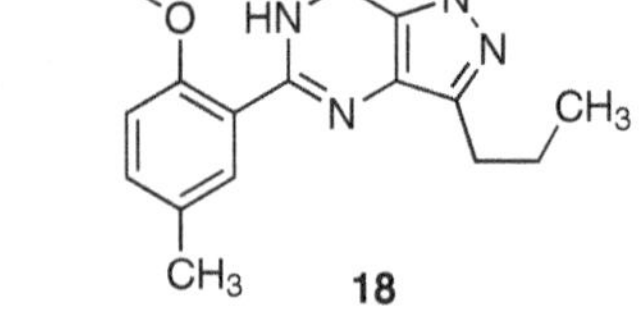

FIGURE 5.6 Sildenafil analogs.

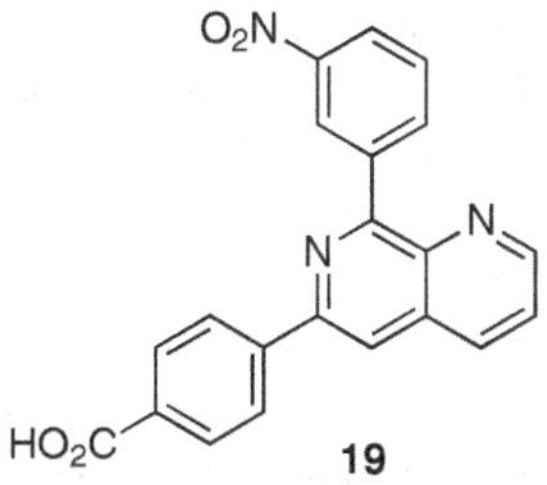

FIGURE 5.7 PDE4 inhibitor NVP.

without the inhibitor present. NVP binds to each PDE in the substrate binding pocket in the same conformation. Comparison of inhibitor-bound structures revealed that there was substantial movement ($\sim$1.1 Å) in PDE4A, compared to PDE4D, of several amino acids (A363–S368) near the key glutamine residue identified above. Less subtle changes were noted when PDE4B and PBE4D with NVP bound were compared. The most substantial difference was associated with the position of a methionine residue that connects helices 14 and 15 in each structure. The authors suggest that achieving selectivity between these two isoforms may be more difficult, although recent results (see below) provide some hope. There are a comparatively large number of differences between unliganded PDE4C and the inhibitor bound PDEs. However, because the substrate binding site for PDE4C is substantially different than the other three family members, achieving selectivity versus this enzyme is less challenging.

ASSAY METHODOLOGY

An excellent summary of *in vitro* enzyme assay protocols for PDE activity was recently published [22]. Although the protocols described are specifically identified with PDE1, PDE2, PDE4, and PDE10, the general procedures should be applicable to other PDEs of interest. These assays rely on radioisotope and fluorescence-based techniques methods to detect the cyclic nucleotide substrates and products of the enzymatic reaction. The latter approach was cited as being more useful for high-throughput screening because of the availability of a commercially available kit and because the fluorescent assay allows activity readouts in much less time than radioisotope methods. The authors advocate using baculovirus-expressed enzyme rather than a tissue source for the desired PDE, possibly because of improved batch-to-batch consistency in terms of specific activity. This chapter also identifies some of the most common problems (e.g., signal to noise ratio, activity signal, and background noise) associated with these *in vitro* assays and proposes solutions that may solve the problem. As an alternative to measuring substrate and/or product of the PDE reaction, a reporter gene assay has been reported that was claimed to be amenable to 384-well formats for high-throughput screening [23]. The assay was exemplified using PDE4B2 expressed in a recombinant stable cell line that was transfected with pCRe-luc and lipofectamine 2000. Validation assays showed a Z' factor of 0.56–0.60 in a 96-well format and 0.65–0.70 in a 384-well format. In each case, the signal-to-noise ratio was ≥ 5. Rolipram, a standard PDE4 inhibitor, was used to help validate the miniaturization from a 96-to a 384-well format. Both assay formats gave a similar IC_{50} value using this readout ($IC_{50} = 7\,nM$).

PDEs are intracellular enzymes; as a result, the development of assay methodology to determine cellular activity of inhibitors is important in a drug discovery effort. Over the last few years, several papers appeared describing cell-based screens for specific PDEs. Herget et al. reported a real-time method to monitor PDE3A, PDE4A1, and PDE5A activity in cells [24]. Using HEK293 cells transfected with a PDE modified with a cAMP-sensing protein (EPAC1) or cGMP sensor (cGES-DE2) and fluorescence resonance energy transfer ligand, it proved to be possible to detect enzyme activity in single cells. The authors showed that regulation of PDE3 activity could be observed based on intracellular cAMP gradients and protein kinase A (PKA)-dependent phosphorylation, and that cGMP could inhibit PDE3A-mediated cAMP hydrolysis.

A cell-based assay to characterize PDE2 inhibitors was reported by Wunder et al. [25]. In this system, PDE2A was stably transfected in Chinese hamster ovary (CHO) cells that also expressed the receptor for atrial natriuretic peptide (ANP) and a cyclic nucleotide gated ion channel that served as a biosensor for cGMP. Luminescence measurements resulting from calcium influx through the ion channel allowed selective quantification of PDE2-mediated cGMP hydrolysis. Enzymatic activity was stimulated by administration of ANP and no activity was detectable in the absence of the peptide. PDE2 selective inhibitors, such as EHNA and BAY 60-7550, were active in this assay, as was the nonselective PDE inhibitor IBMX. Compounds that act selectively on other PDEs, such as rolipram, sildenafil, and milrinone, showed no activity in this assay system. This potentially high-throughput assay could be used to help characterize PDE2 inhibitors for lead optimization. Similar technology was employed to characterize PDE9 inhibitors in a cell-based assay [26].

A PDE4 cell-based assay in 1536-well format was described by Titus et al. [27] These workers used a HEK293 cell line that coexpressed the thyroid-stimulating hormone receptor along with a similar cyclic nucleotide gated ion channel, as reported above. According to the authors of this work, the potential advantages to this methodology, compared to the reporter gene approaches outlined earlier, include a reduced rate of false positives and less of a shift in compound activities associated with possible effects on signal transduction associated with the expression of PDE activity expressed by reporter gene function. Validation assays produced a Z' factor of 0.50, indicating acceptable performance in this format. Over 2100 compounds were screened using this approach, and known PDE4 inhibitors were identified, with IC_{50} values within 10-fold of published results. Compounds that were known not to inhibit PDE4 showed little or no activity in this screen. These results suggest that as a potential cell-based HTS, this technology could be applied.

HEK293 cells were used by Bora et al. for a cell-based PDE10A assay [28]. PDE10A was stably transfected in the cells, along with a luciferase reporter gene under control of a cAMP response element (CRE). The authors note that HEK293 cells express low levels of endogenous cAMP-hydrolyzing PDEs, leading to low background in the assay. Dipyridamole (**8**) was used as a test inhibitor, and as expected, increased luciferase activity correlated with increasing concentration of the inhibitor. The IC_{50} in the assay was 1.16 μM, comparable to a previously reported value [29].

PDE CHEMOTYPES

PDE2 Inhibitors

There has been comparatively little published on new PDE2 inhibitors in the last few years in peer-reviewed literature. A group at Pfizer disclosed oxindole derivative **20** (Figure 5.8) as a chemical tool to explore potential PDE2 applications [30]. This compound showed an IC_{50} of 40 nM against PDE2, and good selectivity in broad receptor and enzyme profiling, kinase, and PDE assays. This molecule has 41% oral bioavailability in rats, and achieves peripheral unbound concentration above the IC_{50} value for >2 h.

PDE4 Inhibitors

As noted above, achieving selectivity between PDE4B and PDE4D is a recognized challenge for medicinal chemists. A high-throughput screen identified a pyrimidine lead, **21** (Figure 5.8), that showed ~10-fold selectivity for PDE4B versus PDE4D and an IC_{50} of 190 nM versus PDE4B [31]. The most fruitful regions for modification of

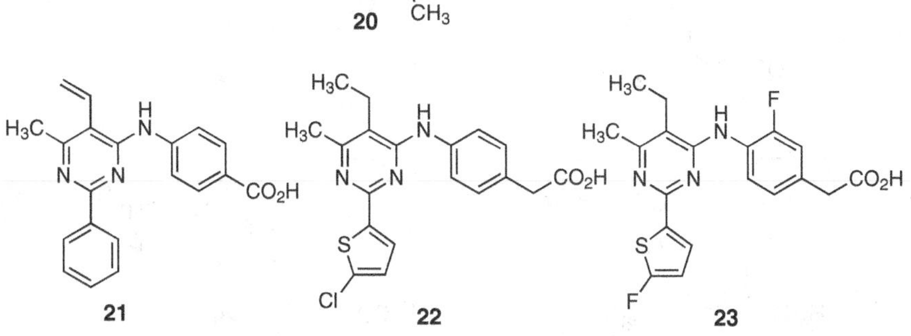

FIGURE 5.8   PDE2- and PDE4B-selective inhibitors.

potency and selectivity involved the phenyl ring attached to the pyrimidine core and the aminobenzoic acid ring. Structure activity studies resulted in compounds **22** and **23**, which showed PDE4B IC_{50} values <15 nM with 100- and 200-fold selectivity, respectively, against PDE4D. Compound **22** was screened *in vivo* for inhibition of TNFα production and it showed an ID_{50} of 14 mg/kg orally without causing emesis in ferrets up to 100 mg/kg. By comparison, cilomilast gave an ID_{50} of 2.2 mg/kg in the same assay and caused two of three ferrets to vomit at an ID_{50} of 10 mg/kg.

Compounds based on rolipram were investigated in an attempt to identify PDE4D-selective inhibitors, and within this group, to study whether it was possible to find analogs that inhibited so-called long, short, and super short isoforms of PDE4D [32]. Selective antagonists might be useful probes to help understand the physiological role(s), subcellular distribution, and regulation of PDE4 isoforms in the CNS [33]. Using a common oxime-derived scaffold, a variety of structural modifications on the substituent attached to the oxime oxygen were investigated (Figure 5.9). The most potent inhibitors contained an amino hydroxy template, and in this group of compounds, the most potent derivative was **24** with an IC_{50} of 1.46 μM. The authors claimed that this analog displayed some selectivity (approximately threefold based on IC_{50} values) for PDE4D3 (long form) compared to PDE4D2 (supershort form). Compound **24** was inactive against other PDE4 isozymes, as well as against PDE1– PDE3, PDE5, and PDE6. In general, inhibitors reported in this work that incorporated a morpholine subunit were preferred compared to other more basic amine derivatives. In some cases, addition of methyl groups next to oxygen on the morpholine ring provided a modest improvement in potency.

As an alternative to active site-directed inhibitors, Burgin et al. reported the discovery and *in vivo* activity of allosteric modulators of PDE4D [34]. Based on observations that some PDE4 inhibitors appeared to show greater activity in cell

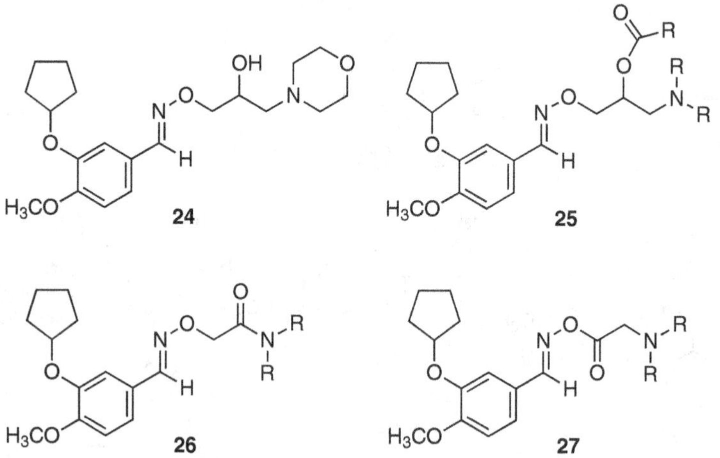

FIGURE 5.9   PDE4D-selective compounds.

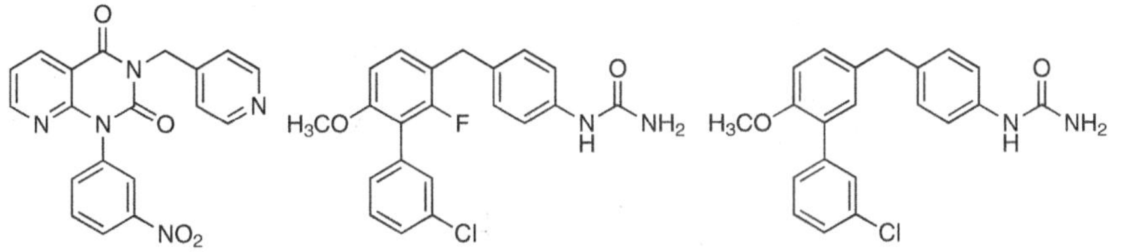

FIGURE 5.10   Allosteric PDE4D modulators.

extracts containing long forms (see above) of PDE4D, the authors hypothesized that this increased activity was associated with interactions with the UCR2 domain. This portion of the enzyme can interact with the substrate binding region and is not present in short forms of PDE4D. X-ray crystal structure determinations with several compounds, for example, R-rolipram, confirmed the existence of a binding site in the UCR2 domain. Using RS-25344, **28** (Figure 5.10) and NVP (**19**) as ligands, crystal structure determinations with PDE4D provided the basis for design of analogs with improved affinity. Key design features included a region with a hydrogen bond acceptor moiety connected by a linker to an aromatic ligand that could interact with a key phenylalanine residue, and a second aromatic group that would provide a second point of interaction with this phenylalanine. These design features imparted good selectivity (60–100-fold) versus PDE4B, and >1000-fold versus other PDEs. In enzymatic assays, these allosteric modulators proved to be partial inhibitors of PDE4D (I_{max} 80–90%). Two compounds, **29** and **30** were evaluated in animal models of cognition such as the scopolamine-impaired Y-maze test and novel object recognition. In the former, the minimal effective dose (MED) of both compounds was $10 \mu g/kg$, and in the novel object assay, the MED was reported to be <10 mg/kg. Both compounds showed acceptable oral bioavailability in mice (28 and 32%, respectively). The dose required to induce emesis for each compound was >1000-fold above the MED for efficacy in animal models of cognition.

Cashman and Ghirmai recently disclosed an interesting dual-target approach that combines PDE4 inhibition with monoamine reuptake inhibition [35]. The hypothesis behind this approach is based on the observation that elevation of serotonin levels in the synapse leads to upregulation of PDE4 activity associated with increased cAMP signaling in neurons. Inhibition of PDE activity could blunt this effect, preserving the activity of the monoamine reuptake antagonist. Furthermore, PDE4 inhibitors such as rolipram are well known to exert antidepressant activity in animal models [1]. The pharmacophore of duloxetine was linked to a phthalazine-based PDE4 inhibitor core to provide compound **31** (Figure 5.11) whose (*R*)- and (*S*)-isomers showed PDE4D2 IC_{50} values of 23 nM and 45 nM, respectively. In monoamine reuptake inhibition assays, these isomers had IC_{50} >1 mM at dopamine and norepinephrine reuptake sites, and ~400 nM against serotonin. Although this level of activity is not as potent as duloxetine, the approach is an interesting one that could be investigated in greater detail.

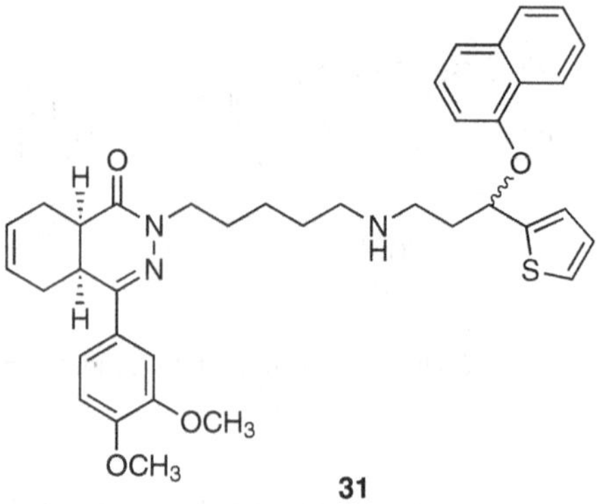

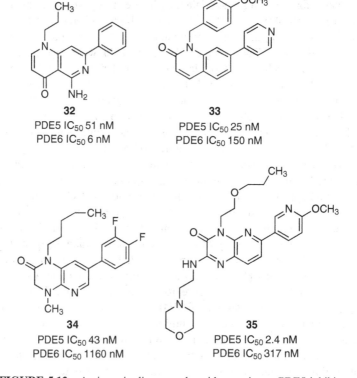

FIGURE 5.11   Dual PDE4-monoamine reuptake inhibitor.

PDE5 Inhibitors

In recent years, there has been increasing interest in identifying additional chemotypes as PDE5 inhibitors beyond those known in the literature[8], possibly with a view toward CNS applications for this class of compounds. Two examples of this work were published recently by researchers at Pfizer. An aminoquinolin-4-one, **32** (Figure 5.12),

32
PDE5 IC_{50} 51 nM
PDE6 IC_{50} 6 nM

33
PDE5 IC_{50} 25 nM
PDE6 IC_{50} 150 nM

34
PDE5 IC_{50} 43 nM
PDE6 IC_{50} 1160 nM

35
PDE5 IC_{50} 2.4 nM
PDE6 IC_{50} 317 nM

FIGURE 5.12   Aminoquinolinone and pyridopyrazinone PDE5 inhibitors.

was identified by high-throughput screening with acceptable PDE5 potency, but no selectivity versus closely related PDE6 [36]. Structural modification of this template was hindered by the comparatively lengthy synthesis of the core and a replacement was sought. It was discovered that 2-quinolinone **33** provided comparable PDE5 potency and moderately improved PDE6 selectivity. Interestingly, the related dihydropyrazinone **34** improved PDE6 selectivity further and was used as the starting point for continued optimization. A preferred compound from this series, **35**, was identified; it showed good PDE5 potency ($IC_{50} = 2.4$ nM) and good selectivity against PDE6 (~150X) and PDE11 (~700X). This particular compound showed high clearance in rats with a 3 h half-life, and according to the authors, continued optimization of pharmacokinetic parameters was desired.

The second paper of this series described optimization of **35** that focused on potency, selectivity, and pharmacokinetic properties [37]. The propoxy- and methoxypyridine substituents on **35** were important for potency, but changes in these regions were not studied in this paper. Effort was focused on evaluation of pyridine regioisomers and the morpholinoethyl substituent. Crystal structure studies indicated that the morpholine ring projected toward solvent, suggesting that modifications here could be used to adjust pharmacokinetic properties without a substantial effect on PDE5 potency. Moving the nitrogen atom of the fused pyridine ring in **35** to so-called southern (**36**) or southeastern (**37**) positions had a significant effect on PDE potency (Figure 5.13). Both **36** and **37** were more potent than **35**, and it was consistently found that southeastern nitrogen regiochemistry was preferred. Although these changes also improved PDE6 potency to a comparable degree, selectivity >200-fold was always observed in the southeastern series. Replacement of the moderately basic morpholine ring with other basic groups or alkyl amide moieties did not improve the pharmacokinetic properties. However, substitution of nonbasic terminal hydroxyl-containing functional groups increased the volume of distribution and reduced the clearance in this series. One example, compound **38** was studied *in vivo* in a spontaneously hypertensive rat pharmacodynamic model. At an oral dose of 3 mg/kg, substantial and prolonged reduction in blood pressure was measured. In dogs, **38** showed 75% oral bioavailability and a half-life of 17.6 h with clearance rate of 3.5 ml/(min kg).

PDE9 Inhibitors

The discovery of orally bioavailable and brain-penetrant PDE9 inhibitors was reported by Verhoest et al. [38]. This series of compounds may have been based, in part on a regioisomeric pyrazolopyrimidinone series (e.g., **39**, Figure 5.14) [39]. Key design criteria included molecular weight <430, a clogP value <3, 1 hydrogen bond donor, and total polar surface area (TPSA) between 50 and 100. By careful filtering of available chemical building blocks based on predicted interactions with P-glycoprotein, cytochrome P450 2D6, and the hERG channel, a set of 2400 compounds was considered and evaluated in docking studies based on the crystal structure of PDE9A with a bound inhibitor. The group attempted to take advantage of interactions with Tyr424, a residue unique to PDE9 in a hydrophobic pocket on the enzyme, and π-stacking with Phe441 to improve potency and selectivity. The hypothetical compound library was evaluated in docking studies and ranked to furnish

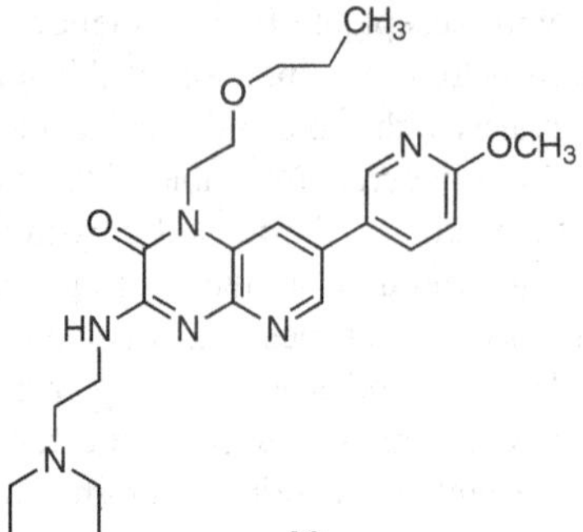

36
PDE5 IC_{50} 0.34 nM

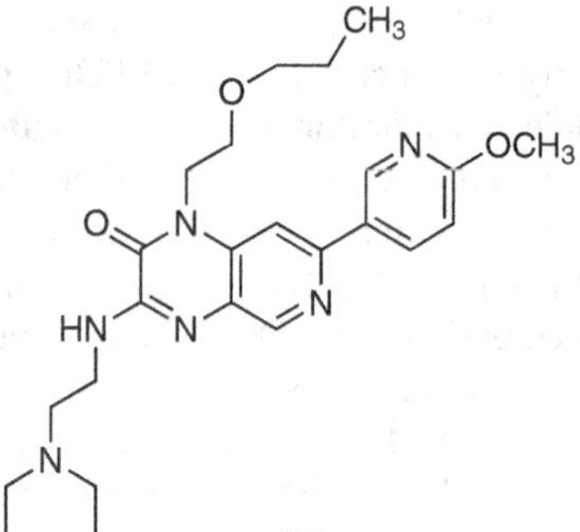

37
PDE5 IC_{50} 0.07 nM

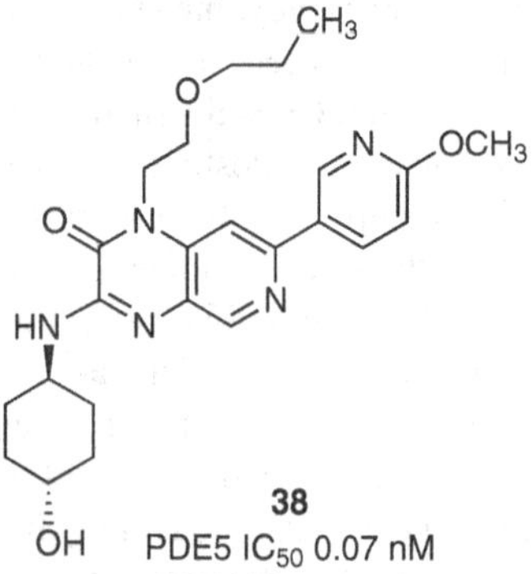

38
PDE5 IC_{50} 0.07 nM
PDE6 IC_{50} 9.8 nM

FIGURE 5.13 Optimized aminopyridazinones.

a subset of 500 analogs for synthesis and evaluation. It was observed that a pyrrolidine ring attached to the pyrazolopyrimidinone core improved potency and substituents on the nitrogen atom could further influence potency and selectivity. Heterocycles attached at this site furnished analogs with good potency (<30 nM) against

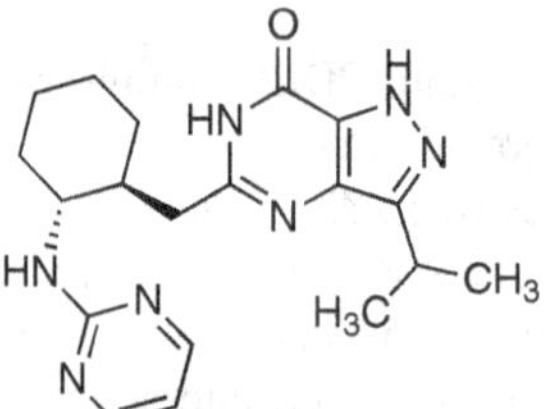

39

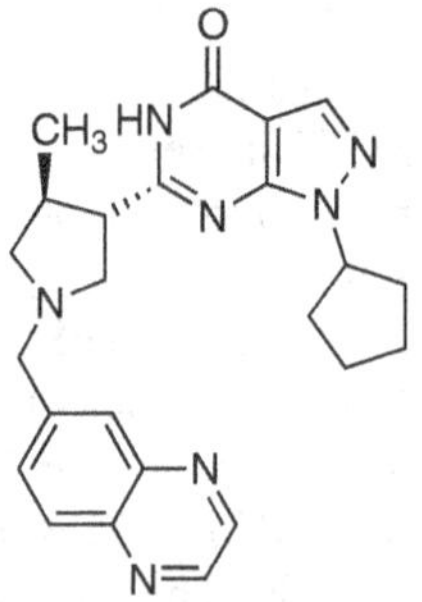

40

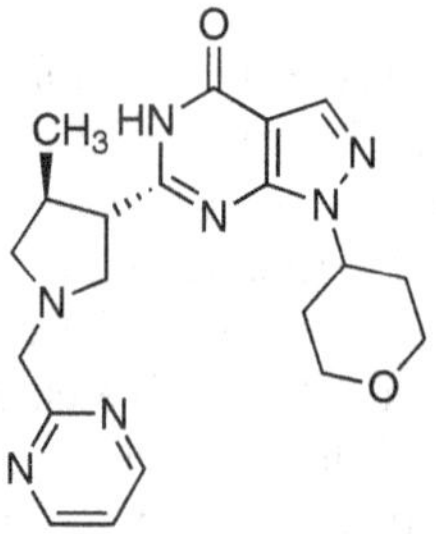

41

FIGURE 5.14 PDE9 inhibitors.

PDE9A and 10–30-fold selectivity against PDE1C. An optimized compound in this group, **40** showed an IC_{50} of 1.8 nM against PDE9A, with >25-fold selectivity versus PDE1C. This compound met the design criteria and when dosed in rats gave a brain-to-plasma ratio of 1.4:1, and elevated cGMP levels in mouse striatum 145 and 215% at doses of 3.2 and 10 mg/kg, respectively.

A related PDE9 inhibitor, PF-04447943, **41** is currently in clinical trials for cognitive disorders. In cognition assays (e.g., mouse spatial recognition, Y-maze, social recognition, and rat novel object recognition), oral administration of 1–3 mg/kg orally resulted in improved performance in all of these screens. The compound also dose dependently elevated cGMP levels in rodent cerebrospinal fluid (CSF) 30 min after administration [40]. PF-04447943 is a derivative of **40** that was optimized for PDE9 potency, selectivity, brain penetration, and pharmaceutical properties. The compound has an IC_{50} value of 8 nM. Like **40**, compound **41** has excellent selectivity versus PDE1C (~150-fold) and good brain penetration in rodents (brain-to-plasma ratio: 0.6–0.9), good oral bioavailability (~75%), and low clearance in dogs and human microsomes [41].

PDE10 Inhibitors

Papaverine, **42** (Figure 5.15), was used as the starting point for optimization of one series of PDE10 inhibitors [42]. This was based on previous observations that papaverine demonstrated efficacy in animal models of schizophrenia [5,43]. Inspection of the crystal structure of a screening hit revealed a key interaction between the methoxy groups of the lead and the glutamine residue within the substrate binding site of PDE10. Using an approach similar to that described by Verhoest et al. for the discovery of PDE9 inhibitors [38], new compound synthesis was guided by chemical and pharmaceutical property calculations. These filters suggested a library of potential PDE10 inhibitors, a subset of which were synthesized and screened for selectivity

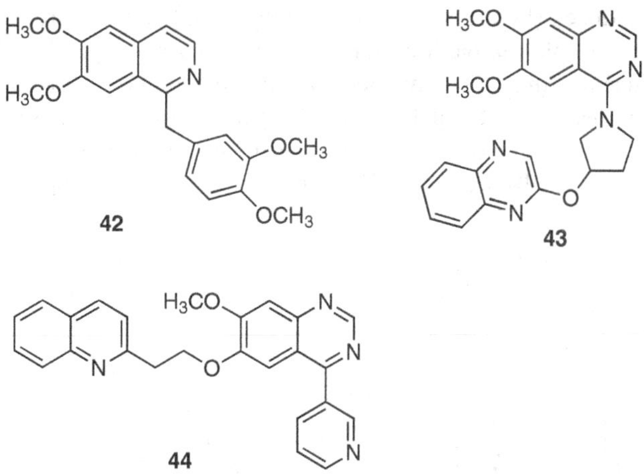

FIGURE 5.15 Papaverine-based PDE10A inhibitors.

versus PDE3A and PDE3B. One compound from this series, **43**, showed an IC_{50} of 4 nM against PDE10A, with >50-fold selectivity against PDE3A and PDE3B. When compound **43** was administered at a dose of 3.2 mg/kg subcutaneously, increased levels of striatal cGMP were measured for at least 90 min after dosing. In addition, increased levels of phosphorylated cAMP response element binding (CREB) were detected for a similar period of time. This line of investigation was extended using structure-based design principles in combination with *in silico* property filters [43,44]. Examination of the PDE10A crystal structure revealed a unique lipophilic pocket near the substrate binding site that if occupied could lead to a highly selective lead series. Using the basic template in **43**, the substituted pyrrolidine was replaced with a pyridine ring and the lipophilic pocket was occupied by extension from one of the phenolic hydroxyl groups to provide **44**. This compound gives an IC_{50} of 5 nM versus PDE10A, shows low clearance in human liver microsomes and elevated cGMP levels in mouse brain following subcutaneous administration, and at a dose of 3.2 mg/kg intraperitoneally demonstrated activity in conditioned avoidance responding in rats.

A subsequent publication from Verhoest et al. described the discovery of PF-2545920, a PDE10A inhibitor that has been in a phase 2 trial for the treatment of schizophrenia and, unfortunately, provided a negative readout [45]. Using structural information derived from an HTS hit (**45**) (Figure 5.16), optimization was carried out to maximize potency, selectivity, and pharmaceutical properties. Lipophilicity was a key property that influenced intrinsic clearance in human liver microsomes, thereby achieving therapeutically relevant concentrations in the brain. It was found that compounds with a clogP <4 were optimal, and it was necessary to balance polarity with brain penetration. This work ultimately led to the selection of PF-2545920 (**46**) as a clinical candidate for treatment of schizophrenia. This compound shows an IC_{50} of 0.37 nM, with >1000-fold PDE selectivity and acceptable pharmacokinetic properties.

Asproni et al. [46] recently reported the initial optimization of phenyl imidazole-based PDE10A inhibitors based on a triazoloquinazolone template (**47**) discovered by workers at Lundbeck (Figure 5.17). This starting point had good oral bioavailability in mice, a brain-to-plasma partition ratio <0.04:1, and water solubility <20 ng/ml. The benzimidazole ring was converted to a phenyl imidazole to provide a more hydrophilic template than **47**, and one of the nitrogen atoms in the triazole ring was removed because it was not involved in any direct interactions with the enzyme. Compounds such as **48** were identified that demonstrated comparable PDE10A

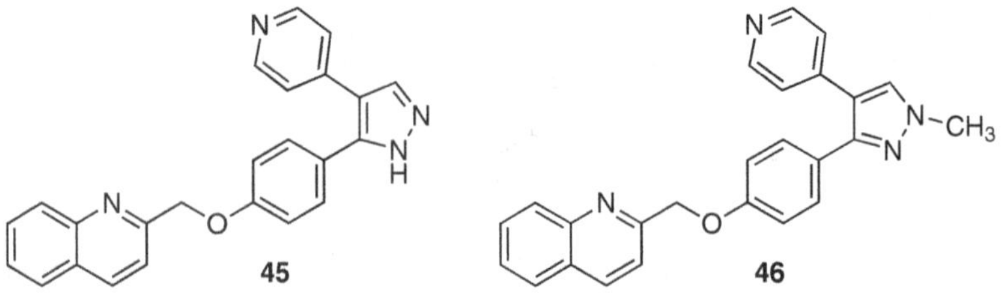

FIGURE 5.16 Discovery of PF-2545920.

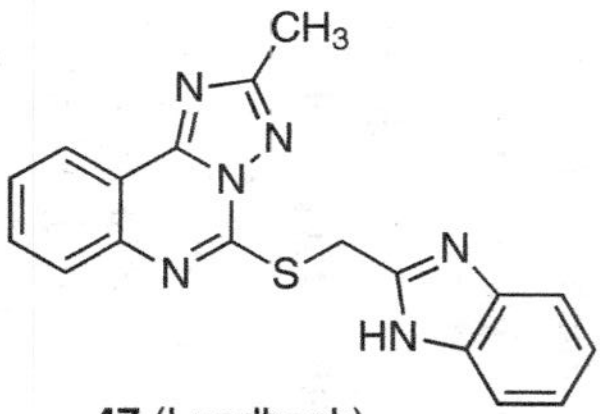

FIGURE 5.17 PDE10A inhibitors.

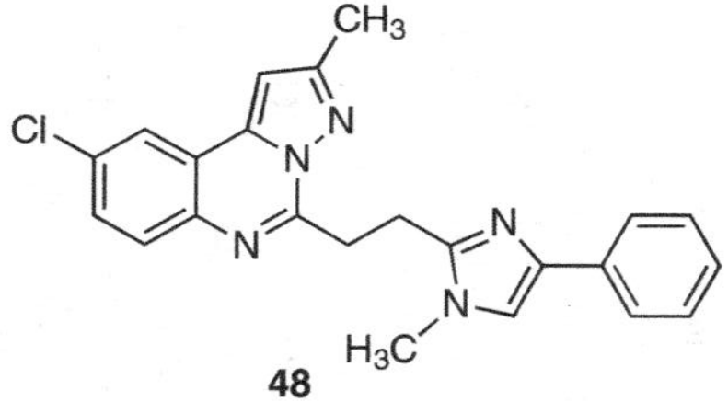

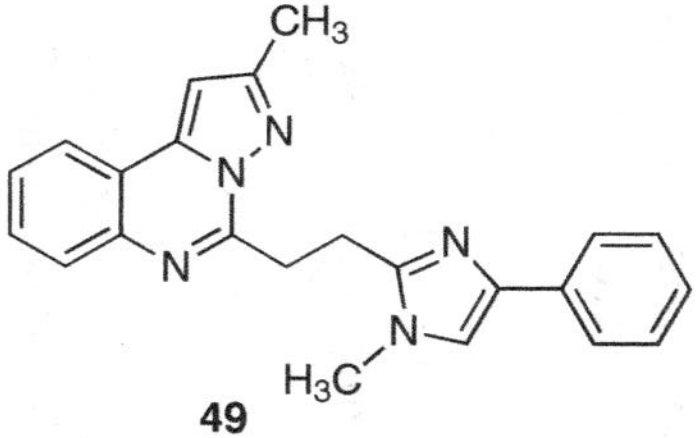

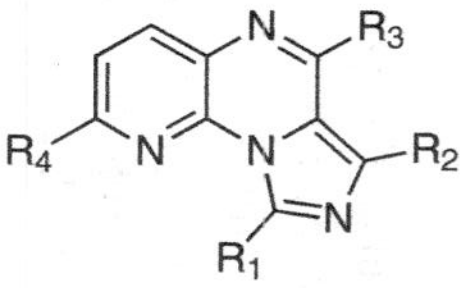

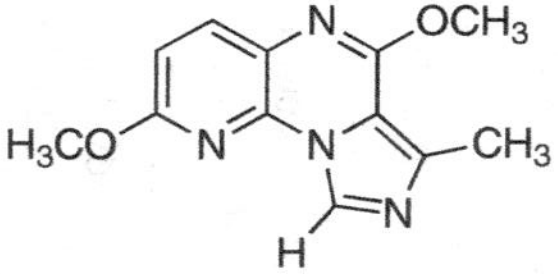

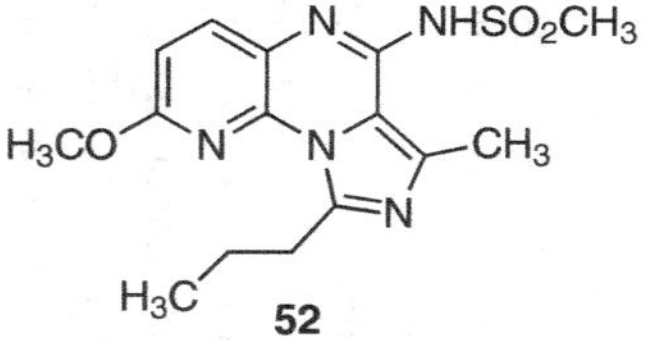

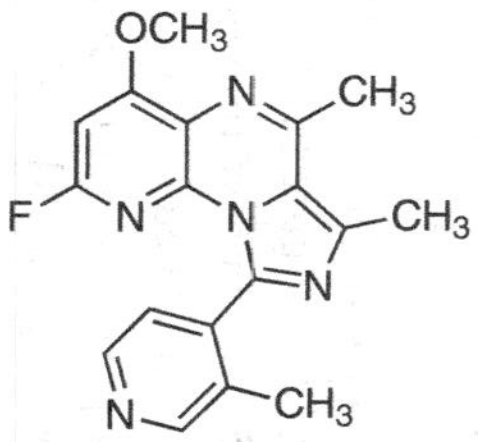

inhibition ($IC_{50} = 16$ nM) to the original Lundbeck lead. The potentially metabolically labile sulfur atom in the linker between the phenyl imidazole and pyrazoloquinazoline could be replaced by a methylene group without losing potency. Halogen or methoxy substituents at the 7- or 8-positions on the quinazoline ring reduced potency 3–10-fold, while in general, substitution at the 9-position proved preferable to the 7- or 8-position. At the 9-position, methyl and chloro (e.g., **48**) were optimal among the limited range of analogs studied. A compound in this series, **49**, showed an improved brain-to-plasma partition ratio of 0.65:1, along with a modest improvement in water solubility (1.8 μg/ml).

Another novel chemotype with potent PDE10A activity was reported by Höfgen et al. (Figure 5.17) [47]. Random screening identified an imidazopyrido pyrazine **50** as an attractive starting point based on structural novelty, activity, and PDE selectivity. This report focused SAR development at the R_1 and R_3 positions. It was found that alkyl and branched alkyl substituents at R_1 improved PDE10A potency significantly ($\sim$100-fold) compared to **51**. Urea, carbamate, amide, and sulfonamide substituents at R_3 had variable effects on potency, while small alkyl groups generally improved activity relative to **51**. Selected compounds were evaluated in an MK-801-induced hyperactivity model in mice and as an example, **52** ($IC_{50} = 7$ nM, 1000-fold selectivity for most PDEs; 77-fold versus PDE5, 4-fold versus PDE11) demonstrated good activity at a dose (intraperitoneal) of 5 mg/kg in this screen. This report was followed by another that employed structure-based design to focus on improved metabolic stability, potency, and selectivity [48]. The alkyl-substituted phenyl moieties in **52** underwent significant metabolism. Replacement of the *n*-propyl chain with a substituted pyridyl ring furnished improved potency because of hydrogen bonding interactions with a bound water molecule. The methane sulfonamide was replaced by a methyl group, and metabolic stability-guided adjustment of phenyl substituents furnished **53**, with a PDE10A IC_{50} of 0.7 nM and excellent selectivity against all other PDEs ($>$100-fold). This optimized compound had good oral bioavailability in dogs (73%) and elevated cyclic nucleotide levels in a dose-dependent manner in mouse striatum. Activity in MK-801 induced hyperactivity, and conditioned avoidance responding following oral administration was demonstrated. The MED was 0.075 mg/kg in hyperactivity and 1 mg/kg in conditioned avoidance responding.

A series of pyrazoloquinolines identified in earlier work was optimized to provide a tool compound **55** (Figure 5.18) [49]. The objective was to optimize a potent and selective lead, **54**, to improve oral bioavailability and, if possible, potency. Attention was initially directed at heteroaromatic replacements for the oxazepane ring. SAR was variable, with 4-pyridyl derivatives providing superior potency compared to substituted phenyl compounds. 2-Pyridine and 3-pyridine regioisomers were significantly less potent than the 4-pyridyl version, and various heterocycles showed a wide range of activity. All of the most potent derivatives ($IC_{50} < 25$ nM) proved to be CYP3A4 inhibitors as well. The best solution to this issue proved to be replacing the pyridine ring with a substituted pyrimidine (**55**) that had low hepatocyte clearance and CYP inhibition ($IC_{50} > 20$ μM) combined with good PDE10A inhibition ($IC_{50} = 3$ nM) and PDE selectivity ($>$500-fold versus all other PDEs). This racemic PDE10A inhibitor

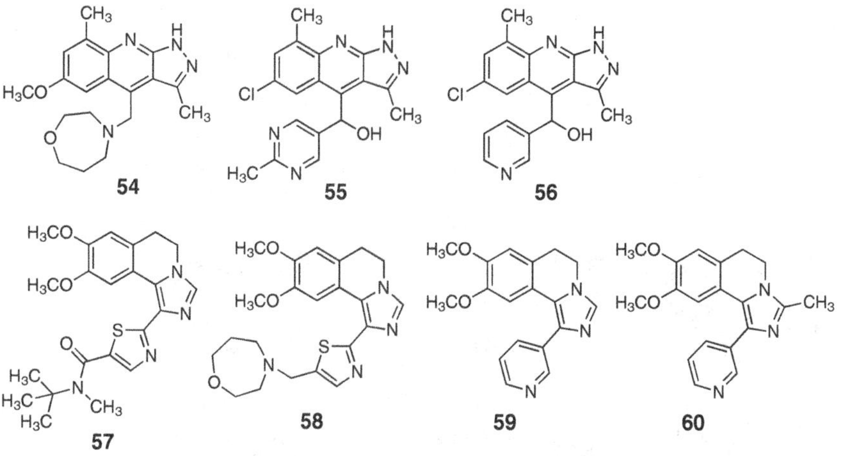

FIGURE 5.18   Pyrazoloquinoline PDE10A inhibitors.

was active in an MK-801 hyperactivity screen with an MED of 10 mg/kg orally. However, **55** showed low brain penetration, prompting the team to evaluate a slightly less potent 3-pyridyl analog **56** that was active in the MK-801 hyperactivity assay and showed a brain-to-plasma ratio of 1.5:1.

Investigators at Merck recently published the discovery and early optimization of an additional PDE10A chemotype [50]. A moderately potent and selective dihydroimidazoisoquinoline scaffold **57** (Figure 5.18) was explored by substitution on the thiazole ring. Amide analogs furnished modest improvement in PDE potency and selectivity, but suffered from low systemic exposure. Removal of the amide and substitution of a heteroalicyclic moiety (e.g., **58**) improved potency and selectivity significantly; however, this series also suffered from low exposure. Replacement of the thiazole with other aromatic rings furnished inhibitors with a wide range of activity. The most potent were 3-substituted phenyl analogs and a 3-pyridyl derivative **59**. Addition of a methyl group to the imidazole ring of the scaffold (to reduce metabolic hydroxylation) provided **60**. This PDE10A inhibitor has a K_i value of 25 nM and shows at least 200-fold selectivity against all PDEs except PDE3, good aqueous solubility, low hERG affinity, and low hepatocyte clearance *in vitro*. In spite of improved plasma exposure, little of the compound was found in rat brain 6 h after oral administration.

A pyridyl cinnoline series was studied by workers at Amgen who discovered that 2-substitution on the pyridine ring in **61** (Figure 5.19) had dramatic effects on PDE10A inhibition [51]. For example, when R is H, PDE10A activity is $>10 \mu M$; when R is methyl, the IC_{50} is 181 nM; and when R is isopropylamino, the IC_{50} is 7.6 nM. Variations in the amine substituent on the pyridyl ring showed that cyclic amines such as piperidine provided compounds with good potency ($IC_{50} \sim 10$–20 nM) and superior PK characteristics compared to acyclic secondary amines, which were generally more potent (<10 nM) but more metabolically labile. Addition of other groups to the pyridine ring (e.g., **62**, $IC_{50} = 2$ nM) had a small effect on potency

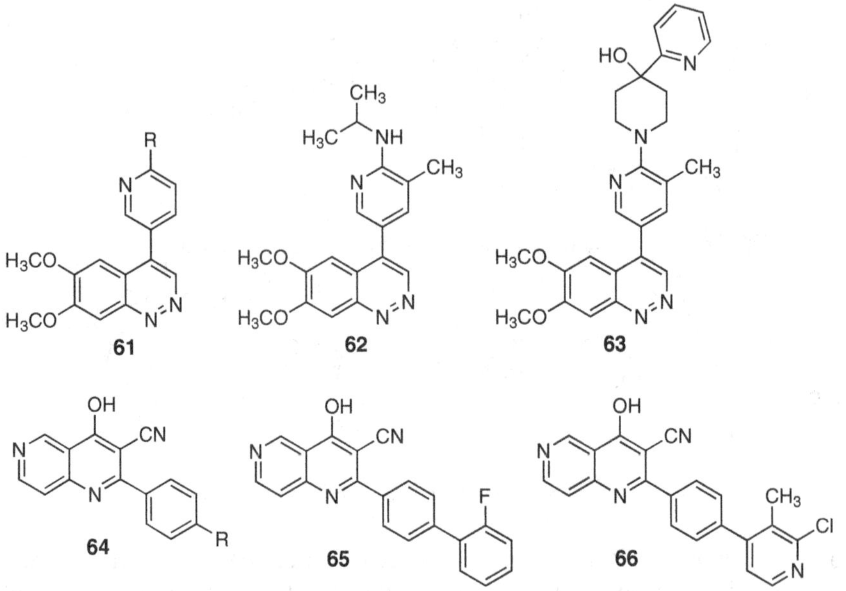

FIGURE 5.19 Recent PDE10A inhibitors.

using an isopropyl amino group at the 2-position. However, when this change was combined with a disubstituted piperidine ring, as in **63**, the resulting PDE10A inhibitors retained good potency with reduced clearance. Compound **63** ($IC_{50} =$ 2.6 nM) was evaluated in conditioned avoidance responding and after oral administration, the MED was 5.6 mg/kg.

Bauer et al. engaged in a scaffold hopping exercise to identify a novel naphthyridine PDE10A inhibitor scaffold **64** (Figure 5.19) [52]. In this template, the nitrile and hydroxyl groups were essential for potency. Fluoro analog **65** with a PDE10A IC_{50} of 22 nM is very selective versus other PDEs, but is structurally related to an inhibitor of dihydroorotate dehydrogenase (DHODH). Because inhibition of this enzyme could result in undesired toxicity in humans, it was essential to reduce DHODH potency. This proved to be possible using substituted pyridine derivatives that also improved PDE10A potency, leading to, for example, **66** with a PDE10A IC_{50} of 1 nM and DHODH IC_{50} of 1300 nM.

The substantial interest in PDE10A as a drug target for schizophrenia also stimulated the development of molecular tools that could be used to provide functional information about the role of the enzyme in real time *in vivo*. As a result, a number of groups have reported the development and use of PET ligands to study PDE10A. An ^{11}C derivative of papaverine, **67** (Figure 5.20), was reported by Tu et al. who also reported rapid (~5 min) biodistribution and accumulation in the striatum of rats with rapid clearance [53]. Two derivatives of the Pfizer clinical candidate PF2545920, an ^{11}C analog **68** [54] and an ^{18}F candidate **69** (^{18}F-JNJ41551047) (Figure 5.20) [55] were also reported. Although **68** rapidly partitioned into rat striatum, where the enzyme is

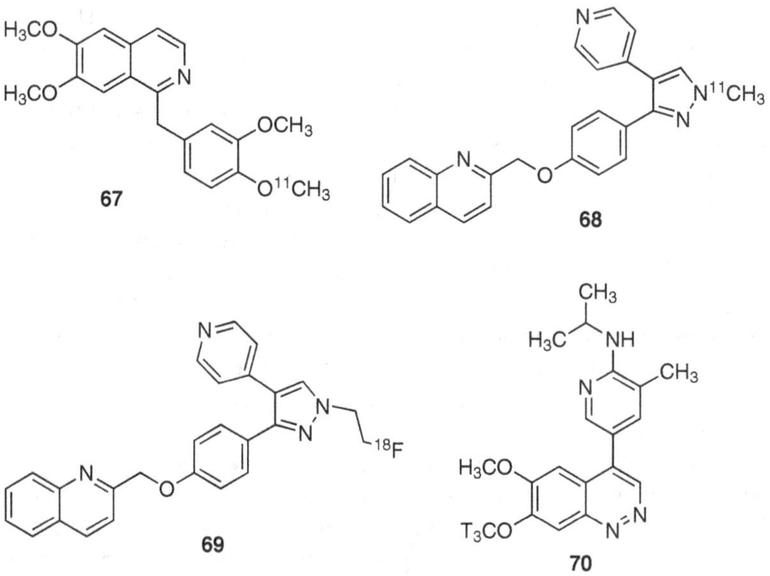

FIGURE 5.20   PDE10A PET ligands.

localized, versus the cerebellum (ratio 6.5:1) within 30 min, an unidentified brain-penetrant, radiolabeled metabolite was formed and detected, complicating analysis. The same general observations were made in rhesus monkeys [54]. Fluoroethyl derivative **69** partitioned rapidly into the brain and striatal tissue without the formation of a similar metabolite. Imaging in PDE10A knockout mice demonstrated that uptake in the striatum was dependent on the presence of the enzyme [47]. Recently, investigators at Amgen synthesized a tritiated derivative, **53a**, and demonstrated its use as a potential PET ligand to monitor PDE10A occupancy [56].

POTENTIAL CNS APPLICATIONS FOR PDE INHIBITORS

The activity of PDE4 inhibitors in preclinical models of cognition is well known and the lead references cited above provide readers with a good survey of the recent literature [1,8,10,31,32,34,35]. A detailed review by Reneerkens et al. summarizes a substantial amount of preclinical work aimed at understanding the potential of PDE inhibitors for cognitive enhancement [57]. Recent papers highlighted the potential for other PDEs to play a role in this process, and in some cases have additional activities that may be beneficial for the treatment of Alzheimer's disease. PDE5 inhibition by sildenafil at a dose of 50 mg/kg has been shown by Puzzo et al. to improve memory and synaptic function and reduce Aβ levels in mice [58]. Memory was measured using a random access water maze and fear-conditioned learning. In a small study in patients with schizophrenia, patients were given 50 or 100 mg doses of sildenafil with a minimum of 48 h between doses. Disease symptoms were measured 1 h after drug administration and memory effects were measured 48 h after dosing. Unfortunately, neither dose group was efficacious for cognitive or psychiatric effects. The authors note that repeated doses may be needed for these effects to be apparent [59].

The PDE4 inhibitor cilomilast was tested in neonatal rats for the ability to improve performance in an odor-preference learning model. Activity in this test is dependent on serotonin levels. At a dose of 3 mg/kg, cilomilast **71** (Figure 5.21) demonstrated positive effects at 24 and 48 h after dosing. These effects correlated with increased levels of cAMP in the striatum. Interestingly, at the 48 h time point, cilomilast was efficacious only in animals that were not treated with isoproterenol, a β-agonist that influences cAMP levels in the olfactory bulb. A PDE9 inhibitor, BAY 73-6691 (**72**), showed positive effects on learning and memory in Wistar rats, following the observation that in hippocampal slices, elevated levels of cGMP were measured and enhanced long-term potentiation (LTP) could be observed [60]. The role of PDE inhibition in a broader sense is discussed in Chapter 7.

Inhibition of PDE2 and PDE5 was shown to attenuate memory deficits caused by serotonin depletion induced by dietary restriction of tryptophan in Wistar rats [61]. The PDE5 inhibitor vardenafil at doses of 1, 3, or 10 mg/kg or the PDE2 inhibitor BAY 60-7550 (**73**) at 0.3, 1, or 3 mg/kg were administered 30 min prior to the start of testing. Vardenafil was efficacious at 3 and 10 mg/kg, whereas the PDE2 inhibitor was active only at 3 mg/kg. Another group has shown that **73** increased basal constitutive nitric oxide synthase activity in the hippocampus and striatum of 3-, 12-, and 24-month-old rats, and this increase in activity correlated with performance object recognition assays [4]. PDE2 inhibitors such as **73** have also exhibited activity in anxiolytic assays [62]. Very recently, **73** showed activity in an extradimensional–intradimensional test of cognitive flexibility in rats. This protocol represents an approach to assessment of executive cognitive function that is most commonly applied to nonhuman primates [63]. Chapter 9 expands on this work.

Previous work demonstrated that sildenafil was efficacious in animal models of stroke [64]. Vardenafil and tadalafil have been examined in rat and mouse middle cerebral occlusion models [65,66]. Vardenafil at a dose of 10 mg/kg twice daily for 14 days did not increase survival or improve behavioral outcome, relative to vehicle-treated animals. In contrast, tadalafil, at doses of 2 or 10 mg/kg administered every 48 h for 6 days 24 h postinsult improved neurological recovery compared to vehicle-treated rats. Animals treated with the PDE5 inhibitor also showed increased cerebral

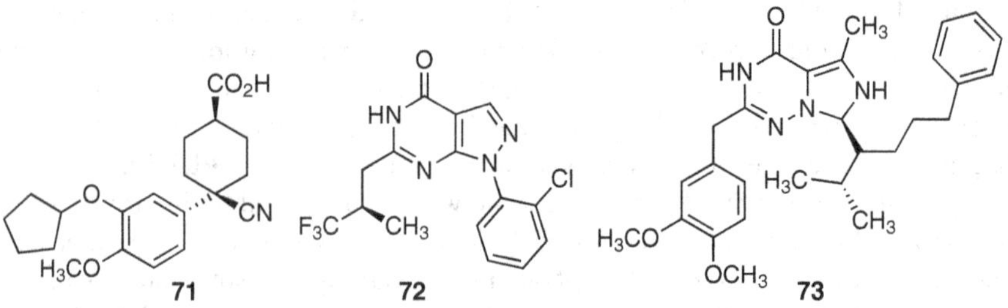

FIGURE 5.21   Cilomilast, BAY 73-6691, and BAY 60-7550.

density and greater ipsilateral SVZ cell proliferation. There was no difference in infarct volume in drug-treated animals. This is also discussed in Chapter 4.

As noted above, PDE10A inhibitors have attracted considerable attention for the potential treatment of schizophrenia and, more recently, Huntington's disease. The enzyme is localized in cell membranes the striatum and plays a role in second messenger signaling via cAMP and cGMP in this key brain region. The striatum is a major regulator of glutaminergic and dopaminergic signaling that influence cognition and behavior. Thus, by influencing second messenger levels, inhibition of PDE10A represents a novel approach for treatment of schizophrenia. Pfizer's PF2545490, **46**, is still in clinical evaluation in patients with the disease, despite a failure in phase 2 in a monotherapy trial. In preclinical models, PDE10A inhibitors demonstrate activity in conditioned avoidance responding [42,43] and MK-801-induced hyperactivity and stereotypy [47], two widely used screens that are considered predictive for antipsychotic efficacy.

Two publications that studied alterations in gene regulation following PDE10A inhibition provide further preclinical support for PDE10A as a target in schizophrenia. Acute inhibition of PDE10A in rats by PF2545490 reduced mRNA for c-fos expression and increased mRNA for neurotensin at 1 h, and measurable effects were noted at 6 h [67]. mRNA levels for enkephalin and substance P were also elevated 3 h after drug administration. Qualitatively similar effects were observed following administration of haloperidol, implying that PDE10A inhibition shares some overlap in terms of signaling pathway effects with an established antipsychotic agent. Chronic effects on gene expression were reported by Kleiman et al. using TP-10 (**45**), a PDE10A inhibitor related to PF2545490 [68]. Upregulation of mRNA encoding glutamate decarboxylase 1, diacylglycerol O-acyl transferase, PDE1C, synaptotagmin 10, choline acetyltransferase, and Kv1.6 was observed, suggesting a role in neurotransmitter synthesis. Downregulation of histone deacetylase 4, follistatin, and claspin mRNA indicates PDE10A inhibition is associated with genes involved in neuroprotection. The neuroprotective potential was supported by *in vivo* studies where mitogen- and stress-activated kinases were activated following drug administration. These effects on gene regulation were absent in PDE10A knockout mice.

In an effort to attempt to develop a clinically useful translational biomarker for PDE10A inhibition in schizophrenia, Dedeurwaerdere et al. investigated the effects of **45** and **46** on induction of region-specific hypermetabolism in the globus pallidus and lateral habenula of C57BL/6 mice [69]. Both PDE10A inhibitors induced a dose-dependent increase in glucose metabolism in the globus pallidus and a bell-shaped curve in the lateral habenula. The dopamine D_2 antagonist haloperidol elevated glucose metabolism in only the lateral habenula. The effects of PDE10 inhibitors were absent in PDE10A knockout mice.

SUMMARY AND OUTLOOK

From a chemical and biological standpoint, the field of PDE inhibition has made significant strides since sildenafil was approved for the treatment of erectile dysfunction in the late 1990s. A wide variety of chemical tools are now available to help understand the role these enzymes play in cell signaling processes and potentially in CNS diseases. Selective, potent inhibitors for most of the known PDE families have been published. These compounds, in combination with other powerful biological techniques, allow scientists to dissect intracellular pathways and study their influence in disease models. Recent work suggests that previously challenging issues, such as isoform selectivity in a family of PDEs, can be addressed using structure-based design and modern synthetic chemistry. The availability of many different crystal structures could permit virtual screening for PDE inhibitors, potentially expanding the range of chemotypes known to have PDE activity. These features combined with increased use of other predictive *in silico* models of pharmacokinetic and pharmaceutical properties increases the probability that the molecules that emerge from lead optimization efforts will have fewer known weaknesses as these compounds move into clinical development.

As additional PDE inhibitors advance into and through clinical development for CNS disorders, pharmaceutical researchers will be able to better answer questions about the role these enzymes play in diseases such as stroke, schizophrenia, Alzheimer's disease, and anxiety. Even if some compounds fail, the lessons learned in the exercise could provide useful information for next-generation compounds, and those compounds that succeed will be approved to treat diseases that are not well addressed with existing agents.

REFERENCES

1. Brandon, N.J. and Rotella, D.P. (2007) *Potential CNS applications for phosphodiesterase enzyme inhibitors*, Vol. 42, Academic Press, pp. 3–12.

2. Wong, M.-L., Whelan, F., Deloukas, P., Whittaker, P., Delgado, M., Cantor, R.M., McCann, S.M., and Licino, J. (2006) Phosphodiesterase genes are associated with susceptibility to major depression and antidepressant treatment response. *Proc. Natl. Acad. Sci. USA*, **103**(41):15124–15129.

3. Rutten, K., Prickaerts, J., Schaenzle, G., Rosenbrock, H., and Brockland, A. (2008) Subchronic rolipram treatment leads to a persistent improvement in long term object memory in rats. *Neurobiol. Learn. Mem.*, **90**, 569–575.

4. Domek-Lopacinska, K. and Strosznajder, J.B. (2008) The effect of selective inhibition of cyclic GMP hydrolyzing phosphodiesterases 2 and 5 on learning and memory processes and nitric oxide synthase activity in brain during aging. *Brain Res.*, **1216**:68–77.

5. Siuciak, J.A., Chapin, D.S., Harms, J.F., Lebel, L.A., McCarthy, S.A., Chambers, L., Shrikhande, A., Wong, S., Menniti, F.S., and Schmidt, C.J. (2006) Inhibition of the striatum-enriched phosphodiesterase PDE10A: a novel approach to the treatment of psychosis. *Neuropharmacology*, **51**:386–396.

6. Zhang, L., Zhang, R.L., Wang, Y., Zhang, C., Zhang, Z.G., Meng, H., and Chopp, M. (2005) Functional recovery in aged and young rats after embolic stroke: treatment with a phosphodiesterase type 5 inhibitor. *Stroke*, **36**:847–852.

7. Supuran, C.T.M.A., Barbaro, G., and Scozzafava, A. (2006) Phosphodiesterase 5 inhibitors-drug design and differentiation based on selectivity, pharmacokinetic and efficacy profiles. *Curr. Pharm. Des.*, **12**(27):3459–3465.

8. Rotella, D.P. (2007) *Phosphodiesterases*, Vol. II, Elsevier Ltd., pp. 920–957.

9. Palmer, M.J., Bell, A.S., Fox, D.N.A., and Brown, D.G. (2007) Design of second generation phosphodiesterase 5 inhibitors. *Curr. Top. Med. Chem.*, **7**:405–419.

10. Menniti, F.S., Faraci, W.S., and Schmidt, C.J. (2006) Phosphodiesterases in the CNS: targets for drug development. *Nat. Rev. Drug Discov.*, **5**(8):660–670.

11. Legraverend, M. and Grierson, D.S. (2006) The purines: potent and versatile small molecule inhibitors and modulators of key biological targets. *Bioorg. Med. Chem.*, **14**(12):3987–4006.

12. Castro, A., Jerez, M.J., Gil, C., Calderon, F., Domenech, T., Nueda, A., and Martinez, A. (2008) CODES, a novel procedure for ligand-based virtual screening: PDE7 inhibitors as an application example. *Eur. J. Med. Chem.*, **43**:1349–1359.

13. Shin, H.J., Kim, H.J., Kwak, J.H., Chun, H.O., Kim, J.H., Park, H., Kim, D.H., and Lee, Y.S. (2002) A prenylated flavonol, sophoflavescenol: a potent and selective inhibitor of cGMP phosphodiesterase 5. *Bioorg. Med. Chem. Lett.*, **12**(17):2313–2316.

14. Duan, H., Zheng, J., Lai, Q., Liu, Z., Tian, G., Wang, Z., Li, J., and Shen, J. (2009) 2-Phenylquinazoline-4(^3H)-one, a class of potent PDE5 inhibitors with high selectivity versus PDE6. *Bioorg. Med. Chem. Lett.*, **19**(10):2777–2779.

15. Ke, H. and Wang, H. (2007) Crystal structures of phosphodiesterases and implications on substrate specificity and inhibitor selectivity. *Curr. Top. Med. Chem.*, **7**:391–403.

16. Xu, R.X., Hassell, A.M., Vanderwall, D., Lambert, M.H., Holmes, W.D., Luther, M.A., Rocque, W.J., Zhao, Y., Ke, H., and Nolte, R.T. (2000) Atomic structure of PDE4: insights into phosphodiesterase mechanism and specificity. *Science*, **288**(5472):1822–1825.

17. Card, G.L., England, B.P., Suzuki, Y., Fong, D., Powell, B., Lee, B., Luu, C., Tabrizizad, M., Gillette, S., Ibrahim, P.N., Artis, D.R., Bollag, G., Milburn, M.V., Kim, S.-H., Schlessinger, J., and Zhang, K.Y.J. (2004) Structural basis for the activity of drugs that inhibit phosphodiesterases. *Structure*, **12**(12):2233–2247.

18. Wang, H., Robinson, H. and Ke, H. (2007) The molecular basis for different recognition of substrates by phosphodiesterase families 4 and 10. *J. Mol. Biol.*, **371**:302–307.

19. Wang, H., Liu, Y., Hou, J., Zheng, M., Robinson, H., and Ke, H. (2007) Structural insight into substrate specificity of phosphodiesterase 10. *Proc. Natl. Acad. Sci. USA*, **104**(14):5782–5787.

20. Chen, G., Wang, H., Robinson, H., Cai, J., Wan, Y., and Ke, H. (2008) An insight into the pharmacophores of phosphodiesterase-5 inhibitors from synthetic and crystal structural studies. *Mol. Pharmacol.*, **75**:1717–1728.

21. Wang, H., Peng, M.-S., Chen, Y., Geng, J., Robinson, J., Houslay, M.D., Cai, J., and Ke, H. (2007) Structures of the four subfamilies of phosphodiesterase-4 provide insight into the selectivity of their inhibitors. *Biochem. J.*, **408**:193–201.

22. Deng, C., Wang, D., Bugaj-Gaweda, B., and De Vivo, M. (2007) Assays for cyclic nucleotide-specific phosphodiesterases (PDEs) in the central nervous system (PDE1, PDE2, PDE4, PDE10). *Curr. Protoc. Neurosci.*, **7**(21):1–21.

23. Nanda, K., Chatterjee, M., Arya, R., Mukherjee, S., Saini, K.S., Dastidar, S., and Ray, A. (2008) Optimization and validation of a reporter gene assay for screening of phosphodiesterase inhibitors in a high throughput system. *Biotechnol. J.*, **3**:1276–1279.

24. Herget, S., Lohse, M.J., and Nikolaev, V.O. (2008) Real-time monitoring of phosphodiesterase inhibition in intact cells. *Cell. Signal.*, **20**:1423–1431.

25. Wunder, F., Gnoth, M.J., Geerts, A., and Barufe, D. (2009) A novel PDE2A reporter cell line: characterization of the cellular activity of PDE inhibitors. *Mol. Pharm.*, **6**(1):326–336.

26. Wunder, F., Tersteegen, A., Rebmann, A., Erb, C., Fahrig, T., and Hendrix, M. (2005) Characterization of the first potent and selective PDE9 inhibitor using a cGMP reporter cell line. *Mol. Pharmacol.*, **68**(6):1775–1781.

27. Titus, S.A., Li, X., Southall, N., Lu, J., Inglese, J., Brasch, M., Austin, C.P., and Zheng, W. (2008) A cell-based PDE4 assay in 1536-well plate format for high throughput screening. *J. Biomol. Screen.*, **13**(7):609–618.

28. Bora, R.S., Gupta, D., Malik, R., Chachra, S., Sharma, P., and Saini, K.S. (2008) Development of a cell-based assay for screening of phosphodiesterase 10A (PDE10A) inhibitors using a stable recombinant HEK-293 cell line expressing high levels of PDE10A. *Biotechnol. Appl. Biochem.*, **49**:129–134.

29. Fujishige, K., Kotera, J., Michibata, H., Yuasa, K., Takebayashi, S., Okumura, K., and Omori, K. (1999) Cloning and characterization of a novel human phosphodiesterase that hydrolyzes both cAMP and cGMP (PDE10A). *J. Biol. Chem.*, **274**(26):18438–18445.

30. Chambers, R.J., Abrams, K., Garceau, N.Y., Kamath, A.V., Manley, C.M., Lilley, S.C., Otte, D.A., Scott, D.O., Sheils, A.L., Tess, D.A., Vellekoop, A.S., Zhang, Y., and Lam, K.T. (2006) A new chemical tool for exploring the physiological function of the PDE2 isozyme. *Bioorg. Med. Chem. Lett.*, **16**:307–310.

31. Naganuma, K., Omura, A., Maekawara, N., Saitoh, M., Ohkawa, N., Kubota, T., Nagumo, H., Kodama, T., Takemura, M., Ohtsuka, Y., Nakamura, J., Tsujita, R., Kawasaki, K., Yokoi, H., and Kawanishi, M. (2009) Discovery of selective PDE4B inhibitors. *Bioorg. Med. Chem. Lett.*, **19**:3174–3176.

32. Bruno, O., Romussi, A., Spallarossa, A., Brullo, C., Schenone, S., Bondavalli, F., Vanthuyne, N., and Roussel, C. (2009) New selective phosphodiesterase 4D inhibitors differently acting on long, short, and supershort isoforms. *J. Med. Chem.*, **52**: 6546–6557.

33. Houslay, M.D. and Adams, D.R. (2003) PDE4 cAMP phosphodiesterases: modular enzymes that orchestrate signalling, crosstalk, desensitization and compartmentalization. *Biochem. J.*, **370**:1–18.

34. Burgin, A.B., Magnusson, O.T., Singh, J., Witte, P., Staker, B.L., Bjornsson, J.M., Thorsteinsdottir, M., Hrafansdottir, S., Hagen, T., Kiselyov, A.S., Stewart, L.J. and Gurney, M.E. (2010) Design of phosphodiesterase 4D (PDE4D) allosteric modulators for enhancing cognition with improved safety. *Nat. Biotechnol.*, **28**(1):63–72.

35. Cashman, J.R. and Ghirmai, S. (2009) Inhibition of serotonin and norepinephrine reuptake and inhibition of phosphodiesterase by multi-target inhibitors as potential agents for depression. *Bioorg. Med. Chem.*, **17**:6890–6897.

36. Owen, D.R., Walker, J.K., Jacobsen, E.J., Freskos, J.N., Hughes, R.O., Brown, D.L., Bell, A.S., Brown, D.G., et al. (2009) Identification, synthesis and SAR of amino substituted pyrido[3,2b]pyrazinones as potent and selective PDE5 inhibitors. *Bioorg. Med. Chem. Lett.*, **19**:4088–4091.

37. Hughes, R.O., Walker, J.K., Rogier, D.J., Heasley, S.E., Blevis-Bal, R.M., Benson, A.G., Jacobsen, E.J., et al. (2009) Optimization of the aminopyridopyrazinone class of PDE5 inhibitors: discovery of 3-[(*trans*-4-hydroxycyclohexyl)amino]-7-(6-methoxypyridin-3-

yl)-1-(2-propoxyethyl)pyrido[3,4b]pyrazin-2(1H)-one. *Bioorg. Med. Chem. Lett.*, **19**: 5209–5213.

38. Verhoest, P.R., Proulx-Lafrance, C., Corman, M., Chenard, L., Helal, C.J., Hou, X., Kleiman, R., Liu, S., Marr, E. et al. (2009) identification of a brain penetrant PDE9A inhibitor utilizing prospective design and chemical enablement as a rapid lead optimization strategy. *J. Med. Chem.*, **52**:7946–7949.

39. DeNinno, M.P., Andrews, M., Bell, A.S., Chen, Y., Eller-Zarbo, C., Eshelby, N., Etienne, J.B., Moore, D.E., Palmer, M.J., Visser, M.S., Yu, L.J., Zavadoski, W.J., and Gibbs, E.M. (2009) The discovery of potent, selective and orally bioavailable PDE9 inhibitors as potential hypoglycemic agents. *Bioorg. Med. Chem. Lett.*, **19**:2537–2541.

40. Hutson, P.H., Finger, E.N., Magliaro, B.C., Smith, S.M., Converso, A., Sanderson, P.E., Mullins, D., Hyde, L.A., Eschle, B.K., Turnbull, Z., et al. (2011) The selective phosphodiesterase 9 inhibitor PF-04447943 (6-[(3S,4S)-4-methyl-1-(pyrimidin-2-ylmethyl) pyrrolidin-3-yl]-1-(tetrahydro-2H-pyran-4-yl)-1,5-dihydro-4H-pyrazolo[3, 4-d]pyrimidin-4-one) enhances synaptic plasticity and cognitive function in rodents. *Neuropharmacology*, **61**:665–676.

41. Verhoest, P.R., Fonseca, K.R., Hou, X., Proulx-LaFrance, C., Corman, M., Helal, C.J., Claffey, M.M., Tuttle, J.B., Coffman, K.J. et al. (2012) Design and discovery of (6-[(3S,4S)-4-methyl-1-(pyrimidin-2-ylmethyl)pyrrolidin-3-yl]-1-(tetrahydro-2H-pyran-4-yl)-1,5-dihydro-4H-pyrazolo[3,4-d]pyrimidin-4-one) (PF-04447943), a selective brain penetrant PDE9A inhibitor for the treatment of cognitive disorders. *J. Med. Chem.*, **55**: 9045–9054.

42. Chappie, T.A., Humphrey, J.M., Allen, M.P., Estep, K.G., Fox, C.G., Lebel, L.A., Liras, S., et al. (2007) Discovery of a series of 6,7-dimethoxy-4-pyrrolidinylquinazoline PDE10A inhibitors. *J. Med. Chem.*, **50**:182–185.

43. Helal, C.J., Kang, Z., Hou, X., Pandit, J., Chappie, T.A., Humphrey, J.M., Marr, E.S., Fennell, K.F., Chenard, L.K., et al. (2011) Use of structure-based design to discover a potent, selective, *in vivo* active phosphodiesterase 10A inhibitor lead series for the treatment of schizophrenia. *J. Med. Chem.*, **54**:4536–4547.

44. Schmidt, C.J., Chapin, D.S., Cianfrogna, J., Corman, M.L., Hajos, M., Harms, J.F., Hoffman, W.E., Lebel, L.A., McCarthy, S.A., Nelson, F.R., Proulx-LaFrance, C., et al. (2008) Preclinical characterization of selective phosphodiesterase 10A inhibitors: a new selective therapeutic approach to the treatment of schizophrenia. *J. Pharmacol. Exp. Ther.*, **325**(2):681–690.

45. Verhoest, P.R., Chapin, D.S., Corman, M., Fonseca, K., Harms, J.F., Hou, X., Marr, E.S., Menniti, F.S., Nelson, F., O'Connor, R., Pandit, J., Proulx-LaFrance, C., Schmidt, A.W., Schmidt, C.J., Suciak, J., and Liras, S. (2009) Discovery of a novel class of phosphodiesterase 10A inhibitors and identification of clinical candidate 2-[4-(1-methyl-4-pyridin-4-yl-1H-pyrazol-3-yl)-phenoxymethyl]-quinoline (PF-2545920) for the treatment of schizophrenia. *J. Med. Chem.*, **52**:5188–5196.

46. Asproni, B., Murineddu, G., Pau, A., Pinna, G.A., Langgard, M., Christoffersen, C.T., Nielsen, J., and Kehler, J. (2011) Synthesis and SAR of new phenylimidazole-pyrazolo [1,5-c]quinazolines as potent phosphodiesterase 10A inhibitors. *Bioorg. Med. Chem.*, **19**:642–649.

47. Höfgen, N., Stange, H., Schindler, R., Lankau, H.-J., Grunwald, C., Langen, B., Egerland, U., Tremmel, P., Pangalos, M.N., Marquis, K.L., Hage, T., Harrison, B.L., Malamas, M.S.,

Brandon, N.J., and Kronbach, T. (2010) Discovery of imidazo[1,5-a]pyrido[3,2-e]pyrazines as a new class of phosphodiesterase 10A inhibitors. *J. Med. Chem.*, **53**:4399–4411.

48. Malamas, M.S., Ni, Y., Erdei, J., Stange, H., Schindler, R., Lankau, H.-J., Grunwald, C., Fan, K.Y., Parris, K., Langen, B., Egerland, U., Hage, T., et al. (2011) Highly potent, selective and orally active phosphodiesterase 10A inhibitors. *J. Med. Chem.*, **54**: 7621–7638.

49. McElroy, W.T., Tan, Z., Basu, K., Yang, S.-W., Smotryski, J., Ho, G.D., Tulshian, D., Greenlee, W.J., Mullins, D., Guzzi, M., Zhang, X., Bleickardt, C. and Hodgson, R. (2012) Pyrazoloquinolines as PDE10A inhibitors: discovery of a tool compound. *Bioorg. Med. Chem. Lett.*, **22**:1335–1339 and references therein.

50. Ho, G.D., Seganish, W.M., Bercovici, A., Tulshian, D., Greenlee, W.J., Van Rijn, R., Hruza, A., Xiao, L., et al. (2012) The SAR development of dihydroimidazoquinoline derivatives as phosphodiesterase 10A inhibitors for the treatment of schizophrenia. *Bioorg. Med. Chem. Lett.*, **22**:2585–2589.

51. Hu, E., Kunz, R.K., Rumfelt, S., Chen, N., Bürli, R., Li, C., Andrews, K.L., Zhang, J., et al. (2012) Discovery of potent, selective and metabolically stable 4-(pyridin-3-yl)cinnolines as novel phosphodiesterase 10A inhibitors. *Bioorg. Med. Chem. Lett.*, **22**:2262–2265.

52. Bauer, U., Giordanetto, F., Bauer, M., O'Mahoney, G., Johansson, K.E., Knecht, W., Hartlieb-Geschwindner, J., Töppner Carlsson, E., and Enroth, C. (2012) Discovery of 4-hydroxy-1,6-naphthyridine-3-carbonitrile derivatives as novel PDE10A inhibitors. *Bioorg. Med. Chem. Lett.*, **22**:1944–1948.

53. Tu, Z., Xu, J., Jones, L.A., Li, S., and Mach, R.H. (2010) Carbon-11 labeled papaverine as a PET tracer for imaging PDE10A: radiosynthesis, *in vitro* and *in vivo* evaluation. *Nucl. Med. Biol.*, **37**:509–516.

54. Tu, Z., Fan, J., Li, S., Jones, L.A., Cui, J., Padakanti, P.K., Xu, J., Zeng, D., Shoghi, K.I., Perlmutter, J.S., and Mach, R.H. (2011) Radiosynthesis and *in vivo* evaluation of [^{11}C]MP-10 as a PET probe for imaging PDE10A in rodent and non-human primate brain. *Bioorg. Med. Chem.*, **19**:1666–1673.

55. Celen, S., Koole, M., De Angelis, M., Sannen, I., Chitneni, S.K., Alcazar, J., Dedeurwaerdere, S., Moechars, D., Schmidt, M., Verbruggen, A., Langlois, X., Van Laere, K., Andres, J.I., and Bormans, G. (2010) Preclinical evaluation of 18F-JNJ41510417 as a radioligand for PET imaging of phosphodiesterase-10A in the brain. *J. Nucl. Med.*, **51**:1584–1591.

56. Hu, E., Ma, J., Biorn, C., Lester-Zeiner, D., Cho, R., Rumfelt, S., Kunz, R.K., Nixey, T., Michelsen, K., et al. (2012) Rapid identification of a novel small molecule phosphodiesterase10A (PDE10A) tracer. *J. Med. Chem.*, **55**:4776–4787.

57. Reneerkens, O.A.H., Rutten, K., Steinbusch, H.W.M., Blokland, A., and Prickaerts, J. (2009) Selective phosphodiesterase inhibitors: a promising target for cognition enhancement. *Psychopharmacology*, **202**:419–443.

58. Puzzo, D., Staniszewski, A., Deng, S.X., Privitera, L., Leznik, E., Liu, S., Zhang, H., Feng, Y., Palmeri, A., Landry, D.W., and Arancio, O. (2009) Phosphodiesterase 5 inhibition improves synaptic function, memory and amyloid-β load in an Alzheimer's disease mouse model. *J. Neurosci.*, **29**(25):8075–8086.

59. Goff, D.C., Cather, C., Freudenreich, O., Henderson, D.C., Evins, A.E., Culhane, M.A., and Walsh, J.P. (2009) A placebo-controlled study of sildenafil effects on cognition in schizophrenia. *Psychopharmacology*, **202**(1–3):411–417.

60. van der Staay, F.J., Rutten, K., Barfacker, L., DeVry, J., Erb, C., Heckroth, H., Karthaus, D., Tersteegen, A., van Kampen, M., Blokland, A., Prickaerts, J., Reymann, K.G., Schroder, U.H., and Hendrix, M. (2008) The novel selective PDE9 inhibitor BAY 73-6691 improves learning and memory in rodents. *Neuropharmacology*, **55**:908–918.

61. van Donkelaar, E.L., Rutten, K., Blokland, A., Akkerman, S., Steinbusch, H.W.M., and Prickaerts, J. (2008) Phosphodiesterase 2 and 5 inhibition attenuates the object memory deficit induced by acute tryptophan depletion. *Eur. J. Pharmacol.*, **600**(1-3):98–104.

62. Masood, A., Huang, Y., Hajjhussein, H., Xiao, L., Li, H., Wang, W., Hamza, A., Zhan, C.-G., and O'Donnell, J.M. (2009) Anxiolytic effects of phosphodiesterase 2 inhibitors associated with increased cGMP signalling. *J. Pharmacol. Exp. Ther.*, **331**(2):690–699.

63. Rodefer, J.S., Saland, S.K., and Eckrich, S.J. (2012) Selective phosphodiesterase inhibitors improve performance on the ED/ID cognitive task in rats. *Neuropharmacology*, **62**: 1182–1190.

64. Bednar, M.M. (2008) The role of sildenafil in the treatment of stroke. *Curr. Opin. Investig. Drugs*, **9**(7):754–759.

65. Royl, G., Balkaya, M., Lehmann, S., Lehnardt, S., Stohlmann, K., Lindauer, U., Endres, M., Dirnagl, U., and Meisel, A. (2009) Effects of the PDE5 inhibitor vardenafil in a mouse stroke model. *Brain Res.*, **1265**:148–157.

66. Zhang, L., Zhang, R.L., Wang, Y., Zhang, C., Zhang, Z.G., Meng, H. and Chopp, M. (2006) Tadalafil, a long-acting type 5 phosphodiesterase isoenzyme inhibitor, improves neurological functional recovery in a rat model of embolic stroke. *Brain Res.*, **1118**:192–198.

67. Strick, C.A., James, L.C., Fox, C.B., Seeger, T.F., Menniti, F.S., and Schmidt, C.J. (2010) Alterations in gene regulation following inhibition of the striatum-enriched phosphodiesterase, PDE10A. *Neuropharmacology*, **58**:444–451.

68. Kleiman, R.J., Kimmel, L.H., Bove, S.E., Lanz, T.A., Harris, J.F., Romegalli, A., Miller, K.S., Willis, A., des Etages, S., Kuhn, M., and Schmidt, C.J. (2011) Chronic suppression of phosphodiesterase 10A alters striatal expression of genes responsible for neurotransmitter synthesis, neurotransmission and signaling pathways implicated in Huntington's disease. *J. Pharmacol. Exp. Ther.*, **336**:64–76.

69. Dedeurwaerdere, S., Wintmolders, C., Vanhoof, G., and Langlois, X. (2011) Patterns of brain glucose metabolism induced by phosphodiesterase 10A inhibitors in the mouse: a potential translational biomarker. *J. Pharmacol. Exp. Ther.*, **339**:210–217.

CHAPTER 6

CRYSTAL STRUCTURES OF PHOSPHODIESTERASES AND IMPLICATION ON DISCOVERY OF INHIBITORS

HENGMING KE, HUANCHEN WANG, and MENGCHUN YE

Department of Biochemistry and Biophysics and Lineberger Comprehensive Cancer Center, The University of North Carolina, Chapel Hill, NC, USA

YINGCHUN HUANG

Department of Biochemistry and Biophysics and Lineberger Comprehensive Cancer Center, The University of North Carolina, Chapel Hill, NC, USA; Biomedical and Pharmaceutical Department, Biochemical Engineering College, Beijing Union University, Beijing, China

INTRODUCTION

Cyclic nucleotide phosphodiesterases (PDEs) are the sole enzymes responsible for hydrolyzing $3',5'$-cyclic adenosine and guanosine monophosphate (cAMP and cGMP) to $5'$-AMP and $5'$-GMP, respectively. The second messengers cAMP and cGMP mediate cell responses to a wide variety of hormones and neurotransmitters, and also modulate many metabolic processes [1–7]. The human genome contains 21 PDE genes that are categorized into 11 families and express approximately 100 isoforms of PDE proteins through alternative mRNA splicing [8–10]. PDE molecules are composed of a variable regulatory domain at the N-terminus and a conserved catalytic domain at the C-terminus. Various structural motifs in the PDE regulatory domains play roles in regulation of the PDE catalytic activities and also in crosstalk with other signaling pathways [8,11]. The conserved catalytic domains of PDEs specifically recognize substrates and selectively bind inhibitors. Family-selective PDE inhibitors have been widely studied as therapeutics for treatment of various human diseases such as schizophrenia and Huntington's disease, and

Cyclic-Nucleotide Phosphodiesterases in the Central Nervous System: From Biology to Drug Discovery, First Edition. Edited by Nicholas J. Brandon and Anthony R. West.
© 2014 John Wiley & Sons, Inc. Published 2014 by John Wiley & Sons, Inc.

improvement of learning and memory [12–19]. The most successful examples are the PDE5 inhibitors sildenafil (Viagra), vardenafil (Levitra), and tadalafil (Cialis), which are drugs for treatment of male erectile dysfunction [12].

The crystal structure of the PDE4B catalytic domain [20] was the milestone breakthrough in the structural studies of PDEs in 2000. Since then, about 190 structures of nine PDE families have been deposited into the RCSB Protein Data Bank (Table 6.1), including the catalytic domains of PDE1B [21], PDE2A [22], PDE3B [23], PDE4A–4D [21,24,25], PDE5A [21,26,27], PDE7A [28], PDE8A [29], PDE9A [30–33], PDE10A [34–37], and the GAF domains of PDE2A [38], PDE5A [39,40], PDE6C [41], and PDE10A [42]. Recently, the structures of a PDE2A fragment containing both GAF and catalytic domains and

TABLE 6.1   Structures of PDEs in Protein Data Bank Until 2010

Protein[a]	Ligand[b]	Resolution (Å)	PDB Code	Reference
PDE1B	Unliganded	1.77	1TAZ	[21]
PDE2A(GAF-A + B)	cGMP	2.86	1MC0	[38]
PDE2A	Unliganded	1.70	1Z1L	[22]
PDE2A(GAF + CAT)	Unliganded	3.02	3IBJ	[43]
PDE2A	Unliganded	2.49	3ITM	[43]
PDE2A	IBMX	1.58	3ITU	[43]
PDE3B	Dihydropyridazine	2.4	1SO2	[23]
PDE3B	IBMX	2.9	1SOJ	[23]
PDE4A10	NPV	2.1	2QYK	[25]
PDE4A	BMPMIL	2.25	3I8V	To be published
PDE4A	Pentoxifylline	2.84	3TVX	[45]
PDE4B	NPV	1.95	2QYL	[25]
PDE4B2B	Unliganded	1.77	1F0J	[20]
PDE4B2B	(R,S)-Rolipram	2.0	1RO6	[46]
PDE4B2B	8-Br-AMP	2.13	1RO9	[46]
PDE4B2B	AMP	2.00	1ROR	[46]
PDE4B2B	Tetrahydrobenzothiophene	2.23	3HMV	[47]
PDE4B	Coumarins	2.60	3LY2	[48]
PDE4B	Quinolines	1.75	3GWT	[49]
PDE4B	Quinolines	1.70	3FRG	[50]
PDE4B	BTMPTP	2.48	3KKT	To be published
PDE4B	Pyrazolopyridines	1.75	3D3P	[51]
PDE4B	AMP	2.15	1TB5	[21]
PDE4B	Cilomilast	2.19	1XLX	[52]
PDE4B	Filaminast	2.06	1XLZ	[52]
PDE4B	Piclamilast	2.31	1XM4	[52]
PDE4B	(R)-Mesopram	1.92	1XM6	[52]
PDE4B	Roflumilast	2.30	1XMU	[52]
PDE4B	(R)-Rolipram	2.40	1XMY	[52]
PDE4B	(R,S)-Rolipram	2.31	1XN0	[52]
PDE4B	Sildenafil	2.28	1XOS	[52]
PDE4B	Vardenafil	2.34	1XOT	[52]
PDE4B	Cl-PDM-PCAEE	2.40	1Y2H	[53]
PDE4B	DMNP-PCAEE	2.55	1Y2J	[53]
PDE4B	D155988	2.63	3G45	[44]
PDE4B	3QJ	1.95	3O0J	To be published

TABLE 6.1   *(Continued)*

Protein[a]	Ligand[b]	Resolution (Å)	PDB Code	Reference
PDE4B2B	ZG1	2.42	3O56	[54]
PDE4B2B	ZG2	2.00	3O57	[54]
PDE4C	Unliganded	1.9	2QYM	[25]
PDE4D (CAT + UCR2)	D155871	2.30	3G4G	[44]
PDE4D	D155871	1.90	3G4I	[44]
PDE4D	Rolipram	1.95	3G4K	[44]
PDE4D	Roflumilast	2.50	3G4L	[44]
PDE4D	D155988	2.05	3G58	[44]
PDE4D	Allosteric modulators	2.65	3IAD	[44]
PDE4D	BMPMIL	2.05	3K4S	To be published
PDE4D	Papaverine	2.80	3IAK	To be published
PDE4D2	NPV	1.57	2QYN	[25]
PDE4D5	Splicing region	NMR	1LOI	[55]
PDE4D	Zardaverine	2.9	1MKD	[56]
PDE4D2	(R,S)-Rolipram	2.0	1OYN	[24]
PDE4D2	(R)-Rolipram	2.3	1Q9M	[24]
PDE4D2	AMP	2.3	1PTW	[57]
PDE4D2	IBMX	2.10	1ZKN	[27]
PDE4D	AMP	1.63	1TB7	[21]
PDE4D	(R,S)-Rolipram	1.6	1TBB	[21]
PDE4D	Cilomilast	1.55	1XOM	[52]
PDE4D	Piclamilast	1.72	1XON	[52]
PDE4D	Roflumilast	1.83	1XOQ	[52]
PDE4D	Zardaverine	1.54	1XOR	[52]
PDE4D	DM-PCAEE	1.40	1Y2B	[53]
PDE4D	DMP-PCAEE	1.67	1Y2C	[53]
PDE4D	MOPDM-PCAEE	1.70	1Y2D	[53]
PDE4D	APDM-PCAEE	2.10	1Y2E	[53]
PDE4D	DMNP-PCAEE	1.36	1Y2K	[53]
PDE4D2	L-869298	2.0	2FM0	[58]
PDE4D2	L869299	2.03	2FM5	[58]
PDE4D2(D201N)	cAMP	1.56	2PW3	[59]
PDE4D2	Apo	2.10	3SL3	[60]
PDE4D2	JN4	1.9	3SL4	[60]
PDE4D2	J25	2.65	3SL5	[60]
PDE4D2	JN8	2.44	3SL6	[60]
PDE4D2	JN7	2.60	3SL8	[60]
PDE4D	IHM	2.10	3V9B	To be published
PDE5A	Sildenafil	2.3	1UDT	[26]
PDE5A	Vardenafil	2.5	1UHO	[26]
PDE5A	Tadalafil	2.83	1UDU	[26]
PDE5A	IBMX	2.05	1RKP	[27]
PDE5A	Unliganded	2.1	1T9R	[21]
PDE5A	GMP	2.0	1T9S	[21]
PDE5A	Sildenafil	1.30	1TBF	[21]
PDE5A	Tadalafil	1.37	1XOZ	[21]
PDE5A	Vardenafil	1.40	1XPO	[21]
PDE5A	Vardenafil	2.07	3B2R	[29]
PDE5A	Unliganded	1.80	2H44	[61]
PDE5A	Icarisid II	1.85	2H40	[61]

(continued)

TABLE 6.1 (*Continued*)

Protein[a]	Ligand[b]	Resolution (Å)	PDB Code	Reference
PDE5A	Sildenafil	2.30	2H42	[61]
PDE5A	Sildenafil derivatives	2.0	3BJC	[62]
PDE5A	PF2	1.60	2CHM	[63]
PDE5A(GAF-A)	cGMP	NMR	2K31	[39]
PDE5A	Aminopyridopyrazinones	1.80	3HDZ	[64]
PDE5A	Aminopyridopyrazinones	1.79	3HC8	[64]
PDE5/6	Sildenafil	2.87	3JWQ	[65]
PDE5/6	IBMX/PDE6γ(70–87)	2.99	3JWR	[65]
PDE6γ		NMR	2JU4	[66]
PDE6C(GAF-A)	cGMP	2.57	3DBA	[41]
PDE7A	IBMX	1.67	1ZKL	[28]
PDE7A	Quinazolines	2.40	3G3N	[67]
PDE8A	Unliganded	1.90	3ECM	[68]
PDE8A	IBMX	2.1	3ECN	[68]
PDE9A	IBMX	2.23	2HD1	[30]
PDE9A	IBMX	2.23	2YY2	[31]
PDE9A	GMP	2.15	3DY8	[31]
PDE9A	cGMP	2.50	3DYQ	[31]
PDE9A	cGMP	2.10	3DYN	[31]
PDE9A	ES complex	2.70	3DYL	[31]
PDE9A	GMP	2.30	3DYS	[31]
PDE9A	JAR	2.72	3JSI	[69]
PDE9A	JAR	2.30	3JSW	[69]
PDE9A(Q453E)	(*R*)-BAY 73-6691	2.70	3K3E	[33]
PDE9A(Q453E)	(*S*)-BAY 73-6691	2.50	3K3H	[33]
PDE9A(Q453E)	BAY 73-6691	2.30	3QI3	[70]
PDE9A(Q453E)	IBMX	2.50	3QI4	[70]
PDE9A(E406A)	IBMX	2.75	3N3Z	To be published
PDE10A	AMP	1.56	2OUN	[34]
PDE10A	Unliganded	1.56	2OUP	[34]
PDE10A	GMP	1.90	2OUQ	[34]
PDE10A(D674A)	cAMP	1.45	2OUR	[34]
PDE10A(D674A)	Unliganded	1.45	2OUS	[34]
PDE10A(D674A)	cGMP	1.52	2OUU	[34]
PDE10A(D564N)	Unliganded	1.56	2OUV	[34]
PDE10A(D564N)	cAMP	1.90	2OUY	[34]
PDE10A (rat)	Quinazolines	1.80	2O8H	[71]
PDE10A (rat)	Quinazolines	2.0	2OVV	[71]
PDE10A (rat)	Quinazolines	2.0	2OVY	[71]
PDE10A(GAF-B)	cAMP	2.1	2ZMF	[42]
PDE10A (rat)	WEB3	2.3	3LXG	[72]
PDE10A	PF4	1.7	3HQW	[73]
PDE10A	PF6	2.0	3HQY	[73]
PDE10A	PF8	1.7	3HQZ	[73]
PDE10A	PF-2545920	1.53	3HR1	[73]
PDE10A	PFK	2.00	3QPN	[74]
PDE10A	PFR	1.80	3QPO	[74]
PDE10A	PFW	1.80	3QPP	[74]
PDE10A	AXC	2.43	2YOJ	[75]
PDE10A	540	1.82	3SN7	[76]
PDE10A	546	1.90	3SNI	[76]
PDE10A	548	2.40	3SNL	[76]

TABLE 6.1 (*Continued*)

Protein[a]	Ligand[b]	Resolution (Å)	PDB Code	Reference
PDE10A	AZ5	2.20	4AEL	To be published
PDE10A	C1L	2.28	3UI7	[77]
PDE10A	OCV	2.11	3UUO	[78]
PDE10A	OJP	2.11	4DFF	[79]
PDE10A	OJQ	2.07	4DDL	[80]
TcrPDEC1	Unliganded	2.00	3V93	[81]
TcrPDEC1	WYQ16	2.33	3V94	[81]
LmjPDEB1	IBMX	1.50	2R8Q	[82]

[a]GAF = N-terminal GAF domain; CAT = catalytic domain; proteins without notation are the catalytic domains.

[b]Ligands are abbreviated as follows: IBMX = 3-isobutyl-1-methylxanthine; NVP = 4-[8-(3-nitrophenyl)-1,7-naphthyridin-6-yl]benzoic acid; BMPMIL = 4-(3-butoxy-4-methoxyphenyl)methyl-2-imidazolidone; BTMPTP = 5-[3-[(1S,2S,4R)-bicyclo[2.2.1]hept-2-yloxy]-4-methoxyphenyl]tetrahydro-2(1H)-pyrimidinone; PCAEE = 1H-pyrazole-4-carboxylic acid ethyl ester; Cl-PDM-PCAEE = 1-(2-chlorophenyl)-3,5-dimethyl-PCAEE; DMNP-PCAEE = 3,5-dimethyl-1-(3-nitrophenyl)-PCAEE; DM-PCAEE = 3,5-dimethyl-PCAEE; DMP-PCAEE = 3,5-dimethyl-1-phenyl-PCAEE; MOPDM-PCAEE = 1-(4-methoxyphenyl)-3,5-dimethyl-PCAEE; APDM-PCAEE = 1-(4-aminophenyl)-3,5-dimethyl-PCAEE; DMNP-PCAEE = 3,5-dimethyl-1-(3-nitrophenyl)-PCAEE; PF2 = 5-(5-acetyl-2-butoxy-3-pyridinyl)-3-ethyl-2-(1-ethyl-3-azetidinyl)-2,6-dihydro-7H-pyrazolo[4,3-d]pyrimidin-7-one; JAR = 6-[(3S,4S)-1-benzyl-4-methylpyrrolidin-3-yl]-1-(1-methylethyl)-1,5-dihydro-4H-pyrazolo[3,4-d]pyrimidin-4-one; WEB3 = imidazo[1,5-a]pyrido[3,2-e]pyrazine; PF4 = 4,5-bis(4-methoxyphenyl)-2-thiophen-2-yl-1H-imidazole; PF6 = 2-({4-[4-(pyridin-4-ylmethyl)-1H-pyrazol-3-yl]phenoxy}methyl)quinoline; PF8 = 2-{[4-(4-pyridin-4-yl-1H-pyrazol-3-yl)phenoxy]methyl}quinoline; PF-2545920 = 2-[4-(1-methyl-4-pyridin-4-yl-1H-pyrazol-3-yl)phenoxymethyl]quinoline; OJQ = 2-{1-[5-(6,7-dimethoxycinnolin-4-yl)-3-methylpyridin-2-yl]piperidin-4-yl}propan-2-ol; OJP = 8,9-dimethoxy-1-(1,3-thiazol-5-yl)-5,6-dihydroimidazo[5,1-a]isoquinoline; OCV = 6-methoxy-3,8-dimethyl-4-(piperazin-1-yl)-1H-pyrazolo[3,4-b]quinolone; AZ5 = 2-(2'-ethoxybiphenyl-4-yl)-4-hydroxy-1,6-naphthyridine-3-carbonitrile; IHM = N-(3-{1-[(1S)-1-[3-(cyclopropylmethoxy)-4-(difluoromethoxy)phenyl]-2-(1-oxidopyridin-4-yl)ethyl]-1H-pyrazol-3-yl}phenyl)acetamide; C1L = 6-methoxy-3,8-dimethyl-4-(morpholin-4-ylmethyl)-1H-pyrazolo[3,4-b]quinoline; JN4 = ethenyl 6-(ethenylcarbamoyl)-2-[(phenylacetyl)amino]-4,5,6,7-tetrahydrothieno[2,3-c]pyridine-3-carboxylate; J25 = diethyl 6-[(thiophen-2-ylacetyl)amino]-4,7-dihydrothieno[2,3-c]pyridine-3,6(5H)-dicarboxylate; JN8 = cyclopentyl 6-(ethylcarbamoyl)-2-[(thiophen-2-ylacetyl)amino]-4,5,6,7-tetrahydrothieno[2,3-c]pyridine-3-carboxylate; JN7 = 3-cyclopentyl 6-ethenyl 2-[(thiophen-2-ylacetyl)amino]-4,7-dihydrothieno[2,3-c]pyridine-3,6(5H)-dicarboxylate; 540 = 8-fluoro-6-methoxy-3,4-dimethyl-1-(3-methylpyridin-4-yl)imidazo[1,5-a]quinoxaline; 546 = 2-methoxy-6,7-dimethyl-9-(4-methylpyridin-3-yl)imidazo[1,5-a]pyrido[3,2-e]pyrazine; 548 = 6-chloro-3,4-dimethyl-1-(3-methylpyridin-4-yl)-8-(trifluoromethyl)imidazo[1,5-a]quinoxaline; 3OJ = 4-[(1-hydroxy-1,3-dihydro-2,1-benzoxaborol-5-yl)oxy]benzene-1,2-dicarbonitrile; ZG2 = 5-[5-benzyl-4-(2-oxo-2-pyrrolidin-1-ylethyl)-1,3-oxazol-2-yl]-1-ethyl-N-(tetrahydro-2H-pyran-4-yl)-1H-pyrazolo[3,4-b]pyridin-4-amine; ZG1 = 1-ethyl-5-[3-(2-oxo-2-pyrrolidin-1-ylethyl)-1,2,4-oxadiazol-5-yl]-N-(tetrahydro-2H-pyran-4-yl)-1H-pyrazolo[3,4-b]pyridin-4-amine; AXC = 5-(1H-benzimidazol-2-ylmethylsulfanyl)-2-methyl-[1,2,4]triazolo[1,5-c]quinazoline; PFK = 6-methoxy-7-[2-(quinolin-2-yl)ethoxy]quinazoline; PFR = 7-methoxy-4-[(3S)-3-phenylpiperidin-1-yl]-6-[2-(pyridin-2-yl)ethoxy]quinazoline; PFW = 7-methoxy-4-[(3S)-3-phenylpiperidin-1-yl]-6-[2-(quinolin-2-yl)ethoxy]quinazoline.

of the PDE4D catalytic domain with a portion of the upstream conserved region 2 (UCR2) have been determined [43,44]. This chapter provides an overview of the three-dimensional structures of PDE families and discusses implications on the design of PDE inhibitors.

OVERVIEW OF STRUCTURES OF PDE CATALYTIC DOMAINS

The conserved catalytic domains of PDE families contain about 350 amino acids that are folded into 16 α-helices, as extensively reviewed in references of Refs [83,84]. The general structural feature of the catalytic domains can be represented by the PDE4D structure (Figure 6.1) and the alignment of the sequences and secondary structures of the PDE families can be found in Chapter 1 and Ref. [84]. Since the first two helices, H1 and H2, demonstrate large conformational and positional differences among PDE families, the core catalytic domain of PDEs can be considered to start from helix H3 [84]. However, there are minor variations in the core catalytic domain: PDE3B contains four very short β-strands before helices H7 and H16 [23]; the H-loop (helices H8 and H9) in PDE5 adopts six different conformations [27,29,61,62]; and PDE8A has two extra helices between H12 and H13 [68]. In addition, an α-helix of C-terminal residues 497–508 was found in PDE4B [20]. This helix is linked with a disordered loop (residues 487–495) to the PDE4B catalytic domain and interacts with the active site of a neighboring molecule in the crystal. It is not clear whether this interaction implies the regulation of PDE4 activity by its C-terminus or represents the lattice contact of the crystal without biological relevance. Further study is needed to clarify this.

The active site of PDEs can be divided into two major subpockets,respectively, for binding of divalent metals and substrates/inhibitors [84]. The metal-binding pocket contains two divalent metal ions. The first metal was identified as zinc by X-ray anomalous scattering experiments [20] and forms an octahedron with two invariant histidines (His164 and His200 in PDE4D2), two invariant aspartic acids (Asp201 and Asp318), and two water molecules. The second metal ion also forms six coordinations with Asp201 and five water molecules in octahedral conformation [84]. The chemical nature of the second metal has not been established, but is generally accepted to be a magnesium ion. However, other divalent ions could possibly be a preferred catalytic metal in some PDE families. For example, manganese was shown to promote the catalytic activities of PDE5, PDE8, and PDE9 twofold better than magnesium [68,85–87]. In addition to the absolute conservation of the metal-binding residues, two histidines at the active site (His160 and His204 in PDE4D2) are identical in all PDE families. The first histidine was proposed to serve as a general acid for the catalysis [31,57], while the role of the second histidine is unclear.

In the nucleotide-binding pocket, residues are well conserved, but only a glutamine (Gln369 in PDE4D2) is invariant across PDE families. Apparently, variation of the residues leads to slightly different sizes and shapes of the nucleotide pockets, thus determining the substrate specificity and inhibitor selectivity. Binding of substrates/inhibitors in the nucleotide pocket shows two common features in all PDEs: the hydrogen bond with the invariant glutamine (Figure 6.1) and the stack against the conserved phenylalanine (Phe372 in PDE4D2, but tryptophan in PDE11). These two characteristics may account for the basic affinity of ligands while other elements in the pocket appear to be important for selective recognition of substrates and inhibitors.

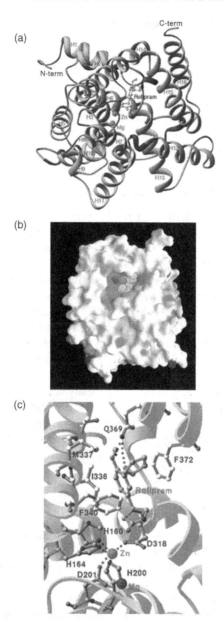

FIGURE 6.1   Structure of the catalytic domain of PDE4D. (a) Ribbon diagram. Alpha-helices are shown in cyan and 3_{10} helices are in blue. The PDE4-selective inhibitor rolipram is represented with golden ball-sticks. (b) Surface presentation of the PDE4D catalytic domain. Red color represents oxygen and blue is for nitrogen. The golden worm represents rolipram. (c) Interaction of rolipram with the active site residues of PDE4D2. Dotted lines represent hydrogen bonds or metal coordinations. (See the color version of the figure in Color Plates section.)

PDE4 STRUCTURES AND IMPLICATION ON THE DESIGN OF ACTIVE SITE INHIBITORS

Cyclic AMP-specific PDE4 can be divided into four subfamilies of 4A to 4D, which have the K_M values of 1–6 μM and the k_{cat} values of 1–7 s^{-1} for cAMP [25]. Molecules of PDE4 contain an N-terminal splicing region, two upstream conserved regions (UCR1 and UCR2), and a catalytic domain. PDE4 is the most extensively studied family of PDEs, and 66 crystal structures and >500 patents on the PDE4s and their inhibitors have been published since 2000.

Rigidity of the PDE4 Binding Pocket

Whether a binding pocket is rigid or not is critical for the success of docking and structure-based design of inhibitors. In general, the rigidity can be judged by comparison of structures bound with different inhibitors and also by examination of B-factors that indicate the thermal movements of atoms. Especially powerful for study of pocket rigidity is comparison between two protein structures that bind two enantiomers of an inhibitor. A good example is a pair of PDE4D inhibitor enantiomers, L869298 (IC$_{50}$ = 0.4 nM) and L869299 (IC$_{50}$ = 43 nM). Intuitively, three large substitution groups that link to the chiral center of L869298/L869299 would be expected to have different orientations and interactions. However, the crystal structures of the PDE4D2 catalytic domain in complex with L869298 or L869299 [58] revealed the similar orientations of two enantiomers in the PDE4D pocket (Figure 6.2a). L869299 had an energetically unfavorable dihedral angle that is close to an "eclipsed" conformation and also lost a hydrogen bond with a bound solvent [58], thus accounting for its 100-fold weaker affinity. This structural study suggests that the PDE4D binding pocket is so rigid as to force the enantiomers to bind in similar conformations. The rigidity of the PDE active sites is also proposed for the structures of PDE8 and PDE9 [30,31,68].

Pharmacophores of PDE4 Inhibitors

Another important step for effective and successful design of inhibitors is to identify pharmacophores. The structural superposition of PDE4s in complex with the catechol-based inhibitors shows that these PDE4 inhibitors can be divided into five pharmacophores, G1 to G5 (Figure 6.2b). The G1 group is the catechol core that stacks against the conserved Phe372 in PDE4D2 and forms two hydrogen bonds with the invariant Gln369 (Figure 6.2b). These interactions have been considered as the common characterizations of binding of substrates and inhibitors in PDE families [57,84]. The G2 group interacts with residues in a small pocket and G3 orients to the "Q" pocket [52]. These two pockets show significant variation of the amino acids across PDE families and are expected to play important roles on selective binding of inhibitors. The G4 group orients toward the metal-binding pocket. Since the metal pocket is made up of the invariant residues and filled with solvents, optimization of G4 might increase affinity, but unlikely selectivity of inhibitors. The G5 group points

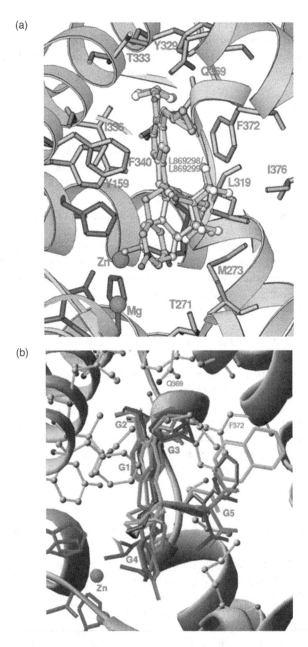

FIGURE 6.2   Binding of PDE4 inhibitors. (a) Binding of enantiomers L869298 (green balls-sticks) and L869299 (golden sticks) to PDE4D2. The metal-binding residues His164, His200, Asp201, and Asp318 are shown in blue. In spite of the similar orientation of the enantiomers, L869299 takes a restricted configuration and thus accounts for its weaker affinity. (b) Superposition of PDE4 inhibitors that are presented in various colors. Five pharmacophores (G1–G5) can be identified. (See the color version of the figure in Color Plates section.)

toward the open space of the molecular surface and interacts with a few residues that vary significantly across PDE families. Thus, the modification of G5 may improve both affinity and selectivity of inhibitors, but its contribution would be dominant due to limited interactions with protein residues in the open space.

Structures of PDE4 Subfamilies and the Subfamily-Selective Inhibitors

PDE4 inhibitors have been extensively studied as anti-inflammatory therapeutics for the treatment of asthma, COPD, and other diseases [13,14,88]. The main obstacles preventing practical application of the PDE4 inhibitors are side effects such as emesis. Although mechanisms for the side effects are unclear, the poor selectivity of the inhibitors against PDE4 subfamilies was thought to be one of the causes [13]. In this regard, it has been suggested that PDE4B may be a target for treatment of asthma [89] while PDE4D may link to the emesis [90] and the cardiac side effects [91]. To illustrate differences in the active sites of PDE4 subfamilies, the crystal structures of four PDE4 subfamilies in complex with a PDE4D-selective inhibitor 4-[8-(3-nitro-phenyl)-1,7-naphthyridin-6-yl]benzoic acid (NVP) [92] were determined [25]. The structures showed that the residues involved in inhibitor binding are identical among the PDE4 subfamilies. However, subtle but substantial conformational differences are observed at the active sites of the PDE4 subfamilies (Figure 6.3). While the active sites of PDE4B and PDE4D resemble each other closely (Figure 6.3c), PDE4A shows significant positional displacements of Ala575 to Ser580 (Figure 6.3b). These residues are N-terminal to the invariant glutamine (Gln581)—a key residue for binding both substrates and inhibitors. PDE4C appears to be distant from other PDE4 members and shows disorder of several residues at the active site, including the invariant Gln491 (Figure 6.3d). Since the major portion of the active sites of PDE4 subfamilies is buried, the subtle structural differences in binding pockets would imply a limitation on development of active site-directed PDE4 subfamily inhibitors, although a few of PDE4 subfamily-selective inhibitors have been reported [93,94].

CONFORMATION VARIATION OF THE PDE5 CATALYTIC DOMAIN

PDE5 is a family of cGMP-specific enzymes with a K_M of 3 μM for cGMP and >40 μM for cAMP [11,95]. Unlike most of other PDE families that have rigid active site pockets, PDE5 shows dramatic conformational variation of its H-loop of the active site. The H-loops of most PDE families contain two short α-helices and have comparable conformation [84], but the H-loop of PDE5 adopts six conformations upon inhibitor binding (Figure 6.4a). The binding of IBMX transforms the coiled H-loop in the unliganded PDE5A1 into two short α-helices and moves its Cα-atoms as much as 7 Å from the positions in the unliganded PDE5A1 structure [27]. Inhibitor wyq3 (Figure 6.5) induces conformational changes of the H-loop in a similar pattern as that in the PDE5A1–IBMX complex [62]. Upon binding of sildenafil, the H-loop is converted to a β-turn and a 3_{10}-helix and the whole loop moves as much as 24 Å to

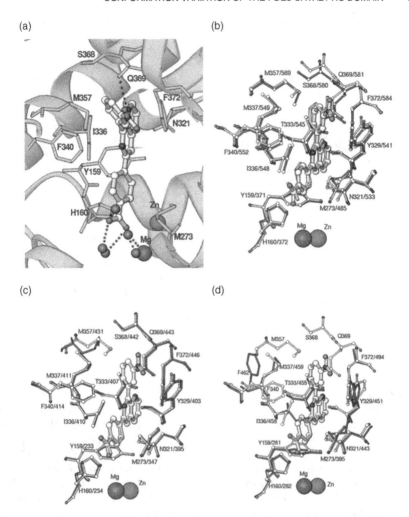

FIGURE 6.3 Binding of NVP to the PDE4 subfamilies. (a) The interaction of NVP (golden bonds) with PDE4D2 residues (green bonds). Dotted lines represent hydrogen bonds or metal coordination. (b) Superposition of PDE4A10–NVP (blue and orange bonds) over PDE4D2–NVP (green bonds). The labels such as M357/569 represent residues of PDE4D2/PDE4A10. Ser368 of PDE4D2 shows a significant shift from Ser580 of PDE4A10. (c) Superposition of PDE4B2B–NVP (blue and orange) over PDE4D2–NVP (green). (d) Superposition of PDE4C2 (blue) over PDE4D2–NVP (green and orange). Significant conformational changes are visible between PDE4C and PDE4D, such as Phe340 of PDE4D2 versus Phe462 of PDE4C2. (See the color version of the figure in Color Plates section.)

cover up the active site [61]. The H-loop in the PDE5A1–vardenafil complex shows a positional shift of as much as 20 Å from that in the unliganded form [29]. The most dramatic change occurs in the structure of PDE5A1–icarisid II (Figure 6.4a), in which the H-loop is converted to two β-strands and moves as much as 35 Å [61].

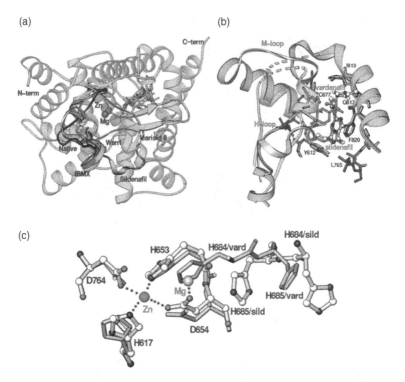

FIGURE 6.4   Structures of PDE5 catalytic domain. (a) Ribbon diagram. The cyan ribbons represent the common structures of the PDE5–inhibitor complexes. The H-loops are shown in green for PDE5–icarisid II, violet for the unliganded PDE5, red for PDE5–IBMX, light purple for PDE5–sildenafil, and gold for PDE5–wyq3 (a derivative of sildenafil). (b) Superimposition of PDE5A1–vardenafil (green and cyan ribbons and green bonds) over PDE5A1–sildenafil (golden ribbons and blue bonds). As a consequence of the different positions of both H- and M-loops between the PDE5A1 complexes of vardenafil and sildenafil, Tyr676, Cys677, and Ile680 interact with vardenafil, while Asn662, Ser663, and Leu804 interact with sildenafil. (c) The superposition of the metal-binding residues between PDE5A1–vardenafil (green) and PDE5A1–sildenafil (gold). The zinc and magnesium ions are drawn from the sildenafil complex, but were absent in the PDE5A1–vardenafil complex. (See the color version of the figure in Color Plates section.)

The conformational changes of the H-loop suggest a complex mechanism of PDE5 inhibition. The H-loops in the wyq3- or IBMX-bound structures have a similar location to that in the unliganded state, but displacements of several angstroms and show different conformations. Since these weak inhibitors do not directly contact the H-loop, the conformational changes must be achieved via an allosteric mode, although the exact pathway needs further study. On the other hand, the direct interactions of the H-loop with the inhibitors of icarisid II, sildenafil, and vardenafil convert the open binding pocket to the closed state. Vardenafil has a similar chemical formula as sildenafil (Figure 6.5). However, the pyrazolopyrimidinone ring in sildenafil is replaced with imidazotriazinone and an extra carbon has been attached

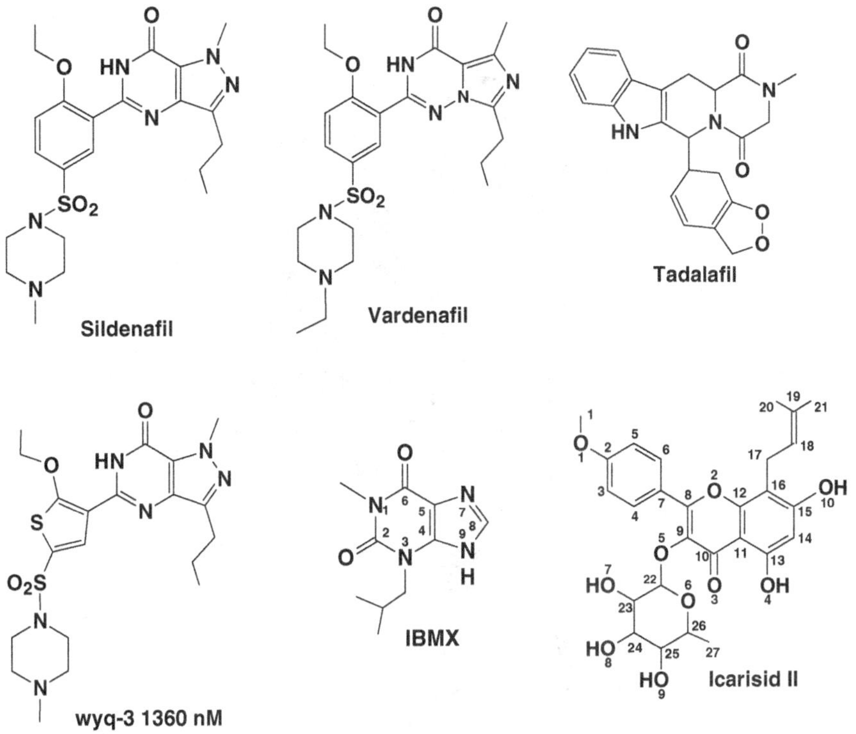

FIGURE 6.5 Chemical structures of some PDE5 inhibitors.

to methylpiperazine. These minor modifications cause about 20-fold difference in affinity, as shown by the IC_{50} values of 3.5–8.5 nM for sildenafil and 0.1–0.8 nM for vardenafil [96]. In addition, the H-loop shows completely different conformations and positions in the structures of PDE5–sildenafil and PDE5–vardenafil (Figure 6.4b). Especially notable is the movement of His684 into the metal-binding pocket (Figure 6.4c), which causes loss of the divalent metal in the PDE5–vardenafil structure [29].

STRUCTURES OF GAF DOMAINS

Introduction to GAF

GAF is a structural motif originally found in three proteins: cGMP-binding phosphodiesterases, cyanobacterial adenylyl cyclases, and a formate-hydrogen-lyase transcription activator from *Escherichia coli* [97]. It contains 140–200 amino acids that are typically folded into a central six-stranded β-sheet flanked with two α-helices on both sides of the sheet (Figure 6.6a). GAF has been reported in more than 7000

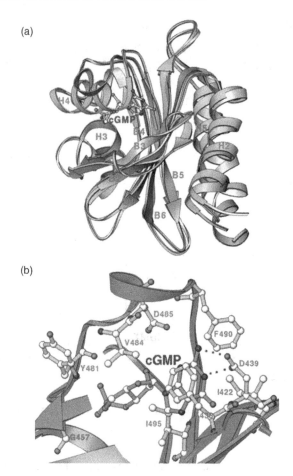

FIGURE 6.6   Ribbon diagram of PDE2A GAF domains. (a) Superposition of GAF-A (green color) over GAF-B (golden) of mouse PDE2A. (b) Binding of cGMP to GAF-B of mouse PDE2A. (See the color version of the figure in Color Plates section.)

proteins [98], but their functions are unclear. PDE2, PDE5, PDE6, PDE10, and PDE11 contain two GAF motifs in their N-terminal regulatory domains. In spite of the similar folding, only one GAF domain, GAF-A or GAF-B, in each PDE family binds a certain type of cyclic nucleotide [98,99]. For example, GAF-B of PDE2 has a K_D of 7–26 nM for cGMP, but 146–300 nM for cAMP [38,100]. GAF-A of mouse PDE5A prefers cGMP (K_D of ~3 nM) over cAMP (K_D of 8–20 μM) [39]. Chicken PDE6C GAF-A has a K_D of 10 nM for cGMP, in comparison with 35 mM for cAMP [101]. The studies on the chimers that contain the cyaB1 adenylyl cyclase catalytic domain and the PDE10 or PDE11 GAF domain suggest the preference of PDE10 GAF-B to cAMP and of PDE11 GAF-A to cGMP [102].

After the first report of the crystal structure of the mouse PDE2A GAF domain [38], three-dimensional structures of the following PDE GAF domains have been determined: a human PDE2A fragment containing both GAF and catalytic domains [43], chicken PDE6C GAF-A in complex with cGMP [41], human PDE10A GAF-B in complex with cAMP [42], the NMR structure of mouse PDE5A GAF-A in complex with cGMP [39], and a fragment of unliganded PDE5A containing N-terminal portion, GAF-A, and GAF-B [40]. These structures showed how the cyclic nucleotides interact with the GAF domains. The strands B1, B2, B5, and B6 of the β-sheet in the PDE2A structure form a hydrophobic core with helices H2 and H5 on one side, and the cGMP-binding pocket with B3, H3, and H4 on another side (Figure 6.6a). In PDE2A, cGMP forms hydrogen bonds with residues Ser424, Asp439, Ile458, Tyr481, Thr492, Glu512, and stacks against Phe480, in addition to van der Waals interactions with Ile422, Cys423, Ser424, Glu512, Asp439, Thr488, and Phe490 (Figure 6.6b). The topological folding and the main characters of the nucleotide binding are conserved in the known structures of PDE2A, PDE5A, PDE6C, and PDE10A.

Conformation Changes upon cGMP Binding

A comparison between the NMR structure of cGMP-bound GAF-A [39] and the X-ray structures of the unliganded PDE5A fragment containing a portion of N-terminus, GAF-A, and GAF-B [40] reveals conformational changes induced by cGMP binding (Figure 6.7). Five regions show positional shifts more than two times the overall RMSD: the β-turn between B1 and B2, a loop after B2 (marked with L2 in Figure 6.7a), a loop after B6 (L6), helix H4, and strand B3. Because L6 is the most variable loop in the NMR conformers [39] and L2 is distant from the binding pocket, the biological meaning of the differences between these loops of the X-ray and NMR structures is unclear. However, other positional changes appear to be biologically relevant. H4 and B3 may serve as "lids" to open and close the cGMP-binding pocket (Figure 6.7b). Upon cGMP binding, the open pocket in the unliganded state is closed due to the movement of helix H4 and strand B3 toward each other by more than 3 Å (Figure 6.7b). On the other hand, the β-turn between B1 and B2 may serve as a relay of cGMP signaling from GAF-A to GAF-B because this is the only region involved in noncovalent interactions between GAF-A and GAF-B, although the cGMP-bound crystal structure is required for confirmation.

Regulation of PDE Activity by Binding of Allosteric Effectors to GAF Domains

The GAF domains are believed to regulate PDE catalytic activity via an allosteric mode [98,103], although it was controversially argued that binding of cyclic nucleotides to the GAF domains of PDE10 and PDE11 does not stimulate the catalytic activity [104]. Regulation of catalytic activity by substrate binding to the GAF sites is not only unique to allosteric enzymes, but also important for development of highly selective inhibitors against the allosteric sites. Although no inhibitors against the GAF site have been reported, the crystal structures of PDE2 and PDE5 have shed light on the allosteric regulation. The structure of the unliganded PDE5A fragment showed a region having noncovalent interactions between GAF-A and

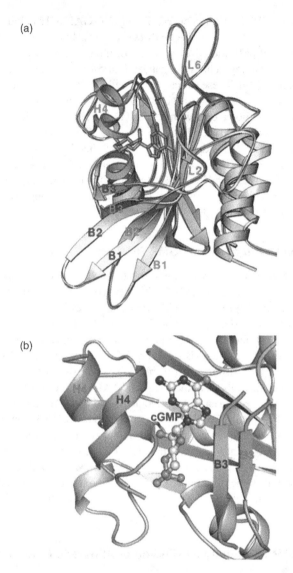

FIGURE 6.7 Conformational changes of PDE5A upon cGMP binding. (a) Superposition between PDE5A1 GAF-A domains of the X-ray (green) and NMR (golden) structures. The purple sticks represent cGMP. (b) Upon cGMP binding, helix H4 and strand B3 moved toward each other and may thus serve as the "lids" to gate the binding pocket. (See the color version of the figure in Color Plates section.)

GAF-B, which was assumed to serve as an allosteric relay of cGMP signaling from GAF-A to GAF-B [40]. On the other hand, PDE2 may have different pathway of the allosteric regulation from that of PDE5 because cGMP binds to GAF-A of PDE5 but GAF-B of PDE2. The structural studies showed that PDE2A can form two different

(a)

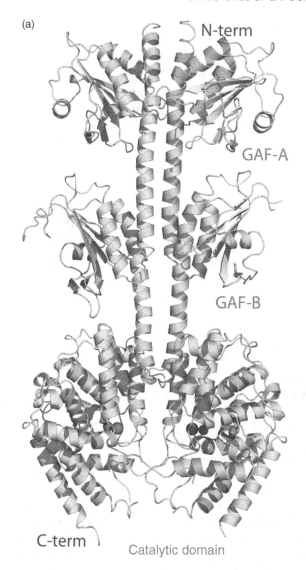

FIGURE 6.8 PDE2 dimers. (a) Parallel dimer of human unliganded PDE2A. (b) Cross dimer of cGMP-bound mouse PDE2A. (See the color version of the figure in Color Plates section.)

types of dimers, depending on the status of cGMP binding (Figure 6.8a and b) . The unliganded PDE2A forms a parallel dimer for its GAF domain [43] while the PDE2–cGMP complex is a dumbbell-shaped "X" dimer [38]. The swap of the GAF-B domains upon cGMP binding may imply a regulation of allosteric conformational changes in PDE2A. However, it could not be ruled out that the positional swap is a crystallization artifact resulting from postmodification during the protein preparation.

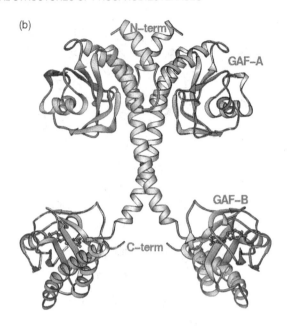

(b)

FIGURE 6.8   *(Continued)*

STRUCTURES OF THE LARGE PDE FRAGMENTS AND IMPLICATION ON DESIGN OF ALLOSTERIC MODULATORS

Recently, a fragment of the unliganded human PDE2A3 (residues 215–900), which contains both regulatory GAF and catalytic domains, has been determined [43]. The structure shows that two GAF domains of PDE2A assemble into a parallel dimer, and two catalytic domains are located underneath the GAF-B domain, but swap their positions (Figure 6.8a). An unusual observation is that the active sites of the PDE2A3 dimer are occupied by the H-loops (residues 702–723) of the opposite catalytic domains. Thus, PDE2A in this crystal form is inactive and the H-loop has to move away from the active site for catalysis. To explain this, an allosteric conformation change on the H-loop and positional separation of the two catalytic domains were proposed for the mechanism of the PDE2A catalysis [43]. This argument is consistent with the observation that the regulatory site in the GAF-B domain is distant from and does not directly interact with the catalytic site.

The assembly of the unliganded PDE2A dimer is the same as that of the unliganded PDE5A1 [40]. However, a superposition between the entire GAF dimers of PDE2A and PDE5A showed significant positional differences in many parts of the structures although their GAF-As and GAF-Bs are separately comparable [40]. This difference may imply that the regulatory mechanisms by cGMP are different between PDE2 and PDE5, because cGMP binds to GAF-A of PDE5 and GAF-B of PDE2, respectively.

Another structure of the large PDE fragments is the PDE4D catalytic domain plus a portion of UCR2 [44]. The structure (Figure 6.9) shows that an α-helix (residues

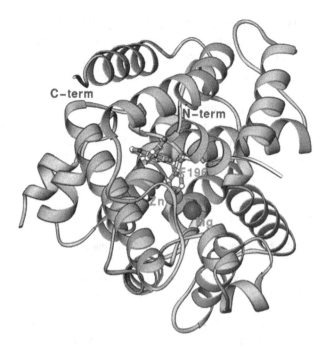

FIGURE 6.9 Interaction of an allosteric modulator with UCR2 helix and the active site of PDE4D.

Asn191 to Asp203) in the UCR2 subdomain interacts with the inhibitors that primarily bind to the active site. This study is helpful for the design of allosteric modulators that interact with both active site and the regulatory UCR2, and leads to the discovery of PDE4 inhibitors that have the potential for treatment of CNS diseases such as schizophrenia [44]. However, since the linker region (about 50 amino acids) between UCR2 and the catalytic domain was disordered in the structures, the real position of the UCR2 helix in biological systems is uncertain, especially in the absence of the inhibitors. Besides, the interaction between the UCR2 residues and the PDE4 inhibitors could be the consequence of the inhibitor binding and thus the structure-based design of allosteric regulators would be complicated.

CONCLUDING REMARKS

1. The active sites of PDE families show subtle differences in the shape and size of the pockets, thus implying that a group of residues are involved in recognition of the cyclic nucleotides and inhibitors.
2. While the active sites of PDE4 subfamilies are rigid, PDE5 shows six different conformations of the H-loop upon inhibitor binding, implying a competitive inhibition mechanism for PDE4 and an allosteric modulation for PDE5.

3. Individual GAF-A or GAF-B domain is comparable among the PDE families, but their assemblies could not be accurately superimposed, suggesting different components involved in the signaling pathways in the PDE families.

4. The structure of the PDE2A fragment containing both GAF and catalytic domains shows that the regulatory site in the GAF domain does not directly interact with the active site, implying an allosteric regulation on catalysis.

5. The interactions between the UCR2 helix and the PDE4 inhibitors provide useful information for the design of PDE4 subfamily-selective inhibitors. However, disorder of the linker region between UCR2 and the catalytic domain may imply a complication on the design of the allosteric modulators.

REFERENCES

1. Antoni, F. (2000) Molecular diversity of cyclic AMP signaling. *Front. Neuroendocrinol.*, **21**:103–132.

2. Zaccolo, M. and Movsesian, M.A.A. (2007) cAMP and cGMP signaling cross-talk: role of phosphodiesterases and implications for cardiac pathophysiology. *Circ. Res.*, **100**:1569–1578.

3. De Felice, F.G., et al. (2007) Cyclic AMP enhancers and Abeta oligomerization blockers as potential therapeutic agents in Alzheimer's disease. *Curr. Alzheimer Res.*, **4**:263–271.

4. Jarnaess, E. and Taskén, K. (2007) Spatiotemporal control of cAMP signalling processes by anchored signalling complexes. *Biochem. Soc. Trans.*, **35**:931–937.

5. Piper, M., van Horck, F., and Holt, C. (2007) The role of cyclic nucleotides in axon guidance. *Adv. Exp. Med. Biol.*, **621**:134–143.

6. O'Neill, J.S., Maywood, E.S., Chesham, J.E., Takahashi, J.S., and Hastings, M.H. (2008) cAMP-dependent signaling as a core component of the mammalian circadian pacemaker. *Science*, **320**:949–953.

7. Horvath, A. and Stratakis, C.A. (2008) Unraveling the molecular basis of micronodular adrenal hyperplasia. *Curr. Opin. Endocrinol. Diabetes Obes.*, **15**:227–233.

8. Bender, A.T. and Beavo, J.A. (2006) Cyclic nucleotide phosphodiesterases: molecular regulation to clinical use. *Pharmacol. Rev.*, **58**:488–520.

9. Omori, K. and Kotera, J. (2007) Overview of PDEs and their regulation. *Circ. Res.*, **100**:309–327.

10. Conti, M. and Beavo, J. (2007) Biochemistry and physiology of cyclic nucleotide phosphodiesterases: essential components in cyclic nucleotide signaling. *Annu. Rev. Biochem.*, **76**:481–511.

11. Mehats, C., Andersen, C.B., Filopanti, M., Jin, S.L., and Conti, M. (2002) Cyclic nucleotide phosphodiesterases and their role in endocrine cell signaling. *Trends Endocrinol. Metab.*, **13**:29–35.

12. Rotella, D.P. (2002) Phosphodiesterase 5 inhibitors: current status and potential applications. *Nat. Rev. Drug Discov.*, **1**:674–682.

13. Lipworth, B.J. (2005) Phosphodiesterase-4 inhibitors for asthma and chronic obstructive pulmonary disease. *Lancet*, **365**:167–175.

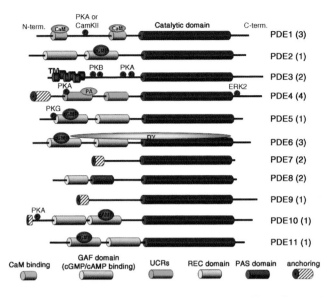

FIGURE 1.2 The domain organization of the different families of phosphodiesterases. Domains are depicted as "barrels" connected by "wires" indicating linker regions. (*See text for full caption.*)

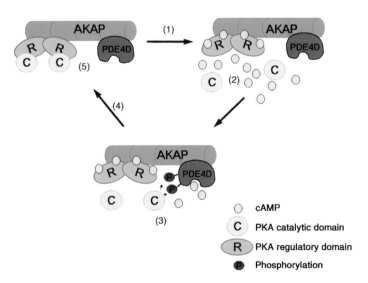

FIGURE 1.4 Localized feedback regulation mediated by a macromolecular complex tethering PKA and PDE4D. (*See text for full caption.*)

Cyclic-Nucleotide Phosphodiesterases in the Central Nervous System: From Biology to Drug Discovery, First Edition. Edited by Nicholas J. Brandon and Anthony R. West.
© 2014 John Wiley & Sons, Inc. Published 2014 by John Wiley & Sons, Inc.

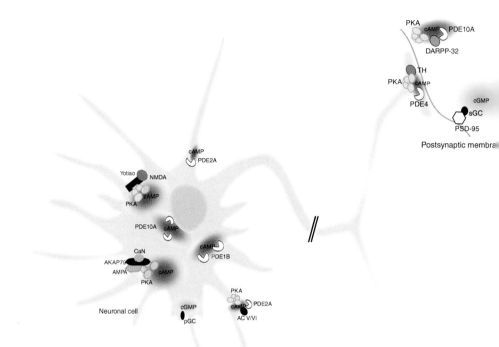

FIGURE 3.1 Compartmentalization of cyclic nucleotides in neurons. (*See text for full caption.*)

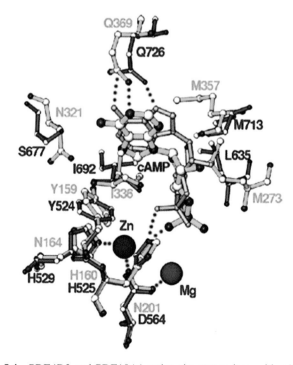

FIGURE 5.4 PDE4D2 and PDE10A1 active site comparison with cAMP bound.

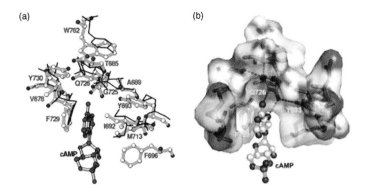

FIGURE 5.5 Hypothetical S-pocket in PDE10A.

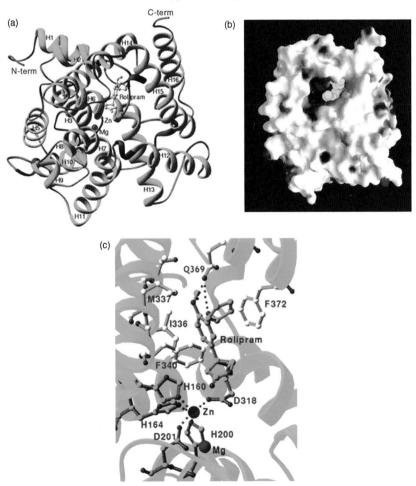

FIGURE 6.1 Structure of the catalytic domain of PDE4D. (*See text for full caption.*)

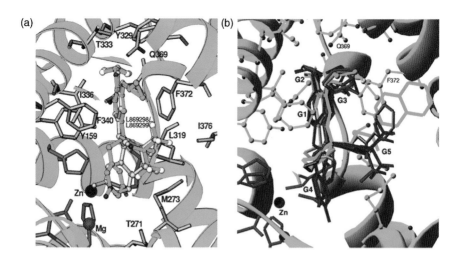

FIGURE 6.2 Binding of PDE4 inhibitors. (*See text for full caption.*)

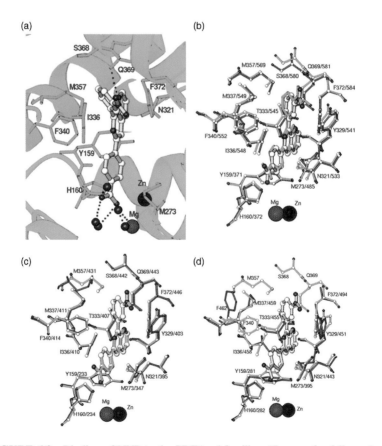

FIGURE 6.3 Binding of NVP to the PDE4 subfamilies. (*See text for full caption.*)

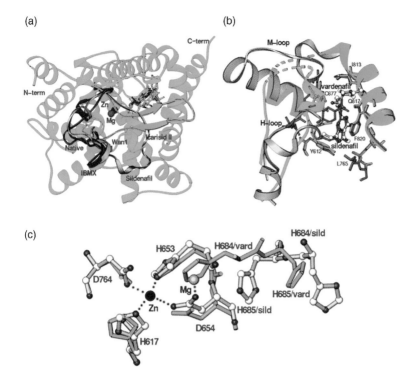

FIGURE 6.4 Structures of PDE5 catalytic domain. (*See text for full caption.*)

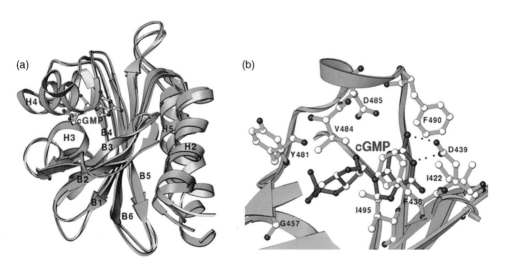

FIGURE 6.6 Ribbon diagram of PDE2A GAF domains. (a) Superposition of GAF-A (green color) over GAF-B (golden) of mouse PDE2A. (b) Binding of cGMP to GAF-B of mouse PDE2A.

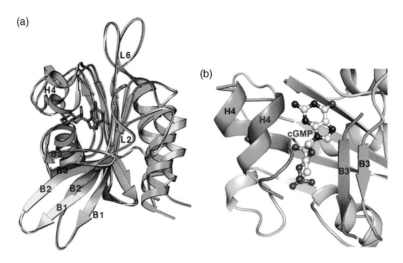

FIGURE 6.7 Conformational changes of PDE5A upon cGMP binding. (*See text for full caption.*)

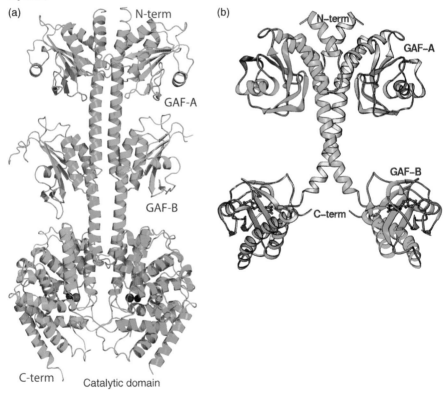

FIGURE 6.8 PDE2 dimers. (*See text for full caption.*)

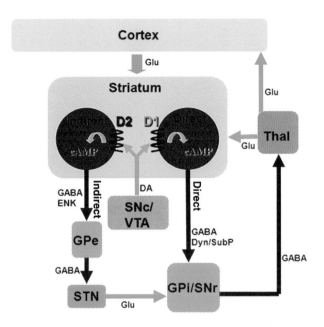

FIGURE 10.3 High PDE10A expression is restricted to both MSN populations of the neostriatum. (*See text for full caption.*)

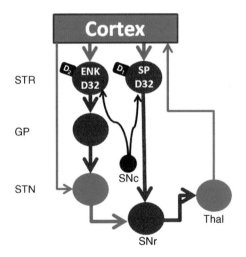

FIGURE 12.1 Schematic showing neuronal pathways that connect nuclei of the basal ganglia. (*See text for full caption.*)

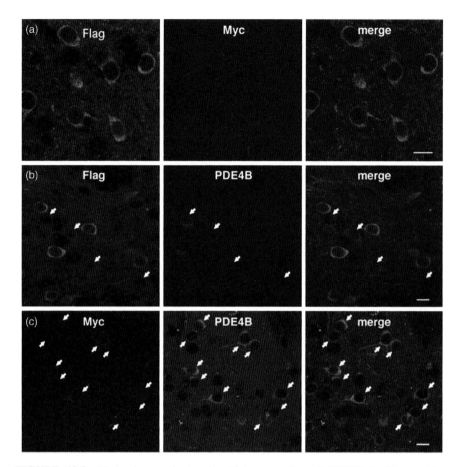

FIGURE 12.2 Photomicrograph showing high expression of PDE4B in Myc-positive, striatopallidal neurons in the striatum of D_1-DARPP-32-Flag/D_2-DARPP-32-Myc mutant mice. (a) Flag- and Myc-tagged DARPP-32 are expressed in striatonigral and striatopallidal neurons, respectively, in the striatum of D_1-DARPP-32-Flag/D_2-DARPP-32-Myc mutant mice. (b) Striatal sections from D_1-DARPP-32-Flag/D_2-DARPP-32-Myc mutant mice double labeled with antibodies against Flag and PDE4B. (c) Striatal sections from D_1-DARPP-32-Flag/D_2-DARPP-32-Myc mutant mice double labeled with antibodies against Myc and PDE4B. Striatal neurons showing strong PDE4B immunoreactivity, indicated by arrows, correspond to Myc-positive, striatopallidal neurons. Scale bars: 10 μm.

14. Houslay, M.D., Schafer, P., and Zhang, K.Y. (2005) Keynote review: phosphodiesterase-4 as a therapeutic target. *Drug Discov. Today*, **10**:1503–1519.

15. Blokland, A., Schreiber, R., and Prickaerts, J. (2006) Improving memory: a role for phosphodiesterases. *Curr. Pharm. Des.*, **12**:2511–2523.

16. Menniti, F.S., Faraci, W.S., and Schmidt, C.J. (2006) Phosphodiesterases in the CNS: targets for drug development. *Nat. Rev. Drug Discov.*, **5**:660–670.

17. Lerner, A. and Epstein, P.M. (2006) Cyclic nucleotide phosphodiesterases as targets for treatment of haematological malignancies. *Biochem. J.*, **393**:21–41.

18. Ghavami, A., Hirst, W.D., and Novak, T.J. (2006) Selective phosphodiesterase (PDE)-4 inhibitors: a novel approach to treating memory deficit? *Drugs R&D*, **7**:63–71.

19. Dastidar, S.G., Rajagopal, D., and Ray, A. (2007) Therapeutic benefit of PDE4 inhibitors in inflammatory diseases. *Curr. Opin. Investig. Drugs*, **8**:364–372.

20. Xu, R.X., et al. (2000) Atomic structure of PDE4: insight into phosphodiesterase mechanism and specificity. *Science*, **288**:1822–1825.

21. Zhang, K.Y., et al. (2004) A glutamine switch mechanism for nucleotide selectivity by phosphodiesterases. *Mol. Cell*, **15**:279–286.

22. Iffland, A., et al. (2005) Structural determinants for inhibitor specificity and selectivity in PDE2A using the wheat germ *in vitro* translation system. *Biochemistry*, **44**:8312–8325.

23. Scapin, G., et al. (2004) Crystal structure of human phosphodiesterase 3B: atomic basis for substrate and inhibitor specificity. *Biochemistry*, **43**:6091–6100.

24. Huai, Q., et al. (2003) Three dimensional structures of PDE4D in complex with roliprams and implication on inhibitor selectivity. *Structure*, **11**:865–873.

25. Wang, H., et al. (2007) Structures of the four subfamilies of phosphodiesterase-4 provide insight into the selectivity of their inhibitors. *Biochem. J.*, **408**:193–201.

26. Sung, B.J., et al. (2003) Structure of the catalytic domain of human phosphodiesterase 5 with bound drug molecules. *Nature*, **425**:98–102.

27. Huai, Q., Liu, Y., Francis, S.H., Corbin, J.D., and Ke, H. (2004) Crystal structures of phosphodiesterases 4 and 5 in complex with inhibitor IBMX suggest a conformation determinant of inhibitor selectivity. *J. Biol. Chem.*, **279**:13095–13101.

28. Wang, H., Liu, Y., Chen, Y., Robinson, H., and Ke, H. (2005) Multiple elements jointly determine inhibitor selectivity of cyclic nucleotide phosphodiesterases 4 and 7. *J. Biol. Chem.*, **280**:30949–30955.

29. Wang, H., Ye, M., Robinson, H., Francis, S.H., and Ke, H. (2008) Conformational variations of both PDE5 and inhibitors provide the structural basis for the physiological effects of vardenafil and sildenafil. *Mol. Pharmacol.*, **73**:104–110.

30. Huai, Q., et al. (2004) Crystal structure of phosphodiesterase 9 in complex with inhibitor IBMX. *Proc. Natl. Acad. Sci. USA*, **101**:9624–9629.

31. Liu, S., et al. (2008) Structural basis for the catalytic mechanism of human phosphodiesterase 9. *Proc. Natl. Acad. Sci. USA*, **105**:13309–13314.

32. Verhoest, P.R., et al. (2009) Identification of a brain penetrant PDE9A inhibitor utilizing prospective design and chemical enablement as a rapid lead optimization strategy. *J. Med. Chem.*, **52**:7946–7949.

33. Wang, H., et al. (2010) Insight into binding of phosphodiesterase-9A selective inhibitors by crystal structures and mutagenesis. *J. Med. Chem.*, **53**:1726–1731.

34. Wang, H., et al. (2007) Structural insight into substrate specificity of phosphodiesterase 10. *Proc. Natl. Acad. Sci. USA*, **104**:5782–5787.

35. Chappie, T.A., et al. (2007) Discovery of a series of 6,7-dimethoxy-4-pyrrolidylquinazoline PDE10A inhibitors. *J. Med. Chem.*, **50**:182–185.

36. Verhoest, P.R., et al. (2009) Discovery of a novel class of phosphodiesterase 10A inhibitors and identification of clinical candidate 2-[4-(1-methyl-4-pyridin-4-yl-1*H*-pyrazol-3-yl)-phenoxymethyl]-quinoline (PF-2545920) for the treatment of schizophrenia. *J. Med. Chem.*, **52**:5188–5196.

37. Höfgen, N., et al. (2010) Discovery of imidazo[1,5-*a*]pyrido[3,2-*e*]pyrazines as a new class of phosphodiesterase 10A inhibitors. *J. Med. Chem.*, **53**:4399–4411.

38. Martinez, S.E., et al. (2002) The two GAF domains in phosphodiesterase 2A have distinct roles in dimerization and in cGMP binding. *Proc. Natl. Acad. Sci. USA*, **99**:13260–13265.

39. Heikaus, C.C., et al. (2008) Solution structure of the cGMP binding GAF domain from phosphodiesterase 5: insights into nucleotide specificity, dimerization, and cGMP-dependent conformational change. *J. Biol. Chem.*, **283**:22749–22759.

40. Wang, H., Robinson, H., and Ke, H. (2010) Conformation changes, N-terminal involvement and cGMP signal relay in phosphodiesterase-5 GAF domain. *J. Biol. Chem.*, **285**:38149–38156.

41. Martinez, S.E., Heikaus, C.C., Klevit, R.E., and Beavo, J.A. (2008) The structure of the GAF A domain from phosphodiesterase 6C reveals determinants of cGMP binding, a conserved binding surface, and a large cGMP-dependent conformational change. *J. Biol. Chem.*, **283**:25913–25919.

42. Handa, N., et al. (2008) Crystal structure of the GAF-B domain from human phosphodiesterase 10A complexed with its ligand, cAMP. *J. Biol. Chem.*, **283**:19657–17664.

43. Pandit, J., Forman, M.D., Fennell, K.F., Dillman, K.S., and Menniti, F.S. (2009) Mechanism for the allosteric regulation of phosphodiesterase 2A deduced from the X-ray structure of a near full-length construct. *Proc. Natl. Acad. Sci. USA*, **106**:18225–18230.

44. Burgin, A.B., et al. (2010) Design of phosphodiesterase 4D (PDE4D) allosteric modulators for enhancing cognition with improved safety. *Nat. Biotechnol.*, **28**:63–70.

45. Recht, M.I., et al. (2012) Fragment-based screening for inhibitors of PDE4A using enthalpy arrays and X-ray crystallography. *J. Biomol. Screen.*, **17**:469–480.

46. Xu, R.X., et al. (2004) Crystal structures of the catalytic domain of phosphodiesterase 4B complexed with AMP, 8-Br-AMP, and rolipram. *J. Mol. Biol.*, **337**:355–365.

47. Kranz, M., et al. (2009) Identification of PDE4B over 4D subtype-selective inhibitors revealing an unprecedented binding mode. *Bioorg. Med. Chem.*, **17**:5336–5341.

48. Govek, S.P., et al. (2010) Water-soluble PDE4 inhibitors for the treatment of dry eye. *Bioorg. Med. Chem. Lett.*, **20**:2928–2932.

49. Woodrow, M.D., et al. (2009) Quinolines as a novel structural class of potent and selective PDE4 inhibitors. Optimisation for inhaled administration. *Bioorg. Med. Chem. Lett.*, **19**:5261–5265.

50. Lunniss, C.J., et al. (2009) Quinolines as a novel structural class of potent and selective PDE4 inhibitors: optimisation for oral administration. *Bioorg. Med. Chem. Lett.*, **19**:1380–1385.

51. Hamblin, J.N., et al. (2008) Pyrazolopyridines as a novel structural class of potent and selective PDE4 inhibitors. *Bioorg. Med. Chem. Lett.*, **18**:4237–4241.

52. Card, G.L., et al. (2004) Structural basis for the activity of drugs that inhibit phosphodiesterases. *Structure*, **12**:2233–2247.

53. Card, G.L., et al. (2005) A family of phosphodiesterase inhibitors discovered by cocrystallography and scaffold-based drug design. *Nat. Biotechnol.*, **23**:201–207.

54. Mitchell, C.J., et al. (2010) Pyrazolopyridines as potent PDE4B inhibitors: 5-heterocycle SAR. *Bioorg. Med. Chem. Lett.*, **20**:5803–5806.

55. Smith, K.J., Scotland, G., Beattie, J., Trayer, I.P., and Houslay, M.D. (1996) Determination of the structure of the N-terminal splice region of the cyclic AMP-specific phosphodiesterase RD1 (RNPDE4A1) by ^{1}H NMR and identification of the membrane association domain using chimeric constructs. *J. Biol. Chem.*, **271**:16703–16711.

56. Lee, M.E., Markowitz, J., Lee, J.O., and Lee, H. (2002) Crystal structure of phosphodiesterase 4D and inhibitor complex (1). *FEBS Lett.*, **530**:53–58.

57. Huai, Q., Colicelli, J., and Ke, H. (2003) The crystal structure of AMP-bound PDE4 suggests a mechanism for phosphodiesterase catalysis. *Biochemistry*, **42**:13220–13226.

58. Huai, Q., et al. (2006) Enantiomer discrimination illustrated by high resolution crystal structures of type 4 phosphodiesterase. *J. Med. Chem.*, **49**:1867–1873.

59. Wang, H., Robinson, H., and Ke, H. (2007) The molecular basis for recognition of different substrates by phosphodiesterase families 4 and 10. *J. Mol. Biol.*, **371**:302–307.

60. Nankervis, J.L., et al. (2011) Thiophene inhibitors of PDE4: crystal structures show a second binding mode at the catalytic domain of PDE4D2. *Bioorg. Med. Chem. Lett.*, **21**:7089–7093.

61. Wang, H., et al. (2006) Multiple conformations of phosphodiesterase-5: implications for enzyme function and drug development. *J. Biol. Chem.*, **281**:21469–21479.

62. Chen, G., et al. (2008) An insight into the pharmacophores of phosphodiesterase-5 inhibitors from synthetic and crystal structural studies. *Biochem. Pharmacol.*, **75**:1717–178.

63. Allerton, C.M., et al. (2006) A novel series of potent and selective PDE5 inhibitors with potential for high and dose-independent oral bioavailability. *J. Med. Chem.*, **49**:3581–3594.

64. Hughes, R.O., et al. (2009) Optimization of the aminopyridopyrazinones class of PDE5 inhibitors: discovery of 3-[(*trans*-4-hydroxycyclohexyl)amino]-7-(6-methoxypyridin-3-yl)-1-(2-propoxyethyl)pyrido[3,4-*b*]pyrazin-2(1*H*)-one. *Bioorg. Med. Chem. Lett.*, **19**:5209–5213.

65. Barren, B., et al. (2009) Structural basis of phosphodiesterase 6 inhibition by the C-terminal region of the gamma-subunit. *EMBO J.*, **28**:3613–3622.

66. Song, J., et al. (2008) Intrinsically disordered gamma-subunit of cGMP phosphodiesterase encodes functionally relevant transient secondary and tertiary structure. *Proc. Natl. Acad. Sci. USA*, **105**:1505–1510.

67. Castaño, T., et al. (2009) Phosphodiesterase 7 inhibitors: understanding their therapeutical relevance. *ChemMedChem*, **4**:866–876.

68. Wang, H., et al. (2008) Kinetic and structural studies of phosphodiesterase-8A and implication on the inhibitor selectivity. *Biochemistry*, **47**:12760–12768.

69. Verhoest, P.R., et al. (2009) Identification of a brain penetrant PDE9A inhibitor utilizing prospective design and chemical enablement as a rapid lead optimization strategy. *J. Med. Chem.*, **52**:7946–7949.

70. Hou, J., et al. (2011) Structural asymmetry of phosphodiesterase-9, potential protonation of a glutamic acid, and role of the invariant glutamine. *PLOS One*, **6**:e18092–e18092.

71. Chappie, T.A., et al. (2007) Discovery of a series of 6,7-dimethoxy-4-pyrrolidylquinazoline PDE10A inhibitors. *J. Med. Chem.*, **50**:182–185.

72. Höfgen, N., et al. (2010) Discovery of imidazo[1,5-*a*]pyrido[3,2-*e*]pyrazines as a new class of phosphodiesterase 10A inhibitors. *J. Med. Chem.*, **53**:4399–4411.

73. Verhoest, P.R., et al. (2009) Discovery of a novel class of phosphodiesterase 10A inhibitors and identification of clinical candidate 2-[4-(1-methyl-4-pyridin-4-yl-1*H*-pyrazol-3-yl)-phenoxymethyl]-quinoline (PF-2545920) for the treatment of schizophrenia. *J. Med. Chem.*, **52**:5188–5196.

74. Helal, C.J., et al. (2011) Use of structure-based design to discover a potent, selective, *in vivo* active phosphodiesterase 10A inhibitor lead series for the treatment of schizophrenia. *J. Med. Chem.*, **54**:4536–4547.

75. Kehler, J., et al. (2011) Triazoloquinazolines as a novel class of phosphodiesterase 10A (PDE10A) inhibitors. *Bioorg. Med. Chem. Lett.*, **21**:3738–3742.

76. Malamas, M.S., et al. (2011) Highly potent, selective, and orally active phosphodiesterase 10A inhibitors. *J. Med. Chem.*, **54**:7621–7638.

77. Yang, S.W., et al. (2012) Discovery of orally active pyrazoloquinolines as potent PDE10 inhibitors for the management of schizophrenia. *Bioorg. Med. Chem. Lett.*, **22**:235–239.

78. Ho, G.D., et al. (2012) The SAR development of dihydroimidazoisoquinoline derivatives as phosphodiesterase 10A inhibitors for the treatment of schizophrenia. *Bioorg. Med. Chem. Lett.*, **22**:2585–2589.

79. Ho, G.D., et al. (2012) The discovery of potent, selective, and orally active pyrazoloquinolines as PDE10A inhibitors for the treatment of schizophrenia. *Bioorg. Med. Chem. Lett.*, **22**:1019–1022.

80. Hu, E., et al. (2012) Discovery of potent, selective, and metabolically stable 4-(pyridin-3-yl)cinnolines as novel phosphodiesterase 10A (PDE10A) inhibitors. *Bioorg. Med. Chem. Lett.*, **22**:2262–2265.

81. Wang, H., et al. (2012) Biological and structural characterization of *Trypanosoma cruzi* phosphodiesterase C and implications for design of parasite selective inhibitors. *J. Biol. Chem.*, **287**:11788–11797.

82. Wang, H., et al. (2007) Crystal structure of the *Leishmania major* phosphodiesterase LmjPDEB1 and insight into the design of the parasite-selective inhibitors. *Mol. Microbiol.*, **66**:1029–1038.

83. Manallack, D.T., Hughes, R.A., and Thompson, P.E. (2005) The next generation of phosphodiesterase inhibitors: structural clues to ligand and substrate selectivity of phosphodiesterases. *J. Med. Chem.*, **48**:3449–3462.

84. Ke, H. and Wang, H. (2007) Crystal structures of phosphodiesterases and implications on substrate specificity and inhibitor selectivity. *Curr. Top. Med. Chem.*, **7**:391–403.

85. Fisher, D.A., Smith, J.F., Pillar, J.S., St. Denis, S.H., and Cheng, J.B. (1998) Isolation and characterization of PDE9A, a novel human cGMP-specific phosphodiesterase. *J. Biol. Chem.*, **273**:15559–15564.

86. Wang, P., Wu, P., Egan, R.W., and Billah, M.M. (2003) Identification and characterization of a new human type 9 cGMP-specific phosphodiesterase splice variant (PDE9A5) differential tissue distribution and subcellular localization of PDE9A variants. *Gene*, **314**:15–27.

87. Francis, S.H., et al. (2006) Phosphodiesterase 5: molecular characteristics relating to structure, function, and regulation. In: Beavo, J.A., Francis, S.H., and Houslay, M.D. (Eds.), *Cyclic Nucleotide Phosphodiesterases in Health and Disease*, Boca Raton, FL: CRC Press, pp. 131–164,

88. Huang, Z., Ducharme, Y., MacDonald, D., and Robinchaud, A. (2001) The next generation of PDE4 inhibitors. *Curr. Opin. Chem. Biol.*, **5**:432–438.

89. Wang, P., Wu, P., Ohleth, K.M., Egan, R.W., and Billah, M.M. (1999) Phosphodiesterase 4B2 is the predominant phosphodiesterase species and undergoes differential regulation of gene expression in human monocytes and neutrophils. *Mol. Pharmacol.*, **56**:170–174.

90. Peter, D., Jin, S.L., Conti, M., Hatzelmann, A., and Zitt, C. (2007) Differential expression and function of phosphodiesterase 4 (PDE4) subtypes in human primary CD4[+] T cells: predominant role of PDE4D. *J. Immunol.*, **178**:4820–4831.

91. Perry, S.J., et al. (2002) Targeting of cyclic AMP degradation to beta 2-adrenergic receptors by beta-arrestins. *Science*, **298**:834–836.

92. Hersperger, R., Bray-French, K., Mazzoni, L., and Muller, T. (2000) Palladium-catalyzed cross-coupling reactions for the synthesis of 6,8-disubstituted 1,7-naphthyridines: a novel class of potent and selective phosphodiesterase type 4D inhibitors. *J. Med. Chem.*, **43**:675–682.

93. Woodrow, M.D., et al. (2009) Quinolines as a novel structural class of potent and selective PDE4 inhibitors. Optimisation for inhaled administration. *Bioorg. Med. Chem. Lett.*, **19**:5261–5265.

94. Mitchell, C.J., et al. (2010) Pyrazolopyridines as potent PDE4B inhibitors: 5-heterocycle SAR. *Bioorg. Med. Chem. Lett.*, **20**:5803–5806.

95. Fink, T.L., Francis, H., Beasley, A., Grimes, K.A., and Corbin, J.D. (1999) Expression of an active, monomeric catalytic domain of the cGMP-binding cGMP-specific phosphodiesterase (PDE5). *J. Biol. Chem.*, **274**:24613–24620.

96. Mehrotra, N., Gupta, M., Kovar, A., and Meibohm, B. (2007) The role of pharmacokinetics and pharmacodynamics in phosphodiesterase-5 inhibitor therapy. *Int. J. Impot. Res.*, **19**:253–264.

97. Aravind, L. and Ponting, C.P. (1997) The GAF domain: an evolutionary link between diverse phototransducing proteins. *Trends Biochem. Sci.*, **22**:458–459.

98. Schultz, J.E. (2009) Structural and biochemical aspects of tandem GAF domains. *Handb. Exp. Pharmacol.*, **191**:93–109.

99. Zoraghi, R., Corbin, J.D., and Francis, S.H. (2004) Properties and functions of GAF domains in cyclic nucleotide phosphodiesterases and other proteins. *Mol. Pharmacol.*, **65**:267–278.

100. Wu, A.Y., Tang, X.B., Martinez, S.E., Ikeda, K., and Beavo, J.A. (2004) Molecular determinants for cyclic nucleotide binding to the regulatory domains of phosphodiesterase 2A. *J. Biol. Chem.*, **279**:37928–37938.

101. Huang, D., Hinds, T.R., Martinez, S.E., Doneanu, C., and Beavo, J.A. (2004) Molecular determinants of cGMP binding to chicken cone photoreceptor phosphodiesterase. *J. Biol. Chem.*, **279**:48143–48151.

102. Gross-Langenhoff, M., Hofbauer, K., Weber, J., Schultz, A., and Schultz, J.E. (2006) cAMP is a ligand for the tandem GAF domain of human phosphodiesterase 10 and

cGMP for the tandem GAF domain of phosphodiesterase 11. *J. Biol. Chem.*, **281**: 2841–2846.

103. Corbin, J.D., Zoraghi, R., and Francis, S.H. (2009) Allosteric-site and catalytic-site ligand effects on PDE5 functions are associated with distinct changes in physical form of the enzyme. *Cell Signal.*, **21**:1768–1774.

104. Matthiesen, K. and Nielsen, J. (2009) Binding of cyclic nucleotides to phosphodiesterase 10A and 11A GAF domains does not stimulate catalytic activity. *Biochem. J.*, **423**: 401–409.

CHAPTER 7

INHIBITION OF CYCLIC NUCLEOTIDE PHOSPHODIESTERASES TO REGULATE MEMORY

HAN-TING ZHANG[*]
Departments of Behavioral Medicine & Psychiatry and Physiology & Pharmacology, West Virginia University Health Sciences Center, Morgantown, WV, USA

YING XU[*] and JAMES O'DONNELL
School of Pharmacy & Pharmaceutical Sciences, The State University of New York at Buffalo, Buffalo, NY, USA

INTRODUCTION

Phosphodiesterases (PDEs), a superfamily of enzymes that catalyze the hydrolysis of the second messengers cyclic AMP (cAMP) and cyclic GMP (cGMP), have been studied for over six decades since the first report by Zittle and Reading on the discovery of PDE in rats and rabbits [1]. Sixteen years later, the first PDE was isolated and purified by Butcher and Sutherland [2]. In 1970, Beavo et al. isolated and characterized PDEs in rat and bovine tissues [3]. Since these initial pioneering studies, 11 PDE families (PDE1-11) have been identified based on structural similarities, ability to hydrolyze cyclic nucleotides (cAMP and cGMP), cellular and subcellular distributions, sensitivity to endogenous modulators (cGMP, Ca^{2+}, and calmodulin), and pharmacological inhibitors [4–8]. PDEs are classified into three categories based on different substrates: cAMP-specific PDEs (PDE4, PDE7, and PDE8), cGMP-specific PDEs (PDE5, PDE6, and PDE9), and dual substrate PDEs (PDE1–PDE3, PDE10, and PDE11). Each family is encoded by 1–4 distinct genes, leading to a total of 21 PDE isoforms (see Chapter 1 for more details). All PDEs have three functional domains: the regulatory N-terminus, the conserved catalytic domain, and the C-terminus [9–11]. The N-terminal domain varies among PDEs and may play a

[*] Authors contributed equally.

Cyclic-Nucleotide Phosphodiesterases in the Central Nervous System: From Biology to Drug Discovery, First Edition. Edited by Nicholas J. Brandon and Anthony R. West.
© 2014 John Wiley & Sons, Inc. Published 2014 by John Wiley & Sons, Inc.

role in the regulation of catalytic activity and/or subcellular localization of PDEs [12,13]. N-terminal domains may contain different sequences, including calmodulin binding protein in PDE1, cGMP binding domain in PDE2, and a transducin binding domain in PDE6. They may also exert similar structural and functional features such as phosphorylation sites for protein kinases, such as PKA, PKG, in PDE1–PDE5, which regulate activity of the PDE enzymes.

Most PDEs are expressed in the brain at high levels. The relative expression levels vary depending on specific brain regions and PDE families, or PDE isoforms from the same family, indicating different contributions of individual PDEs to CNS functions (see Chapter 2). For instance, PDE2, PDE4, and PDE9 are highly expressed in the cerebral cortex, hippocampus, amygdala, and the olfactory system [14–19]; PDE1, PDE7, and PDE10 are expressed at high levels in the striatum, whereas PDE11 is abundant in the dorsal root ganglia [18,20–22] and hippocampus [23]. In contrast, PDE3, PDE5, and PDE8 are expressed at relatively low levels in the brain, although the expression may vary depending on the specific PDE isoforms in question [20,24,25]; PDE3 is primarily expressed in peripheral tissues. The distribution patterns of PDE2, PDE4, and PDE9 are consistent with the role of these enzymes in the mediation of factors associated with depression [8,26], anxiety [27–29], and/or memory [30–35]. Furthermore, PDE1, PDE10, and PDE11 have been shown to contribute to cognition and factors associated with schizophrenia [23,36–38]. Overall, given that cAMP and cGMP are important second messengers in the mediation of memory processes, PDEs highly or relatively highly expressed in the brain (all except PDE3 and PDE6) are potential targets for mediating memory. These PDE enzymes are involved in cognition associated with neuropsychiatric disorders such as schizophrenia (PDE1 and PDE10) and neurodegenerative disorders such as Alzheimer's disease (PDE4 and PDE5). Inhibitors of these PDEs typically enhance memory and may reverse cognitive dysfunction or memory deficits associated with these diseases.

PDE1 AND MEMORY

PDE1 Distribution in the Brain and Implication in Memory

PDE1 was first identified in 1970 in the rat brain [39]. It is characterized by Ca^{2+}-dependent stimulation via the Ca^{2+} binding protein calmodulin (CaM). PDE1 is the only PDE family that directly links Ca^{2+} signaling to the cyclic nucleotide signaling pathways [40]. It has three subtypes – PDE1A, PDE1B, and PDE1C – encoded by distinct genes; they are different in their regulatory properties, specific activities, substrate affinities, activation constants for CaM, and tissue distributions [40–43]. The three PDE1 subtypes are differentially expressed in the brain. Specifically, PDE1A, which is distributed throughout the brain, is highly expressed in the hippocampus and cerebellum, but is found in low levels in the striatum [44,45]. PDE1B is predominantly expressed in brain regions with high levels of dopaminergic innervations, including the striatum, olfactory tubercle, and dentate gyrus [45–47]. This determines the particular role of PDE1B in behaviors mediated by the dopaminergic system. PDE1C is primarily expressed in the olfactory epithelium, cerebellum, and striatum [41,42]. Overall, based on the high expression of PDE1 in the striatum, hippocampus, and olfactory tract and its ability to interact with cAMP/ cGMP signaling, PDE1 may be involved in motor activity, learning and memory, and olfaction. Using pharmacological and genetic approaches, it has been demonstrated that PDE1 plays a role in the mediation of spatial memory and dopamine-associated cognition [38,48].

Effect of PDE1 Inhibitors on Memory

Vinpocetine, a synthetic derivative of vincamine that is extracted from the leaves of the lesser periwinkle plant (Vinca minor), is a potent PDE1 inhibitor. Studies have shown that vinpocetine facilitates long-term potentiation (LTP) [49,50], increases structural dynamics of dendritic spines [51], enhances memory retrieval [52], and reverses memory deficits produced by scopolamine [53] and streptozotocin [48], which are used to generate animal models associated with Alzheimer's disease. The memory-enhancing effect of vinpocetine is primarily attributed to its inhibition of PDE1, leading to increases in levels of cAMP and cGMP [54] and subsequent activation of the cAMP/ protein kinase A (PKA)/cAMP response element-binding protein (CREB) and cGMP/ protein kinase G (PKG)/CREB signaling cascades, respectively. Indeed, both pathways play an important role in memory processes [55–58]. This is supported by the findings that vinpocetine increases phosphorylation of CREB [59] and enhances memory [60]. In addition, the increase in cGMP levels improves cerebral blood flow [61], which may also contribute to memory enhancement. However, it has also been shown that PDE5–cGMP-mediated memory consolidation is independent of cerebrovascular effects [33]. Other mechanisms involved in the memory-enhancing effect of vinpocetine may include increases in cholinergic function [48], prevention of neuronal damage [62], and its antioxidant properties [48,63].

A novel, selective PDE1B inhibitor has been developed by Intra-Cellular Therapies Inc. [38]. Preclinical studies showed that this compound (name unreleased) enhanced memory in the novel object recognition task and attenuated catalepsy produced by antipsychotic drugs such as risperidone and haloperidol. The major advantage of this novel PDE1B inhibitor is that, compared to dopamine D1 receptor agonists, it increased wakefulness, but did not exhibit psychostimulant activity. The compound did not alter prepulse inhibition (PPI) or startle magnitude. Overall, PDE1B inhibitors may provide a novel approach with unique advantages over direct-acting D1 receptor agonists for treatment of cognitive dysfunction associated with schizophrenia (see Chapter 12).

Memory Changes in Mice Deficient in PDE1B

The role of PDE1B in the mediation of cognition has been confirmed using mice deficient in PDE1B, which is predominantly expressed in the striatum, in particular, in neurons expressing D1 receptors [45,47,64]. Consistent with this, mice deficient in PDE1B display increased striatal dopamine turnover (DOPAC/DA and HVA/DA ratios), increased basal motor activity, and exaggerated locomotor responses to dopaminergic stimulants such as amphetamine and methamphetamine [65,66], suggesting an important role of PDE1B in striatal functions. This may account for unaltered depression- and anxiety-like behaviors in PDE1B knockout mice [36,67].

The results of memory tests using PDE1B knockout mice are controversial. Reed et al. found that mice deficient in PDE1B displayed impairment of spatial memory in the Morris water maze test [65]. However, this was not confirmed by a follow-up study using the same test, although mice deficient in both PDE1B and DARPP32 (dopamine- and cAMP-regulated phosphoprotein, MW 32 kDa) displayed memory deficits in the Morris water maze test [67]. Consistent with the latter study, PDE1B knockout mice did not show any difference from their wild-type controls in passive avoidance and conditioned avoidance responding tasks [66]. No acquisition differences as seen in PDE1B inhibitor-treated mice have been observed in PDE1B knockout mice in the Morris water maze test [67]. One explanation for this discrepancy might be the difference in the memory tasks and/or procedures. Since PDE1B is particularly enriched in the striatum [45,47], the use of striatum-dependent, instead of hippocampus-dependent, memory tasks may help clarify this issue.

While PDE1A knockout mice are not available, it has been reported that mice deficient in PDE1C display reduced sensitivity and attenuated adaptation to repeated odor stimulation, as revealed using electro-olfactogram analysis [68]. These observations suggest a specific role of PDE1C in the regulation of olfactory transduction. Together, the three PDE1 isoforms are differentially expressed in the brain, leading to different roles in CNS functions. Given that PDE1B is predominantly expressed in striatal neurons expressing dopamine D1 receptors, PDE1B may play a unique role in D1 receptor signaling-mediated cognition. Therefore, PDE1B inhibitors may be a novel class of cognition enhancers for treatment of cognition dysfunction in schizophrenia [38].

PDE2 AND MEMORY

PDE2 Structure

PDE2 was first identified by Beavo et al. [3]. It is a dual-substrate PDE family that is activated by cGMP binding to its specific N-terminal domain. PDE2 is a 105 kDa homodimer with a relatively conserved C-terminal catalytic domain and a more variable N-terminal region; the latter contains different regulatory modules involved in the regulation of enzymatic activity, subcellular localization, and interactions with other proteins [69,70]. PDE2 activity is normally regulated through the N-terminal GAF domains, a small-molecule binding domain identified in more than 7400 proteins [71]. More specifically, the GAF domain interacts with cGMP, leading to the regulation of catalytic activity. Mammalian PDE2 contains a tandem of GAF domains in the N-terminal region designated GAF-A and GAF-B, which can functionally bind to cGMP. Studies have indicated that the structure of mammalian PDE GAF domains is a parallel dimer, whereas the structure of *Anabaena* cyclase GAF domains is an antiparallel dimer [72]. Despite the structural difference, mammalian GAF domains can replace those in the *Anabaena* adenylyl cyclase [70,73,74]. It is interesting that the chimeric protein is stimulated by cGMP when the PDE2 GAF domains are used to substitute for the GAF domains of cyclase using molecular biological techniques [75]. Studies on the structure of mammalian PDE GAF domains may make it possible to design drugs to target allosteric regulatory nucleotide binding sites of PDEs that contain GAF domains [72]. A similar design strategy has led to the synthesis of PDE4D allosteric modulators, which produce memory enhancement, but lack an emetic-like effect [32].

PDE2 Distribution in the Brain

PDE2 is found in the limbic system, such as cortical areas, hippocampus, amygdala, and olfactory bulb, and the adrenal cortex [76]. Since the limbic system is important for mood and cognition, expression of PDE2 in these brain regions suggests the potential role of PDE2 in signaling pathways involved in these processes. Neuronal projections from the olfactory bulb to the limbic system have a major influence on emotional behavior. PDE2 in olfactory neurons could modulate the opening of cyclic-nucleotide-gated ion channels by controlling intracellular cGMP concentrations [14,77], affecting neuronal activity downstream.

Role of PDE2 in Memory and Related Mood Disorders

It is well known that the rodent adrenal cortex synthesizes and releases stress hormones, such as corticosterone and aldosterone, which mediate physiological responses to stress. The hallmark of stress responses is feedback inhibition of the limbic-hypothalamic–pituitary–adrenal (HPA) axis, which is also reported in depressed patients [78,79]. The adrenal cortex and limbic system contain relevant concentrations of PDE2, indicating that increased cGMP by PDE2 inhibition may influence stress-related disorders, such as learning and memory impairment, depression, and anxiety [80].

In primary neuronal cultures, inhibition of PDE2 by the nonspecific inhibitor EHNA enhances NMDA receptor-mediated cGMP up to 2000% of control, but does not produce a significant increase in cAMP [81]. BAY 60-7550, another highly selective PDE2 inhibitor, also increases cGMP in the presence of NMDA or guanylate cyclase activators. In contrast to EHNA, BAY 60-7550 also increases cAMP formation induced by activation of NMDA receptors, but the magnitude of this effect is markedly less than the increase in cGMP [28].

Studies have shown that EHNA is also a potent inhibitor of adenosine deaminase. This property of EHNA has limited its use *in vivo* as a PDE2 inhibitor [82]. ND7001, a more recently developed PDE2 inhibitor, exhibits at least 100-fold selectivity for inhibition of PDE2 relative to other PDE families [83]. Recent studies from Masood et al. demonstrate that both ND7001 and Bay 60-7550 are effective in reversing anxiogenic-like effects induced by chronic stress [28]. In addition, Bay 60-7550 also reverses oxidative stress-induced anxiety-like behaviors in the elevated plus-maze and hole-board tests [29]. Furthermore, Bay 60-7550 improves memory performance, including memory storage and retrieval, in both novel object and social recognition tasks; it also reverses memory deficits produced in rats by the NMDA receptor antagonist MK-801 in spontaneous alternation tasks [31,84,85]. Finally, Bay 60-7550 not only enhances both early and late phase consolidation of spatial object memory [86], but also reverses object memory deficits induced by acute tryptophan depletion [87,88].

Mechanisms Whereby PDE2 Mediates Memory

It is reported that cyclic nucleotide pathways can crosstalk to modulate each other's synthesis, degradation, and actions. Increased cGMP can potentiate PDE2 activity and enhance the hydrolysis of cAMP [89]. However, such an effect was not found for BAY 60-7550 in primary cultures of cerebral cortical neurons in the presence of NMDA [28]. It is possible that there is no direct interaction between cAMP and cGMP pathways associated with NMDA receptor-mediated signaling in neuronal cultures.

The underlying mechanism may also include the nitric oxide (NO)–cGMP signaling pathway, given that PDE2 inhibition increases the activity of cGMP-dependent protein kinase, including cGMP kinase II, which mediates emotion-related behavioral changes [90]. Memory enhancement may be mediated by subsequent increases in intracellular cGMP and/or cAMP levels after PDE2 inhibition since both cGMP and cAMP are involved in consolidation processes [86,91,92].

PDE4 AND MEMORY

PDE4 Structure and Diversity

PDE4 is the most complicated PDE family (see Chapter 1). Mammalian PDE4 is encoded by four distinct genes, comprising four subtypes: PDE4A–PDE4D [93,94]. Each PDE4 subtype has 3-11 splice variants, which are generated by truncating differentially at the unique N-termini [7,95–100]. To date, 25 distinct splice variants have been identified, including 6 PDE4As, 5 PDE4Bs, 3 PDE4Cs, and 11 PDE4Ds (Figure 7.1) [8,94,96,101]. These variants can be classified into four categories based on the consequences of N-terminal truncation: long-form PDE4s, which contain two highly conserved N-terminal regions (i.e., upstream conserved regions 1 and 2 (UCR1 and UCR2)); short-form PDE4s (PDE4B2, PDE4D1, and PDE4D2), which lack UCR1; supershort-form PDE4s (PDE4A1, PDE4B5, PDE4D6, and PDE4D10), which lack UCR1 and a portion of UCR2; and dead-short-form PDE4 (i.e., PDE4A7, thus far), which has the N- and C-terminally truncated catalytic domain and is inactive, without either UCR1 or UCR2 [8,97,102]. Long-form PDE4s (all but those specified) uniquely contain a conserved phosphorylation site for PKA at the N-termini, which provides feedback and stimulatory regulation of PDE4 activity [93,103,104]. The catalytic domains of all the PDE4 isoforms, except for PDE4As, contain an extracellular signal-regulated kinase (ERK) phosphorylation site, which provides inhibitory (for long-form PDE4s and, to a less extent, supershort-form PDE4s) or stimulatory (for short-form PDE4s) regulation of PDE4 activity in a variant-specific manner [97,105–107].

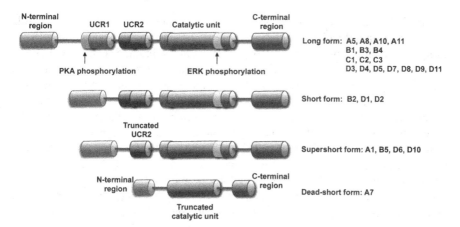

FIGURE 7.1 PDE4 diversity. Four distinct genes encode PDE4 subtypes (PDE4A–PDE4D), each of which can be truncated alternatively at the N-terminus into multiple forms, leading to at least 25 different splice variants. These variants can be divided into four categories based on the structure of the unique N-terminus: long-form, short-form, supershort-form, and dead-short -form PDE4s. Long-form PDE4s are characterized by the unique upstream conserved regions 1 and 2 (UCR1 and UCR2) and a PKA phosphorylation site, while short-form PDE4s simply have intact UCR2. Supershort-form PDE4s only have a portion of UCR2 formed by truncation of UCR2 at the N-terminus, whereas dead-short-form PDE4s only have an inactive catalytic unit formed from truncating of the catalytic domain at both N- and C-termini. There is an ERK phosphorylation site at the catalytic unit, with the exception of PDE4A variants and dead-short-form PDE4s.

PDE4 Distribution in the Brain

PDE4 is widely distributed in the brain, but in a region-specific manner. It is particularly high in the cerebral cortex, olfactory bulb, hippocampus, thalamus, cerebellum, and caudate putamen, as demonstrated by *in vitro* and *in vivo* auto-radiography of [³H]rolipram binding in the rat brain [16,108]. *In situ* hybridization histochemistry has also revealed a similar expression pattern of PDE4 mRNA in the rat brain (see Chapter 2) [109]. The brain distribution patterns of PDE4 have been confirmed by PET scans of the brain in conscious and anesthetized rats, although the latter display an overall lower levels of PDE4 compared to conscious animals [110,111]. The wide and rich expression of PDE4 in the brain is consistent with various implications for the function and dysfunction of PDE4 in the CNS, including memory [30,34,112,113], depression [8,26,35], anxiety [26,29,114], schizophrenia [66,115], stroke [116,117], alcohol consumption and seeking behavior [118,119], and drug abuse [120]. In addition, the distribution of PDE4 in the brain does not vary significantly across species, including rats [16,108,109], mice [15], primates [121], and humans [16,77,122]. This helps readily translate findings from rodents to humans.

The distribution pattern of PDE4 also varies in a subtype-specific manner. While PDE4C is primarily distributed in peripheral tissues, all the other three PDE4 subtypes are highly but differentially expressed in the brain. Specifically, PDE4A and PDE4D are expressed at high levels in the hippocampus, the cerebral cortex, the olfactory system, and the brain stem in a similar pattern, whereas PDE4B is predominantly expressed in basal ganglia and related areas, including the caudate putamen, nucleus accumbens, amygdala, and certain thalamic nuclei [15,16]. The differential distributions of PDE4 subtypes in the brain suggest that they may play different roles in CNS functions. This is supported by findings that PDE4D is important in the mediation of synaptic plasticity and memory [32,87,112,123] and antidepressant activity [124], while PDE4B is involved in anxiety [29,125] and antipsychotic activity [36,126]. In addition, PDE4 expression levels can be changed in an isoform-specific manner by chronic treatment with rolipram or classic anti-depressant drugs [16,127–130] or repeated electroconvulsive shock [131,132]. These observations suggest that adaptive changes in PDE4 subtypes may occur under certain conditions. This phenomenon provides the neurochemical basis for PDE4 inhibitors for treatment of neuropsychiatric and neurodegenerative disorders (see Chapter 8 for more details).

Role of PDE4 in Memory

PDE4 plays a critical role in the control of intracellular cAMP concentrations [12,133] and is responsible for a major component of cAMP hydrolysis in neurons [124,134]. Inhibition of PDE4 results in cAMP accumulation in the brain, leading to activation of downstream signaling, including activation of PKA and phosphorylation of CREB. This signaling pathway is important in the mediation of memory [30,112], as well as in the antidepressant activity and other central functions [7,26,35,135–137]. Therefore, it is reasonable to believe that PDE4 may be an important target in the mediation of cognitive function and memory.

Studies on the function of PDE4 in memory began in the early 1980s, demonstrating that selective PDE4 inhibitors rolipram and Ro 20-1742 enhanced memory or reversed protein synthesis inhibitor-induced memory deficits in the passive avoidance task in mice [138,139]. Randt et al. also demonstrated that the memory-enhancing effect of rolipram was related to rolipram-induced increases in cAMP levels in the brain [138]. Extensive studies since then support the idea that PDE4 plays an important role in the mediation of synaptic plasticity and memory. First, inhibition of PDE4 enhances memory. PDE4 inhibitors such as rolipram increase cAMP in the brain and facilitate memory formation in untreated animals [30,87,112,140]. The memory-enhancing effect of rolipram is independent of the cerebrovascular effects associated with rolipram treatment [33]. Repeated drug administration is usually necessary to obtain a significant memory-enhancing effect. This may be explained by the requirement of repeated rolipram treatment for upregulation of cAMP signaling components such as pCREB and for increased adult neurogenesis in the hippocampus, which are important for memory processes [26,112,140–142]. Consistent with the memory-enhancing effect, inhibition of PDE4 by rolipram facilitates the induction of hippocampal LTP, a cellular model of memory [30,143]. Second, rolipram reverses memory deficits induced by a variety of approaches (Table 7.1). Specifically, rolipram reverses memory deficits produced by various pharmacological agents, including the NMDA receptor antagonist MK-801 [34,144], the muscarinic cholinergic receptor antagonist scopolamine [76,145,146], the MEK inhibitor U0126 [35], the protein synthesis inhibitor anisomycin [138], and amyloid-β peptides (Aβ) [147,148]. Rolipram's reversal of Aβ-induced memory deficits is consistent with its blockade of deficits of LTP and memory in a transgenic mouse model of Alzheimer's disease [149]. It is also contributed by the anti-inflammatory and antiapoptotic effects of rolipram. These findings suggest that PDE4 may be a target for treatment of memory loss associated with Alzheimer's disease. In addition, rolipram reverses memory impairment induced by physical manipulations, such as cerebral ischemia [112,150,151]. Given that PDE4D is associated with ischemic stroke [116], PDE4 may be a target for treatment of some aspects of stroke. Furthermore, rolipram ameliorates memory deficits induced by genetic manipulations, such as CREB binding protein (CBP) +/− mutant mice, a mouse model of Rubinstein–Taybi syndrome [152]. Animals whose memory is impaired are usually more responsive to rolipram treatment compared to untreated animals [153]. Third, PDE4 is highly expressed in the hippocampus, an important mediator of memory [16,111,154]. Inhibition of PDE4 in this region improves memory [140] and reverses memory deficits induced by pharmacological means [35,147,148]. Overall, the findings to date have provided strong evidence for an important role of PDE4 in mediating memory; inhibition of PDE4 improves memory and reverses memory deficits produced by various approaches.

Mechanisms Whereby PDE4 Mediates Memory

Cyclic AMP has three targets: cAMP-dependent PKA, Epac (exchange protein directly activated by cAMP), and cAMP-regulated ion channels. The cAMP/PKA/CREB cascade is the most extensively studied and most important pathway in the mediation of memory. PDE4 specifically hydrolyzes cAMP, indicating that stimulation of the cAMP/PKA/CREB signaling cascade may be the major mechanism by which PDE4 mediates memory and PDE4 inhibitors enhance memory. This can account for the high correlation of memory enhancement and increased cAMP/CREB signaling in the hippocampus after rolipram administration [112,140,147]. Given that NMDA receptors are important in regulating cAMP/CREB signaling in the hippocampus [146,160,161], PDE4 could be a link between NMDA receptor-mediated cAMP signaling and memory. This is supported by findings from a variety of *in vivo* and *in vitro* studies. First of all, inhibition of PDE4 blocks memory deficits induced by NMDA receptor antagonism. Pretreatment with rolipram reverses memory deficits induced by the NMDA receptor antagonist MK-801 in the radial-arm maze and step-through passive avoidance tasks in rats [144,146]. This is in agreement with results from another study in which rolipram was found to attenuate MK-801-induced deficits in latent inhibition, a measure of cognition associated with animal models of schizophrenia [162]. In addition, *in vitro* studies have consistently reported that

TABLE 7.1 Effects of PDE Inhibitors in Animal Models of Impaired Memory[a]

Animal Model	Treatment for Amnesia	Drug	Dose (mg/kg)	Efficacy	References
Morris water maze and step-through passive avoidance tests (rats)	Streptozotocin	PDE1 inhibitor vinpocetine	5, 10, 20 i.p. for 21 days	+, WM, LTM	[48]
T-maze task (mice)	MK-801	PDE2 inhibitor BAY 60-7550	3.0, i.p.	+, WM	[31,87]
Object recognition task (rats)	Aging	PDE2 inhibitor BAY 60-7550	3.0, i.p.	+, Acquisition; +, consolidation	
Object recognition task (rats)	Acute tryptophan depletion	PDE2 inhibitor BAY 60-7550	3.0	+, STM	
Passive avoidance (mice)	Anisomycin	PDE4 inhibitor rolipram	10, i.p.	+, LTM	[32,34,35,76,138, 144–147,148,149, 151,152]
Morris water maze (rats/mice)	Cerebral ischemia (rats) Aβ25–Aβ35 or Aβ1–Aβ40 (rats)	PDE4 inhibitor rolipram PDE4 inhibitor rolipram	3.0, i.p., repeated 0.1–0.5, i.p., repeated	+, LTM; +, acquisition +, LTM; +, acquisition	
Object recognition test (mice)	APP/PS1 transgenic mice Scopolamine	PDE4 inhibitor rolipram PDE4 inhibitor rolipram PDE4 inhibitor rolipram	0.03, SC, repeated 0.1, i.p. 0.1, i.p.	+, LTM +, WM	

Test	Model	Inhibitor	Dose	Effect
Y-maze test (mice)	CBP+/− mutant mice	PDE4D partial inhibitors D159404, D159687	0.1, 0.3, and 1.0 μg/kg, i.v.	+, LTM
	Scopolamine			+, WM
Radial arm maze test (rats)		PDE4 inhibitor rolipram	0.1, i.p.	+, WM/RM
			0.1, i.p.	+, WM/RM
			0.02–0.2, P.O.	+, WM
	MK-801		0.1, i.p.	+, RM
				Acquisition/STM
Radial arm maze test (mice)	Scopolamine	PDE4 inhibitor rolipram	0.03, SC, repeated	+, LTM
	U0126	PDE4 inhibitor rolipram	0.1, i.p.	+, LTM
			0.1, i.p.	+, LTM
Step-through passive avoidance test (rats)	APP/PS1 transgenic mice		15 μg, CA1	+, LTM
	MK-801	PDE4 inhibitor rolipram	0.02–0.1, PO	+, LTM
			0.1–0.5, i.p., repeated	+, LTM
	U0126			
	Scopolamine	PDE4 inhibitor rolipram	0.03, SC, repeated	
Fear conditioning (mice)	FRFS			
	Aβ25–Aβ35 or Aβ1–Aβ40			
	APP/PS1 transgenic mice			

(continued)

TABLE 7.1 (*Continued*)

Animal Model	Treatment for Amnesia	Drug	Dose (mg/kg)	Efficacy	References
T-maze task (mice)	APP/PS1 transgenic mice	PDE5 inhibitor sildenafil	3.0, i.p.	+, WM	[87,155,156]
Fear conditioning and Morris water maze tasks (mice)	APP/PS1 transgenic mice	PDE5 inhibitor sildenafil	3.0, i.p.	+, WM	
Object recognition test (rats)	Acute tryptophan depletion	PDE5 inhibitor vardenafil	3.0, 10	+, STM	
Passive avoidance test (rats)	Scopolamine	PDE9 inhibitor BAY 73-6691	1.0, i.p.	+, Acquisition; +, consolidation; +, LTM	[157]
T-maze and object recognition tests (rats)	MK-801	PDE9 inhibitor BAY 73-6691	10, i.p.	+, WM, LTM	
Passive avoidance (mice)	PDE10A knockout	–	–	+, Acquisition	[158,159]
Social odor recognition (mice)	MK-801	PDE10 inhibitor MP-10	3.0, P.O.	+, LTM	

[a]Animals were treated with drugs or manipulated physically, genetically, or by aging to produce memory impairment. +, Enhancement. WM, working memory; RM, reference memory; LTM, long-term memory; STM, short-term memory. i.p., intraperitoneal injection; PO, per os; SC, subcutaneous. FRFS, frequently repetitive febrile seizures. CBP, cAMP-responsive element binding protein (CREB)-binding protein. PO, per os; SC, subcutaneous. FRFS, frequently repetitive febrile seizures. CBP, cAMP-responsive element binding protein (CREB)-binding protein.

rolipram potentiates NMDA-induced facilitation of cAMP in cultured neurons or slices of the rat frontal cortex and hippocampus [34,81,163]. Furthermore, rolipram converts tetanus-stimulated early-phase LTP (E-LTP) to late-phase LTP (L-LTP), which correspond to short- and long-term memory, respectively (see Chapter 11 for more details). In slices of rat hippocampal CA1, this conversion is blocked by the NMDA receptor antagonist AP-5 [164]. Together, the effects of rolipram appear to depend on activation of NMDA receptors; PDE4 is an integral component of NMDA receptor-mediated cAMP signaling in the mediation of memory (Figure 7.2).

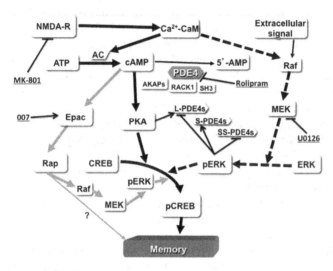

FIGURE 7.2   Phosphodiesterase-4 (PDE4)-mediated intracellular signaling in the mediation of memory. Activation of NMDA receptors (NMDA-R) stimulates adenylyl cyclase (AC) via calcium-calmodulin (CaM), leading to increased formation of cAMP, which triggers downstream signaling, including activation of protein kinase A (PKA) and subsequent phosphorylation and activation of CREB (dark wide solid arrows). NMDA receptor activation also initiates extracellular signal-regulated kinase (ERK) signaling by activation of Raf kinase, which phosphorylates mitogen-activated protein kinase (MAPK)/ERK kinase (MEK). Activated MEK phosphorylates ERK1 and ERK2, which phosphorylate CREB (dark wide dashed arrows). Phospho-CREB (pCREB) mediates memory via related gene expression. On the other hand, cAMP activates Epac (exchange protein directly activated by cAMP), which activates Rap and subsequently CREB via MEK/ERK signaling (gray wide solid arrows). Epac/Rap may also mediate memory through other mechanisms. PDE4 catalyzes the hydrolysis of cAMP and plays a critical role in controlling intracellular cAMP concentrations. The intracellular localization and function of PDE4 are affected by its interaction with other proteins, such as receptors for activated C kinases 1 (RACK1), A-kinase-anchoring proteins (AKAPs), and SH3 proteins (i.e., proteins containing SH3 domains). PDE4 activity can be regulated by activated PKA and/or ERK, that is, PKA stimulates long-form PDE4s (L-PDE4s), while phospho-ERK inhibits long-form and supershort-form PDE4s but stimulates short-form PDE4s. Inhibition of PDE4 by rolipram increases cAMP-mediated signaling in these pathways and subsequently enhances memory. Activation of Epac/Rap signaling by the cAMP analog 8-pCPT-2-O-Me-cAMP (007) that specifically activates Epac also enhances memory. Blockade of NMDA receptors by MK-801 or inhibition of MEK by U0126 impairs memory.

While many of the downstream effects of cAMP are mediated by cAMP-dependent PKA, more recently, Epac proteins (Epac1 and Epac2), also known as cAMP-guanine nucleotide exchange factor (GEF) I and cAMP-GEF II, have been shown to mediate functions of cAMP signaling in a manner that is independent of PKA [165,166]. Epac proteins act as GEFs for Rap1 and Rap2, which belong to the Ras family of small G proteins cycling between the inactive GDP- and active GTP-bound states. GEFs activate Rap via the formation of RapGTP. The major role of Epac/Rap signaling has been considered to be the regulation of cell adhesion [167] and insulin secretion [168,169]. More recently, it has been found that both Epac1 and Epac2 are expressed in most parts of the nervous system and are involved in the mediation of GABAergic neurotransmission and neurotransmitter release [170,171]. Activation of Epac increases synaptic plasticity [172,173], enhances memory [174], and improves cognitive dysfunction in a schizophrenia-related model [33]. However, direct evidence for the role of PDE4 in Epac signaling in the mediation of memory is lacking. Nevertheless, since PDE4 subtypes such as PDE4D are sequestered with Epac to a discrete functional signaling complex [175,176], inhibition of these PDE4 isoforms in the complex should increase cAMP and subsequently activate Epac signaling, which may contribute to the mediation of memory (Figure 7.2). Further studies are required to demonstrate the role of PDE4-cAMP-Epac signaling in memory.

The other mechanism by which PDE4 mediates memory may be through activation of the ERK pathway, which is also regulated by NMDA receptors [177] and is involved in mediating memory [178,179]. Inhibition of PDE4 by rolipram converts an early form of long-term depression (LTD), which is associated with learning and memory requiring object–place configuration [180] and normally decays within 2–3 h, to a long-lasting LTD in slices of rat hippocampal CA1 [181]. This conversion is prevented by the specific MEK inhibitor U0126, indicating that the effect of rolipram is mediated by the MEK/ERK signaling pathway. Consistent with this, rolipram potentiates phosphorylation of ERK induced by activation of β-adrenergic receptors in a human embryonic kidney cell line [182]. Interestingly, ERK signaling appears to activate CREB via inhibitory effects on PDE4, in addition to direct activation of CREB (Figure 7.2). Phosphorylation of ERK inhibits long-form PDE4s except PDE4A [12,105], leading to an increase in cAMP and subsequent triggering of the downstream signaling cascade. This has been demonstrated to exhibit an effect on behavior in our studies [35]. Intrahippocampal CA1 infusion of U0126 reduces phospho-ERK1/ERK2 expression via MEK inhibition and impairs long-term memory in the passive avoidance and radial arm maze tests. Pretreatment of rats with rolipram reverses memory deficits induced by U0126, without altering expression of phosphorylated ERK [35]. Given that U0126-induced memory deficits and ERK inhibition are highly correlated with increased PDE4 activity, the results strongly support the idea that PDE4 in hippocampal CA1 plays an essential role in ERK signaling-mediated memory.

PDE4 Subtypes and Memory

While the role of PDE4 in memory has been well established, it has been particularly difficult to identify the specific subtypes responsible for PDE4–cAMP signaling-mediated memory regulation because of the lack of highly selective inhibitors of specific PDE4 subtypes. Fortunately, this situation has been changed since the generation of mice deficient in specific PDE4 subtypes [183,184]. Studies to date have provided initial indications regarding the contributions of individual PDE4 subtypes to memory, including the distribution patterns and structural features of PDE4s and the behavioral phenotype of mice deficient in a specific PDE4 subtype.

PDE4 subtypes are differentially distributed in brain regions associated with memory. For instance, PDE4A and PDE4D are highly expressed in the CA1 subregions of the rat hippocampus, whereas PDE4B is expressed at a relatively low level; PDE4C, which is primarily localized to peripheral tissues, is not expressed in the hippocampus [15,16]. The distribution pattern indicates that PDE4 subtypes may have different contributions to the mediation of memory. More specifically, PDE4A and PDE4D may be the major PDE4 subtypes involved in mediating memory. Comparably, PDE4B may not be important in hippocampal-dependent memory. This appears to be supported by the negative results in the water maze memory test using mice deficient in PDE4B [29,84,185].

PDE4 has been shown to be a component of ERK signaling in the mediation of memory [35]. Since PDE4A does not have an ERK phosphorylation site [105] and PDE4B is minimally expressed in the hippocampus [16], PDE4D appears to be the major PDE4 in memory mediated by ERK signaling. This is supported by *in vitro* studies showing the regulation of PDE4D activity by activated ERK [182,186]. It is also consistent with our findings that PDE4D is the primary component contributing to PDE4 activity in the hippocampus [29,124]. However, the role of PDE4A in memory that is not mediated by ERK signaling cannot be excluded, given the high expression of this subtype in memory-related brain regions such as the hippocampus [15,16].

Consistent with the major contribution of PDE4D to PDE4 activity in the hippocampus, cerebral cortical slices from PDE4D-deficient mice exhibit basal cAMP concentrations ~200% of those of wild-type controls [124]; pCREB is also significantly increased in the hippocampus of PDE4D-deficient mice [112]. In addition, mice deficient in PDE4D, but not PDE4B, display enhanced memory performance in hippocampal-dependent memory tests, including Morris water maze, radial arm maze, and object recognition tests [29,84,187]. The behavioral results appear to be in agreement with enhanced induction of hippocampal LTP in PDE4D-deficient mice, although memory of these mice is "unexpectedly" impaired in fear conditioning [188], a test primarily for amygdala-dependent memory. Other studies also support the role of PDE4D in the mediation of memory. PDE4 inhibitors with high or modest selectivity for PDE4D enhance memory in the object recognition test [123] or reverse amnesic effects of MK-801 in the rat radial arm maze and passive avoidance tests [144]. In addition, overexpression of PDE4D induced by administration of high doses of PDE4 inhibitors impairs spatial memory in the water maze test in rats [189]. Most recently, Burgin et al. have demonstrated that allosteric modulators of PDE4D, which only partially inhibit PDE4 activity, enhance memory and reverse scopolamine-induced memory deficits in the Y-maze and object recognition tests, but do not cause emetic-like effects [32]. These results suggest that the PDE4D subtype plays the major role in the regulation of cAMP hydrolysis and in the mediation of memory, although the role for PDE4A and PDE4B, in particular the former, cannot be excluded.

PDE4 Splice Variants and Memory

The importance of PDE4D in memory [32,112] has provided the possibility to narrow the 25 PDE4 splice variants in identification of their roles in the mediation of memory. This is necessary because gene knockout techniques that have been used successfully to identify the function of specific PDE4 subtypes [29,36,112,124,185,188] result in gene deficiency throughout the whole body and may lead to abnormal development and mortality [183], in addition to high costs of generating and breeding animals. The gene knockout approach cannot be used to identify the role of PDE4 variants because of the multiple variants for each subtype. Most recently, Zhang and coworkers have applied microRNAs (miRNAs), an efficient approach of RNA interference (RNAi) for gene silencing, to target long-form PDE4Ds in mice and have reported memory-enhancing effects of these genetic manipulations [112].

MicroRNAs are 19–22 nucleotide endogenous, noncoding RNA sequences, which are formed from the precursor pre-miRNAs. MicroRNAs are incorporated into the RNA-induced silencing complex (RISC), leading to degradation of mRNA or suppression of mRNA translation [190,191]. Gene silencing of PDE4 subtypes or other PDEs has been achieved *in vitro* in nonneuronal cells using small interfering RNAs (siRNAs) that target the corresponding PDE enzymes [182,192–197]. However, the effect of synthetic siRNAs is unstable, transient (several hours to days), and neuron insensitive [198]. For stable, long-term, and neuron-sensitive suppression, the viral vector-mediated miRNA has been developed and applied in brain research [199,200]. Therefore, we applied lentiviral vectors, which are replication defective and self-inactivating [201–203], to harbor miRNAs that target long-form PDE4Ds [112]. The consideration of targeting long-form PDE4Ds was based on the important role of PDE4D in synaptic plasticity and memory [32,112,144,188,189] and the essential contribution of long-form PDE4s to cAMP hydrolysis [105]. MicroRNA-mediated knockdown of long-form PDE4Ds, in particular PDE4D4 and PDE4D5, in the hippocampus increases pCREB and adult neurogenesis and enhances memory in passive avoidance and object recognition tests [112]. In the same study, these effects are also observed in mice treated with rolipram and mice deficient in PDE4D (Figure 7.3). Interestingly, while rolipram and PDE4D deficiency produce emetic-like response, as demonstrated by shortened α_2-adrenergic receptor-mediated anesthesia, a surrogate measure of emesis in rodents [204,205], mice administered PDE4D-miRNAs in the hippocampus do not exhibit any emetic-like behavior [112]. This could be due to the fact that the hippocampus is not involved in emesis; these results suggest that appropriate manipulations of PDE4D can achieve memory enhancement without causing emesis, although PDE4D plays a substantial role in emetic responses [112,205].

The above observations indicate that PDE4 inhibitors produce consistent memory-enhancing effects in animals whose memory is impaired by pharmacological, physical, or genetic approaches. This may involve PDE4D, in particular long-form PDE4D

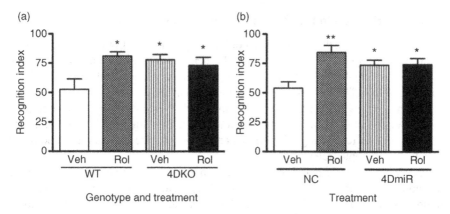

FIGURE 7.3 Memory enhancement in the object recognition test in mice of which PDE4D was mutant or downregulated. (a) Increased recognition index in mice deficient in PDE4D (4DKO) or treated with rolipram (Rol). (b) Recognition index changes in ICR mice treated with lentiviral vectors containing either miRNAs that target long-form PDE4Ds (4DmiR) or the negative control (NC) sequence that does not target any known vertebrate gene, with or without rolipram. Animals were trained to discriminate a novel object from the object that was exposed 24 h earlier (i.e., the familiar object). The recognition index was the exploration time in the novel object (T_n) divided by the total exploration time (i.e., T_n plus the exploration time in the familiar object (T_f)). Rolipram (1.25 mg/kg) or vehicle was injected (i.p.) once daily for 22 days (a) or 12 days (b) prior to the test, which was performed 1 h after the rolipram injection. 4DmiR or NC was microinfused into bilateral dentate gyri using a single dose of 4×10^6 tu per side 12 days before the test. Values shown are means $\pm$ SEM; $n = 7$–10; $^*P < 0.05$, $^{**}P < 0.01$ versus corresponding WT/NC + Veh. Redrawn from Ref. [112].

variants such as PDE4D4 and PDE4D5. However, depending on specific memory tasks, the role of PDE4A and PDE4B, in particular the former, cannot be excluded. Regardless of the subtype contribution, PDE4 appears to be not only a key connection of the crosstalk between cAMP and ERK signaling mediated by NMDA receptors but also an important component of the cAMP and ERK signaling pathways in the mediation of memory. Inhibition of PDE4 improves memory primarily via stimulation of cAMP signaling, which involves PKA and probably, to a certain extent, Epac. Therefore, PDE4 may be a target for therapeutic agents for treatment of disorders affecting memory and cognitive functions, such as Alzheimer's disease.

PDE5 AND MEMORY

PDE5 is a cGMP-specific phosphodiesterase encoded by a single gene. It has a GAF-A and a GAF-B domain at its N-terminus, which determines substrate specificity and functional differentiation (see Chapter 1). Binding of cGMP to at least one GAF domain activates catalytic activity of PDE5 [206].

PDE5 Distribution in the Brain and Involvement in Memory

PDE5 is expressed in the brain, spinal cord, lung, platelets, and smooth muscle. In the brain, moderate/weak PDE5 signals can be detected in the cerebral cortex, hippocampus (pyramidal cells), olfactory bulb, basal ganglia, and cerebellum [18,207,208]. Expression of PDE5 is particularly significant in cerebellar Purkinje cells; this contributes greatly to the processes of cognition associated with the cerebellum [209] via the regulation of cGMP [210]. Therefore, PDE5 inhibition has been brought into consideration in rodent cognition measurements during recent years. For additional details on this topic, refer to Chapter 9.

Zaprinast, an earlier PDE5 inhibitor that is not highly selective, was first observed to improve memory performance in 1997 [211]. Later on, as more PDE5 selective inhibitors were developed, animal studies indicated that PDE5 inhibition could potentially improve working memory, reference memory, and mood-related memory in animals. PDE5 inhibitors such as sildenafil and tadalafil, which easily cross the blood–brain barrier, have been shown to exhibit promising effects on learning and memory. In the rat object recognition task, both drugs improve memory consolidation [92,212,213]. Recent findings suggest that PDE5 inhibitors improve the early-phase consolidation of memory performance in rats [33]. Moreover, in an adapted version of the modified elevated plus maze that tests anxiety-related learning and memory, sildenafil produces dose-dependent improvement in short-term memory in both rats and mice [214,215]. Using the nitric oxide inhibitor L-NAME in a similar behavioral test, Devan et al. further demonstrate that the PDE5-related NO–cGMP pathway may be a target for drug development to offset cognitive changes during aging [216]. The same pathway has been reported to play a role in sildenafil-induced anxiogenic-like effect [217]. With regard to emotion-related memory, sildenafil and zaprinast induce enhancement of memory in active and passive avoidance learning in shuttle box tasks [155,215,218–220]. However, sildenafil has not been proved to enhance short-term memory in humans [221,222].

In addition to its role in memory, PDE5 is involved in the regulation of mood disorders such as depression, although the results to date have been controversial. Increasing cGMP levels via PDE5 inhibitors such as sildenafil may block antidepressant-like effects in some animal models, including forced-swim and tail-suspension tests [223,224]. In contrast, while there is evidence for cognitive-enhancing effects of sildenafil in animal models including monkeys [92], attempts using sildenafil to ameliorate schizophrenia-related deficits of cognition have not been successful in patients [225].

Mechanisms Whereby PDE5 Mediates Memory

Both sildenafil and vardenafil increase cGMP, but not cAMP [212], indicating that the memory-enhancing effect of PDE5 inhibitors may be primarily mediated by the cGMP/PKG/CREB pathway, which is important in synaptic plasticity such as LTP [226,227] and cognition associated with Alzheimer's disease [156]. The previous studies suggest that the memory-enhancing effect of PDE5 inhibitors may not involve the direct vascular action of cGMP, that is, increases in cerebral blood flow and delivery of glucose and oxygen to the brain. However, it cannot be excluded that the indirect effect of PDE5 inhibitors on cerebral blood vessels, such as activating downstream signal molecules related to blood vessel dilation, may contribute to their memory-enhancing effects. Furthermore, PDE5 inhibitors can activate K^+ channels, leading to blockade of the antidepressant-like effect of K^+ channel inhibitors on forced-swim behavior [223]. This mechanism may also be involved in the memory-enhancing effect of PDE5 inhibitors.

Several groups have reported that sildenafil has antidepressant-like effects only when central muscarinic acetylcholine receptors are blocked [228–230]. This appears to be in agreement with the findings that sildenafil bolsters cholinergic responses in the brain [216,231] and that increases in central cholinergic activity result in depressogenic profiles [232]. This also accounts for the fact that, when coadministered with an antimuscarinic agent, NO-cGMP activation by sildenafil results in antidepressant-like effects.

Other studies have revealed additional mechanisms whereby PDE5 may be associated with CNS diseases or their treatment. Studies have suggested a link between MDMA-induced 5-HT toxicity and increased free radical formation [233]; this is prevented by activation of PKG [234,235]. Based on these findings, Puerta et al. have inferred that sildenafil protects against MDMA-induced 5-HT depletion through activation of PKG and mitochondrial ATP-sensitive K^+ channels [236]. The PI3-K/Akt/eNOS/sGC pathway may also be involved in the effect of sildenafil [237], although this was not confirmed in the study [236]. Thus, the neuroprotective effect of PDE5 inhibition may contribute differentially in different models of cytotoxicity.

Another central regulator of behavioral activity is the mitogen-activated protein kinase (MAPK) signaling pathway. In an ischemia–reperfusion renal injury model, pretreatment with sildenafil increases iNOS/eNOS and cleavage of caspase-3, decreases TUNEL-positive cells and the Bax/Bcl-2 ratio via phosphorylating the upstream regulator ERK, and thus prevents apoptosis [238]. Further studies have extended this signaling pathway to more of its downstream signaling components, including the mitochondrial ATP-sensitive K^+ channels [239].

PDE9 AND MEMORY

PDE9 Structure and Distribution in the Brain

PDE9 is encoded by two genes (PDE9A and PDE9B) and is cGMP specific. To date, at least 20 different splice variants have been discovered (PDE9A1–PDE9A20) in human and mouse [240,241]. Structural studies of PDE9A have shown that its cDNAs of different splice variants share a high percentage of amino acid identity in the catalytic domain. However, despite its highest specificity for cGMP among all the PDEs [242,243], PDE9A lacks a GAF domain, whose binding of cGMP usually activates catalytic activity [244].

Besides its expression in the kidney, spleen, and other peripheral organs, PDE9 is widespread throughout the brain in mice, rats, and humans with a high similarity of expression, namely, in the striatum, cerebellum, olfactory bulb, amygdala, and midbrain [18]. The expression is mainly detected in neurons and subpopulations of astrocytes, with different sensitivity to PDE9 inhibitors observed in different brain regions [212].

Role of PDE9 in Memory

The location of PDEs may predict their roles in mediating certain cognitive functions that are related to specific brain regions. Studies of the healthy human brain have shown that a moderate distribution of the mRNA for PDE9 is observed in the cortical areas, including insular, entorhinal, and visual cortices, and in the Ammon's horn (CA1, CA2, and CA3) and dentate gyrus of the hippocampus [245]. There is a general tendency for PDE9 mRNA to be decreased in most brain regions of patients with Alzheimer's disease [246]. These results suggest that PDE9 may be a target for treatment of symptoms of Alzheimer's disease, although little is known about the side effect profiles and isoform specificity of PDE9 inhibitors [86]. Gene studies have also revealed a potential link between the functional status of PDE9 and affective disorders [247,248]. PDE9 is found to be associated with both the diagnosis of major depressive disorder and the treatment response. Behavioral research also showed that PDE9 inhibitor PF-04447943 reverses object memory deficits induced by scopolamine and enhances synaptic plasticity and cognitive function in rodents [245,249].

Mechanisms Whereby PDE9 Mediates Memory

Only a few PDE9 selective inhibitors have been developed so far, including BAY 73-6691, SCH 51866, and PF-04447943 [157,249,250]. In a stably transfected PDE9 CHO cell line, Wunder et al. have reported that BAY 73-6691 is able to dose-dependently induce intracellular cGMP accumulation, indicating its ability to penetrate cells and inhibit PDE9 activity. However, the PDE9 inhibitor is effective only when the cell line is prestimulated with a submaximal concentration of the sGC stimulator BAY 58-2667 [251]. Van der Staay et al. demonstrate that Bay 73-6691 enhances early LTP and improves acquisition, consolidation, and retention of long-term memory in the social recognition task and tends to enhance long-term memory in the object recognition task in old rats [157]. BAY 73-6691 attenuates deficits caused by other compounds such as scopolamine and MK-801 in passive avoidance and T-maze alternation tasks. These effects appear to involve elevation of cGMP presynaptically during early LTP. SCH 51866, another PDE9 inhibitor, produces a slight increase in cGMP when coadministered with the NO donor DEANO [250]. The Pfizer drug PF_04447943 is currently under phase II clinical trials. Overall, the results to date suggest that inhibition of PDE9 may be a novel and effective approach to improve memory consolidation via the NO–cGMP pathway.

PDE10 AND MEMORY

PDE10 Structure

The PDE10 family was initially discovered by three laboratories simultaneously [22,252,253]. There is only one gene encoding PDE10, that is, PDE10A, which hydrolyzes both cAMP and cGMP about equally well. In humans, there are four known PDE10A isoforms, PDE10A1–PDE10A4, which differ at their N- and C-termini, while high amino acid sequence similarity is found for the catalytic region and the GAF domains (see Chapter 10 for more details). PDE10A is allosterically activated by both cAMP and cGMP via binding to GAF in some neurons, particularly in the medium spiny neurons [254]. Some studies indicate that the tandem GAF domain of PDE10A uses cAMP, rather than cGMP, as an allosteric ligand [255,256].

PDE10 Distribution in the Brain

PDE10 is found in the testes, thyroid, and brain. It exhibits particularly high expression in the caudate and putamen nuclei (i.e., striatum), and much more modest expression in cortex, hippocampus, thalamus, and the spinal cord [22,72]. PDE10A mRNA and protein are expressed in γ-aminobutyric acid (GABA)-containing medium spiny projection neurons of the mammalian striatum, which is the core site for regulation of motoric, appetitive, and cognitive processes associated with planning and sequencing of movement [21,85].

Role of PDE10 in Memory

Dysfunction of dopaminergic/GABAergic systems may contribute to a variety of CNS disorders, such as Parkinson's disease, Huntington's disease, and psychiatric disorders. The expression of PDE10 in the hippocampal pyramidal cell layer, the granule cell layer of the hippocampal dentate gyrus, and cells throughout the cerebral cortex potentially implicates a role of PDE10 in the mediation of cognition and mood disorders [158,257]. The expression of PDE10 in nonstriatal neuronal populations indicates that PDE10 may play a broader role in the regulation of synaptic input/output and neuronal homeostasis [21]. This is supported by the finding that stimulation that induces stable LTP increases mRNA levels of PDE10A splice variants in the hippocampus of rats [258]. As a cellular model of memory, LTP is usually used as an indication of memory changes. Animal models with enhanced LTP usually also show improvement in hippocampal-dependent learning and memory [259].

Reduction or inhibition of PDE10 may cause changes in cyclic nucleotide signaling that contribute to the regulation of some aspects of behavior relevant to schizophrenia [37]. Studies using mice deficient in PDE10A suggest that it regulates striatal output, possibly by decreasing the sensitivity of medium spiny neurons to glutamatergic excitation by corticostriatal inputs [159]. This has recently been demonstrated by West and colleagues using both the nonselective PDE10 inhibitor papaverine and the selective PDE10A inhibitor TP-10 (see Chapter 11 for more details) [260]. These findings are supported further by studies that show that papaverine is effective in improving deficits in executive function [159,261]. Since phosphorylation of CREB (pCREB) is increased by inhibition of PDE10A in primary cultures of cortical and hippocampal neurons [85], cellular signaling involving CREB may be a major mechanism underlying the role of PDE10 in the mediation of memory.

PDE11 AND MEMORY

PDE11 Structure and Diversity

PDE11 is the most recently identified member of the PDE superfamily. The first report on PDE11A was published about a decade ago by Fawcett et al. [262], who obtained a partial sequence of PDE11A from a commercially available, expressed sequence tag database based on homology with other mammalian PDEs [263]; PDE11 is coded by a single gene [264]. Alternate splicing is responsible for further variation of the PDE11 family. To date, four N-terminal splicing variants (PDE11A1–PDE11A14) have been identified. PDE11A1–PDE11A3 have an N-terminus GAF domain and a catalytic domain, while PDE11A4, the longest PDE11A variant, has two N-terminal GAF domains, in addition to the catalytic domain [72]. The GAF domain in PDE11A is homologous to that in PDE2, PDE5, PDE6, and PDE10.

PDE11 Distribution in the Brain and Role in Memory

Initial studies revealed that PDE11A is expressed in the skeletal muscle, prostate, testes, pancreas, and salivary glands. This distribution pattern is supported by subsequent studies, which have also shown expression of PDE11A in the brain, pituitary, and spinal cord [262,264]. Recent findings suggest that PDE11A mRNA and protein in brain are largely restricted to the hippocampus [23].

There is evidence to date showing a contribution of PDE11 to social memory [23]. The expression of PDE11A in the brain suggests that PDE11A may be involved in functions related to psychiatric diseases [23]. PDE11A knockout mice show subtle psychiatric disease-related deficits, including hyperactivity in the open field test, increased sensitivity to the NMDA receptor antagonist MK-801, and impaired social behavior. In addition, PDE11A haplotypes are associated with major depressive disorders and response to treatment [248,265]. Therefore, PDE11A may be involved in the mediation of cognition associated with depression and schizophrenia.

Mechanisms Whereby PDE11 Mediates Memory

PDE11A hydrolyzes both cAMP and cGMP and has a catalytic site most similar to PDE5 (50% identity and 71% similarity) [262]. One of the major PDE5 inhibitors tadalafil has been found to have a relatively high affinity for PDE11 with reported IC_{50} of ~73 nM [113]. Tadalafil is widely used to treat erectile dysfunction, so its effects on PDE11 need to be kept in mind. While there is a lack of selective inhibitors of

PDE11 that are publicly available, the nonselective PDE inhibitors IBMX, zaprinast, and dipyridamole have all been shown to inhibit PDE11 [263].

FUTURE DIRECTIONS

During the last six decades, the progress in PDE research has been dramatic, leading to direct production of two classes of drugs for clinical uses, that is, the PDE5 inhibitors such as sildenafil (Viagra) for treatment of erectile dysfunction and the PDE4 inhibitor roflumilast (Daxas) for asthma and chronic obstructive pulmonary disease (COPD). However, while there are some PDE inhibitors available for use in the clinic [266], no PDE inhibitors to date have been approved for treatment of CNS disorders, although they have been demonstrated to exert therapeutic potential such as memory-enhancing and antidepressant actions. This may be due to the complexity of the PDE superfamily, which includes 11 families with 21 subtypes and more than 100 splice variants, and the lack of selective inhibitors of individual PDE subtypes or splice variants. Most recently, there have been two important advances in PDE4 studies: the identification of the important role of PDE4D and its long-form isoforms in the mediation of memory [112]; and the successful synthesis of PDE4D allosteric modulators, which enhance memory and reverse scopolamine-induced memory deficits, but do not exhibit emetic-like effects [32]. These findings appear to provide important directions for future PDE studies. Using gene knockout and RNAi-mediated gene silencing (e.g., miRNAs) techniques, we can identify the role of individual PDE isoforms in CNS functions, including memory and anti-depressant activity. Using small-molecule allosteric modulators of certain PDE subtypes such as PDE4D, we can achieve therapeutic effects but avoid potential major side effects. The strategy will aid in the development of selective inhibitors of specific PDE isoforms for treatment of neurodegenerative disorders such as Alzheimer's disease and neuropsychiatric disorders such as depression and schizophrenia.

REFERENCES

1. Zittle, C.A. and Reading, E.H. (1946) Ribonucleinase and non-specific phosphodiesterase in rat and rabbit blood and tissues. *J. Franklin Inst.*, **242**:424–428.
2. Butcher, R.W. and Sutherland, E.W. (1962) Adenosine 3′,5′-phosphate in biological materials: I. Purification and properties of cyclic 3′,5′-nucleotide phosphodiesterase and use of this enzyme to characterize adenosine 3′,5′-phosphate in human urine. *J. Biol. Chem.*, **237**:1244–1250.
3. Beavo, J.A., Hardman, J.G., and Sutherland, E.W. (1970) Hydrolysis of cyclic guanosine and adenosine 3′,5′-monophosphates by rat and bovine tissues. *J. Biol. Chem.*, **245**:5649–5655.
4. Beavo, J.A. (1995) Cyclic nucleotide phosphodiesterases: functional implications of multiple isoforms. *Physiol. Rev.*, **75**:725–748.

5. Conti, M. and Beavo, J. (2007) Biochemistry and physiology of cyclic nucleotide phosphodiesterases: essential components in cyclic nucleotide signaling. *Annu. Rev. Biochem.*, **76**:481–511.

6. Soderling, S.H. and Beavo, J.A. (2000) Regulation of cAMP and cGMP signaling: new phosphodiesterases and new functions. *Curr. Opin. Cell Biol.*, **12**:174–179.

7. O'Donnell, J.M. and Zhang, H.-T. (2004) Antidepressant effects of inhibitors of cAMP phosphodiesterase (PDE4). *Trends Pharmacol. Sci.*, **25**:158–163.

8. Zhang, H.-T. (2009) Cyclic AMP-specific phosphodiesterase-4 as a target for the development of antidepressant drugs. *Curr. Pharm. Des.*, **15**:1688–1698.

9. Thompson, W.J. (1991) Cyclic nucleotide phosphodiesterases: pharmacology, biochemistry and function. *Pharmacol. Ther.*, **51**:13–33.

10. Bolger, G.B., Rodgers, L., and Riggs, M. (1994) Differential CNS expression of alternative mRNA isoforms of the mammalian genes encoding cAMP-specific phosphodiesterases. *Gene*, **149**:237–244.

11. Keravis, T. and Lugnier, C. (2010) Cyclic nucleotide phosphodiesterases (PDE) and peptide motifs. *Curr. Pharm. Des.*, **16**:1114–1125.

12. Houslay, M.D. and Adams, D.R. (2003) PDE4 cAMP phosphodiesterases: modular enzymes that orchestrate signalling cross-talk, desensitization and compartmentalization. *Biochem. J.*, **370**:1–318.

13. Sonnenburg, W.K., et al. (1995) Identification of inhibitory and calmodulin-binding domains of the PDE1A1 and PDE1A2 calmodulin-stimulated cyclic nucleotide phosphodiesterases. *J. Biol. Chem.*, **270**:30989–31000.

14. Juilfs, D.M., et al. (1997) A subset of olfactory neurons that selectively express cGMP-stimulated phosphodiesterase (PDE2) and guanylyl cyclase-D define a unique olfactory signal transduction pathway. *Proc. Natl. Acad. Sci. USA*, **94**:3388–3395.

15. Cherry, J.A. and Davis, R.L. (1999) Cyclic AMP phosphodiesterases are localized in regions of the mouse brain associated with reinforcement, movement, and affect. *J. Comp. Neurol.*, **407**:287–301.

16. Miró, X., et al. (2002) Regulation of cAMP phosphodiesterase mRNAs expression in rat brain by acute and chronic fluoxetine treatment: an *in situ* hybridization study. *Neuropharmacology*, **43**:1148–1157.

17. Repaske, D.R., Corbin, J.G., Conti, M., and Goy, M.F. (1993) A cyclic GMP-stimulated cyclic nucleotide phosphodiesterase gene is highly expressed in the limbic system of the rat brain. *Neuroscience*, **56**:673–686.

18. Van Staveren, W.C.G., et al. (2003) mRNA expression patterns of the cGMP-hydrolyzing phosphodiesterases types 2, 5, and 9 during development of the rat brain. *J. Comp. Neurol.*, **467**:566–580.

19. van Staveren, W.C.G. and Markerink-van Ittersum, M. (2005) Localization of cyclic guanosine 3′,5′-monophosphate-hydrolyzing phosphodiesterase type 9 in rat brain by nonradioactive *in situ* hybridization. *Methods Mol. Biol.* **307**:75–84.

20. Lakics, V., Karran, E.H., and Boess, F.G. (2010) Quantitative comparison of phosphodiesterase mRNA distribution in human brain and peripheral tissues. *Neuropharmacology*, **59**:367–374.

21. Seeger, T.F., et al. (2003) Immunohistochemical localization of PDE10A in the rat brain. *Brain Res.* **985**:113–126.

22. Soderling, S.H., Bayuga, S.J., and Beavo, J.A. (1999) Isolation and characterization of a dual-substrate phosphodiesterase gene family: PDE10A. *Proc. Natl. Acad. Sci. USA*, **96**:7071–7076.

23. Kelly, M.P., et al. (2010) Phosphodiesterase 11A in brain is enriched in ventral hippocampus and deletion causes psychiatric disease-related phenotypes. *Proc. Natl. Acad. Sci. USA*, **107**:8457–8462.

24. Degerman, E., Belfrage, P., and Manganiello, V.C. (1997) Structure, localization, and regulation of cGMP-inhibited phosphodiesterase (PDE3). *J. Biol. Chem.*, **272**:6823–6826.

25. Kobayashi, T., et al. (2003) Molecular comparison of rat cyclic nucleotide phosphodiesterase 8 family: unique expression of PDE8B in rat brain. *Gene*, **319**:21–31.

26. Li, Y.-F., et al. (2009) Antidepressant- and anxiolytic-like effects of the phosphodiesterase-4 inhibitor rolipram on behavior depend on cyclic AMP response element binding protein-mediated neurogenesis in the hippocampus. *Neuropsychopharmacology*, **34**:2404–2419.

27. Masood, A., Nadeem, A., Mustafa, S.J., and O'Donnell, J.M. (2008) Reversal of oxidative stress-induced anxiety by inhibition of phosphodiesterase-2 in mice. *J. Pharmacol. Exp. Ther.* **326**:369–379.

28. Masood, A., et al. (2009) Anxiolytic effects of phosphodiesterase-2 inhibitors associated with increased cGMP signaling. *J. Pharmacol. Exp. Ther.*, **331**:690–699.

29. Zhang, H.T., et al. (2008) Anxiogenic-like behavioral phenotype of mice deficient in phosphodiesterase 4B (PDE4B). *Neuropsychopharmacology*, **33**:1611–1623.

30. Barad, M., Bourtchouladze, R., Winder, D.G., Golan, H., and Kandel, E. (1998) Rolipram, a type IV-specific phosphodiesterase inhibitor, facilitates the establishment of long-lasting long-term potentiation and improves memory. *Proc. Natl. Acad. Sci. USA*, **95**:15020–15025.

31. Boess, F.G., et al. (2004) Inhibition of phosphodiesterase 2 increases neuronal cGMP, synaptic plasticity and memory performance. *Neuropharmacology*, **47**:1081–1092.

32. Burgin, A.B., et al. (2010) Design of phosphodiesterase 4D (PDE4D) allosteric modulators for enhancing cognition with improved safety. *Nat. Biotechnol.*, **28**:63–70.

33. Rutten, K., et al. (2009) Phosphodiesterase inhibitors enhance object memory independent of cerebral blood flow and glucose utilization in rats. *Neuropsychopharmacology*, **34**:1914–1925.

34. Zhang, H.T., Crissman, A.M., Dorairaj, N.R., Chandler, L.J., and O'Donnell, J.M. (2000) Inhibition of cyclic AMP phosphodiesterase (PDE4) reverses memory deficits associated with NMDA receptor antagonism. *Neuropsychopharmacology*, **23**:198–204.

35. Zhang, H.-T., et al. (2004) Inhibition of the phosphodiesterase 4 (PDE4) enzyme reverses memory deficits produced by infusion of the MEK inhibitor U0126 into the CA1 subregion of the rat hippocampus. *Neuropsychopharmacology*, **29**:1432–1439.

36. Siuciak, J.A., et al. (2007) Behavioral and neurochemical characterization of mice deficient in the phosphodiesterase-1B (PDE1B) enzyme. *Neuropharmacology*, **53**:113–124.

37. Siuciak, J.A., et al. (2008) Behavioral characterization of mice deficient in the phosphodiesterase-10A (PDE10A) enzyme on a C57/Bl6N congenic background. *Neuropharmacology*, **54**:417–427.

38. Zhang, H.T. (2010) Phosphodiesterase targets for cognitive dysfunction and schizophrenia: a New York Academy of Sciences Meeting. *IDrugs*, **13**:166–168.

39. Kakiuchi, S. and Yamazaki, R. (1970) Calcium dependent phosphodiesterase activity and its activating factor (PAF) from brain studies on cyclic 3′,5′-nucleotide phosphodiesterase (3). *Biochem. Biophys. Res. Commun.*, **41**:1104–1110.

40. Goraya, T.A. and Cooper, D.M.F. (2005) Ca2+-calmodulin-dependent phosphodiesterase (PDE1): current perspectives. *Cell Signal*, **17**:789–797.

41. Yan, C., et al. (1995) Molecular cloning and characterization of a calmodulin-dependent phosphodiesterase enriched in olfactory sensory neurons. *Proc. Natl. Acad. Sci. USA*, **92**:9677–9681.

42. Yan, C., Zhao, A.Z., Bentley, J.K., and Beavo, J.A. (1996) The calmodulin-dependent phosphodiesterase gene PDE1C encodes several functionally different splice variants in a tissue-specific manner. *J. Biol. Chem.*, **271**:25699–25706.

43. Yu, J., et al. (1997) Identification and characterisation of a human calmodulin-stimulated phosphodiesterase PDE1B1. *Cell Signal*, **9**:519–529.

44. Borisy, F.F., et al. (1992) Calcium/calmodulin-activated phosphodiesterase expressed in olfactory receptor neurons. *J. Neurosci.*, **12**:915–923.

45. Yan, C., Bentley, J.K., Sonnenburg, W.K., and Beavo, J.A. (1994) Differential expression of the 61 kDa and 63 kDa calmodulin-dependent phosphodiesterases in the mouse brain. *J. Neurosci.*, **14**:973–984.

46. Furuyama, T., Iwahashi, Y., Tano, Y., Takagi, H., and Inagaki, S. (1994) Localization of 63-kDa calmodulin-stimulated phosphodiesterase mRNA in the rat brain by *in situ* hybridization histochemistry. *Mol. Brain Res.*, **26**:331–336.

47. Polli, J.W. and Kincaid, R.L. (1994) Expression of a calmodulin-dependent phosphodiesterase isoform (PDE1B1) correlates with brain regions having extensive dopaminergic innervation. *J. Neurosci.*, **14**:1251–1261.

48. Deshmukh, R., Sharma, V., Mehan, S., Sharma, N., and Bedi, K.L. (2009) Amelioration of intracerebroventricular streptozotocin induced cognitive dysfunction and oxidative stress by vinpocetine: a PDE1 inhibitor. *Eur. J. Pharmacol.*, **620**:49–56.

49. Ishihara, K., Katsuki, H., Sugimura, M., and Satoh, M. (1989) Idebenone and vinpocetine augment long-term potentiation in hippocampal slices in the guinea pig. *Neuropharmacology*, **28**:569–573.

50. Molnár, P. and Gaál, L. (1992) Effect of different subtypes of cognition enhancers on long-term potentiation in the rat dentate gyrus *in vivo*. *Eur. J. Pharmacol.*, **215**:17–22.

51. Lendvai, B., Zelles, T., Rozsa, B., and Vizi, E.S. (2003) A vinca alkaloid enhances morphological dynamics of dendritic spines of neocortical layer 2/3 pyramidal cells. *Brain Res. Bull.*, **59**:257–260.

52. DeNoble, V.J. (1987) Vinpocetine enhances retrieval of a step-through passive avoidance response in rats. *Pharmacol. Biochem. Behav.*, **26**:183–186.

53. DeNoble, V.J., Repetti, S.J., Gelpke, L.W., Wood, L.M., and Keim, K.L. (1986) Vinpocetine: nootropic effects on scopolamine-induced and hypoxia-induced retrieval deficits of a step-through passive avoidance response in rats. *Pharmacol. Biochem. Behav.*, **24**:1123–1128.

54. van Staveren, W.C., Markerink-van Ittersum, M., Steinbusch, H.W., and de Vente, J. (2001) The effects of phosphodiesterase inhibition on cyclic GMP and cyclic AMP accumulation in the hippocampus of the rat. *Brain Res.*, **888**:275–286.

55. Barco, A., Pittenger, C., and Kandel, E.R. (2003) CREB, memory enhancement and the treatment of memory disorders: promises, pitfalls and prospects. *Expert Opin. Ther. Targets*, **7**:101–114.

56. Benito, E. and Barco, A. (2010) CREB's control of intrinsic and synaptic plasticity: implications for CREB-dependent memory models. *Trends Neurosci.*, **33**:230–240.

57. Josselyn, S.A. and Nguyen, P.V. (2005) CREB, synapses and memory disorders: past progress and future challenges. *Curr. Drug Targets CNS Neurol. Disord.* **4**:481–497.

58. Puzzo, D., Palmeri, A., and Arancio, O. (2006) Involvement of the nitric oxide pathway in synaptic dysfunction following amyloid elevation in Alzheimer's disease. *Rev. Neurosci.*, **17**:497–523.

59. Krahe, T.E., Wang, W., and Medina, A.E. (2009) Phosphodiesterase inhibition increases CREB phosphorylation and restores orientation selectivity in a model of fetal alcohol spectrum disorders. *PLoS One*, **4**:e6643.

60. Filgueiras, C.C., Krahe, T.E., and Medina, A.E. (2010) Phosphodiesterase type 1 inhibition improves learning in rats exposed to alcohol during the third trimester equivalent of human gestation. *Neurosci. Lett.*, **473**:202–207.

61. West, G.A., et al. (2003) cGMP-dependent and not cAMP-dependent kinase is required for adenosine-induced dilation of intracerebral arterioles. *J. Cardiovasc. Pharmacol.*, **41**:444–451.

62. Rischke, R. and Krieglstein, J. (1991) Protective effect of vinpocetine against brain damage caused by ischemia. *Jpn. J. Pharmacol.*, **56**:349–356.

63. Horvath, B., Vekasi, J., Kesmarky, G., and Toth, K. (2008) *In vitro* antioxidant properties of pentoxifylline and vinpocetine in a rheological model. *Clin. Hemorheol. Microcirc.*, **40**:165–166.

64. Nishi, A. and Snyder, G.L. (2010) Advanced research on dopamine signaling to develop drugs for the treatment of mental disorders: biochemical and behavioral profiles of phosphodiesterase inhibition in dopaminergic neurotransmission. *J. Pharmacol. Sci.*, **114**:6–16.

65. Reed, T.M., Repaske, D.R., Snyder, G.L., Greengard, P., and Vorhees, C.V. (2002) Phosphodiesterase 1B knock-out mice exhibit exaggerated locomotor hyperactivity and DARPP-32 phosphorylation in response to dopamine agonists and display impaired spatial learning. *J. Neurosci.*, **22**:5188–5197.

66. Siuciak, J.A., Chapin, D.S., McCarthy, S.A., and Martin, A.N. (2007) Antipsychotic profile of rolipram: efficacy in rats and reduced sensitivity in mice deficient in the phosphodiesterase-4B (PDE4B) enzyme. *Psychopharmacology*, **192**:415–424.

67. Ehrman, L.A., et al. (2006) Phosphodiesterase 1B differentially modulates the effects of methamphetamine on locomotor activity and spatial learning through DARPP32-dependent pathways: evidence from PDE1B-DARPP32 double-knockout mice. *Genes Brain Behav.*, **5**:540–551.

68. Cygnar, K.D. and Zhao, H. (2009) Phosphodiesterase 1C is dispensable for rapid response termination of olfactory sensory neurons. *Nat. Neurosci.*, **12**:454–462.

69. Jäger, R., Schwede, F., Genieser, H.-G., Koesling, D., and Russwurm, M. (2010) Activation of PDE2 and PDE5 by specific GAF ligands: delayed activation of PDE5. *Br. J. Pharmacol.*, **161**:1645–1660.

70. Pandit, J., Forman, M.D., Fennell, K.F., Dillman, K.S., and Menniti, F.S. (2009) Mechanism for the allosteric regulation of phosphodiesterase 2A deduced from the

X-ray structure of a near full-length construct. *Proc. Natl. Acad. Sci. USA*, **106**:18225–18230.

71. Aravind, L. and Ponting, C.P. (1997) The GAF domain: an evolutionary link between diverse phototransducing proteins. *Trends Biochem. Sci.*, **22**:458–459.

72. Bender, A.T. and Beavo, J.A. (2006) Cyclic nucleotide phosphodiesterases: molecular regulation to clinical use. *Pharmacol. Rev.*, **58**:488–520.

73. Kanacher, T., Schultz, A., Linder, J.U., and Schultz, J.E. (2002) A GAF-domain-regulated adenylyl cyclase from *Anabaena* is a self-activating cAMP switch. *EMBO J.* **21**:3672–3680.

74. Schultz, J.E. (2009) Structural and biochemical aspects of tandem GAF domains. *Handb. Exp. Pharmacol.*, **191**:93–109.

75. Knacker, T., et al. (2005) The EU-project ERAPharm: incentives for the further development of guidance documents? *Environ. Sci. Pollut. Res. Int.*, **12**:62–65.

76. Rutten, K., Prickaerts, J., and Blokland, A. (2006) Rolipram reverses scopolamine-induced and time-dependent memory deficits in object recognition by different mechanisms of action. *Neurobiol. Learn. Mem.*, **85**:132–138.

77. DaSilva, J.N., et al. (2002) Imaging cAMP-specific phosphodiesterase-4 in human brain with R-[11C]rolipram and positron emission tomography. *Eur. J. Nucl. Med. Mol. Imaging*, **29**:1680–1683.

78. Morley-Fletcher, S., et al. (2004) Chronic treatment with imipramine reverses immobility behaviour, hippocampal corticosteroid receptors and cortical 5-HT(1A) receptor mRNA in prenatally stressed rats. *Neuropharmacology*, **47**:841–847.

79. Tafet, G.E. and Bernardini, R. (2003) Psychoneuroendocrinological links between chronic stress and depression. *Prog. Neuropsychopharmacol. Biol. Psychiatry*, **27**:893–903.

80. Xu, Y., Barish, P.A., Pan, J., Ogle, W.O., and O'Donnell, J.M. (2012) Animal models of depression and neuroplasticity: assessing drug action in relation to behavior and neurogenesis. *Methods Mol. Biol.*, **829**:103–124.

81. Suvarna, N.U. and O'Donnell, J.M. (2002) Hydrolysis of N-methyl-D-aspartate receptor-stimulated cAMP and cGMP by PDE4 and PDE2 phosphodiesterases in primary neuronal cultures of rat cerebral cortex and hippocampus. *J. Pharmacol. Exp. Ther.*, **302**:249–256.

82. Rivet-Bastide, M., et al. (1997) cGMP-stimulated cyclic nucleotide phosphodiesterase regulates the basal calcium current in human atrial myocytes. *J. Clin. Invest.*, **99**:2710–2718.

83. Abarghaz, M., Biondi, S., Duranton, J., Limanton, E., and Moddadori, C.W.P. (2005) Benzo [1,4] diazepin-2-one derivatives as phosphodiesterase PDE2 inhibitors, preparation and therapeutics use thereof. *World Patent WO/2005/063723*, pp. 1–38.

84. Reneerkens, O.A.H., et al. (2012) Inhibition of phoshodiesterase type 2 or type 10 reverses object memory deficits induced by scopolamine or MK-801. *Behav. Brain Res.* **236C**:16–22.

85. Menniti, F.S., Faraci, W.S., and Schmidt, C.J. (2006) Phosphodiesterases in the CNS: targets for drug development. *Nat. Rev. Drug Discov.* **5**:660–670.

86. Reneerkens, O.A.H., Rutten, K., Steinbusch, H.W.M., Blokland, A., and Prickaerts, J. (2009) Selective phosphodiesterase inhibitors: a promising target for cognition enhancement. *Psychopharmacology*, **202**:419–443.

87. van Donkelaar, E.L., et al. (2008) Phosphodiesterase 2 and 5 inhibition attenuates the object memory deficit induced by acute tryptophan depletion. *Eur. J. Pharmacol.*, **600**:98–104.

88. Klaassen, T., Riedel, W.J., Deutz, N.E., van Someren, A., and van Praag, H.M. (1999) Specificity of the tryptophan depletion method. *Psychopharmacology*, **141**:279–286.

89. Pelligrino, D.A. and Wang, Q. (1998) Cyclic nucleotide crosstalk and the regulation of cerebral vasodilation. *Prog. Neurobiol.*, **56**:1–18.

90. Werner, C., et al. (2004) Importance of NO/cGMP signalling via cGMP-dependent protein kinase II for controlling emotionality and neurobehavioural effects of alcohol. *Eur. J. Neurosci.*, **20**:3498–3506.

91. Izquierdo, L.A., et al. (2002) Molecular pharmacological dissection of short- and long-term memory. *Cell Mol. Neurobiol.*, **22**:269–287.

92. Rutten, K., et al. (2007) Time-dependent involvement of cAMP and cGMP in consolidation of object memory: studies using selective phosphodiesterase type 2, 4 and 5 inhibitors. *Eur. J. Pharmacol.*, **558**:107–112.

93. Conti, M., et al. (2003) Cyclic AMP-specific PDE4 phosphodiesterases as critical components of cyclic AMP signaling. *J. Biol. Chem.*, **278**:5493–5496.

94. Houslay, M.D. (2001) PDE4 cAMP-specific phosphodiesterases. *Prog. Nucleic Acid Res. Mol. Biol.*, **69**:249–315.

95. Chandrasekaran, A., et al. (2008) Identification and characterization of novel mouse PDE4D isoforms: molecular cloning, subcellular distribution and detection of isoform-specific intracellular localization signals. *Cell Signal*, **20**:139–153.

96. Cheung, Y.-F., et al. (2007) PDE4B5, a novel, super-short, brain-specific cAMP phosphodiesterase-4 variant whose isoform-specifying N-terminal region is identical to that of cAMP phosphodiesterase-4D6 (PDE4D6). *J. Pharmacol. Exp. Ther.*, **322**:600–609.

97. Houslay, M.D., Baillie, G.S., and Maurice, D.H. (2007) cAMP-specific phosphodiesterase-4 enzymes in the cardiovascular system: a molecular toolbox for generating compartmentalized cAMP signaling. *Circ. Res.*, **100**:950–966.

98. Lynex, C.N., et al. (2008) Identification and molecular characterization of a novel PDE4D11 cAMP-specific phosphodiesterase isoform. *Cell Signal*, **20**:2247–2255.

99. Mackenzie, K.F., et al. (2008) Human PDE4A8, a novel brain-expressed PDE4 cAMP-specific phosphodiesterase that has undergone rapid evolutionary change. *Biochem. J.*, **411**:361–369.

100. Wallace, D.A., et al. (2005) Identification and characterization of PDE4A11, a novel, widely expressed long isoform encoded by the human PDE4A cAMP phosphodiesterase gene. *Mol. Pharmacol.*, **67**:1920–1934.

101. Richter, W., Hermsdorf, T., and Dettmer, D. (2005) Renaturation of the catalytic domain of PDE4A expressed in *Escherichia coli* as inclusion bodies. *Methods Mol. Biol.*, **307**:155–165.

102. Johnston, L.A., et al. (2004) Expression, intracellular distribution and basis for lack of catalytic activity of the PDE4A7 isoform encoded by the human PDE4A cAMP-specific phosphodiesterase gene. *Biochem. J.*, **380**:371–384.

103. MacKenzie, S.J., et al. (2002) Long PDE4 cAMP specific phosphodiesterases are activated by protein kinase A-mediated phosphorylation of a single serine residue in Upstream Conserved Region 1 (UCR1). *Br. J. Pharmacol.*, **136**:421–433.

104. Sette, C. and Conti, M. (1996) Phosphorylation and activation of a cAMP-specific phosphodiesterase by the cAMP-dependent protein kinase: involvement of serine 54 in the enzyme activation. *J. Biol. Chem.*, **271**:16526–16534.

105. Baillie, G.S., MacKenzie, S.J., McPhee, I., and Houslay, M.D. (2000) Sub-family selective actions in the ability of Erk2 MAP kinase to phosphorylate and regulate the activity of PDE4 cyclic AMP-specific phosphodiesterases. *Br. J. Pharmacol.*, **131**:811–819.

106. Hoffmann, R., Baillie, G.S., MacKenzie, S.J., Yarwood, S.J., and Houslay, M.D. (1999) The MAP kinase ERK2 inhibits the cyclic AMP-specific phosphodiesterase HSPDE4D3 by phosphorylating it at Ser579. *EMBO J.*, **18**:893–903.

107. MacKenzie, S.J., Baillie, G.S., McPhee, I., Bolger, G.B., and Houslay, M.D. (2000) ERK2 mitogen-activated protein kinase binding, phosphorylation, and regulation of the PDE4D cAMP-specific phosphodiesterases: the involvement of COOH-terminal docking sites and NH2-terminal UCR regions. *J. Biol. Chem.*, **275**:16609–16617.

108. Kaulen, P., Brüning, G., Schneider, H.H., Sarter, M., and Baumgarten, H.G. (1989) Autoradiographic mapping of a selective cyclic adenosine monophosphate phosphodiesterase in rat brain with the antidepressant [3H]rolipram. *Brain Res.*, **503**:229–245.

109. Iwahashi, Y., et al. (1996) Differential distribution of mRNA encoding cAMP-specific phosphodiesterase isoforms in the rat brain. *Brain Res. Mol. Brain Res.*, **38**:14–24.

110. Fujita, M., et al. (2005) Quantification of brain phosphodiesterase 4 in rat with (R)-[11C] rolipram-PET. *NeuroImage*, **26**:1201–1210.

111. Itoh, T., et al. (2009) PET measurement of the *in vivo* affinity of 11C-(R)-rolipram and the density of its target, phosphodiesterase-4, in the brains of conscious and anesthetized rats. *J. Nucl. Med.*, **50**:749–756.

112. Li, Y.-F., et al. (2011) Phosphodiesterase-4D knock-out and RNA interference-mediated knock-down enhance memory and increase hippocampal neurogenesis via increased cAMP signaling. *J. Neurosci.*, **31**:172–183.

113. Card, G.L., et al. (2004) Structural basis for the activity of drugs that inhibit phosphodiesterases. *Structure (London, England: 1993)*, **12**:2233–2247.

114. Silvestre, J.S., Fernández, A.G., and Palacios, J.M. (1999) Effects of rolipram on the elevated plus-maze test in rats: a preliminary study. *J. Psychopharmacol.*, **13**:274–277.

115. Halene, T.B. and Siegel, S.J. (2008) Antipsychotic-like properties of phosphodiesterase 4 inhibitors: evaluation of 4-(3-butoxy-4-methoxybenzyl)-2-imidazolidinone (RO-20-1724) with auditory event-related potentials and prepulse inhibition of startle. *J. Pharmacol. Exp. Ther.*, **326**:230–239.

116. Gretarsdottir, S., et al. (2003) The gene encoding phosphodiesterase 4D confers risk of ischemic stroke. *Nat. Genet.*, **35**:131–138.

117. Sasaki, T., et al. (2007) The phosphodiesterase inhibitor rolipram promotes survival of newborn hippocampal neurons after ischemia. *Stroke*, **38**:1597–1605.

118. Hu, W., et al. (2011) Inhibition of phosphodiesterase-4 decreases ethanol intake in mice. *Psychopharmacology*, **218**:331–339.

119. Wen, R.-T., et al. (2012) The phosphodiesterase-4 (PDE4) inhibitor rolipram decreases ethanol seeking and consumption in alcohol-preferring fawn-hooded rats. *Alcohol Clin. Exp. Res.*, **36**:2157–2167.

120. Mamiya, T., et al. (2001) Involvement of cyclic AMP systems in morphine physical dependence in mice: prevention of development of morphine dependence by rolipram, a phosphodiesterase 4 inhibitor. *Br. J. Pharmacol.*, **132**:1111–1117.

121. Lamontagne, S., et al. (2001) Localization of phosphodiesterase-4 isoforms in the medulla and nodose ganglion of the squirrel monkey. *Brain Res.*, **920**:84–96.

122. Schneider, H.H., Schmiechen, R., Brezinski, M., and Seidler, J. (1986) Stereospecific binding of the antidepressant rolipram to brain protein structures. *Eur. J. Pharmacol.*, **127**:105–115.

123. Bruno, O., et al. (2011) GEBR-7b, a novel PDE4D selective inhibitor that improves memory in rodents at non-emetic doses. *Br. J. Pharmacol.*, **164**:2054–2063.

124. Zhang, H.-T., et al. (2002) Antidepressant-like profile and reduced sensitivity to rolipram in mice deficient in the PDE4D phosphodiesterase enzyme. *Neuropsychopharmacology*, **27**:587–595.

125. Otowa, T., et al. (2011) Association study of PDE4B with panic disorder in the Japanese population. *Prog. Neuropsychopharmacol. Biol. Psychiatry*, **35**:545–549.

126. Fatemi, S.H., et al. (2008) PDE4B polymorphisms and decreased PDE4B expression are associated with schizophrenia. *Schizophr. Res.*, **101**:36–49.

127. Dlaboga, D., Hajjhussein, H., and O'Donnell, J.M. (2006) Regulation of phosphodiesterase-4 (PDE4) expression in mouse brain by repeated antidepressant treatment: comparison with rolipram. *Brain Res.*, **1096**:104–112.

128. Fujita, M., et al. (2007) *In vivo* and *in vitro* measurement of brain phosphodiesterase 4 in rats after antidepressant administration. *Synapse*, **61**:78–86.

129. Takahashi, M., et al. (1999) Chronic antidepressant administration increases the expression of cAMP-specific phosphodiesterase 4A and 4B isoforms. *J. Neurosci.*, **19**:610–618.

130. Ye, Y., Jackson, K., and O'Donnell, J.M. (2000) Effects of repeated antidepressant treatment of type 4A phosphodiesterase (PDE4A) in rat brain. *J. Neurochem.*, **74**:1257–1262.

131. Cho, C.H., Cho, D.H., Seo, M.R., and Juhnn, Y.S. (2000) Differential changes in the expression of cyclic nucleotide phosphodiesterase isoforms in rat brains by chronic treatment with electroconvulsive shock. *Exp. Mol. Med.*, **32**:110–114.

132. D'Sa, C., Eisch, A.J., Bolger, G.B., and Duman, R.S. (2005) Differential expression and regulation of the cAMP-selective phosphodiesterase type 4A splice variants in rat brain by chronic antidepressant administration. *Eur. J. Neurosci.*, **22**:1463–1475.

133. Schneider, H.H. (1984) Brain cAMP response to phosphodiesterase inhibitors in rats killed by microwave irradiation or decapitation. *Biochem. Pharmacol.*, **33**:1690–1693.

134. Ye, Y. and O'Donnell, J.M. (1996) Diminished noradrenergic stimulation reduces the activity of rolipram-sensitive, high-affinity cyclic AMP phosphodiesterase in rat cerebral cortex. *J. Neurochem.*, **66**:1894–1902.

135. Houslay, M.D., Schafer, P., and Zhang, K.Y.J. (2005) Keynote review: phosphodiesterase-4 as a therapeutic target. *Drug Discov. Today*, **10**:1503–1519.

136. Millar, J.K., et al. (2005) DISC1 and PDE4B are interacting genetic factors in schizophrenia that regulate cAMP signaling. *Science*, **310**:1187–1191.

137. Zhu, J., Mix, E., and Winblad, B. (2001) The antidepressant and antiinflammatory effects of rolipram in the central nervous system. *CNS Drug Rev.*, **7**:387–398.

138. Randt, C.T., Judge, M.E., Bonnet, K.A., and Quartermain, D. (1982) Brain cyclic AMP and memory in mice. *Pharmacol. Biochem. Behav.*, **17**:677–680.

139. Villiger, J.W. and Dunn, A.J. (1981) Phosphodiesterase inhibitors facilitate memory for passive avoidance conditioning. *Behav. Neural. Biol.*, **31**:354–359.

140. Monti, B., Berteotti, C., and Contestabile, A. (2006) Subchronic rolipram delivery activates hippocampal CREB and arc, enhances retention and slows down extinction of conditioned fear. *Neuropsychopharmacology*, **31**:278–286.

141. Nakagawa, S., et al. (2002) Localization of phosphorylated cAMP response element-binding protein in immature neurons of adult hippocampus. *J. Neurosci.*, **22**:9868–9876.

142. Saxe, M.D., et al. (2006) Ablation of hippocampal neurogenesis impairs contextual fear conditioning and synaptic plasticity in the dentate gyrus. *Proc. Natl. Acad. Sci. USA*, **103**:17501–17506.

143. Otmakhov, N., et al. (2004) Forskolin-induced LTP in the CA1 hippocampal region is NMDA receptor dependent. *J. Neurophysiol.*, **91**:1955–1962.

144. Zhang, H.-T., et al. (2005) Effects of the novel PDE4 inhibitors MEM1018 and MEM1091 on memory in the radial-arm maze and inhibitory avoidance tests in rats. *Psychopharmacology*, **179**:613–619.

145. Egawa, T., Mishima, K., Matsumoto, Y., Iwasaki, K., and Fujiwara, M. (1997) Rolipram and its optical isomers, phosphodiesterase 4 inhibitors, attenuated the scopolamine-induced impairments of learning and memory in rats. *Jpn. J. Pharmacol.*, **75**:275–281.

146. Zhang, H.T. and O'Donnell, J.M. (2000) Effects of rolipram on scopolamine-induced impairment of working and reference memory in the radial-arm maze tests in rats. *Psychopharmacology*, **150**:311–316.

147. Cheng, Y.-F., et al. (2010) Inhibition of phosphodiesterase-4 reverses memory deficits produced by Aβ25–35 or Aβ1–40 peptide in rats. *Psychopharmacology*, **212**:181–191.

148. Wang, C., et al. (2012) The phosphodiesterase-4 inhibitor rolipram reverses Aβ-induced cognitive impairment and neuroinflammatory and apoptotic responses in rats. *Int. J. Neuropsychopharmacol.*, **15**:749–766.

149. Gong, B., et al. (2004) Persistent improvement in synaptic and cognitive functions in an Alzheimer mouse model after rolipram treatment. *J. Clin. Invest.*, **114**:1624–1634.

150. Imanishi, T., et al. (1997) Ameliorating effects of rolipram on experimentally induced impairments of learning and memory in rodents. *Eur. J. Pharmacol.*, **321**:273–278.

151. Nagakura, A., Niimura, M., and Takeo, S. (2002) Effects of a phosphodiesterase IV inhibitor rolipram on microsphere embolism-induced defects in memory function and cerebral cyclic AMP signal transduction system in rats. *Br. J. Pharmacol.*, **135**:1783–1793.

152. Bourtchouladze, R., et al. (2003) A mouse model of Rubinstein–Taybi syndrome: defective long-term memory is ameliorated by inhibitors of phosphodiesterase 4. *Proc. Natl. Acad. Sci. USA*, **100**:10518–10522.

153. Zhang, H.T. and O'Donnell, J.M. (2007) In: Beavo, J.A., Francis, S.H., and Houslay, M.D. (Eds), *Phosphodiesterase-4 as a pharmacological target mediating antidepressant and cognitive effects on behavior*, Boca Raton, FL: CRC Press, pp. 539–558.

154. Engels, P., Abdel'Al, S., Hulley, P., and Lübbert, H. (1995) Brain distribution of four rat homologues of the *Drosophila* dunce cAMP phosphodiesterase. *J. Neurosci.Res.*, **41**:169–178.

155. Devan, B.D., et al. (2004) Phosphodiesterase inhibition by sildenafil citrate attenuates the learning impairment induced by blockade of cholinergic muscarinic receptors in rats. *Pharmacol. Biochem. Behav.*, **79**:691–699.

156. Puzzo, D., et al. (2009) Phosphodiesterase 5 inhibition improves synaptic function, memory, and amyloid-beta load in an Alzheimer's disease mouse model. *J. Neurosci.*, **29**:8075–8086.

157. van der Staay, F.J., et al. (2008) The novel selective PDE9 inhibitor BAY 73-6691 improves learning and memory in rodents. *Neuropharmacology*, **55**:908–918.

158. Grauer, S.M., et al. (2009) Phosphodiesterase 10A inhibitor activity in preclinical models of the positive, cognitive, and negative symptoms of schizophrenia. *J. Pharmacol. Exp. Ther.*, **331**:574–590.

159. Siuciak, J.A., et al. (2006) Inhibition of the striatum-enriched phosphodiesterase PDE10A: a novel approach to the treatment of psychosis. *Neuropharmacology*, **51**:386–396.

160. Chetkovich, D.M., Gray, R., Johnston, D., and Sweatt, J.D. (1991) N-Methyl-D-aspartate receptor activation increases cAMP levels and voltage-gated Ca^{2+} channel activity in area CA1 of hippocampus. *Proc. Natl. Acad. Sci. USA*, **88**:6467–6471.

161. Sala, C., Rudolph-Correia, S., and Sheng, M. (2000) Developmentally regulated NMDA receptor-dependent dephosphorylation of cAMP response element-binding protein (CREB) in hippocampal neurons. *J. Neurosci.*, **20**:3529–3536.

162. Davis, J.A. and Gould, T.J. (2005) Rolipram attenuates MK-801-induced deficits in latent inhibition. *Behav. Neurosci.* **119**:595–602.

163. Hajjhussein, H., Suvarna, N.U., Gremillion, C., Chandler, L.J., and O'Donnell, J.M. (2007) Changes in NMDA receptor-induced cyclic nucleotide synthesis regulate the age-dependent increase in PDE4A expression in primary cortical cultures. *Brain Res.*, **1149**:58–68.

164. Navakkode, S., Sajikumar, S., and Frey, J.U. (2004) The type IV-specific phosphodiesterase inhibitor rolipram and its effect on hippocampal long-term potentiation and synaptic tagging. *J. Neurosci.*, **24**:7740–7744.

165. Bos, J.L. (2006) Epac proteins: multi-purpose cAMP targets. *Trends Biochem. Sci.*, **31**:680–686.

166. Gloerich, M. and Bos, J.L. (2010) Epac: defining a new mechanism for cAMP action. *Annu. Rev. Pharmacol. Toxicol.* **50**:355–375.

167. Rangarajan, S., et al. (2003) Cyclic AMP induces integrin-mediated cell adhesion through Epac and Rap1 upon stimulation of the beta 2-adrenergic receptor. *J. Cell Biol.*, **160**:487–493.

168. Kashima, Y., et al. (2001) Critical role of cAMP-GEFII–Rim2 complex in incretin-potentiated insulin secretion. *J. Biol. Chem.*, **276**:46046–46053.

169. Ozaki, N., et al. (2000) cAMP-GEFII is a direct target of cAMP in regulated exocytosis. *Nat. Cell Biol.*, **2**:805–811.

170. Gekel, I. and Neher, E. (2008) Application of an Epac activator enhances neurotransmitter release at excitatory central synapses. *J. Neurosci.*, **28**:7991–8002.

171. Sakaba, T. and Neher, E. (2003) Direct modulation of synaptic vesicle priming by GABA (B) receptor activation at a glutamatergic synapse. *Nature*, **424**:775–778.

172. Ster, J., et al. (2009) Epac mediates PACAP-dependent long-term depression in the hippocampus. *J. Physiol.*, **587**:101–113.

173. Zhong, N. and Zucker, R.S. (2004) Roles of Ca^{2+}, hyperpolarization and cyclic nucleotide-activated channel activation, and actin in temporal synaptic tagging. *J. Neurosci.*, **24**:4205–4212.

174. Ouyang, M., Zhang, L., Zhu, J.J., Schwede, F., and Thomas, S.A. (2008) Epac signaling is required for hippocampus-dependent memory retrieval. *Proc. Natl. Acad. Sci. USA*, **105**:11993–11997.

175. Houslay, M.D. (2010) Underpinning compartmentalised cAMP signalling through targeted cAMP breakdown. *Trends Biochem. Sci.*, **35**:91–100.

176. Rampersad, S.N., et al. (2010) Cyclic AMP phosphodiesterase 4D (PDE4D) Tethers EPAC1 in a vascular endothelial cadherin (VE-Cad)-based signaling complex and controls cAMP-mediated vascular permeability. *J. Biol. Chem.*, **285**:33614–33622.

177. Chandler, L.J., Sutton, G., Dorairaj, N.R., and Norwood, D. (2001) *N*-Methyl D-aspartate receptor-mediated bidirectional control of extracellular signal-regulated kinase activity in cortical neuronal cultures. *J. Biol. Chem.*, **276**:2627–2636.

178. Giovannini, M.G. (2006) The role of the extracellular signal-regulated kinase pathway in memory encoding. *Rev. Neurosci.*, **17**:619–634.

179. Sweatt, J.D. (2001) The neuronal MAP kinase cascade: a biochemical signal integration system subserving synaptic plasticity and memory. *J. Neurochem.*, **76**:1–10.

180. Massey, P.V. and Bashir, Z.I. (2007) Long-term depression: multiple forms and implications for brain function. *Trends Neurosci.*, **30**:176–184.

181. Navakkode, S., Sajikumar, S., and Frey, J.U. (2005) Mitogen-activated protein kinase-mediated reinforcement of hippocampal early long-term depression by the type IV-specific phosphodiesterase inhibitor rolipram and its effect on synaptic tagging. *J. Neurosci.*, **25**:10664–10670.

182. Lynch, M.J., et al. (2005) RNA silencing identifies PDE4D5 as the functionally relevant cAMP phosphodiesterase interacting with beta arrestin to control the protein kinase A/AKAP79-mediated switching of the beta2-adrenergic receptor to activation of ERK in HEK293B2 cells. *J. Biol. Chem.*, **280**:33178–33189.

183. Jin, S.L., Richard, F.J., Kuo, W.P., D'Ercole, A.J., and Conti, M. (1999) Impaired growth and fertility of cAMP-specific phosphodiesterase PDE4D-deficient mice. *Proc. Natl. Acad. Sci. USA*, **96**:11998–12003.

184. Jin, S.-L.C. and Conti, M. (2002) Induction of the cyclic nucleotide phosphodiesterase PDE4B is essential for LPS-activated TNF-alpha responses. *Proc. Natl. Acad. Sci. USA*, **99**:7628–7633.

185. Siuciak, J.A. (2008) The role of phosphodiesterases in schizophrenia: therapeutic implications. *CNS Drugs*, **22**:983–993.

186. Liu, H., et al. (2000) Expression of phosphodiesterase 4D (PDE4D) is regulated by both the cyclic AMP-dependent protein kinase and mitogen-activated protein kinase signaling pathways: a potential mechanism allowing for the coordinated regulation of PDE4D activity and expression. *J. Biol. Chem.*, **275**:26615–26624.

187. Li, L.-X., et al. (2011) Prevention of cerebral ischemia-induced memory deficits by inhibition of phosphodiesterase-4 in rats. *Metab. Brain Dis.*, **26**:37–47.

188. Rutten, K., Prickaerts, J., Schaenzle, G., Rosenbrock, H., and Blokland, A. (2008) Sub-chronic rolipram treatment leads to a persistent improvement in long-term object memory in rats. *Neurobiol. Learn. Mem.* **90**:569–575.

189. Giorgi, M., Modica, A., Pompili, A., Pacitti, C., and Gasbarri, A. (2004) The induction of cyclic nucleotide phosphodiesterase 4 gene (PDE4D) impairs memory in a water maze task. *Behav. Brain Res.*, **154**:99–106.

190. Cullen, B.R. (2005) RNAi the natural way. *Nat. Genet.*, **37**:1163–1165.

191. Zeringue, H.C. and Constantine-Paton, M. (2004) Post-transcriptional gene silencing in neurons. *Curr. Opin. Neurobiol.*, **14**:654–659.

192. Day, J.P., Houslay, M.D., and Davies, S.-A. (2006) A novel role for a *Drosophila* homologue of cGMP-specific phosphodiesterase in the active transport of cGMP. *Biochem. J.*, **393**:481–488.

193. Kolosionek, E., et al. (2009) Expression and activity of phosphodiesterase isoforms during epithelial mesenchymal transition: the role of phosphodiesterase 4. *Mol. Biol. Cell*, **20**:4751–4765.

194. Li, X., Huston, E., Lynch, M.J., Houslay, M.D., and Baillie, G.S. (2006) Phosphodiesterase-4 influences the PKA phosphorylation status and membrane translocation of G-protein receptor kinase 2 (GRK2) in HEK-293beta2 cells and cardiac myocytes. *Biochem. J.*, **394**:427–435.

195. Nagel, D.J., et al. (2006) Role of nuclear Ca^{2+}/calmodulin-stimulated phosphodiesterase 1A in vascular smooth muscle cell growth and survival. *Circ. Res.*, **98**:777–784.

196. Pekkinen, M., Ahlström, M.E.B., Riehle, U., Huttunen, M.M., and Lamberg-Allardt, C.J. E. (2008) Effects of phosphodiesterase 7 inhibition by RNA interference on the gene expression and differentiation of human mesenchymal stem cell-derived osteoblasts. *Bone*, **43**:84–91.

197. Waddleton, D., et al. (2008) Phosphodiesterase 3 and 4 comprise the major cAMP metabolizing enzymes responsible for insulin secretion in INS-1 (832/13) cells and rat islets. *Biochem. Pharmacol.*, **76**:884–893.

198. Zentilin, L. and Giacca, M. (2004) *In vivo* transfer and expression of genes coding for short interfering RNAs. *Curr. Pharm. Biotechnol.*, **5**:341–347.

199. Schratt, G.M., et al. (2006) A brain-specific microRNA regulates dendritic spine development. *Nature*, **439**:283–289.

200. Stegmeier, F., Hu, G., Rickles, R.J., Hannon, G.J., and Elledge, S.J. (2005) A lentiviral microRNA-based system for single-copy polymerase II-regulated RNA interference in mammalian cells. *Proc. Natl. Acad. Sci. USA*, **102**:13212–13217.

201. Dull, T., et al. (1998) A third-generation lentivirus vector with a conditional packaging system. *J. Virol.*, **72**:8463–8471.

202. Grillo, C.A., et al. (2007) Lentivirus-mediated downregulation of hypothalamic insulin receptor expression. *Physiol. Behav.*, **92**:691–701.

203. Zufferey, R., et al. (1998) Self-inactivating lentivirus vector for safe and efficient *in vivo* gene delivery. *J. Virol.*, **72**:9873–9880.

204. Robichaud, A., Savoie, C., Stamatiou, P.B., Tattersall, F.D., and Chan, C.C. (2001) PDE4 inhibitors induce emesis in ferrets via a noradrenergic pathway. *Neuropharmacology*, **40**:262–269.

205. Robichaud, A., et al. (2002) Assessing the emetic potential of PDE4 inhibitors in rats. *Br. J. Pharmacol.*, **135**:113–118.

206. Martinez, S.E., Beavo, J.A., and Hol, W.G.J. (2002) GAF domains: two-billion-year-old molecular switches that bind cyclic nucleotides. *Mol. Interv.*, **2**:317–323.

207. Kleppisch, T. (2009) Phosphodiesterases in the central nervous system. *Handb. Exp. Pharmacol.*, **191**:71–92.

208. Garthwaite, J. and Boulton, C.L. (1995) Nitric oxide signaling in the central nervous system. *Annu. Rev. Physiol.*, **57**:683–706.

209. Rapoport, M., van Reekum, R., and Mayberg, H. (2000) The role of the cerebellum in cognition and behavior: a selective review. *J. Neuropsychiatry Clin. Neurosci.*, **12**:193–198.

210. Hartell, N.A. (1996) Inhibition of cGMP breakdown promotes the induction of cerebellar long-term depression. *J. Neurosci.*, **16**:2881–2890.

211. Prickaerts, J., Steinbusch, H.W., Smits, J.F., and de Vente, J. (1997) Possible role of nitric oxide-cyclic GMP pathway in object recognition memory: effects of 7-nitroindazole and zaprinast. *Eur. J. Pharmacol.*, **337**:125–136.

212. Prickaerts, J., et al. (2002) Effects of two selective phosphodiesterase type 5 inhibitors, sildenafil and vardenafil, on object recognition memory and hippocampal cyclic GMP levels in the rat. *Neuroscience*, **113**:351–361.

213. Rutten, K., et al. (2005) The selective PDE5 inhibitor, sildenafil, improves object memory in Swiss mice and increases cGMP levels in hippocampal slices. *Behav. Brain Res.*, **164**:11–16.

214. Singh, N. and Parle, M. (2003) Sildenafil improves acquisition and retention of memory in mice. *Indian J. Physiol. Pharmacol.*, **47**:318–324.

215. Patil, C.S., Singh, V.P., and Kulkarni, S.K. (2006) Modulatory effect of sildenafil in diabetes and electroconvulsive shock-induced cognitive dysfunction in rats. *Pharmacol. Rep.*, **58**:373–380.

216. Devan, B.D., et al. (2006) Phosphodiesterase inhibition by sildenafil citrate attenuates a maze learning impairment in rats induced by nitric oxide synthase inhibition. *Psychopharmacology*, **183**:439–445.

217. Kurt, M., et al. (2004) Effect of sildenafil on anxiety in the plus-maze test in mice. *Pol. J. Pharmacol.* **56**:353–357.

218. Baratti, C.M. and Boccia, M.M. (1999) Effects of sildenafil on long-term retention of an inhibitory avoidance response in mice. *Behav. Pharmacol.*, **10**:731–737.

219. Campbell, E. and Edwards, T. (2006) Zaprinast consolidates long-term memory when administered to neonate chicks trained using a weakly reinforced single trial passive avoidance task. *Behav. Brain Res.*, **169**:181–185.

220. Edwards, T.M. and Lindley, N. (2007) Phosphodiesterase type 5 inhibition coupled to strong reinforcement results in two periods of transient retention loss in the young chick. *Behav. Brain Res.*, **183**:231–235.

221. Grass, H., et al. (2001) Sildenafil (Viagra): is there an influence on psychological performance? *Int. Urol. Nephrol.*, **32**:409–412.

222. Schultheiss, D., et al. (2001) Central effects of sildenafil (Viagra) on auditory selective attention and verbal recognition memory in humans: a study with event-related brain potentials. *World J. Urol.*, **19**:46–50.

223. Kaster, M.P., Rosa, A.O., Santos, A.R.S., and Rodrigues, A.L.S. (2005) Involvement of nitric oxide–cGMP pathway in the antidepressant-like effects of adenosine in the forced swimming test. *Int. J. Neuropsychopharmacol.*, **8**:601–606.

224. Almeida, R.C., Felisbino, C.S., López, M.G., Rodrigues, A.L.S., and Gabilan, N.H. (2006) Evidence for the involvement of L-arginine–nitric oxide–cyclic guanosine

monophosphate pathway in the antidepressant-like effect of memantine in mice. *Behav. Brain Res.*, **168**:318–322.

225. Goff, D.C., et al. (2009) A placebo-controlled study of sildenafil effects on cognition in schizophrenia. *Psychopharmacology*, **202**:411–417.

226. Monfort, P., et al. (2004) Sequential activation of soluble guanylate cyclase, protein kinase G and cGMP-degrading phosphodiesterase is necessary for proper induction of long-term potentiation in CA1 of hippocampus: alterations in hyperammonemia. *Neurochem. Int.*, **45**:895–901.

227. Shimizu-Albergine, M., et al. (2003) Individual cerebellar Purkinje cells express different cGMP phosphodiesterases (PDEs): *in vivo* phosphorylation of cGMP-specific PDE (PDE5) as an indicator of cGMP-dependent protein kinase (PKG) activation. *J. Neurosci.*, **23**:6452–6459.

228. Brink, C.B., Clapton, J.D., Eagar, B.E., and Harvey, B.H. (2008) Appearance of antidepressant-like effect by sildenafil in rats after central muscarinic receptor blockade: evidence from behavioural and neuro-receptor studies. *J. Neural. Transm.*, **115**:117–125.

229. Liebenberg, N., Harvey, B.H., Brand, L., and Brink, C.B. (2010) Antidepressant-like properties of phosphodiesterase type 5 inhibitors and cholinergic dependency in a genetic rat model of depression. *Behav. Pharmacol.*, **21**:540–547.

230. Ribeiro, J.A. and Sebastião, A.M. (2010) Caffeine and adenosine. *J. Alzheimers Dis.*, **20** (Suppl. 1):S3–S15.

231. Patil, C.S., Jain, N.K., Singh, V.P., and Kulkarni, S.K. (2004) Cholinergic-NO-cGMP mediation of sildenafil-induced antinociception. *Indian J. Exp. Biol.*, **42**:361–367.

232. Janowsky, D.S., el-Yousef, M.K., Davis, J.M., and Sekerke, H.J. (1972) A cholinergic–adrenergic hypothesis of mania and depression. *Lancet*, **2**:632–635.

233. Aguirre, N., Barrionuevo, M., Ramírez, M.J., Del Río, J., and Lasheras, B. (1999) Alpha-lipoic acid prevents 3,4-methylenedioxy-methamphetamine (MDMA)-induced neurotoxicity. *Neuroreport*, **10**:3675–3680.

234. Abdul, H.M. and Butterfield, D.A. (2007) Involvement of PI3K/PKG/ERK1/2 signaling pathways in cortical neurons to trigger protection by cotreatment of acetyl-L-carnitine and alpha-lipoic acid against HNE-mediated oxidative stress and neurotoxicity: implications for Alzheimer's disease. *Free Radic. Biol. Med.*, **42**:371–384.

235. Lee, S.Y., Andoh, T., Murphy, D.L., and Chiueh, C.C. (2003) 17beta-estradiol activates ICI 182,780-sensitive estrogen receptors and cyclic GMP-dependent thioredoxin expression for neuroprotection. *FASEB J.*, **17**:947–948.

236. Puerta, E., et al. (2009) Phosphodiesterase 5 inhibitors prevent 3,4-methylenedioxymethamphetamine-induced 5-HT deficits in the rat. *J. Neurochem.*, **108**:755–766.

237. Wang, L., Gang Zhang, Z., Lan Zhang, R., and Chopp, M. (2005) Activation of the PI3-K/Akt pathway mediates cGMP enhanced-neurogenesis in the adult progenitor cells derived from the subventricular zone. *J. Cereb. Blood Flow Metab.*, **25**:1150–1158.

238. Choi, D.E., et al. (2009) Pretreatment of sildenafil attenuates ischemia–reperfusion renal injury in rats. *Am. J. Physiol. Renal. Physiol.*, **297**:F362–F370.

239. Yuan, Z., Hein, T.W., Rosa, R.H., and Kuo, L. (2008) Sildenafil (Viagra) evokes retinal arteriolar dilation: dual pathways via NOS activation and phosphodiesterase inhibition. *Invest. Ophthalmol. Vis. Sci.*, **49**:720–725.

240. Fisher, D.A., Smith, J.F., Pillar, J.S., St Denis, S.H., and Cheng, J.B. (1998) Isolation and characterization of PDE9A, a novel human cGMP-specific phosphodiesterase. *J. Biol. Chem.*, **273**:15559–15564.

241. Guipponi, M., et al. (1998) Identification and characterization of a novel cyclic nucleotide phosphodiesterase gene (PDE9A) that maps to 21q22.3: alternative splicing of mRNA transcripts, genomic structure and sequence. *Hum. Genet.*, **103**:386–392.

242. Kleiman, R.J., et al. (2012) Phosphodiesterase 9A regulates central cGMP and modulates responses to cholinergic and monoaminergic perturbation *in vivo*. *J. Pharmacol. Exp. Ther.*, **341**:396–409.

243. Soderling, S.H., Bayuga, S.J., and Beavo, J.A. (1998) Identification and characterization of a novel family of cyclic nucleotide phosphodiesterases. *J. Biol. Chem.*, **273**:15553–15558.

244. Esposito, M., et al. (2009) The efficacy of horizontal and vertical bone augmentation procedures for dental implants: a Cochrane systematic review. *Eur. J. Oral Implantol.*, **2**:167–184.

245. Hutson, P.H., et al. (2011) The selective phosphodiesterase 9 (PDE9) inhibitor PF-04447943 (6-[(3S,4S)-4-methyl-1-(pyrimidin-2-ylmethyl)pyrrolidin-3-yl]-1-(tetrahydro-2H-pyran-4-yl)-1,5-dihydro-4H-pyrazolo[3,4-d]pyrimidin-4-one) enhances synaptic plasticity and cognitive function in rodents. *Neuropharmacology*, **61**:665–676.

246. Reyes-Irisarri, E., Markerink-Van Ittersum, M., Mengod, G., and de Vente, J. (2007) Expression of the cGMP-specific phosphodiesterases 2 and 9 in normal and Alzheimer's disease human brains. *Eur. J. Neurosci.*, **25**:3332–3338.

247. Straub, R.E., et al. (1994) A possible vulnerability locus for bipolar affective disorder on chromosome 21q22.3. *Nat. Genet.*, **8**:291–296.

248. Wong, M.-L., et al. (2006) Phosphodiesterase genes are associated with susceptibility to major depression and antidepressant treatment response. *Proc. Natl. Acad. Sci. USA*, **103**:15124–15129.

249. Vardigan, J.D., Converso, A., Hutson, P.H., and Uslaner, J.M. (2011) The selective phosphodiesterase 9 (PDE9) inhibitor PF-04447943 attenuates a scopolamine-induced deficit in a novel rodent attention task. *J. Neurogenet.*, **25**:120–126.

250. de Vente, J., Markerink-van Ittersum, M., and Vles, J.S.H. (2006) The role of phosphodiesterase isoforms 2, 5, and 9 in the regulation of NO-dependent and NO-independent cGMP production in the rat cervical spinal cord. *J. Chem. Neuroanat.*, **31**:275–303.

251. Wunder, F., et al. (2005) Characterization of the first potent and selective PDE9 inhibitor using a cGMP reporter cell line. *Mol. Pharmacol.*, **68**:1775–1781.

252. Fujishige, K., Kotera, J., and Omori, K. (1999) Striatum- and testis-specific phosphodiesterase PDE10A isolation and characterization of a rat PDE10A. *Eur. J. Biochem.*, **266**:1118–1127.

253. Loughney, K., et al. (1999) Isolation and characterization of PDE10A, a novel human 3′,5′-cyclic nucleotide phosphodiesterase. *Gene*, **234**:109–117.

254. Giampà, C., et al. (2010) Inhibition of the striatal specific phosphodiesterase PDE10A ameliorates striatal and cortical pathology in R6/2 mouse model of Huntington's disease. *PloS One*, **5**:e13417.

255. Hofbauer, K., Schultz, A., and Schultz, J.E. (2008) Functional chimeras of the phosphodiesterase 5 and 10 tandem GAF domains. *J. Biol. Chem.*, **283**:25164–25170.

256. Gross-Langenhoff, M., Hofbauer, K., Weber, J., Schultz, A., and Schultz, J.E. (2006) cAMP is a ligand for the tandem GAF domain of human phosphodiesterase 10 and cGMP for the tandem GAF domain of phosphodiesterase 11. *J. Biol. Chem.*, **281**:2841–2846.

257. Smith, S.M., et al. (2013) The novel phosphodiesterase 10A inhibitor THPP-1 has antipsychotic-like effects in rat and improves cognition in rat and rhesus monkey. *Neuropharmacology*, **64**:215–223.

258. O'Connor, V., et al. (2004) Differential amplification of intron-containing transcripts reveals long term potentiation-associated upregulation of specific PDE10A phosphodiesterase splice variants. *J. Biol. Chem.*, **279**:15841–15849.

259. Migaud, M., et al. (1998) Enhanced long-term potentiation and impaired learning in mice with mutant postsynaptic density-95 protein. *Nature*, **396**:433–439.

260. Threlfell, S., Sammut, S., Menniti, F.S., Schmidt, C.J., and West, A.R. (2009) Inhibition of phosphodiesterase 10A increases the responsiveness of striatal projection neurons to cortical stimulation. *J. Pharmacol. Exp. Ther.*, **328**:785–795.

261. Rodefer, J.S., Murphy, E.R., and Baxter, M.G. (2005) PDE10 inhibition reverses subchronic PCP-induced deficits in attentional set-shifting in rats. *Eur. J. Neurosci.*, **21**:1070–1076.

262. Fawcett, L., et al. (2000) Molecular cloning and characterization of a distinct human phosphodiesterase gene family: PDE11A. *Proc. Natl. Acad. Sci. USA*, **97**:3702–3707.

263. Makhlouf, A., Kshirsagar, A., and Niederberger, C. (2006) Phosphodiesterase 11: a brief review of structure, expression and function. *Int. J. Impot. Res.* **18**:501–509.

264. Yuasa, K., Kanoh, Y., Okumura, K., and Omori, K. (2001) Genomic organization of the human phosphodiesterase PDE11A gene: evolutionary relatedness with other PDEs containing GAF domains. *Eur. J. Biochem.*, **268**:168–178.

265. Luo, H.-R., et al. (2009) Association of PDE11A global haplotype with major depression and antidepressant drug response. *Neuropsychiatr. Dis. Treat.*, **5**:163–170.

266. Prickaerts, J. (2010) Phosphodiesterase inhibitors. In: Stolerman, I. (Ed), *Encyclopedia of Psychopharmacology*, Hamburg: Springer, pp. 1022–1028.

CHAPTER 8

EMERGING ROLE FOR PDE4 IN NEUROPSYCHIATRIC DISORDERS: TRANSLATING ADVANCES FROM GENETIC STUDIES INTO RELEVANT THERAPEUTIC STRATEGIES

SANDRA P. ZOUBOVSKY and AKIRA SAWA
Department of Psychiatry, Johns Hopkins University School of Medicine, Baltimore, MD, USA

NICHOLAS J. BRANDON
AstraZeneca Neuroscience iMED, Cambridge, MA, USA

INTRODUCTION

There is growing awareness in our society about the alarming burden on health and economies attributable to major psychiatric disorders, such as schizophrenia (SZ) and mood disorders [1]. Unfortunately, recent progress in drug development to treat these disorders has been very unproductive. Patients have had to be satisfied with therapies that provide modest improvement over medicines discovered serendipitously many decades ago [1]. More effective therapeutic strategies are likely to be identified by achieving a better understanding of molecular mechanisms underlying these neuro-psychiatric disorders. Our knowledge of major mental illnesses has been enriched at the molecular level as a result of newer, more powerful methodologies in the field of human genetics. The role of phosphodiesterase (PDE) 4 as a genetic risk factor for SZ and depression has been supported by genetic association studies and incredibly insightful cytogenetic approaches [2,3]. A role for PDE4 in disease has been further elucidated through the availability of PDE4 inhibitors as pharmacological tools. Indeed, PDE4-dependent processes have been implicated in SZ and depression, and more recently Huntington's disease (HD). Many of these pharmacological studies

Cyclic-Nucleotide Phosphodiesterases in the Central Nervous System: From Biology to Drug Discovery, First Edition. Edited by Nicholas J. Brandon and Anthony R. West.
© 2014 John Wiley & Sons, Inc. Published 2014 by John Wiley & Sons, Inc.

were performed principally in rodents using the "prototypical" PDE4 inhibitor rolipram [4]. Postmortem expression data have provided further support for the role of PDE4 molecular pathways contributing to disease pathophysiology [4]. Additionally, PDE4 is a downstream target of dopamine and glutamate receptor signaling, two of the critical neurotransmitter systems implicated in mental illness [5].

Collectively these observations have driven the reevaluation of PDE4 inhibition–modulation as an approach to treat major mental and neurological disorders such as SZ, HD, and depression. We will highlight the emerging data supporting these observations within this chapter. In particular, we will focus on human genetic findings, but will not dwell on the pharmacology based around rolipram, as this is covered in Chapters 7 and 12.

PDE4 SIGNALING IN SCHIZOPHRENIA

SZ is a complex disorder with onset normally in young adulthood. Several lines of evidence indicate that disturbances elicited by the interaction of multiple risk genes and environmental factors during early neurodevelopment disturb postnatal brain maturation and set the stage for the emergence of the disorder many years later, possibly through interactions with major developmental stressors such as puberty. Defects during cortical development, resulting in dysfunctional neuronal circuitries, are highly likely to be associated with SZ [6–8]. Many studies have provided supportive evidence for deficits in the cortico–striatal–pallido–thalamo–cortical associative loop and cortico–hippocampal circuit underlying the pathophysiology [9]. Because disturbances in dopaminergic neurotransmission have also been robustly implicated in SZ via clinicopharmacological studies, cyclic adenosine monophosphate (cAMP) signaling, a key intracellular cascade downstream of dopamine receptors, has been postulated as a drug target for SZ [4].

The earliest and most robust data linking PDE4 to SZ-like behaviors were generated in studies with rolipram, which seems to have the profile of an antipsychotic (see Chapter 12 for more details). Suggestive human genetic evidence of the involvement of PDE4 in SZ was originally reported in Ref. [10]. Their group identified a patient with SZ with a balanced t(1;16)(p31.2q21) translocation. A cousin of the proband who also carried the translocation had a psychotic illness with prolonged hospital admission. It was shown by traditional but elegant approaches that the breakpoint on chromosome 1 disrupted *PDE4B1*. Intriguingly, the expression of the B1 isoform of PDE4B was significantly reduced (by ~50%) in lymphoblastoid cell lines derived from patients with the t(1;16) translocation. Nevertheless, the study was unable to determine whether downregulation of PDE4 levels may be causally involved in the etiology of SZ in the probands.

Within the same study [11], there was also the report that PDE4B binds to the protein product of the *disrupted in schizophrenia 1* (*DISC1*) gene. *DISC1*, a major candidate risk gene for SZ, was originally identified at the site of a breakpoint on chromosome 1 as a result of a balanced chromosomal translocation t(1;11) segregating with major mental illness in a large Scottish pedigree [10]. The DISC1 protein interacts with PDE4B1 in resting cells. However, dissociation of the DISC1–PDE4B1 protein complex was observed after augmentation of cAMP levels [11]. Because a selective inhibitor of protein kinase A reverses this dissociation of the DISC1– PDE4B1 protein complex, interference with the binding of these two proteins can be associated with phosphorylation of PDE4B by cAMP-dependent protein kinase (protein kinase A [PKA]). Moreover, blocking the DISC1–PDE4B1 interaction increases phosphodiesterase activity, suggesting that DISC1 appears to bind primarily to dephosphorylated PDE4B, a form with reduced catalytic activity. Signaling events that elevate cellular levels of cAMP, such as those triggered by neurotransmitters, could consequently lead to phosphorylation of PDE4B, loss of DISC1 anchoring, and increased PDE activity. Houslay and colleagues [12] went further and elaborated on the cAMP regulation of the DISC1–PDE4 interaction by showing that PDE4D3 and PDE4C2 isoform interaction with DISC1 is disrupted by elevated cAMP while the binding of PDE4B1 and PDE4A5 isoforms was unaffected. The molecular explanation for this dichotomy in interactions was teased apart by mapping the interaction domains between DISC1 and the different PDE4 variants. Those variants resistant to cAMP manipulation were suggested to have additional sites of binding to DISC1 [12].

A recent report added a further layer of complexity to the DISC1–PDE4 relationship, by showing that PDE4D9 is regulated negatively at the level of transcription by a DISC1-activating transcription factor 4 (ATF4) complex [13]. Interestingly, loss of either DISC1 or ATF4 leads to increased levels of PDE4D9, whereas PKA phosphorylation of DISC1 regulates the DISC1–ATF4 complex integrity that is needed for PDE4D9 repression. These molecular insights likely explain the upregulation of PDE4D9 by dopamine D_1 receptor (D1R) stimulation, where PKA activation leads to DISC1 phosphorylation and disassociation from the PDE4D9 locus. Together, this provides a simple feedback loop for regulating D_1-dependent signaling.

Genetic studies on rare variants of *PDE4* in conjunction with cell biology of DISC1 and PDE4 are supportive of the notion that molecular pathways involving PDE4 may play a role in SZ [10]. However, the influence of common genetic variants in PDE4 on the risk for developing SZ remains uncertain. Thus far, the outcome of association studies focusing on the *PDE4* genes is controversial, as several studies have observed positive association, whereas others did not. For example, one group reported on the association of *PDE4B* polymorphisms and decreased PDE4B expression with SZ [14], and several studies reported positive association between the *PDE4B* gene and SZ in the Japanese population [15]. Most notably, in studies performed by the Peltonen lab, 11 genes coding for DISC1 protein interactors were selected and studied in 476 families, including 1857 genotyped individuals [16]. Single-nucleotide polymorphism (SNP) markers and haplotypes for their association with SZ were analyzed in two independent sets of families. For markers and haplotypes found to be consistently associated in both sets, the overall significance was tested with both family sample sets combined. Peltonen's group identified SNPs to be associated with SZ in *PDE4D* (rs1120303, $p = 0.021$), *PDE4B* (rs7412571, $p = 0.018$), and *NDEL1* (rs17806986, $p = 0.0038$). Greater significance was observed with allelic haplotypes of *PDE4D* ($p = 0.00084$), *PDE4B* ($p = 0.0022$ and $p = 0.029$), and *NDEL1* ($p = 0.0027$) that were either underrepresented or overrepresented in families ascertained for SZ. Taken together, these findings suggest that a cluster of genes in the DISC1–PDE4 cascade may contribute to the risk for SZ.

Data obtained from DISC1-based genetic animal models also seem to support the role for DISC1–PDE4 pathways in SZ. For example, behavioral and anatomical abnormalities relevant to SZ, including changes in prepulse inhibition, locomotion, working memory, and ventricle size, can be observed in animal models with genetically modified or perturbed DISC1 [1]. It is not perfectly clear yet whether such changes are mediated by PDE4–DISC1 interactions. However, there are two DISC1 mouse models generated by N-ethyl-N-nitrosourea (ENU) mutagenesis that have provided tantalizing data for a key role for the DISC1–PDE4 complex [17]. The DISC1–Q31L line shows a range of phenotypes, such as increased immobility in forced swim test, that can potentially classify it as a "depression" model. Although, this can be debated, of interest is the fact that the line shows a reduction of PDE4B activity. The decrease in activity was shown to be functionally important as attempts to reverse any of the phenotypes of the line with rolipram were unsuccessful. The second line, L100P, has been classified as a "SZ model," with deficits, for example, in prepulse inhibition, which were modulated by rolipram. More recently, these mouse lines have confirmed the close relationship between glycogen synthase kinase 3 (GSK3) and PDE4. In the Q31L line, a reduction in binding of GSK3 to mutated DISC1 is observed, along with increased enzymatic activity of GSK3. Additionally, DISC1–Q31L mutant mice show increased sensitivity to GSK3 inhibitor TDZD-8, which was able to reverse decreased prepulse inhibition and reduce immobility in forced swim test, thus suggesting synergy between PDE4 and GSK3 in a DISC1-dependent manner. Additionally, in the L100P line, combined treatment with rolipram and TDZD-8 reverses behavioral phenotypes observed, also indicating synergistic interactions between PDE4, GSK3, and DISC1 [17–19]. Another important DISC1 mouse model, DISC1$^{TM1/KARA}$, has a naturally occurring mutation (stop codon) in exon 7 in the 129Sv/Ev strain. Thus, in a simplistic manner this DISC1 truncation may effectively model the mutation found in the Scottish pedigree [20]. This line has been shown to exhibit significant decreases in PDE4B and PDE4D protein isoforms, concomitant decreased PDE4 activity, and elevated cAMP levels. These changes in cAMP regulation have been suggested to underlie deficits in axonal targeting and dendritic outgrowth, which may contribute to the robust working memory deficit seen in this line [21]. Finally, a behavioral study with PDE4B knockout mice reports a "prepsychotic" phenotype with a decrease in both prepulse inhibition and immobility in forced swim test, which potentially might be of relevance to SZ and depression [22]. Together, the data from mouse models, summarized in Table 8.1, provides compelling, but complex, evidence that the DISC1–PDE4 axis is important in regulating behaviors in animal assays relevant for SZ.

It is important to note that several other PDEs, in addition to PDE4, have been implicated in the pathophysiology of SZ. For example, PDE10A, a PDE with a highly enriched expression in the medium spiny neurons of the striatum [23], has been suggested to be a novel treatment target for SZ [24] (see also Chapters 10 and 12). Additionally, PDE11A knockout mice show subtle SZ-related phenotypes, including hyperactivity in an open field, deficits in social behaviors, enlarged lateral ventricles,

TABLE 8.1 Comparison of Behavioral Changes in PDE4BKO, PDE4DKO, Wild-Type with Rolipram, R6/2, and Several DISC1 Models

Relevance to Disease	Schizophrenia	Depression	Anxiety	PDE4
Assays	PPI	FST	EPM	Enzyme activity
DISC1 (Q31L)	↓	↑	–	↓
DISC1 (L100P)	↓	–	–	–
DISC1$^{TM1/KARA}$	–	NA	NA	↓
DISC1 (DN-Tg)	↓	↑	–	?
PDE4B (−/−)	↓	↓	NA	↓
PDE4D (−/−)	NA	↓	–	↓
WT + Rolipram	NA	↓	↓	↓
R6/2	NA	↓	↓	↑

The *down arrow* indicates a reduction, the *up arrow* indicates augmentation in activity, the hyphen indicates no change, and *N/A* indicates not available. In some models PDE4 activity has been examined, but it remains to be studied in others, such as DISC1 (DN-Tg). *Q31L*, mouse model with a point mutation in *DISC1*; *L100P*, mouse model with a point mutation in DISC1; *DISC1$^{TM1/KARA}$*, mouse model with a truncating lesion in the endogenous murine DISC1 ortholog; *PDE4B(−/−)*, phosphodiesterase 4 B knockout; *PDE4D (−/−)*, phosphodiesterase 4D knockout; *R6/2*, mouse model transgenic for exon 1 of the mutant *htt* gene expressing approximately 150 CAG repeats.

and increased activity in CA1 area of hippocampus [25]. Furthermore, PDE1 inhibitors are also being studied for treating cognitive impairments associated with SZ [26].

PDE4 SIGNALING IN DEPRESSION AND ANXIETY

Depression is a highly prevalent condition associated with significant morbidity and mortality. Approximately 10% of the population suffers from depression and it is twice more common in women than in men. For example, the 12-month prevalence of major depression in the adult population in the United States was quoted as 6.7% [27]. The role of PDE4 in depression has been studied for many years. Suggestions of antidepressant activity of rolipram originally led to a series of clinical trials for depression in the 1980s. For example, investigators carried out a 4-week study with patients hospitalized with major depressive disorder. The study demonstrated the antidepressant efficacy of rolipram with three different doses (3×0.25, 3×0.50, and 3×1.00 mg), although, interestingly, an inferior performance was obtained at higher dosages [28–30]. Overall the efficacy of rolipram in these studies was very encouraging, with similar efficacy to comparators such as imipramine. Unfortunately, its side effect profile, notably its emetic and sedative actions, has limited its clinical utility and prevented this compound from becoming a commercially viable product. As summarized in Table 8.1, consistent with the clinical evidence, mice deficient in PDE4D exhibit an antidepressant-like profile in the forced swim test, as shown by reduced immobility relative to wild-type controls [31]. This reduced immobility is similar to that observed if the mice are administered marketed antidepressants. In parallel, as seen in Table 8.1, pharmacological studies show that chronic administration of rolipram in mice reduces immobility in the forced swim test and tail suspension test [32,33], and reverses reserpine-induced hypothermia, another classic test for antidepressant-like activity [34]. Although it is unclear how PDE4 is involved in the pathology of depression, its role in this disorder is likely to be complex. This is highlighted by the human imaging finding by positron emission tomography (PET) using a ^{11}C-labeled version of rolipram that showed a decrease in PDE4 across brain regions in a group of unmedicated patients with major depressive disorder [35]. The previous medication history and current disease status at the time of scan could be important in the interpretation of these data.

Depression is often associated with anxiety, with ~85% of depressed patients reporting symptoms of anxiety of at least moderate severity [36]. Interestingly, signaling proteins such as cAMP response element binding (CREB), implicated in the action of chronic antidepressants, have also been found to activate intracellular signaling cascades that are implicated in anxiety [37], suggesting similarities in the molecular mechanisms underlying depressive and anxiety disorders. Consistent with this idea, repeated rolipram treatment also produced anxiolytic-like effects on behavior in mice (see Table 8.1), as evidenced by increased percentage of entries and time spent in the open arms in the elevated plus maze test [33]. Taken together, these findings might indicate that the PDE4 pathway plays a role in the pathophysiology of comorbid depression and anxiety.

In the unique Scottish pedigree from which the *DISC1* gene was identified, the family members with the balanced translocation manifest major mental illness with a 50-fold greater incidence than the general population. Intriguingly, the major diagnoses in the original description of the pedigree turned out to be major depression and not SZ as the name implies [1,38]. Thus, although DISC1 obviously confers risk for SZ, this family can also support a role for DISC1 signaling (possibly including the DISC1–PDE4 cascade) in depression.

It is important again to note that other PDEs have been linked with depressive disorder and different domains of these complex disorders in humans, such as motivation. These are discussed in more detail in Chapter 12. Notably, polymorphisms in the PDE9A and PDE11A genes have been significantly associated with the diagnosis of depression and remission in response to antidepressants [39].

PDE4 SIGNALING IN HUNTINGTON'S DISEASE

HD is characterized by the triad of motor, cognitive, and psychiatric symptoms [40,41]. Identification of the causal gene, *huntingtin(htt)*, has helped to build etiologically relevant rodent models for HD and facilitate understanding of molecular pathways that may mediate the disease pathology. In HD pedigrees, subjects with mutation in the *htt* gene display depression more frequently than family members without the mutation [42]. Associated with psychiatric symptoms, the completed suicide rate in HD is about 10-fold higher than that observed in the general population [40]. Indeed, the rate of suicide in HD is much higher than that in other medical and neurodegenerative conditions [41], suggesting that specific biological mechanisms, associated with HD, may underlie this unfortunate manifestation linked to mood disturbance. Mutant Htt protein is prone to form insoluble intracellular aggregates. Nevertheless, the role for aggregate formation in HD pathology remains unclear. This process may be, in part, a protective reaction against toxic mutant Htt. Nevertheless, this process also results in sequestering many Htt-associated molecules and interferes with their physiological functions [43,44]. Taken together, although HD is regarded as a neurological condition, its psychiatric manifestations are an even more important target for medication and indeed HD is an organic model for depression and perhaps other psychiatric conditions.

In HD, levels of cAMP are reportedly decreased in cerebrospinal fluid, lymphoblasts, and autopsied brains from patients, as well as in brains from a HD animal model, Hdh^{Q111} knock-in mouse expressing 111-glutamine mutant Htt [45]. There have been reports that rolipram can ameliorate HD pathology in cell and animal models, such as R6/2 mice containing exon 1 of Htt with an expanded CAG repeat [46], which exhibit decreased immobility in forced swim test and decreased anxiety in elevated plus maze, as well as elevated levels of PDE4. Rolipram treatment in R6/2 mice not only resulted in longer survival but also less severe and delayed onset of neurological impairment [47,48]. However, the underlying mechanism remains to be elucidated, although roles for CREB and extracellular signal regulated kinase (ERK) signaling have been proposed as neuroprotective mechanisms [47,49]. A possible breakthrough in understanding the mechanism may come from the knowledge of Htt–DISC1 protein interactions. This idea was initially suggested by abundant numbers of proteins that commonly interact with both Htt and DISC1, which include dynactin P150, N-CoR, and Karilin/Trio, as seen in Figure 8.1 [50–52]. This idea has

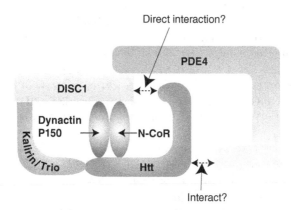

FIGURE 8.1   DISC1–PDE4 protein interaction (and hypothetical interaction of DISC1–Htt with common interactors). This figure illustrates examples of a number of proteins that commonly interact with both Htt and DISC1, which include dynactin P150, N-CoR, and Karilin/Trio. Plausible DISC1–Htt interaction could elucidate how PDE-mediated pathways are implicated in the pathology of Huntington's disease, thus contributing to novel treatment interventions.

been further elaborated by a bioinformatic approach [53]. Because PDE4–DISC1 protein interactions have already been established, this DISC1–Htt protein interaction can, if it is experimentally proven, provide a clue to elucidate how PDE4 function may be impaired in the presence of mutant Htt, probably via modulation of DISC1 complexes.

In addition to PDE4, involvement of other PDEs, especially those being expressed selectively in striatal medium spiny neurons (PDE10A and PDE1B), has been suggested in HD [54,55]. Dosing with the potent and selective PDE10A inhibitor TP-10 also ameliorated HD-associated pathology in the R6/2 model [56].

FUTURE PERSPECTIVES

cAMP is a key intracellular signaling molecule found downstream of neurotransmitter receptors implicated in neuropsychiatric conditions, such as the dopamine D_2 receptor. Meanwhile, cAMP–PKA are upstream of key cellular executors and transcription factors, such as CREB, critical for neuronal plasticity and neurogenesis. Modulation of PDE4 activity may be one of the best means to regulate cAMP. The importance of scaffolding proteins that interact with PDE4 in modulating PDE4 activity has been suggested. This chapter emphasizes that DISC1 may be a promising scaffold to regulate PDE4 in the context of various neuropsychiatric conditions.

REFERENCES

1. Brandon, N.J. and Sawa A. (2011) Linking neurodevelopmental and synaptic theories of mental illness through DISC1. *Nat. Rev. Neurosci.*, **12**:707–722.
2. Porteous, D. (2008) Genetic causality in schizophrenia and bipolar disorder: out with the old and in with the new. *Curr. Opin. Genet. Dev.*, **18**:229–234.
3. Zhang, H.T. (2009) Cyclic AMP-specific phosphodiesterase-4 as a target for the development of antidepressant drugs. *Curr. Pharm. Des.*, **15**:1688–1698.
4. Halene, T.B. and Siegel, S.J. (2007) PDE inhibitors in psychiatry: future options for dementia, depression and schizophrenia? *Drug Discov. Today*, **12**:870–878.
5. Soares, D.C., Carlyle, B.C., Bradshaw, N.J., and Porteous, D.J. (2011) DISC1: structure, function, and therapeutic potential for major mental illness. *ACS Chem. Neurosci.*, **2**:609–632.
6. Kamiya, A., Kitabatake, Y., and Sawa, A. (2008) Neurodevelopmental disturbance in the pathogenesis of major mental disorders. *Brain Nerve*, **60**:445–452.
7. Weinberger, D.R. (1987) Implications of normal brain development for the pathogenesis of schizophrenia. *Arch. Gen. Psychiatry*, **44**:660–669.
8. Cannon, T.D., et al. (2003) Early and late neurodevelopmental influences in the prodrome to schizophrenia: contributions of genes, environment, and their interactions. *Schizophr. Bull.*, **29**:653–669.
9. Simpson, E.H., Kellendonk, C., and Kandel, E. (2010) A possible role for the striatum in the pathogenesis of the cognitive symptoms of schizophrenia. *Neuron*, **65**:585–596.

10. Millar, J.K., et al. (2000) Disruption of two novel genes by a translocation co-segregating with schizophrenia. *Hum. Mol. Genet.*, **9**:1415–1423.

11. Millar, J.K., et al. (2005) DISC1 and PDE4B are interacting genetic factors in schizophrenia that regulate cAMP signaling. *Science*, **310**:1187–1191.

12. Murdoch, H., et al. (2007) Isoform-selective susceptibility of DISC1/phosphodiesterase-4 complexes to dissociation by elevated intracellular cAMP levels. *J. Neurosci.*, **27**: 9513–9524.

13. Soda, T., et al. (2013) DISC1-ATF4 transcriptional repression complex: dual regulation of the cAMP-PDE4 cascade by DISC1. *Mol. Psychiatry*, **18**:898–908.

14. Fatemi, S.H., et al. (2008) PDE4B polymorphisms and decreased PDE4B expression are associated with schizophrenia. *Schizophr. Res.*, **101**:36–49.

15. Numata, S., et al. (2008) Positive association of the PDE4B (phosphodiesterase 4B) gene with schizophrenia in the Japanese population. *J. Psychiatr. Res.*, **43**:7–12.

16. Tomppo, L., et al. (2009) Association between genes of disrupted in schizophrenia 1 (DISC1) interactors and schizophrenia supports the role of the DISC1 pathway in the etiology of major mental illnesses. *Biol. Psychiatry*, **65**:1055–1062.

17. Clapcote, S.J., et al. (2007) Behavioral phenotypes of Disc1 missense mutations in mice. *Neuron*, **54**:387–402.

18. Lipina, T.V., et al. (2011) Genetic and pharmacological evidence for schizophrenia-related Disc1 interaction with GSK-3. *Synapse*, **65**:234–248.

19. Lipina, T.V., Wang, M., Liu, F., and Roder, J.C. (2012) Synergistic interactions between PDE4B and GSK-3: DISC1 mutant mice. *Neuropharmacology*, **62**:1252–1262.

20. Koike, H., Arguello, P.A., Kvajo, M., Karayiorgou, M., and Gogos, J.A. (2006) Disc1 is mutated in the 129S6/SvEv strain and modulates working memory in mice. *Proc. Natl. Acad. Sci. USA*, **103**:3693–3697.

21. Kvajo, M., et al. (2011) Altered axonal targeting and short-term plasticity in the hippocampus of Disc1 mutant mice. *Proc. Natl. Acad. Sci. USA*, **108**:E1349–1358.

22. Siuciak, J.A., McCarthy, S.A., Chapin, D.S., and Martin A.N. (2008) Behavioral and neurochemical characterization of mice deficient in the phosphodiesterase-4B (PDE4B) enzyme. *Psychopharmacology (Berl.)*, **197**:115–126.

23. Xie, Z., et al. (2006) Cellular and subcellular localization of PDE10A, a striatum-enriched phosphodiesterase. *Neuroscience*, **139**:597–607.

24. Kehler, J. and Nielsen, J. (2011) PDE10A inhibitors: novel therapeutic drugs for schizophrenia. *Curr. Pharm. Des.*, **17**:137–150.

25. Kelly, M.P., et al. (2010) Phosphodiesterase 11A in brain is enriched in ventral hippocampus and deletion causes psychiatric disease-related phenotypes. *Proc. Natl. Acad. Sci. USA*, **107**:8457–8462.

26. Gray, J.A. and Roth, B.L. (2007) Molecular targets for treating cognitive dysfunction in schizophrenia. *Schizophr. Bull.*, **33**:1100–1119.

27. Kessler, R.C., Chiu, W.T., Demler, O., Merikangas, K.R., and Walters, E.E. (2005) Prevalence, severity, and comorbidity of 12-month DSM-IV disorders in the National Comorbidity Survey Replication. *Arch. Gen. Psychiatry*, **62**:617–627.

28. Hebenstreit, G.F., et al. (1989) Rolipram in major depressive disorder: results of a double-blind comparative study with imipramine. *Pharmacopsychiatry*, **22**:156–160.

29. Zeller, E., Stief, H.J., Pflug, B., and Sastre-y-Hernandez, M. (1984) Results of a phase II study of the antidepressant effect of rolipram. *Pharmacopsychiatry*, **17**:188–190.

30. Fleischhacker, W.W., et al. (1992) A multicenter double-blind study of three different doses of the new cAMP-phosphodiesterase inhibitor rolipram in patients with major depressive disorder. *Neuropsychobiology*, **26**:59–64.

31. Zhang, H.T., et al. (2002) Antidepressant-like profile and reduced sensitivity to rolipram in mice deficient in the PDE4D phosphodiesterase enzyme. *Neuropsychopharmacology*, **27**:587–595.

32. Przegalinski, E. and Bigajska, K. (1983) Antidepressant properties of some phosphodiesterase inhibitors. *Pol. J. Pharmacol. Pharm.*, **35**:233–240.

33. Li, Y.F., et al. (2009) Antidepressant- and anxiolytic-like effects of the phosphodiesterase-4 inhibitor rolipram on behavior depend on cyclic AMP response element binding protein-mediated neurogenesis in the hippocampus. *Neuropsychopharmacology*, **34**:2404–2419.

34. Wachtel, H. (1983) Potential antidepressant activity of rolipram and other selective cyclic adenosine $3',5'$-monophosphate phosphodiesterase inhibitors. *Neuropharmacology*, **22**:267–272.

35. Fujita, M., et al. (2012) Downregulation of brain phosphodiesterase type IV measured with (11)C-(*R*)-rolipram positron emission tomography in major depressive disorder. *Biol. Psychiatry*, **72**:548–554.

36. Gorman, J.M. (1996) Comorbid depression and anxiety spectrum disorders. *Depress. Anxiety*, **4**:160–168.

37. Carlezon, W.A., Jr., Duman, R.S., and Nestler, E.J. (2005) The many faces of CREB. *Trends Neurosci.*, **28**:436–445.

38. Blackwood, D.H., et al. (2001) Schizophrenia and affective disorders—cosegregation with a translocation at chromosome 1q42 that directly disrupts brain-expressed genes: clinical and P300 findings in a family. *Am. J. Hum. Genet.*, **69**:428–433.

39. Wong, M.L., et al. (2006) Phosphodiesterase genes are associated with susceptibility to major depression and antidepressant treatment response. *Proc. Natl. Acad. Sci. USA*, **103**:15124–15129.

40. Craufurd, D. Snowde, J. (2002) Neuropsychological and neuropsychiatric aspects of Huntington's disease. In: Bates, G., Harper, P., and Jones, L. (Eds.), *Huntington's Disease*, 3rd ed., Oxford University Press, Oxford, UK, pp. 62–94.

41. Paulsen, J.S., et al. (2006) Preparing for preventive clinical trials: the Predict-HD study. *Arch. Neurol.*, **63**:883–890.

42. Duff, K., Paulsen, J.S., Beglinger, L.J., Langbehn, D.R., and Stout, J.C. (2007) Psychiatric symptoms in Huntington's disease before diagnosis: the Predict-HD Study. *Biol. Psychiatry*, **62**:1341–1346.

43. Li, H., Li, S.H., Johnston, H., Shelbourne, P.F., and Li, X.J. (2000) Amino-terminal fragments of mutant huntingtin show selective accumulation in striatal neurons and synaptic toxicity. *Nat. Genet.*, **25**:385–389.

44. Martindale, D., et al. (1998) Length of huntingtin and its polyglutamine tract influences localization and frequency of intracellular aggregates. *Nat. Genet.*, **18**:150–154.

45. Gines, S., et al. (2003) Specific progressive cAMP reduction implicates energy deficit in presymptomatic Huntington's disease knock-in mice. *Hum. Mol. Genet.*, **12**: 497–508.

46. Mangiarini, L., et al. (1996) Exon 1 of the HD gene with an expanded CAG repeat is sufficient to cause a progressive neurological phenotype in transgenic mice. *Cell*, **87**:493–506.

47. Giampa, C., et al. (2009) Phosphodiesterase type IV inhibition prevents sequestration of CREB binding protein, protects striatal parvalbumin interneurons and rescues motor deficits in the R6/2 mouse model of Huntington's disease. *Eur. J. Neurosci.*, **29**:902–910.

48. DeMarch, Z., Giampa, C., Patassini, S., Bernardi, G., and Fusco, F.R. (2008) Beneficial effects of rolipram in the R6/2 mouse model of Huntington's disease. *Neurobiol. Dis.*, **30**:375–387.

49. Fusco, F.R., et al. (2012) Changes in the expression of extracellular regulated kinase (ERK 1/2) in the R6/2 mouse model of Huntington's disease after phosphodiesterase IV inhibition. *Neurobiol. Dis.*, **46**:225–233.

50. Kamiya, A., et al. (2005) A schizophrenia-associated mutation of DISC1 perturbs cerebral cortex development. *Nat. Cell. Biol.*, **7**:1167–1178.

51. Sawamura, N., et al. (2008) Nuclear DISC1 regulates CRE-mediated gene transcription and sleep homeostasis in the fruit fly. *Mol. Psychiatry*, **13**:1138–1148, 1069.

52. Hayashi-Takagi, A., et al. (2010) Disrupted-in-schizophrenia 1 (DISC1) regulates spines of the glutamate synapse via Rac1. *Nat. Neurosci.*, **13**:327–332.

53. Boxall, R., Porteous, D.J., and Thomson, P.A. (2011) DISC1 and Huntington's disease: overlapping pathways of vulnerability to neurological disorder? *PLoS One*, **6**:e16263.

54. Kleiman, R.J., et al. (2011) Chronic suppression of phosphodiesterase 10A alters striatal expression of genes responsible for neurotransmitter synthesis, neurotransmission, and signaling pathways implicated in Huntington's disease. *J. Pharmacol. Exp. Ther.*, **336**:64–76.

55. Hebb, A.L., Robertson, H.A., and Denovan-Wright, E.M. (2004) Striatal phosphodiesterase mRNA and protein levels are reduced in Huntington's disease transgenic mice prior to the onset of motor symptoms. *Neuroscience*, **123**:967–981.

56. Giampa, C, et al. (2010) Inhibition of the striatal specific phosphodiesterase PDE10A ameliorates striatal and cortical pathology in R6/2 mouse model of Huntington's disease. *PLoS One*, **5**:e13417.

CHAPTER 9

BEYOND ERECTILE DYSFUNCTION: UNDERSTANDING PDE5 ACTIVITY IN THE CENTRAL NERVOUS SYSTEM

EVA P.P. BOLLEN, KRIS RUTTEN, OLGA A.H. RENEERKENS, HARRY W.M. STEINBUSCH, and JOS PRICKAERTS

Department of Psychiatry and Neuropsychology, School for Mental Health and Neuroscience, Maastricht University, Maastricht, The Netherlands

INTRODUCTION

Phosphodiesterases (PDEs) are widely distributed enzymes in the body that break down cyclic adenosine monophosphate (cAMP) and cyclic guanosine monophosphate (cGMP) into their inactive form, $5'$-AMP and $5'$-GMP, respectively [1]. cAMP and cGMP are cyclic nucleotides that are essential for signal transduction in several physiological functions. PDEs are represented as a superfamily, as they have been classified into 11 subtypes (PDE1–PDE11) based largely on their sequence homology. Most of these subtypes have more than one gene product (e.g., PDE4A, PDE4B, PDE4C, and PDE4D). In addition, each gene product may have multiple splice variants or isozymes (e.g., PDE4D1–PDE4D9). In total there are more than 100 specific PDEs [1]. The different subtypes are discriminated using several criteria, such as localization, subcellular distribution, mechanism of regulation, and enzymatic and kinetic properties (see Chapter 1 for more details). One fundamental distinction between subfamilies comprises the difference in affinity for the two distinct cyclic nucleotides. A differentiation is possible between cAMP-specific enzymes (PDE4, PDE7, and PDE8), cGMP-specific enzymes (PDE5, PDE6, and PDE9), and the so-called dual-substrate PDEs that have affinity for both cyclic nucleotides (PDE1–PDE3, PDE10, and PDE11).

Because of the regulatory role of PDEs in essential cyclic nucleotide-dependent signaling in various physiological systems, they have been identified as interesting drug targets for treatment of a wide array of disorders. Bender and Beavo described a

Cyclic-Nucleotide Phosphodiesterases in the Central Nervous System: From Biology to Drug Discovery, First Edition. Edited by Nicholas J. Brandon and Anthony R. West.

number of possible factors that contribute to the current interest in PDEs as a drug target [1]. One factor is the remarkably high number of members of the PDE family. An initial problem of the therapeutic application of PDE inhibitors was the lack of specificity. Early PDE inhibitors blocked virtually all PDE activity in the body, which often resulted in narrow therapeutic windows for these agents, which limited their applicability in humans. However, the identification of different PDE subfamilies and isozymes allowed the development of more specific pharmacological agents, which can often be targeted to only one subfamily or isozyme. Different subtypes and isozymes are thought to be involved in diverse physiological functions in the body as demonstrated by their unique distribution and specificity (see Chapters 2 and 6). As such, specific functions can be selectively targeted without causing off-target side effects. In addition, the variety of PDEs and their different physiological roles promise multiple therapeutical purposes, which are not restricted to one particular pathology or medical application.

Two additional pharmacological properties make PDEs interesting targets for drug development. As enzymes, they participate in the degradation process of their substrates cAMP and cGMP. It is known that alteration in ligand levels is more effective when pharmacological intervention is directed at degradation processes compared to synthesis processes [1]. Furthermore, cyclic nucleotides are only available in relatively small amounts in the cell (<1–$10\,\mu M$). Therefore, PDE inhibitors have little competition in binding the targeted PDE enzyme.

The first report of clinical properties of a PDE inhibitor dates back to 1886 when the effects of caffeine on bronchodilation were described [2]. However, it was only discovered several years later that these effects were caused by the caffeine-induced inhibition of cAMP-specific PDEs. In 1970, PDEs were identified in rat and bovine tissue, and it was demonstrated that they hydrolyze the phosphodiesteric bonds of cGMP and cAMP [3].

A first commercial success for the clinical application of PDE inhibitors was sildenafil, a selective inhibitor of PDE5. Although initially developed for the treatment of arterial hypertension and angina pectoris, sildenafil was approved by the U.S. Food and Drug Administration (FDA) in 1998 for the treatment of erectile dysfunction (ED) and marketed by Pfizer under the name Viagra® [4].

The discovery and success of sildenafil boosted the research and development of several other inhibitors of PDE5. At the same time, it stimulated researchers to explore the therapeutic potential of other classes of PDEs in different disorders of the body and the brain. In addition, previously explored PDEs, such as PDE4, were reevaluated after first being dismissed as putative targets because of side effects and lack of specificity or efficacy. For instance, in 1984 the PDE4 inhibitor rolipram was being developed as a putative antidepressant, but it never made it to the market owing to severe emetic side effects (e.g., nausea, vomiting). At present, several PDE inhibitors are in different phases of development for the treatment of a wide variety of pathologies. However, today only two more PDE3 inhibitors, for the treatment of congestive heart failure or intermittent claudication, respectively, and PDE4 inhibitor roflumilast for reducing exacerbations in chronic obstructive pulmonary disorder [5,6] have been approved by the FDA [2,7,8]. At this moment almost 1000 clinical trials involving PDE inhibitors are listed in the database of the U.S. National Institutes of Health [9].

The commercial success of the three FDA-approved PDE5 inhibitors for ED (sildenafil as Viagra, tadalafil as Cialis®, and vardenafil as Levitra® [4,10,11]) has been enormous, with the three major players (Pfizer, Eli Lilly, and Bayer, respectively) posting sales of US$3.1 billion in 2006. These compounds exert their effects throughout the entire body, and are therefore also considered for disorders other than ED. In 2005, for example, sildenafil was additionally approved by the FDA under the name of Revatio® for therapy against pulmonary arterial hypertension (PAH), recently followed by the approval of tadalafil (Adcirca®) for the same purpose [12–15].

The availability and relative safety of PDE5 inhibitors make it worthwhile to evaluate their effects in many different species, conditions, and disorders. This has led to a vast amount of preclinical and clinical data on the central effects of these drugs. This chapter considers the possibilities for application of PDE5 inhibitors in disorders of the central nervous system (CNS). First, the properties of PDE5 in the CNS and the forthcoming physiological changes by PDE5 inhibition are evaluated. Second, we review the relevance of PDE5 inhibition treatment for CNS disorders gathered in preclinical and clinical studies.

Properties of PDE5 Inhibition

PDE5. PDE5 catalyzes specifically the breakdown of cGMP to 5′GMP. PDE5 has high affinity binding sites for cGMP and is therefore able to hydrolyze cGMP even in low substrate levels. PDE5 consists of two identical monomers, which consist of both a C-terminal catalytic domain and an N-terminal regulatory domain [16,17]. The catalytic domain of the enzyme is responsible for the actual degradation of cGMP. However, cGMP can also bind to one of the noncatalytic allosteric domains available on the N-terminal region [18]. The occupation of this so-called GAF-A binding site by cGMP is critical for specific phosphorylation of Ser92 by cGMP-dependent protein kinase (PKG) (and to a lesser extent cAMP-dependent protein kinase (PKA)). When bound, PKG activates the degradation of the catalytic domain and increases affinity for cGMP [19]. This negative feedback mechanism closely regulates PDE5 activity (Figure 9.1). The higher the available levels of cGMP and PKG, the more likely the PDE5 will be activated to reduce cGMP levels.

Thus far, a single gene, PDE5A, has been identified to produce PDE5 [20]. PDE5A is considered a cytosolic protein, with rodent studies showing the highest levels of PDE5A mRNA in the kidneys, pancreas, cerebellum, lung, and heart [21,22]. However, significant expression has also been observed in human vascular smooth muscle, placenta, platelets, several gastrointestinal tissues, and the brain [20,23,24]. When focusing on the CNS, the highest levels of PDE5 mRNA expression are detected in the cerebellum, with large quantities found in the Purkinje cells [25]. Additionally, in the cortex and hippocampus considerable amounts of PDE5A mRNA have been detected [22,26,27]. Three splice variants of the gene are known to exist (PDE5A1, PDE5A2, and PDE5A3) that differ in their N-terminal sequence [28]. These variants show different localization patterns. Whereas PDE5A1 and PDE5A2 are ubiquitous, PDE5A3 is specifically located in smooth muscle tissue [28].

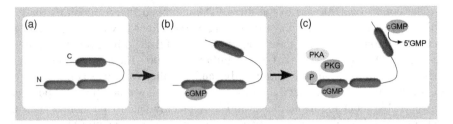

FIGURE 9.1 Enzymatic regulation of PDE5. (a) PDE5 consists of a C-terminal catalytic domain and an N-terminal regulatory domain. (b) cGMP binds to the regulatory domain. (c) This promotes phosphorylation of the N-terminal region and subsequent binding of PKG (and to a lesser extent PKA). These changes alter the shape of the enzyme, and thereby enhance the enzymatic activity at the catalytic domain. Adapted from [17].

PDE5 Inhibitors. The commercially available PDE5 inhibitors compete with cGMP to bind to the catalytic site of PDE5 [29]. Their molecular structures are therefore mostly based on that of cGMP (see Chapter 5). Sildenafil and vardenafil have a similar cGMP-based molecular makeup, while the structure of the latest FDA-approved PDE5 inhibitor tadalafil is rather different and is derived from β-carboline (Figure 9.2). This structural difference has implications for the selectivity of the inhibitor to PDE5. Whereas vardenafil and sildenafil are selective for PDE5 isozymes with some affinity to PDE6, tadalafil has additional affinity for PDE11 [30]. The functional relevance of PDE11 is relatively unknown in contrast to PDE6, which is involved in visual processes. The differences in chemical makeup between the PDE5 inhibitors also have consequences for their pharmacokinetic properties. The half-life of sildenafil and vardenafil is ~3–4 h. In contrast, tadalafil has a longer half-life of ~18 h [31]. Vardenafil has a five-times greater inhibitory potential toward PDE5 than sildenafil and is the most potent of the three commercially available drugs [29,30]. Tadalafil is a second-generation drug, which is reflected in its long half-life and long-lasting effects. All three compounds are rapidly absorbed in the gastrointestinal tract at the level of the small intestine. They are predominantly metabolized in the liver by CYP3A4 [32,33]. All three metabolized PDE5 inhibitors are excreted via both the liver and the kidneys, but predominantly in feces compared to urine.

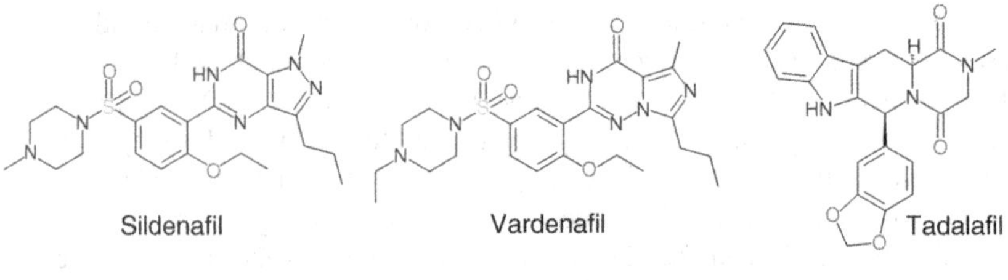

Sildenafil Vardenafil Tadalafil

FIGURE 9.2 Molecular structure of the FDA-approved PDE5 inhibitors sildenafil, vardenafil, and tadalafil.

When assessing the efficacy of PDE5 inhibitors in CNS-related disorders, an important issue is the ability of these compounds to cross the blood–brain barrier. Considering the lipophilic structure of sildenafil and vardenafil, they can be expected to penetrate into the brain. For both sildenafil and vardenafil, this was confirmed in preclinical animal studies [34,35]. Tadalafil has been shown to not cross the blood–brain barrier [34].

Mechanism. Because the effects of PDE5 inhibition are generated through changes in cGMP levels, we now focus on the functions of this cyclic nucleotide to further comprehend the effects of PDE5 inhibitors in the body and the brain. cGMP is synthesized from guanosine triphosphate (GTP) by guanylyl cyclase (GC) [36]. The latter enzyme is an important target of nitric oxide (NO) [37]. NO is formed of L-arginine by catalysis by nitric oxide synthase (NOS). NO functions as a messenger in signal transduction, most notably in retrograde signaling. As a gaseous molecule, NO is able to diffuse through the cell membrane and pass on a signal to neighboring cells, where it activates GC functioning, thereby promoting cGMP synthesis. Downstream, cGMP regulates PKG, PDEs, and cGMP-gated ion channels [37]. All these downstream targets of the NO–cGMP pathway mediate activity of different proteins involved in cell signal transduction, which eventually leads to a biological system-dependent cellular and physiological response. NO has been studied for decades for its vasodilatory effects and its involvement in *N*-methyl-D-aspartate (NMDA)-mediated cell signaling in the brain [36,38]. Because PDE5 inhibitors and NO both increase cGMP levels, it is most likely that they also exert their effects by the same downstream mechanisms. As such, two main mechanisms are likely at the basis of the effects induced by PDE5 inhibition: changes in blood flow or neuronal signal transduction.

Blood Flow. The most studied function of PDE5 is its regulatory role in hemodynamics. PDE5 modulates vascular smooth muscle contraction by regulation of cGMP levels (Figure 9.3) [1]. This characteristic lies at the base of the development of the commercially available PDE5 inhibitors. Inhibition of PDE5 increases levels of cGMP in the cavernosal smooth muscles, which, in turn, via PKG can act in diverse ways to promote erectile function, including activation of ion channels and contractile regulatory proteins that promote relaxation of the penile smooth muscle tissue [39]. NO functions as a neurotransmitter, which links signals of central or peripheral sexual stimulation to smooth muscle relaxation, leading to penile erection [40]. Treatment of ED with PDE5 inhibitors still requires presence of NO-mediated signals, that is, sexual stimulation, since PDE5 is downstream of NO.

In addition to the treatment of ED, the FDA also approved sildenafil and tadalafil for the treatment of pulmonary hypertension [14,15]. In pulmonary hypertension, the small blood vessels in the lungs are more resistant to blood flow, causing increased stress on the right ventricle of the heart [41]. Also in this condition, PDE5 inhibition can attenuate the stress on the heart by counteracting vasoconstriction and thereby lowering the resistance for blood flow in the lungs [42].

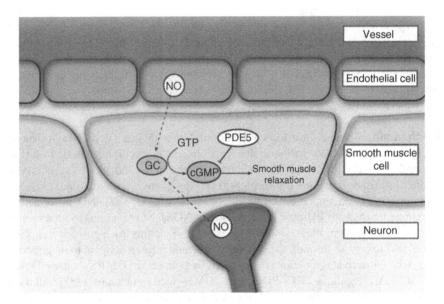

FIGURE 9.3   Role of PDE5 in blood flow regulation. Nitric oxide can diffuse from both the endothelial cells and neurons to the smooth muscle cell layer of vessels. NO activates GC, which produces cGMP from GTP. Elevated levels of cGMP lead to relaxation of smooth muscle cells surrounding the vessel, thereby enhancing blood flow in that vessel. PDE5 counteracts this process by degrading cGMP and preventing cGMP levels from rising. Inhibitors of PDE5 prevent the breakdown of cGMP and thus have vasodilating properties.

PDE5 inhibition induces changes in the central vascular system by similar mechanisms (see Figure 9.3). Administration of zaprinast, an early, relatively nonselective PDE5 inhibitor, dilates basilar and cerebral arteries in various animal models [43,44], and a temporary increase in local cerebral blood flow (CBF) was also found after sildenafil administration [45]. In humans, sildenafil also enhances cerebral vascular reactivity [46]. Although PDE5 inhibitors can exhibit cerebrovascular effects, this effect cannot be generalized. Several studies could not find changes in CBF or vasodilation of cerebral arteries after inhibition of PDE5 [47–50]. It is unclear what causes this discrepancy between studies, although the different techniques and dosages used as well as the interspecies differences may play a role.

Neuronal Signaling.   The NO–cGMP pathway is important in synaptic plasticity. Synaptic plasticity is the ability of neurons to actively enhance or depress their connectivity with neighboring neurons in order to induce changes in signal transduction efficiency. Connections between neurons can be strengthened by functional neuronal changes, such as enhanced neurotransmitter release and receptor efficacy and number, and by structural changes, such as the outgrowth of new synapses. Synaptic plasticity has been observed in various brain structures, such as the hippocampus, cerebral cortex, and cerebellum. The most recognized form of synaptic

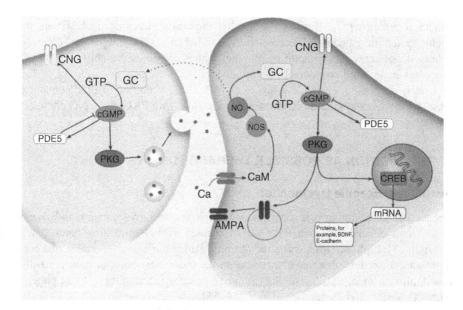

FIGURE 9.4   Role of PDE5 in synaptic plasticity. After the entrance of Ca^{2+} in the postsynapse, NO activates GC either postsynaptically or presynaptically. cGMP synthesis is increased, which then leads to the opening of CNG and to the activation of PKG. The influx of Ca^{2+} through CNG channels further depolarizes synaptic membranes. PKG acts in different ways. Presynaptically, it enhances neurotransmitter release into the synaptic cleft. In the postsynapse, activation of PKG can lead to the insertion of functional AMPA receptors into the synapse, and to the synthesis of relevant proteins, mediated through transcription factors, such as CREB. All these changes bring about an enhanced signal transduction, and thus synaptic plasticity. PDE5 acts as a brake on these processes by controlling cGMP levels. Therefore, inhibition of PDE5 may boost synaptic plasticity.

plasticity is long-term potentiation (LTP), which is strongly associated with learning and memory. LTP was first discovered in the hippocampus, a structure renowned for its involvement in memory processes [51]. Studies show that the NO–cGMP pathway is involved in at least the NMDA-dependent form of this specific type of synaptic plasticity (see also Chapter 11). NO is activated by influx of calcium (Ca^{2+}) in the postsynapse, following binding of glutamate to NMDA-receptors [37]. Next, the efficiency of signal transduction can be enhanced in several ways (Figure 9.4).

First, the gaseous nature of NO allows it to act as a retrograde messenger from the postsynapse back to the presynapse. There, the NO–cGMP–PKG pathway is activated, which then causes increased presynaptic glutamate release following depolarization of the postsynapse [52]. In addition, in the postsynapse, the same pathway increases the insertion of AMPA receptors at the synaptic membrane and mediates the activation of transcription factors such as cAMP-responsive element binding protein (CREB) [53]. CREB facilitates the synthesis of proteins, such as membrane channels and transporters, other transcription factors, cytokines, and structural proteins [37]. Besides PKG, other important effectors of cGMP are the cyclic nucleotide-gated (CNG) channels. The influx of Ca^{2+} through CNG channels further depolarizes the synaptic membrane, which promotes the above-mentioned processes in pre- and postsynaptic membranes. PDE5 acts as a natural brake on this system by decreasing available levels of cGMP. Thus, inhibition of PDE5 enhances these processes, allowing more interaction of cGMP with PKG and CNG channels.

PDE5 INHIBITION AS POSSIBLE THERAPEUTIC CNS TARGET

Sexual and Erectile Dysfunction

The use of PDE5 inhibitors is recognized as a first-line therapy for men with ED, and is safe and effective in men with various causes of ED, including neurological disorders. Multiple sclerosis, systemic atrophy, Parkinson's disease, and spinal cord injury are disorders that have their origin in the CNS, but can affect erectile functioning in men. PDE5 inhibitors improve erectile functioning in these patients without causing significant side effects [54–58].

In psychogenic ED, causes of the disorder are psychological in nature, rather than physical [59]. ED is often observed in men facing stress or depression. Although most studies find an improvement after PDE5 inhibitor treatment in this group of patients [59,60], the diagnostic criteria in these clinical studies for labeling ED as psychogenic are mostly unclear. It is difficult for clinicians to extricate psychological from physiological problems in ED. One study that focused on a specific population of psychogenic ED patients, that is, soldiers with ED caused by posttraumatic stress disorder, could not find any beneficial effects of PDE5 inhibitor treatment [61].

Sildenafil has been successfully used to treat serotonin reuptake inhibitor (SSRI)-associated ED in men [62–64]. In addition, the impact of treatment with sildenafil on the depressive symptoms and quality of life in male patients with ED who have untreated depressive symptoms was recently measured in a phase IV study [65]. It was found that compared to placebo, sildenafil led to significant improvement in depressive symptoms. However, it will be difficult to disentangle whether a possible beneficial effect on mood is a consequence of treatment of the sexual dysfunction or whether PDE5 inhibition directly leads to a reduction of depressive symptoms and thereby attenuates the erectile/sexual dysfunction. In a 6-week open-label study in patients with erectile function, an improvement in depression was observed without changes in sexual function, indicating that in men, PDE5 inhibitor treatment has antidepressant effects [66]. Moreover, the antidepressant potential of sildenafil was recently shown in preclinical rodent studies [67,68].

The same holds for female sexual arousal disorder (FSAD). Women who experience a lack of sexual arousal have been treated with PDE5 inhibitors to enhance sexual functioning [69–71]. Again, PDE5 inhibitor treatment has been reported to be effective in FSAD secondary to neurological dysfunctions such as multiple sclerosis, spinal cord injury, and SSRI-treatment because of major depression [72–74]. Whether this improvement is a result of genital or neurological changes remains controversial. PDE5 can be found in the clitoris, vagina, and labia minora [75,76]. This implies that the improvement in sexual functioning and increase in arousal induced by PDE5 inhibition could be a consequence of relaxation of vascular smooth muscles, causing enhanced genital blood flow, as well as of nonvascular smooth muscles, which causes increased genital arousal. In women, however, there seems to be a discrepancy between genital and subjective arousal [77]. Genital arousal is believed to have a minor influence on subjective feelings of arousal [78]. Thus, increasing genital arousal would not, *per se*, stimulate feelings of desire or sexual pleasure [69–71]. This suggests that, besides increased genital sensation, PDE5 inhibition induces changes in mental arousal processes in women. Oxytocin might be of importance in this regard, as a recent study showed that vardenafil treatment enhances the expression of this hormone that is associated with affiliation and sexual behavior in the brain of rats [79]. But, again, the disentanglement of bodily versus mental processes sets hurdles for researchers, and up until now, it remains difficult to estimate the contribution of the CNS in enhanced sexual functioning caused by PDE5 inhibitor treatment.

Ischemic Stroke

Preclinical research in rats demonstrates beneficial effects of PDE5 inhibitor treatment on functional outcome after ischemic stroke. Ischemic stroke is characterized by a disturbance of blood and glucose supply to certain regions in the brain, which leads to infarction. As mentioned before, PDE5 inhibitors modulate vasorelaxation. Not only the peripheral, but most likely also the central vascular system is susceptible to PDE5-induced changes in CBF. As mentioned above, PDE5 inhibitors can increase perfusion rates in the cerebral vessels in rats. Changes in local CBF were observed after sildenafil and tadalafil treatment, as well as after administration of PF-5, a new PDE5 inhibitor developed by Pfizer [80–82].

However, it is most unlikely that PDE5 inhibitors act only as protectors against vascular occlusion, as administration of a PDE5 inhibitor 24 h or more after an ischemic insult still generates beneficial effects on functional outcomes, while at that time point the main cerebral damage has already occurred [82]. In accordance with this, the volume of the induced infarcts did not change with PDE5 inhibitor treatment after an ischemic insult [82,83]. Moreover, considerable debate exists when it comes to CBF changes caused by PDE5 inhibitor treatment. For example, a recent study using a mouse stroke model could not find any changes in CBF after vardenafil treatment, which may imply that CBF effects are species-specific [48]. Furthermore, in healthy humans, single-photon emission computed tomography (SPECT) and ultrasound measures after sildenafil treatment also did not reveal any changes in cerebral perfusion [84].

Several studies show that besides CBF, PDE5 inhibitor treatment (sildenafil, tadalafil, and zaprinast) affects angiogenesis and neurogenesis after stroke [82,85–88]. Thus, improved functional recovery could be caused by enhanced angio- and/or neurogenesis, rather than temporary changes in CBF. Angiogenesis is regulated in endothelial cells by vascular endothelial growth factor (VEGF) and angiopoietin

(Ang-1) [89,90]. It is known that sildenafil promotes the VEGF–Ang-1 system. Blockage of VEGF disrupts sildenafil-induced angiogenesis [88]. Moreover, PDE5 inhibitor treatment triggers angiogenic regulatory proteins and promotes phosphorylation of e-NOS and Akt [91]. Also in neural progenitor cells, phosphorylation and activation of Akt is increased following sildenafil treatment [92]. More specifically, activation of the PI3-K–Akt–GSK-3 pathway in response to elevated cGMP levels seems to underlie sildenafil-induced enhancements in neural proliferation [92]. It was recently demonstrated that sildenafil, as well as tadalafil, increased neurogenesis in the ischemic brain [81,82]. However, not only survival of new cells, but also structural changes in preexisting neurons can facilitate functional recovery. Experimental studies showed synaptic sprouting and axonal remodeling in response to PDE5 inhibition after embolic stroke [45,87]. Elevated levels of cGMP in neurons stimulate transcription factors that produce proteins necessary for these structural neuronal changes. Taken together, PDE5 inhibitors have been shown to enhance angiogenesis, neurogenesis, and synaptogenesis after stroke, thereby promoting functional recovery.

The observation that changes induced by tadalafil treatment are not only vascular, but also neuronal in nature is in apparent contrast to the finding that tadalafil is unable to cross the blood–brain barrier [34]. Two possible explanations can be given. On the one hand, it could be that all neuronal changes result from an increased blood supply caused by enhanced CBF and angiogenesis. On the other hand, it is possible that induction of an ischemic insult compromises the efficacy of the blood–brain barrier. In a preclinical study, Ko et al. demonstrated that neuronal cGMP levels were enhanced by tadalafil only in gerbils that underwent an ischemic insult, and not in control animals [93]. These findings are in favor of a disrupted blood–brain barrier after vascular occlusion.

Paradoxically, sildenafil has been associated with transient ischemic attack and ischemic stroke in patients with preexisting cerebrovascular disease [94,95]. These effects were described in case reports, but in experimental conditions no association between sildenafil and risk for stroke has been reported [96]. Seemingly, it is rather a rare side effect that might be caused by an interaction of sildenafil treatment with damaged blood vessels as a result of the preexisting cerebrovascular disease [94,96].

Hemorrhagic Stroke

Besides stroke induced by lack of blood and glucose supply to the brain, a second category of stroke is caused by intracranial hemorrhage. One specific type of hemorrhagic stroke is subarachnoid hemorrhage, in which blood accumulates between the arachnoid membrane and pia mater. A frequent complication in this type of stroke is vasospasm. The constriction of blood vessels can cause an ischemic insult in these patients, secondary to the initial hemorrhage. The spasms are likely caused by hemoglobin and its breakdown products [97]. Hemoglobin as well as the by-products disrupts NO signaling, which can lead to decreases in cGMP and, consequently, to vasoconstriction. In an experimental model of this specific type of stroke, sildenafil prevented the occurrence of vasospasms [98]. It seems that in this case, PDE5 inhibitors are very potent relaxers of the cerebral vasculature. In clinical practice, papaverine is often used as vasodilator for vasospasm. Although a non-selective PDE10 inhibitor, papaverine is likely to mainly target cGMP breakdown. Whereas both PDE inhibitor treatments have similar mechanisms, inhibition of PDE5 seems to be more potent. Papaverine requires intraarterial injections, whereas oral administration of PDE5 inhibitors is at least as effective [99,100]. For this specific subtype of stroke, PDE5 inhibitor treatment is therefore very promising.

Hemorrhage, and more specifically intracerebral hemorrhage, has been associated with use of all three commercially available PDE5 inhibitors in several case reports [101–106]. The reason for this remains unclear, but it has been suggested that sildenafil redistributes CBF by dilating the cerebral (micro) vasculature [107]. This redistribution could ultimately lead to intracerebral hemorrhage [101,108]. Nevertheless, this seems to be a fairly rare adverse effect of PDE5 inhibitor treatment.

Cognitive Deficits

The procognitive effect of PDE5 inhibition is probably the most-well-studied possible CNS application in preclinical research. The evidence from these studies in several animal models is substantial [109–112]. Zaprinast was the first PDE5 inhibitor that was reported to enhance memory [113]. Because of additional affinity of zaprinast for PDE9, PDE10, and PDE11, no solid conclusions regarding the memory-enhancing effects of PDE5 inhibitor treatment could be drawn, especially since PDE9 and PDE10 are also considered interesting targets for cognition enhancement (see Chapter 4). However, the more selective PDE5 inhibitors sildenafil and vardenafil show similar procognitive effects in several learning and memory paradigms and across species [35,110,112,114–117]. Although prefrontal functions, such as executive functioning, are also improved by PDE5 inhibition [110,118], its memory-enhancing effects have received most attention in this regard. As such, PDE5 inhibitors mediate memory formation in spatial, fear, and object memory, during acquisition as well as consolidation of the memory trace [109]. These procognitive effects have also been studied in animal deficit models, which mimic memory decline due to various conditions. Memory deficits caused by experimental models of diabetes, hyperammonemia, stroke, and electroconvulsive shock have been reversed by PDE5 inhibitor treatment [93,119,120]. Several studies have assessed the effects of PDE5 inhibitors in human subjects, but until now attempts to demonstrate cognition-enhancing effects in humans were rather disappointing [121–123]. However, changes in cognition-associated event-related potentials (ERPs) suggest that PDE5 inhibitors do affect cognitive processes to some degree [123]. Further studies with larger sample sizes and in patient populations are required to evaluate whether procognitive effects of PDE inhibitors observed in preclinical studies can be translated toward the human situation.

It is unlikely that the procognitive effects of PDE5 inhibitor treatment are related to cerebrovascular changes. A preclinical study found no changes in blood flow or glucose utilization in cognition-related brain regions at doses of vardenafil that effectively enhanced memory performance in rats [49]. Compounds that do not cross the blood–brain barrier, for example, tadalafil, should therefore not affect cognition. Whereas procognitive effects of sildenafil and vardenafil have repeatedly been reported in preclinical studies, to our knowledge, only one study found an increased memory performance after tadalafil treatment [93]. In this study, tadalafil treatment reversed a memory deficit caused by an ischemic insult. It might be speculated that stroke affected the integrity of the blood–brain barrier, thus allowing tadalafil to expand its effects into the brain. Further research in this area is required.

The effects of PDE inhibitors on cognition are thus most likely caused by increased synaptic connectivity. In the most established cellular model of memory, that is, LTP, plasticity of the synapses is essential. It is theorized that the strengthening of the synaptic connections by facilitating signal transduction leads to a memory trace [124]. The synaptic changes can be temporary or structural in nature, depending on the strength of the signal or memory. The NO–cGMP pathway has an important role in this process. First, cGMP has a role in temporarily boosting presynaptic release of neurotransmitters upon depolarization of the postsynaptic cell via the retrograde action of NO. Second, cGMP promotes structural synaptic changes by activating transcription factors, which are responsible for the synthesis of proteins such as growth factors.

It was shown that the NO–cGMP pathway is disrupted in an animal model of Alzheimer's disease [125]. The latter finding together with the consistency of the observed memory improvements increased hopes for PDE5 inhibition efficacy in Alzheimer's disease. Indeed, in transgenic mice (APP/PS1) that have Alzheimer-like amyloid depositions, chronic treatment of sildenafil improved memory and synaptic function and decreased amyloid-β load [34]. However, a recent study revealed that the expression of PDE5 is strongly reduced in the brains of healthy elderly persons and of patients with Alzheimer's disease [126], which was confirmed by a lack of efficacy of PDE5 inhibitor treatment in a memory paradigm in old animals [127]. This suggests that PDE5 inhibition treatment might be less suited as a candidate for treatment of Alzheimer's disease. Several other PDEs that target cGMP, including PDE1, PDE2, PDE9, and PDE10, are present in the brains of aged subjects, and inhibitors of these PDEs have also been shown to improve memory [109]. Therefore, inhibitors of PDE1, PDE2, PDE9, or PDE10 might be more successful than PDE5 inhibitors in treating memory decline caused by Alzheimer's disease (see also Chapter 11).

Neuropathic Pain

PDE5 inhibitors have potential for the treatment of pain as antinociceptive effects have been reported after local peripheral and systemic administration in animal models [41,128–135]. Several lines of evidence point to an involvement of the NO–cGMP pathway in pain perception. However, there is still debate ongoing about the functional role of the NO–cGMP pathway in nociception. This can in part be explained by the seemingly contradictory results of nociceptive studies with sildenafil. As such, several preclinical studies have found antinociceptive effects of sildenafil, others have reported hyperalgesia [135,136].

Whether PDE5 inhibition induces analgesic or hyperalgesic effects, could be dependent on location of administration. In one study, intrathecal and intracerebro-ventricular administration of L-arginine, the precursor of NO, led to hyperalgesic and antinociceptive effects, respectively [135]. Because PDE5 inhibitors and L-arginine both affect the NO–cGMP pathway, similar effects should occur when inhibiting PDE5. Not only the administration site, but also the dosage of the PDE5 inhibitor applied may influence the outcome, as cGMP analogs produce hyperalgesia at high doses, while causing antinociception at low doses. Thus, the neuronal balance of cGMP concentrations is important for the up- or downregulation of nociceptors [137]. In animal models of neuropathic pain caused by sciatic nerve transection, increases in neuronal NOS (nNOS) were observed [138]. In contrast, the activity of nNOS is reduced in chronic neuropathic pain because of diabetes mellitus [139,140]. There-fore, the latter type of neuropathic pain seems to be a good candidate to be treated with PDE5 inhibitors, as it is likely that in this condition cGMP levels are declined in accordance with the decreased nNOS activity. Indeed, sildenafil treatment had analgesic effects in a mouse model for diabetes [129,133]. Although the cause of pain symptoms in diabetic patients remains to be clarified, dysfunctional neuronal NO generation, and consequent decreases in cGMP levels in the dorsal root ganglion (DRG), have been suggested as factors involved in the pathogenesis of this specific type of neuropathy [140]. PDE5 inhibitors increase cGMP levels, which leads to the activation of PKG. K+ channels are targets of PKG, and are known to be involved in peripheral antinociception [141,142]. Several lines of evidence suggest a similar involvement of K+ channels in the DRG [129,143]. The influx of K+ hyperpolarizes the central terminal of primary afferent neurons, which causes antinociception [129]. Given the above, it can be concluded that for specific types of pain, such as chronic neuropathy caused by diabetes mellitus, PDE5 inhibition might provide a new therapeutic option.

Side Effects

Unwanted side effects can severely limit the therapeutic potential of drug treatment candidates. The available PDE5 inhibitors are generally regarded as safe drugs, with some minor side effects. Among the most commonly reported are headache, flushing, and dyspepsia. A number of common and less-common side effects caused by PDE5 inhibition have a neurological origin. Headache or cephalgia is mostly reported as mild and transient; however, there is also an association between PDE5 inhibitors and migraine [144]. In general, headaches are thought to be caused by a complex interaction of factors, including excitatory thresholds, neurotransmitter levels, and vascular dynamics [108]. The induction of headaches and migraine after PDE5 inhibition were mostly regarded as being a result of cerebral vasodilation. However, recent studies showed that the migraine attacks induced by PDE5 inhibition are independent of CBF changes in the main cerebral arteries, and electrophysiological measures did not reveal increases of neuronal excitability [47,84,145]. An alternative hypothesis is that an enhanced NO–cGMP pathway causes cortical spreading depression (CSD), in which a wave of hyperactivity and vasodilation in the brain is followed by a wave of inhibition and vasoconstriction [146]. CSD has been strongly associated with migraine attacks preceded by an aura. PDE5 inhibition could be responsible for the initial increased CBF and cerebral hyperactivity, thereby instigating the consequent inhibition. Migraine is a risk factor for transient global amnesia, which has been associated with PDE5 inhibition treatment in two case studies [147,148]. Again, CSD could explain transient global amnesia by vasoconstriction of vessels in the hippocampal area.

Blurred vision is another side effect reported in sildenafil and vardenafil treatment, but not in tadalafil treatment. The former PDE5 inhibitors have additional affinity to PDE6 [30], which is important for visual signal transduction. PDE6 is mainly found in the rod and cone cells in the retina. Inhibition of PDE6 disturbs normal visual processing at the level of the retina, and thereby induces visual deficiencies. Tadalafil does not affect PDE6, and is therefore not associated with this side effect.

Previously, we discussed the contraindications of PDE5 inhibition in patients with preexisting cerebrovascular diseases, considering the increased risk for stroke. Increased risk for seizures has been reported as another rare, although severe, side effect of PDE5 inhibition. Case reports described tonic–clonic seizures after intake of sildenafil and vardenafil [149,150]. Also in a seizure mouse model, sildenafil reduced the threshold for clonic seizures, which according to Riazi et al. was caused by hyperexcitability in the brain because of cGMP accumulation [151]. It is known that endogenous NO is involved in NMDA-dependent excitatory neurotransmission. In addition, there are indications that the activity of GABAergic receptors is reduced by NO. A disturbance in the NO–cGMP pathway could in that manner lead to excessive glutamate neurotransmitter release and proconvulsant effects [152].

CONCLUSIONS

PDE5 inhibition has been shown successful in treating ED and pulmonary hypertension. The function and widespread localization of PDE5 in the body, as well as relative safety and availability of its inhibitors, has attracted a lot of researchers in different medical fields to evaluate PDE5 as a drug target. This has led to the identification of a number of CNS applications in which PDE5 inhibitors could produce beneficial effects, including stroke and cerebral vasospasm, specific types of neuropathy, and memory decline. In ED and FSAD, PDE5 inhibition also effectively attenuates disease symptoms, even if the cause of the problem is neurological. However, for these disorders the actual contribution of CNS might be questioned.

It should be noted that several interspecies differences and several seemingly contrasting findings were encountered in the literature. PDE5 inhibition can have positive as well as negative effects on pain perception, and is associated with increased risk for stroke, but at the same time can enhance functional outcome after stroke. This clearly shows the complexity of the underlying mechanisms. cGMP is involved in many bodily functions, and it is important to keep in mind that many of these functions will be affected when administering PDE5 inhibitors. Nevertheless, we believe that PDE5 inhibitors have potential in treating CNS-related disorders.

Future research should aim to confirm the gathered findings in clinical studies and further elucidate the underlying mechanisms of action.

REFERENCES

1. Bender, A.T. and Beavo, J.A. (2006) Cyclic nucleotide phosphodiesterases: molecular regulation to clinical use. *Pharmacol. Rev.*, **58**(3):488–520.

2. Boswell-Smith, V., Spina, D., and Page, C.P. (2006) Phosphodiesterase inhibitors. *Br. J. Pharmacol.*, **147**(Suppl. 1):S252–S257.

3. Beavo, J.A., Hardman, J.G., and Sutherland, E.W. (1970) Hydrolysis of cyclic guanosine and adenosine 3',5'-monophosphates by rat and bovine tissues. *J. Biol. Chem.*, **245**(21):5649–5655.

4. U.S. Food and Drug Administration (1998) Viagra®, Center for Drug Evaluation and Research, Department of Health and Human Services.

5. Boswell-Smith, V. and Spina, D. (2007) PDE4 inhibitors as potential therapeutic agents in the treatment of COPD-focus on roflumilast. *Int. J. Chronic Obstruct. Pulm. Dis.*, **2**(2): 121–129.

6. U.S. Food and Drug Administration (2011) Daliresp, Center for Drug Evaluation and Research, Departement of Health and Human Services.

7. Petersen, J.W. and Felker, G.M. (2008) Inotropes in the management of acute heart failure. *Crit. Care Med.*, **36**(1 Suppl.):S106–S111.

8. Chapman, T.M. and Goa, K.L. (2003) Cilostazol: a review of its use in intermittent claudication. *Am. J. Cardiovasc. Drugs*, **3**(2):117–138.

9. U.S. National Institutes of Health, ClinicalTrials.gov (accessed January 13, 2014).

10. U.S. Food and Drug Administration (2003) Cialis®, Center for Drug Evaluation and Research, Department of Health and Human Services.

11. U.S. Food and Drug Administration (2003) Levitra®, Center for Drug Evaluation and Research, Department of Health and Human Services.

12. Patel, M.D. and Katz, S.D. (2005) Phosphodiesterase 5 inhibition in chronic heart failure and pulmonary hypertension. *Am. J. Cardiol.*, **96**(12B):47M–51M.

13. Lewis, G.D. and Semigran, M.J. (2004) Type 5 phosphodiesterase inhibition in heart failure and pulmonary hypertension. *Curr. Heart Fail. Rep.*, **1**(4):183–189.

14. U.S. Food and Drug Administration (2005) Revatio®, Center for Drug Evaluation and Research, Department of Health and Human Services.

15. U.S. Food and Drug Administration (2009) Adcirca®, Center for Drug Evaluation and Research, Department of Health and Human Services.

16. Corbin, J.D. and Francis, S.H. (1999) Cyclic GMP phosphodiesterase-5: target of sildenafil. *J. Biol. Chem.*, **274**(20):13729–13732.

17. Omori, K. and Kotera, J. (2007) Overview of PDEs and their regulation. *Circ. Res.*, **100**(3):309–327.

18. Liu, L., Underwood, T., Li, H., Pamukcu, R., and Thompson, W.J. (2002) Specific cGMP binding by the cGMP binding domains of cGMP-binding cGMP specific phosphodiesterase. *Cell. Signal.*, **14**(1):45–51.

19. Corbin, J.D., Turko, I.V., Beasley, A., and Francis, S.H. (2000) Phosphorylation of phosphodiesterase-5 by cyclic nucleotide-dependent protein kinase alters its catalytic and allosteric cGMP-binding activities. *Eur. J. Biochem.*, **267**(9):2760–2767.

20. Yanaka, N., et al. (1998) Expression, structure and chromosomal localization of the human cGMP-binding cGMP-specific phosphodiesterase PDE5A gene. *Eur. J. Biochem.*, **255**(2):391–399.

21. Giordano, D., De Stefano, M.E., Citro, G., Modica, A., and Giorgi, M. (2001) Expression of cGMP-binding cGMP-specific phosphodiesterase (PDE5) in mouse tissues and cell lines using an antibody against the enzyme amino-terminal domain. *Biochim. Biophys. Acta*, **1539**(1–2):16–27.

22. Kotera, J., Fujishige, K., and Omori, K. (2000) Immunohistochemical localization of cGMP-binding cGMP-specific phosphodiesterase (PDE5) in rat tissues. *J. Histochem. Cytochem.*, **48**(5):685–693.

23. Loughney, K., et al. (1998) Isolation and characterization of cDNAs encoding PDE5A, a human cGMP-binding, cGMP-specific 3′,5′-cyclic nucleotide phosphodiesterase. *Gene*, **216**(1):139–147.

24. Stacey, P., Rulten, S., Dapling, A., and Phillips, S.C. (1998) Molecular cloning and expression of human cGMP-binding cGMP-specific phosphodiesterase (PDE5). *Biochem. Biophys. Res. Commun.*, **247**(2):249–254.

25. Shimizu-Albergine, M., et al. (2003) Individual cerebellar Purkinje cells express different cGMP phosphodiesterases (PDEs): *in vivo* phosphorylation of cGMP-specific PDE (PDE5) as an indicator of cGMP-dependent protein kinase (PKG) activation. *J. Neurosci.*, **23**(16):6452–6459.

26. Marte, A., Pepicelli, O., Cavallero, A., Raiteri, M., and Fedele, E. (2008) *In vivo* effects of phosphodiesterase inhibition on basal cyclic guanosine monophosphate levels in the prefrontal cortex, hippocampus and cerebellum of freely moving rats. *J. Neurosci. Res.*, **86**(15):3338–3347.

27. Van Staveren, W.C., et al. (2003) mRNA expression patterns of the cGMP-hydrolyzing phosphodiesterases types 2, 5, and 9 during development of the rat brain. *J. Comp. Neurol.*, **467**(4):566–580.

28. Lin, C.S., Lau, A., Tu, R., and Lue, T.F. (2000) Expression of three isoforms of cGMP-binding cGMP-specific phosphodiesterase (PDE5) in human penile cavernosum. *Biochem. Biophys. Res. Commun.*, **268**(2):628–635.

29. Blount, M.A., et al. (2004) Binding of tritiated sildenafil, tadalafil, or vardenafil to the phosphodiesterase-5 catalytic site displays potency, specificity, heterogeneity, and cGMP stimulation. *Mol. Pharmacol.*, **66**(1):144–152.

30. Bischoff, E. (2004) Potency, selectivity, and consequences of nonselectivity of PDE inhibition. *Int. J. Impot. Res.*, **16**(Suppl. 1):S11–S14.

31. Kulkarni, S.K. and Patil, C.S. (2004) Phosphodiesterase 5 enzyme and its inhibitors: update on pharmacological and therapeutic aspects. *Methods Find. Exp. Clin. Pharmacol.*, **26**(10):789–799.

32. Bischoff, E. (2004) Vardenafil preclinical trial data: potency, pharmacodynamics, pharmacokinetics, and adverse events. *Int. J. Impot. Res.*, **16**(Suppl. 1):S34–S37.

33. Cirino, G., Fusco, F., Imbimbo, C., and Mirone, V. (2006) Pharmacology of erectile dysfunction in man. *Pharmacol. Ther.*, **111**(2):400–423.

34. Puzzo, D., et al. (2009) Phosphodiesterase 5 inhibition improves synaptic function, memory, and amyloid-beta load in an Alzheimer's disease mouse model. *J. Neurosci.*, **29** (25):8075–8086.

35. Reneerkens, O.A., et al. (2012) Phosphodiesterase type 5 (PDE5) inhibition improves object recognition memory: indications for central and peripheral mechanisms. *Neurobiol. Learn. Mem.*, **97**(4):370–379.

36. Friebe, A. and Koesling, D. (2003) Regulation of nitric oxide-sensitive guanylyl cyclase. *Circ. Res.*, **93**(2):96–105.

37. Kleppisch, T. and Feil, R. (2009) cGMP signalling in the mammalian brain: role in synaptic plasticity and behaviour. *Handb. Exp. Pharmacol.*, (191):549–579.

38. Puzzo, D., Sapienza, S., Arancio, O., and Palmeri, A. (2008) Role of phosphodiesterase 5 in synaptic plasticity and memory. *Neuropsychiatr. Dis. Treat.*, **4**(2):371–387.

39. Corbin, J.D. (2004) Mechanisms of action of PDE5 inhibition in erectile dysfunction. *Int. J. Impot. Res.*, **16**(Suppl. 1):S4–S7.

40. Burnett, A.L. (2005) Phosphodiesterase 5 mechanisms and therapeutic applications. *Am. J. Cardiol.*, **96**(12B):29M–31M.

41. Uthayathas, S., et al. (2007) Versatile effects of sildenafil: recent pharmacological applications. *Pharmacol. Rep.*, **59**(2):150–163.

42. Archer, S.L. and Michelakis, E.D. (2009) Phosphodiesterase type 5 inhibitors for pulmonary arterial hypertension. *N. Engl. J. Med.*, **361**(19):1864–1871.

43. Sobey, C.G. and Quan, L. (1999) Impaired cerebral vasodilator responses to NO and PDE V inhibition after subarachnoid hemorrhage. *Am. J. Physiol.*, **277**(5 Part 2):H1718–H1724.

44. Kim, P., Schini, V.B., Sundt, T.M., Jr., and Vanhoutte, P.M. (1992) Reduced production of cGMP underlies the loss of endothelium-dependent relaxations in the canine basilar artery after subarachnoid hemorrhage. *Circ. Res.*, **70**(2):248–256.

45. Ding, G., et al. (2008) Magnetic resonance imaging investigation of axonal remodeling and angiogenesis after embolic stroke in sildenafil-treated rats. *J. Cereb. Blood Flow Metab.*, **28**(8):1440–1448.

46. Diomedi, M., et al. (2005) Sildenafil increases cerebrovascular reactivity: a transcranial Doppler study. *Neurology*, **65**(6):919–921.

47. Kruuse, C., Thomsen, L.L., Birk, S., and Olesen, J. (2003) Migraine can be induced by sildenafil without changes in middle cerebral artery diameter. *Brain*, **126**(Part 1):241–247.

48. Royl, G., et al. (2009) Effects of the PDE5-inhibitor vardenafil in a mouse stroke model. *Brain Res.*, **1265**:148–157.

49. Rutten, K., et al. (2009) Phosphodiesterase inhibitors enhance object memory independent of cerebral blood flow and glucose utilization in rats. *Neuropsychopharmacology*, **34**(8):1914–1925.

50. Kruuse, C., Gupta, S., Nilsson, E., Kruse, L., and Edvinsson, L. (2012) Differential vasoactive effects of sildenafil and tadalafil on cerebral arteries. *Eur. J. Pharmacol.*, **674**(2–3):345–351.

51. Bliss, T.V. and Lomo, T. (1973) Long-lasting potentiation of synaptic transmission in the dentate area of the anaesthetized rabbit following stimulation of the perforant path. *J. Physiol.*, **232**(2):331–356.

52. Arancio, O., et al. (1996) Nitric oxide acts directly in the presynaptic neuron to produce long-term potentiation in cultured hippocampal neurons. *Cell*, **87**(6):1025–1035.

53. Lu, Y.F., Kandel, E.R., and Hawkins, R.D. (1999) Nitric oxide signaling contributes to late-phase LTP and CREB phosphorylation in the hippocampus. *J. Neurosci.*, **19**(23):10250–10261.

54. Fowler, C.J., et al. (2005) A double blind, randomised study of sildenafil citrate for erectile dysfunction in men with multiple sclerosis. *J. Neurol. Neurosurg. Psychiatry*, **76**(5):700–705.

55. Nehra, A. and Moreland, R.B. (2001) Neurologic erectile dysfunction. *Urol. Clin. North Am.*, **28**(2):289–308.

56. Giammusso, B., et al. (2002) Sildenafil in the treatment of erectile dysfunction in elderly depressed patients with idiopathic Parkinson's disease. *Arch. Gerontol. Geriatr. Suppl.*, **8**:157–163.

57. Giuliano, F., et al. (1999) Randomized trial of sildenafil for the treatment of erectile dysfunction in spinal cord injury. Sildenafil Study Group. *Ann. Neurol.*, **46**(1):15–21.

58. Zesiewicz, T.A., Helal, M., and Hauser, R.A. (2000) Sildenafil citrate (Viagra) for the treatment of erectile dysfunction in men with Parkinson's disease. *Mov. Disord.*, **15**(2):305–308.

59. Donatucci, C., et al. (2004) Vardenafil improves erectile function in men with erectile dysfunction irrespective of disease severity and disease classification. *J. Sex. Med.*, **1**(3):301–309.

60. Goldenberg, M.M. (1998) Safety and efficacy of sildenafil citrate in the treatment of male erectile dysfunction. *Clin. Ther.*, **20**(6):1033–1048.

61. Safarinejad, M.R., Kolahi, A.A., and Ghaedi, G. (2009) Safety and efficacy of sildenafil citrate in treating erectile dysfunction in patients with combat-related post-traumatic stress disorder: a double-blind, randomized and placebo-controlled study. *BJU Int.*, **104** (3):376–383.

62. Damis, M., Patel, Y., and Simpson, G.M. (1999) Sildenafil in the treatment of SSRI-induced sexual dysfunction: a pilot study. *Prim. Care Companion J. Clin. Psychiatry*, **1**(6):184–187.

63. Nurnberg, H.G., et al. (2001) Efficacy of sildenafil citrate for the treatment of erectile dysfunction in men taking serotonin reuptake inhibitors. *Am. J. Psychiatry*, **158** (11):1926–1928.

64. Nurnberg, H.G., et al. (2003) Treatment of antidepressant-associated sexual dysfunction with sildenafil: a randomized controlled trial. *JAMA*, **289**(1):56–64.

65. Pfizer (2008) *Efficacy study measuring the impact of treatment with Viagra on the depressive symptoms of men with erectile dysfunction.* Bethesda (MD): National Library of Medicine. Available at: http://clinicaltrials.gov/ct2/show/NCT00159809, NLM identifier: NCT00159809.

66. Orr, G., Seidman, S.N., Weiser, M., Gershon, A.A., Dubrov, Y., and Klein, D.F. (2008) An open-label pilot study to evaluate the efficacy of sildenafil citrate in middle-aged men with late-onset dysthymia. *J. Nerv. Ment. Dis.*, **196**(6):496–500.

67. Brink, C.B., Clapton, J.D., Eagar, B.E., and Harvey, B.H. (2008) Appearance of anti-depressant-like effect by sildenafil in rats after central muscarinic receptor blockade: evidence from behavioural and neuro-receptor studies. *J. Neural Transm.*, **115**(1):117–125.

68. Liebenberg, N., Harvey, B.H., Brand, L., and Brink, C.B. (2010) Antidepressant-like properties of phosphodiesterase type 5 inhibitors and cholinergic dependency in a genetic rat model of depression. *Behav. Pharmacol.*, **21**(5–6):540–547.

69. Foster, R., Mears, A., and Goldmeier, D. (2009) A literature review and case reports series on the use of phosphodiesterase inhibitors in the treatment of female sexual dysfunction. *Int. J. STD AIDS*, **20**(3):152–157.

70. Caruso, S., Intelisano, G., Lupo, L., and Agnello, C. (2001) Premenopausal women affected by sexual arousal disorder treated with sildenafil: a double-blind, cross-over, placebo-controlled study. *BJOG*, **108**(6):623–628.

71. Berman, J.R., Berman, L.A., Toler, S.M., Gill, J., and Haughie, S. (2003) Safety and efficacy of sildenafil citrate for the treatment of female sexual arousal disorder: a double-blind, placebo controlled study. *J. Urol.*, **170**(6 Part 1):2333–2338.

72. Sipski, M.L., Rosen, R.C., Alexander, C.J., and Hamer, R.M. (2000) Sildenafil effects on sexual and cardiovascular responses in women with spinal cord injury. *Urology*, **55**(6):812–815.

73. Dasgupta, R., Wiseman, O.J., Kanabar, G., Fowler, C.J., and Mikol, D.D. (2004) Efficacy of sildenafil in the treatment of female sexual dysfunction due to multiple sclerosis. *J. Urol.*, **171**(3):1189–1193; discussion 1193.

74. Nurnberg, H.G., et al. (2008) Sildenafil treatment of women with antidepressant-associated sexual dysfunction: a randomized controlled trial. *JAMA*, **300**(4):395–404.

75. Park, K., Moreland, R.B., Goldstein, I., Atala, A., and Traish, A. (1998) Sildenafil inhibits phosphodiesterase type 5 in human clitoral corpus cavernosum smooth muscle. *Biochem. Biophys. Res. Commun.*, **249**(3):612–617.

76. D'Amati, G., et al. (2002) Type 5 phosphodiesterase expression in the human vagina. *Urology*, **60**(1):191–195.

77. Leiblum, S.R. and Chivers, M.L. (2007) Normal and persistent genital arousal in women: new perspectives. *J. Sex Marital Ther.*, **33**(4):357–373.

78. Basson, R., et al. (2004) Revised definitions of women's sexual dysfunction. *J. Sex. Med.*, **1**(1):40–48.

79. Shin, M.S., et al. (2010) Vardenafil enhances oxytocin expression in the paraventricular nucleus without sexual stimulation. *Int. Neurourol. J.*, **14**(4):213–219.

80. Menniti, F.S., et al. (2009) PDE5A inhibitors improve functional recovery after stroke in rats: optimized dosing regimen with implications for mechanism. *J. Pharmacol. Exp. Ther.*, **331**(3):842–850.

81. Zhang, L., et al. (2006) Tadalafil, a long-acting type 5 phosphodiesterase isoenzyme inhibitor, improves neurological functional recovery in a rat model of embolic stroke. *Brain Res.*, **1118**(1):192–198.

82. Zhang, R., et al. (2002) Sildenafil (Viagra) induces neurogenesis and promotes functional recovery after stroke in rats. *Stroke*, **33**(11):2675–2680.

83. Li, L., et al. (2007) Angiogenesis and improved cerebral blood flow in the ischemic boundary area detected by MRI after administration of sildenafil to rats with embolic stroke. *Brain Res.*, **1132**(1):185–192.

84. Kruuse, C., Thomsen, L.L., Jacobsen, T.B., and Olesen, J. (2002) The phosphodiesterase 5 inhibitor sildenafil has no effect on cerebral blood flow or blood velocity, but nevertheless induces headache in healthy subjects. *J. Cereb. Blood Flow Metab.*, **22**(9):1124–1131.

85. Ding, G., et al. (2008) Angiogenesis detected after embolic stroke in rat brain using magnetic resonance T2*WI. *Stroke*, **39**(5):1563–1568.

86. Gao, F., Sugita, M., and Nukui, H. (2005) Phosphodiesterase 5 inhibitor, zaprinast, selectively increases cerebral blood flow in the ischemic penumbra in the rat brain. *Neurol. Res.*, **27**(6):638–643.

87. Zhang, L., et al. (2005) Functional recovery in aged and young rats after embolic stroke: treatment with a phosphodiesterase type 5 inhibitor. *Stroke*, **36**(4):847–852.

88. Zhang, R., et al. (2003) Nitric oxide enhances angiogenesis via the synthesis of vascular endothelial growth factor and cGMP after stroke in the rat. *Circ. Res.*, **92**(3):308–313.

89. Hanahan, D. (1997) Signaling vascular morphogenesis and maintenance. *Science*, **277**(5322):48–50.

90. Risau, W. (1997) Mechanisms of angiogenesis. *Nature*, **386**(6626):671–674.

91. Koneru, S., et al. (2008) Sildenafil-mediated neovascularization and protection against myocardial ischaemia reperfusion injury in rats: role of VEGF/angiopoietin-1. *J. Cell. Mol. Med.*, **12**(6B):2651–2664.

92. Wang, L., Gang Zhang, Z., Lan Zhang, R., and Chopp, M. (2005) Activation of the PI3-K/Akt pathway mediates cGMP enhanced-neurogenesis in the adult progenitor cells derived from the subventricular zone. *J. Cereb. Blood Flow Metab.*, **25**(9):1150–1158.

93. Ko, I.G., et al. (2009) Tadalafil improves short-term memory by suppressing ischemia-induced apoptosis of hippocampal neuronal cells in gerbils. *Pharmacol. Biochem. Behav.*, **91**(4):629–635.

94. Habek, M. and Petravic, D. (2006) Stroke: an adverse reaction to sildenafil. *Clin. Neuropharmacol.*, **29**(3):165–167.

95. Morgan, J.C., Alhatou, M., Oberlies, J., and Johnston, K.C. (2001) Transient ischemic attack and stroke associated with sildenafil (Viagra) use. *Neurology*, **57**(9):1730–1731.

96. Morales, A., Gingell, C., Collins, M., Wicker, P.A., and Osterloh, I.H. (1998) Clinical safety of oral sildenafil citrate (Viagra) in the treatment of erectile dysfunction. *Int. J. Impot. Res.*, **10**(2):69–73; discussion 73–64.

97. Pluta, R.M. (2005) Delayed cerebral vasospasm and nitric oxide: review, new hypothesis, and proposed treatment. *Pharmacol. Ther.*, **105**(1):23–56.

98. Atalay, B., et al. (2006) Systemic administration of phosphodiesterase V inhibitor, sildenafil citrate, for attenuation of cerebral vasospasm after experimental subarachnoid hemorrhage. *Neurosurgery*, **59**(5):1102–1107; discussion 1107–1108.

99. Inoha, S., et al. (2002) Type V phosphodiesterase expression in cerebral arteries with vasospasm after subarachnoid hemorrhage in a canine model. *Neurol. Res.*, **24**(6):607–612.

100. Firlik, K.S., Kaufmann, A.M., Firlik, A.D., Jungreis, C.A., and Yonas, H. (1999) Intra-arterial papaverine for the treatment of cerebral vasospasm following aneurysmal subarachnoid hemorrhage. *Surg. Neurol.*, **51**(1):66–74.

101. Alpsan, M.H., et al. (2008) Intracerebral hemorrhage associated with sildenafil use: a case report. *J. Neurol.*, **255**(6):932–933.

102. Buxton, N., Flannery, T., Wild, D., and Bassi, S. (2001) Sildenafil (Viagra)-induced spontaneous intracerebral haemorrhage. *Br. J. Neurosurg.*, **15**(4):347–349.

103. Gazzeri, R., Neroni, M., Galarza, M., and Esposito, S. (2008) Intracerebral hemorrhage associated with use of tadalafil (Cialis). *Neurology*, **70**(15):1289–1290.

104. Marti, I. and Marti Masso, J.F. (2004) Hemiballism due to sildenafil use. *Neurology*, **63**(3):534.

105. McGee, H.T., Egan, R.A., and Clark, W.M. (2005) Visual field defect and intracerebral hemorrhage associated with use of vardenafil (Levitra). *Neurology*, **64**(6):1095–1096.

106. Monastero, R., Pipia, C., Camarda, L.K., and Camarda, R. (2001) Intracerebral haemorrhage associated with sildenafil citrate. *J. Neurol.*, **248**(2):141–142.

107. Ballard, S.A., et al. (1998) Effects of sildenafil on the relaxation of human corpus cavernosum tissue *in vitro* and on the activities of cyclic nucleotide phosphodiesterase isozymes. *J. Urol.*, **159**(6):2164–2171.

108. Farooq, M.U., et al. (2008) Role of sildenafil in neurological disorders. *Clin. Neuropharmacol.*, **31**(6):353–362.

109. Reneerkens, O.A., Rutten, K., Steinbusch, H.W., Blokland, A., and Prickaerts, J. (2009) Selective phosphodiesterase inhibitors: a promising target for cognition enhancement. *Psychopharmacology (Berl.)*, **202**(1–3):419–443.

110. Rutten, K., Basile, J.L., Prickaerts, J., Blokland, A., and Vivian, J.A. (2008) Selective PDE inhibitors rolipram and sildenafil improve object retrieval performance in adult cynomolgus macaques. *Psychopharmacology (Berl.)*, **196**(4):643–648.

111. Rutten, K., et al. (2007) Time-dependent involvement of cAMP and cGMP in consolidation of object memory: studies using selective phosphodiesterase type 2, 4 and 5 inhibitors. *Eur. J. Pharmacol.*, **558**(1–3):107–112.

112. Rutten, K., et al. (2005) The selective PDE5 inhibitor, sildenafil, improves object memory in Swiss mice and increases cGMP levels in hippocampal slices. *Behav. Brain Res.*, **164**(1):11–16.

113. Prickaerts, J., Steinbusch, H.W., Smits, J.F., and de Vente, J. (1997) Possible role of nitric oxide–cyclic GMP pathway in object recognition memory: effects of 7-nitroindazole and zaprinast. *Eur. J. Pharmacol.*, **337**(2–3):125–136.

114. Campbell, E. and Edwards, T. (2006) Zaprinast consolidates long-term memory when administered to neonate chicks trained using a weakly reinforced single trial passive avoidance task. *Behav. Brain Res.*, **169**(1):181–185.

115. Shafiei, M., Mahmoudian, M., Rostami, P., and Nemati, F. (2006) Effect of sildenafil (Viagra) on memory retention of a passive avoidance response in rats. *Acta Physiol. Acad. Sci. Hung.*, **93**(1):53–59.

116. Baratti, C.M. and Boccia, M.M. (1999) Effects of sildenafil on long-term retention of an inhibitory avoidance response in mice. *Behav. Pharmacol.*, **10**(8):731–737.

117. Boccia, M.M., Blake, M.G., Krawczyk, M.C., and Baratti, C.M. (2011) Sildenafil, a selective phosphodiesterase type 5 inhibitor, enhances memory reconsolidation of an inhibitory avoidance task in mice. *Behav. Brain Res.*, **220**(2):319–324.

118. Rodefer, J.S., Saland, S.K., and Eckrich, S.J. (2012) Selective phosphodiesterase inhibitors improve performance on the ED/ID cognitive task in rats. *Neuropharmacology*, **62**(3):1182–1190.

119. Erceg, S., et al. (2006) Role of extracellular cGMP and of hyperammonemia in the impairment of learning in rats with chronic hepatic failure: therapeutic implications. *Neurochem. Int.*, **48**(6–7):441–446.

120. Patil, C.S., Singh, V.P., and Kulkarni, S.K. (2006) Modulatory effect of sildenafil in diabetes and electroconvulsive shock-induced cognitive dysfunction in rats. *Pharmacol. Rep.*, **58**(3):373–380.

121. Goff, D.C., et al. (2009) A placebo-controlled study of sildenafil effects on cognition in schizophrenia. *Psychopharmacology (Berl.)*, **202**(1–3):411–417.

122. Grass, H., et al. (2001) Sildenafil (Viagra): is there an influence on psychological performance? *Int. Urol. Nephrol.*, **32**(3):409–412.

123. Schultheiss, D., et al. (2001) Central effects of sildenafil (Viagra) on auditory selective attention and verbal recognition memory in humans: a study with event-related brain potentials. *World J. Urol.*, **19**(1):46–50.

124. Hebb, D.O. (1949) *The Organization of Behavior*, New York: John Wiley & Sons, Inc.

125. Puzzo, D., Palmeri, A., and Arancio, O. (2006) Involvement of the nitric oxide pathway in synaptic dysfunction following amyloid elevation in Alzheimer's disease. *Rev. Neurosci.*, **17**(5):497–523.

126. Reyes-Irisarri, E., Markerink-Van Ittersum, M., Mengod, G., and de Vente, J. (2007) Expression of the cGMP-specific phosphodiesterases 2 and 9 in normal and Alzheimer's disease human brains. *Eur. J. Neurosci.*, **25**(11):3332–3338.

127. Domek-Lopacinska, K. and Strosznajder, J.B. (2008) The effect of selective inhibition of cyclic GMP hydrolyzing phosphodiesterases 2 and 5 on learning and memory processes and nitric oxide synthase activity in brain during aging. *Brain Res.*, **1216**:68–77.

128. Jain, N.K., Patil, C.S., Singh, A., and Kulkarni, S.K. (2001) Sildenafil-induced peripheral analgesia and activation of the nitric oxide-cyclic GMP pathway. *Brain Res.*, **909**(1–2):170–178.

129. Araiza-Saldana, C.I., Reyes-Garcia, G., Bermudez-Ocana, D.Y., Perez-Severiano, F., and Granados-Soto, V. (2005) Effect of diabetes on the mechanisms of intrathecal antinociception of sildenafil in rats. *Eur. J. Pharmacol.*, **527**(1–3):60–70.

130. Patil, C.S., Jain, N.K., Singh, V.P., and Kulkarni, S.K. (2004) Cholinergic-NO-cGMP mediation of sildenafil-induced antinociception. *Indian J. Exp. Biol.*, **42**(4):361–367.

131. Jain, N.K., Patil, C.S., Singh, A., and Kulkarni, S.K. (2003) Sildenafil, a phosphodiesterase-5 inhibitor, enhances the antinociceptive effect of morphine. *Pharmacology*, **67**(3):150–156.

132. Asomoza-Espinosa, R., et al. (2001) Sildenafil increases diclofenac antinociception in the formalin test. *Eur. J. Pharmacol.*, **418**(3):195–200.

133. Patil, C.S., Singh, V.P., Singh, S., and Kulkarni, S.K. (2004) Modulatory effect of the PDE-5 inhibitor sildenafil in diabetic neuropathy. *Pharmacology*, **72**(3):190–195.

134. Mixcoatl-Zecuatl, T., Aguirre-Banuelos, P., and Granados-Soto, V. (2000) Sildenafil produces antinociception and increases morphine antinociception in the formalin test. *Eur. J. Pharmacol.*, **400**(1):81–87.

135. Kitto, K.F., Haley, J.E., and Wilcox, G.L. (1992) Involvement of nitric oxide in spinally mediated hyperalgesia in the mouse. *Neurosci. Lett.*, **148**(1–2):1–5.

136. Patil, C.S., Padi, S.V., Singh, V.P., and Kulkarni, S.K. (2006) Sildenafil induces hyperalgesia via activation of the NO–cGMP pathway in the rat neuropathic pain model. *Inflammopharmacology*, **14**(1–2):22–27.

137. Torres-Lopez, J.E., Arguelles, C.F., and Granados-Soto, V. (2002) Participation of peripheral and spinal phosphodiesterases 4 and 5 in inflammatory pain. *Proc. West. Pharmacol. Soc.*, **45**:141–143.

138. Guedes, R.P., et al. (2009) Sciatic nerve transection increases glutathione antioxidant system activity and neuronal nitric oxide synthase expression in the spinal cord. *Brain Res. Bull.*, **80**(6):422–427.

139. Rodella, L., Rezzani, R., Corsetti, G., and Bianchi, R. (2000) Nitric oxide involvement in the trigeminal hyperalgesia in diabetic rats. *Brain Res.*, **865**(1):112–115.

140. Sasaki, T., Yasuda, H., Maeda, K., and Kikkawa, R. (1998) Hyperalgesia and decreased neuronal nitric oxide synthase in diabetic rats. *Neuroreport*, **9**(2):243–247.

141. Ambriz-Tututi, M., Velazquez-Zamora, D.A., Urquiza-Marin, H., and Granados-Soto, V. (2005) Analysis of the mechanism underlying the peripheral antinociceptive action of sildenafil in the formalin test. *Eur. J. Pharmacol.*, **512**(2–3):121–127.

142. Sachs, D., Cunha, F.Q., and Ferreira, S.H. (2004) Peripheral analgesic blockade of hypernociception: activation of arginine/NO/cGMP/protein kinase G/ATP-sensitive K+ channel pathway. *Proc. Natl. Acad. Sci. USA*, **101**(10):3680–3685.

143. Yamazumi, I., Okuda, T., and Koga, Y. (2001) Involvement of potassium channels in spinal antinociceptions induced by fentanyl, clonidine and bethanechol in rats. *Jpn. J. Pharmacol.*, **87**(4):268–276.

144. Evans, R.W. and Kruuse, C. (2004) Phosphodiesterase-5 inhibitors and migraine. *Headache*, **44**(9):925–926.

145. Kruuse, C., Hansen, A.E., Larsson, H.B., Lauritzen, M., and Rostrup, E. (2009) Cerebral haemodynamic response or excitability is not affected by sildenafil. *J. Cereb. Blood Flow Metab.*, **29**(4):830–839.

146. Tfelt-Hansen, P.C. (2010) History of migraine with aura and cortical spreading depression from 1941 and onwards. *Cephalalgia*, **30**(7):780–792.

147. Savitz, S.A. and Caplan, L.R. (2002) Transient global amnesia after sildenafil (Viagra) use. *Neurology*, **59**(5):778.

148. Gandolfo, C., Sugo, A., and Del Sette, M. (2003) Sildenafil and transient global amnesia. *Neurol. Sci.*, **24**(3):145–146.

149. Gilad, R., Lampl, Y., Eshel, Y., and Sadeh, M. (2002) Tonic–clonic seizures in patients taking sildenafil. *BMJ*, **325**(7369):869.

150. Striano, P., Zara, F., Minetti, C., and Striano, S. (2006) Epileptic seizures can follow high doses of oral vardenafil. *BMJ*, **333**(7572):785.

151. Riazi, K., et al. (2006) The proconvulsant effect of sildenafil in mice: role of nitric oxide-cGMP pathway. *Br. J. Pharmacol.*, **147**(8):935–943.

152. Robello, M., et al. (1996) Nitric oxide and GABAA receptor function in the rat cerebral cortex and cerebellar granule cells. *Neuroscience*, **74**(1):99–105.

CHAPTER 10

MOLECULAR AND CELLULAR UNDERSTANDING OF PDE10A: A DUAL-SUBSTRATE PHOSPHODIESTERASE WITH THERAPEUTIC POTENTIAL TO MODULATE BASAL GANGLIA FUNCTION

ERIK I. CHARYCH
Lundbeck Research USA, Paramus, NJ, USA

NICHOLAS J. BRANDON
AstraZeneca Neuroscience iMED, Cambridge, MA, USA

INTRODUCTION

Of all the 11 phosphodiesterase (PDE) gene families, the PDE10 family (consisting of PDE10A as the sole member identified to date) has one of the most restricted regional and cellular distribution patterns, with the majority of protein expression confined to the medium-sized spiny projection neurons (MSNs) of the striatum [1–3]. These neurons are the principal input sites of the basal ganglia, a group of interconnected subcortical nuclei, the coordinated activity of which is crucial for initiating the planning and execution of behaviorally relevant cognitive, emotional, and motor patterns while suppressing irrelevant ones [4]. Given this, it has been predicted that inhibition of PDE10A enzymatic activity may augment MSN output, affecting basal ganglia circuit activity in a manner that is consistent with the potential for anti-psychotic efficacy. Indeed, this notion has since gained strong preclinical support and, as a result, there is growing interest in PDE10A as a potential therapeutic target for human disorders such as schizophrenia and Huntington's disease (HD), in which

Cyclic-Nucleotide Phosphodiesterases in the Central Nervous System: From Biology to Drug Discovery,
First Edition. Edited by Nicholas J. Brandon and Anthony R. West.
© 2014 John Wiley & Sons, Inc. Published 2014 by John Wiley & Sons, Inc.

basal ganglia dysfunction is heavily implicated. This chapter reviews the major findings that have contributed to our present understanding of the molecular and cellular biology of the PDE10A enzyme in the context of its therapeutic potential for these diseases, as well as identifies gaps in our understanding where further insights into the function of PDE10A may yet be gained.

PDE10A IS A MEMBER OF THE SUPERFAMILY OF CYCLIC NUCLEOTIDE PHOSPHODIESTERASES

Domain Structure and Isoform Diversity of PDE10A

Mammalian PDEs, possessing class I catalytic domains (shared by the subkingdoms protozoa and metazoa), are encoded by 21 genes identified to date (see Chapter 1 for more details) [5]. The PDE-encoding genes identified herein are grouped into 11 gene families (PDE1–PDE11) based on comparative structural and functional analysis, with an amino acid sequence identity within the conserved catalytic domain ranging from 35 to 50% across PDE families [6,7]. Members of a given PDE gene family (PDE4A–PDE4D) share identical protein functional domains and exhibit greater than 70% amino acid sequence identity within the catalytic domain [7]. Moreover, each member of a PDE gene family can be further subdivided into multiple transcript variants arising from alternative transcriptional start sites and alternative splicing (i.e., PDE10A1–PDE10A18) (Figure 10.1) [6]. To date, the PDE10 gene family consists of just one member, PDE10A, that maps onto chromosomes 17A1, 1q11, and 6q26 of the mouse, rat, and human genomes, respectively (NM_011866.2, NM_022236.1, and NM_006661.2, respectively). Multiple PDE10A transcript variants have also been identified across mouse, rat, and human (Figure 10.1), which is discussed in greater detail in the following sections.

Domain Structure and Function of PDE10A: Regulatory Domain

Upon initial cloning and characterization [8–10], the structural organization of the PDE10A gene product was deduced, as with other PDEs, to consist of (i) an N-terminal region presumed to serve a regulatory role on catalytic activity, and (ii) a C-terminal region containing a catalytic domain known to be highly conserved across all PDEs (Figure 10.1). As with PDE2, PDE5, PDE6, and PDE11, the putative N-terminal regulatory region consists of a pair of highly homologous GAF domains arranged in tandem (GAF-A and GAF-B; so named because they were first recognized as sequence motifs in cGMP-specific and cGMP-stimulated PDEs, *Anabaena* *a*denylate cyclases, and *Escherichia coli* (*F*hlA)) [11]. GAF domains are present in a large and diverse set of protein classes across most phyla, possessing small-molecule binding properties and providing an interface for protein–protein interactions [12]. The GAF domains of PDE2, PDE5, PDE6, PDE10, and PDE11 are sites for high affinity and high selectivity cyclic nucleotide binding, and also provide an interface for PDE parallel dimerization, the latter being consistent with the notion that the

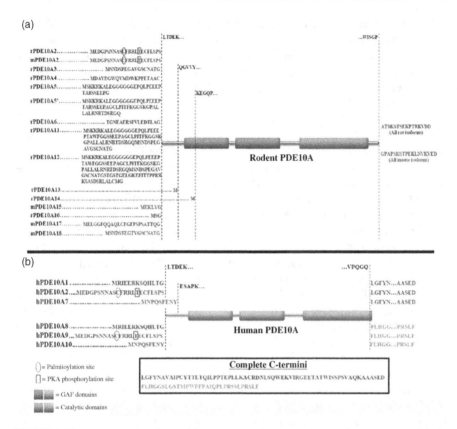

FIGURE 10.1 Predicted amino acid sequences of PDE10A transcript variants. With the exception of rat PDE10A14, which is missing the N-terminal portion of the GAF-A domain, all known PDE10A transcript variants exhibit alternate peptide sequences only at the extreme N- or C-termini, while the core polypeptide containing the tandem GAF domain region, the catalytic domain, and intermediate regions remains intact. (a) Predicted amino acid sequences of rodent PDE10A transcript variants. (b) Predicted amino acid sequences of human PDE10A transcript variants. Only PDE10A2 protein expression has been verified in rodent or human striatum. Furthermore, the function of the PDE10A2-specific N-terminus has been characterized experimentally, where membrane attachment is mediated by palmitoylation of cysteine 11 (circle), a modification that is regulated by cAMP/PKA-mediated phosphorylation of threonine 16 (square). Note that human PDE10A9 is predicted to possess the same N-terminus as PDE10A2 (b), although translation of this transcript versus PDE10A2 has not been verified using C-terminal-specific antibodies. Adapted from Ref. [42].

known PDEs are thought to exist and function as dimers [6,12]. Moreover, experimental evidence suggests that, upon cyclic nucleotide binding, PDE GAF domains have the potential to allosterically modulate the activity of the adjoining catalytic domain. For example, the GAF-A domain of PDE5A binds cGMP with >100-fold selectivity over cyclic adenosine monophosphate (cAMP) [13]. Upon binding of cGMP to GAF-A of PDE5A, the activity of the adjoining catalytic domain is greatly enhanced, as it was demonstrated that PDE5A catalytic activity was stimulated 9–11-fold over basal activity by cGMP, and this increase in activity could be abrogated using a monoclonal antibody directed against the GAF-A domain of PDE5A [14]. At the same time, both GAF domains of PDE5A appear to participate in parallel homodimerization, as shown by truncation and mutation analysis of the recombinant PDE5A protein [13].

Although the cyclic nucleotide ligand for the GAF domains of PDE2, PDE5, PDE6, and PDE11 is cGMP [13–20], that for PDE10A is cAMP [16,21–23]. Taking advantage of the similarity and interchangeability of the PDE GAF domains with those from cyanobacteria adenylyl cyclase (CyaB1 AC) [24], in 2005 Gross-Langenhoff et al. first explored the function of this region within PDE10A by replacing the tandem GAF domain region of CyaB1 AC with that of PDE10A1. This allowed these investigators to use the AC activity of the resulting chimera as a convenient reporter assay system that avoids the difficulties posed by the fact that the allosteric activator and substrate (cyclic nucleotides) of the native PDE may be identical [16]. Unexpectedly, this group found that, in contrast to all other GAF domain-containing PDEs, the AC activity of the PDE10A1-AC chimera was stimulated only by cAMP ($EC_{50} = 19.8\ \mu M$), but not by cGMP [16]. Furthermore, when the so-called NKFDE signature sequence, previously implicated in forming the cyclic nucleotide binding pocket in the GAF domains of other PDEs [13,17,25–27] and only present in GAF-B of PDE10A, was disrupted by a D397A mutation, the AC activity of the resulting mutated chimera was reduced by 97% of the original activity. These findings suggest that GAF-B of PDE10A is responsible for cAMP binding [16]. This was further corroborated by a report of the crystal structure of PDE10A GAF-B in complex with cAMP, a study that also showed that in addition to cAMP binding, GAF-B was a key determinant of PDE10A dimerization [21]. Another study employing functional chimeric constructs fused to AC demonstrated the importance of the linker region between GAF-A and GAF-B of PDE10A, in addition to GAF-B itself, in cAMP binding and intramolecular signaling [22].

The stimulation of AC activity by cAMP in the above PDE10A1-AC chimeras was confirmed to be a reliable proxy for determining the specificity of the PDE10A GAF-B domain for cAMP over cGMP. However, whether or not this finding implies that cAMP binding to PDE10A GAF-B modulates the phosphodiesterase activity of the native holoenzyme is not entirely clear. Matthiesen and Nielsen used cyclic nucleotide analogs, exhibiting a higher affinity for PDE GAF domains than for the catalytic domains, in a scintillation proximity-based assay to test the effect of cyclic nucleotide binding on the PDE activity of the PDE10A holoenzyme, using PDE2A as a methodological control [23]. The authors showed that while cyclic nucleotide analog binding to the PDE2A GAF domain resulted in strong activation of its PDE activity, no alteration of any kind was observed for PDE10A PDE activity, suggesting that the GAF domains of PDE10A do not modulate the activity of the adjoining catalytic domain [23]. These data must be interpreted with caution; however, as the GAF domain ligand used to make this determination was not cAMP, the natural ligand for the PDE10A GAF-B domain, and this analog may owe its high selectivity for PDE10A GAF-B over the catalytic domain to a distinct mode of docking that might not transduce the same conformational change that would otherwise allow cAMP to allosterically modulate PDE10A catalytic activity. The binding of cAMP to GAF-B of PDE10A could alternatively inhibit PDE10A phosphodiesterase activity, which may be consistent with early reports that cAMP potently inhibits the hydrolysis of cGMP by PDE10A [8,10]. However, it is not clear whether this inhibition is a result of either (i) allosteric modulation of the catalytic domain by cAMP-GAF-B domain binding, (ii) competitive inhibition at the catalytic domain, because of a substantially lower K_m for cAMP compared to that for cGMP [8,10], or (iii) some combination of the above (see the following section for more details on enzyme kinetics and substrate specificity). Moreover, the regulation of PDE10A phosphodiesterase activity by cAMP-GAF-B binding may yet occur by an alternative mechanism *in vivo*, through induced changes in phosphorylation state, membrane attachment, or protein–protein interactions [23].

Domain Structure and Function of PDE10A: Catalytic Domain

The initial classification of PDE10A as a novel PDE family was made, for the most part, on the basis of the amino acid sequence homology of its catalytic domain to that of other known PDEs [8–10]. Thus, the catalytic domain of PDE10A shares the highest amino acid sequence identity with PDE2A, PDE5A, PDE6A, and PDE11A (41, 46, 37, and 41%, respectively) [9,28]. Interestingly, the PDEs that share the highest amino acid sequence identity with PDE10A in their catalytic domains (PDE2A, PDE5A, PDE6A, and PDE11A) are also the only known PDEs to also contain tandem GAF domains in the N-terminal regulatory region, suggesting that these GAF domains may have coevolved with their corresponding catalytic domains under pressure to provide allosteric modulation. Unlike that for PDE5A, and PDE6A that are selective for cGMP, the catalytic domain of PDE10A can, however, hydrolyze both cAMP and cGMP at physiological concentrations [8,10,29] and is thus classified as a dual-substrate PDE. Upon initial characterization of PDE10A, Fujishige et al. reported K_m values of 0.26 and 7.2 μM, for cAMP and cGMP, respectively, and a V_{max} ratio (cGMP:cAMP) of 2.2, using cell extracts of COS-7 cells expressing human PDE10A1 [8]. At the same time, Soderling et al. reported K_m values of 0.05 and 3 μM for cAMP and cGMP, respectively, and a V_{max} ratio (cGMP:cAMP) of 4.7, using mouse PDE10A15 (Figure 10.1a) purified from baculovirus-infected Sf9 insect cells [10]. Later, Wang et al. reported K_m values of 0.056 and 4.4 μM for cAMP and cGMP, respectively, and a V_{max} ratio (cGMP:cAMP) of 3.7, using human PDE10A catalytic domain purified from plasmid-transformed *E. coli* [30]. In all cases, the K_m for cAMP was substantially lower than that for cGMP, whereas the reverse was true for the V_{max}. Combined with this, both Fujishige et al. and Soderling et al. also found that the hydrolysis of cGMP by PDE10A was potently inhibited by cAMP [8,10], leading to the hypothesis that PDE10A may function as a cAMP-inhibited cGMP PDE; however, this has not yet been demonstrated *in vivo*. Furthermore, the extent to which the inhibition of PDE10A PDE activity by cAMP is caused by allosteric modulation through GAF-B binding, if any, is also not clear (see section above).

The dual-specificity of the PDE10A catalytic domain for both cAMP and cGMP was confirmed by an X-ray crystallography study [30]. However, the reported crystallography data support a mechanism for dual-specificity that is distinct from the "glutamine switch" model for substrate specificity [31]. This model, based on the crystal structures of ligand-bound PDE4 and PDE5 catalytic domains, assumes that an invariant glutamine in the substrate binding pocket of cAMP-specific PDEs forms two hydrogen bonds with cAMP but only one hydrogen bond with cGMP, while the reverse is true for the invariant glutamine of cGMP-specific PDEs [31]. The "glutamine switch" model also assumes that for dual-specific PDEs, such as PDE10A and PDE1B, the invariant glutamine is free to rotate to form two hydrogen bonds with either cAMP or cGMP. This was supported, for example, by the crystal structure of PDE1B, a dual-specific PDE, where this glutamine could be freely rotated in the unbound state [31]. For the catalytic domain of PDE10A, however, the invariant glutamine side chain was shown to be locked through interactions with a nearby tyrosine residue and with nearby water molecules [30]. Therefore, the recently reported crystal structure of the PDE10A catalytic domain suggests instead that cAMP and cGMP bind to the same substrate pocket within the PDE10A catalytic domain, but with different orientations and interactions [30], rather than in the same orientation accommodated by a rotating glutamine side chain [31].

The biological significance of the dual-substrate specificity of PDE10A observed *in vitro* also eluded investigators until recently, with the discovery of highly potent and selective PDE10A inhibitors (including MP-10 and TP-10) [32]. Administration of either MP-10 or TP-10 causes significant, dose-dependent elevations in the levels of both cAMP and cGMP in the striatum of mice, suggesting that striatal PDE10A hydrolyzes both cAMP and cGMP *in vivo* [32,33]. This *in vivo* confirmation of the dual-substrate specificity of PDE10A has implications for the functional consequences of PDE10A inhibition in the modulation of basal ganglia activity, as will be discussed in a later section.

Domain Structure and Function of PDE10A: Transcript Variation

As shown in Figure 10.1, multiple transcript variants derived from the PDE10A gene have been described for both human and rodent. With the exception of rat PDE10A14, which lacks the N-terminal portion of the GAF-A domain (Figure 10.1a), all known PDE10A transcript variants exhibit alternate peptide sequences only at the extreme N- or C-termini, while the core polypeptide containing the tandem GAF domain region, the catalytic domain, and intermediate regions, remains intact. Analysis of the genomic structure of the human PDE10A gene compared with cloned cDNA sequences has confirmed that the six known human PDE10A transcript variants— PDE10A1, PDE10A2, and PDE10A7–PDE10A10—are products of exon skipping [34]. Although human PDE10A1, human/mouse/rat PDE10A2, and rat PDE10A3 (Figure 10.1) mRNA are most readily detected in multiple tissues by hybridization and polymerase chain reaction (PCR) amplification approaches [8–10,29,34–36], to date only the PDE10A2 protein has been positively identified *in vivo* [36]. Moreover, assuming that all the PDE10A transcript variants illustrated in

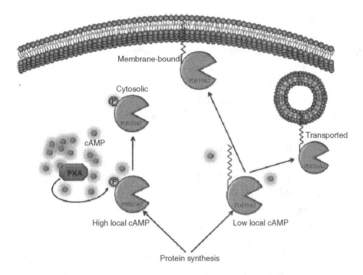

FIGURE 10.2 Proposed model for the regulation of PDE10A localization in response to local fluctuations in cAMP levels. Under conditions of high local cAMP levels near the site of synthesis, PDE10A2 becomes phosphorylated at Thr16 as a consequence of increased cAMP-mediated activation of PKA. Thr16 phosphorylation interferes with palmitoylation at Cys11, resulting in the local, cytosolic accumulation of PDE10A2 where it can normalize cAMP levels through its catalytic activity. Under conditions of low local cAMP at the site of PDE10A2 synthesis, PDE10A2 can become palmitoylated, facilitating the association with intracellular transport vesicles, permitting distal transport and plasma membrane targeting where it may serve to regulate intracellular signaling cascades associated with dopaminergic and glutamatergic synapses. Reproduced with permission from Ref. [50].

Figure 10.1 are functionally expressed, the full significance of PDE10A isoform diversity and the functional properties conferred by each of the alternate N- and C-terminal peptide sequences is presently not known. One exception is the PDE10A2-specific N-terminus, which might be required for tight spatial regulation of PDE10A2 (Figures 10.1 and 10.2, and discussed in greater detail in the following), is presently not known. Another exception might be the apparent long-term potentiation (LTP)-induced increase in transcription of specific PDE10A splice variants in the rat dentate gyrus [37]. For these studies, O'Connor et al. used tetanic stimulation of the perforant pathway in freely moving rats to induce a stable form of LTP (which persists for 6 h), followed by differential display of expressed mRNA to identify LTP-modulated transcripts in the dentate gyrus [37,38]. Surprisingly, the authors found an increase in the abundance of 5 reverse-transcription polymerase chain reaction (RT-PCR) amplicons 1 h after LTP induction. All of these changes corresponded to an intron within the PDE10A gene, consistent with an LTP-induced increase in PDE10A primary RNA transcripts [37]. Following this, the authors showed by *in situ* hybridization, using antisense probes to multiple PDE10A-specific exons, that the abundance of mRNA transcripts corresponding to PDE10A3/PDE10A6 and

PDE10A5/PDE10A11, but not to PDE10A2 or PDE10A4, was elevated in the dentate gyrus 1 h after LTP-induction [37]. Interestingly, the authors determined that the LTP-associated splice variants shared a common leading exon, and speculated that the transcription of these splice variants might be driven by a common activity-related promoter [37]. These studies raise the interesting possibility that at least some degree of PDE10A transcript variation, particularly for N-terminal sequence variants, derives from the use of alternative promoters. For example, genomic studies reveal that greater than half of all human genes are associated with more than one promoter, and these alternative promoters can be selectively utilized to regulate tissue-specific or temporal expression, as well as to regulate length or exon content of the corresponding transcript open reading frame [39,40]. Because alternative promoters for many genes are associated with different transcriptional start sites, one specific consequence of the use of alternative promoters is the incorporation of distinct 5′ leading exons [39,40]. Thus, the diversity of PDE10A N-terminal variants may reflect a high degree of alternative promoter usage, including activity-related promoters in the case of the above LTP studies [37]. This notion is also supported by the finding that PDE10A1 and PDE10A2 expression is driven by two distinct promoters associated with different transcriptional start sites [41], and these two isoforms differ only in their 5′ leading exons [29,34]. Taken together, it is possible that not all PDE10A isoform-specific N-terminal peptide sequences (Figure 10.1) necessarily confer distinct functional properties, but may in some cases only represent a secondary consequence of alternative promoter usage. The latter of these possibilities may prove to be important for the spatial or temporal control of PDE10A expression, or perhaps also for the regulation of PDE10A expression under specific physiological or pathological conditions. However, the full extent of alternative promoter usage for PDE10A and its functional implications are currently not well understood.

PDE10A IS POSITIONED TO PLAY A CENTRAL ROLE IN THE MODULATION OF THE CORTICOBASAL GANGLIA–THALAMOCORTICAL LOOP

Expression of PDE10A in Brain and Peripheral Tissues

PDE10A mRNA and protein are highly enriched in the brain across multiple mammalian species, with the highest peripheral expression found in testes at levels approximately 20-fold lower than that in brain [1,2,8–10,35,42]. Within the rodent brain, *in situ* hybridization analysis has shown that PDE10A mRNA is most robustly expressed in the striatum (caudate putamen, nucleus accumbens, and olfactory tubercle), a central component of the basal ganglia, at levels at least one order of magnitude higher than in any other brain region [2]. At higher resolution, it was shown that the mRNA encoding PDE10A is entirely confined to the cell bodies of MSNs, but absent from the dendrites and axons of these neurons [2]. MSNs are the major output neurons of the striatum, comprising >95% of all striatal neuron classes. MSNs receive glutamatergic input on their dendritic spines from the cerebral cortex and thalamus, dopaminergic input from the substantia nigra pars compacta and ventral tegmental area, and send GABAergic projections to the globus pallidus and substantia nigra pars reticulata, the major output nuclei of the basal ganglia [43]. Thus, MSNs represent a major point of integration in the corticobasal ganglia–thalamocortical loop, a circuit strongly implicated in the underlying biology of major mental illness such as schizophrenia [42,44,45]. MSNs can be segregated into two distinct populations based on their neurochemical properties and neuroanatomical connections: (i) those that preferentially express dopamine D_1 receptors, which stimulate cAMP–protein kinase A (PKA) signaling [46], and participate in the so-called direct or striatonigral output pathway, and (ii) those that preferentially express dopamine D_2 receptors, which inhibit cAMP–PKA signaling [47], and participate in the so-called indirect or striatopallidal output pathway. Activation of MSNs belonging to the striatonigral output pathway is generally thought to promote the initiation of relevant motor or cognitive patterns, whereas activation of MSNs belonging to the striato-pallidal output pathway is generally thought to suppress irrelevant motor or cognitive patterns. Thus, given the preferential expression of dopamine D_1 (activate cAMP–PKA signaling) and D_2 (inhibit cAMP–PKA signaling) receptors in striatonigral and striatopallidal pathway MSNs, respectively, dopamine is thought to increase stria-tonigral MSN output and suppress striatopallidal MSN output in response to glutamatergic input (Figure 10.3). Because PDE10A is likely to be expressed at similar levels in these two populations of striatal MSNs, selective PDE10A inhibition, elevating cyclic nucleotide levels in both MSN populations, was predicted to increase the output of both pathways in response to glutamatergic input [48] (see also Chapter 11 for more details).

Similar to the distribution of the mRNA encoding PDE10A within the brain, immunohistochemical studies have confirmed that the highest brain PDE10A protein levels are also found in MSN cell bodies of the caudate putamen, nucleus accumbens and olfactory tubercle [2]. In contrast to PDE10A mRNA, however, these studies clearly demonstrated that, in addition to MSN cell bodies, comparatively strong immunoreactivity is also observed in the striatal neuropil, and as observed at the electron microscopy (EM) level, in dendrites and spines. Robust immunolabeling was also evident in striatonigral and striatopallidal white matter tracts, and in the neuropil of the substantia nigra and globus pallidus, the two principle targets of striatal MSN axons, with an absence of PDE10A immunoreactivity in the cell bodies of these two target nuclei [1,2]. This distribution pattern of striatal PDE10A protein is strikingly different from that observed in pyramidal neurons of the cerebral cortex, hippocam-pus, and dentate gyrus, as well as from that in the granule cell layer of the cerebellum, where PDE10A immunoreactivity was shown to be significantly weaker and confined to the cell bodies of these extrastriatal neurons [1,2]. Thus, the dramatic contrast between PDE10A mRNA and protein distribution strongly implies a scenario in which PDE10A mRNA is translated only in the striatal MSN cell bodies, while PDE10A protein is transported from the cell bodies throughout MSN dendrites (which can span ~ 300–$500\,\mu m$ in length), axons, and axon terminals. Possible mechanisms by which PDE10A protein transport occurs are discussed in the following section.

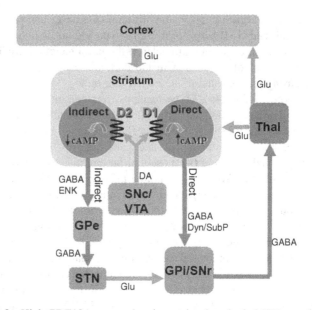

FIGURE 10.3   High PDE10A expression is restricted to both MSN populations of the neostriatum. Medium spiny neurons (filled red circles) of the neostriatum (caudate putamen and nucleus accumbens) represent a major point of integration in the corticobasal ganglia–thalamocortical loop, a circuit strongly implicated in the underlying biology of major mental illnesses such as schizophrenia. MSNs receive glutamatergic input (green arrows) on their dendritic spines from the cerebral cortex and thalamic nuclei. Dopaminergic input from the substantia nigra pars compacta and ventral tegmental area (purple arrows) is thought to modulate the responsiveness of MSNs to cortical input. MSNs, in turn, send GABAergic projections (red arrows) to the globus pallidus and substantia nigra pars reticulata, the major output nuclei of the basal ganglia. MSNs can be segregated into two distinct populations based on their neurochemical properties and neuroanatomical connections, *Left*: Those that preferentially express enkephalin and dopamine D_2 receptors, the latter inhibiting cAMP–PKA signaling through direct coupling to $G\alpha_i$, and participate in the so-called indirect or striatopallidal output pathway, *Right*: Those that preferentially express dynorphin–substance P and dopamine D_1 receptors, the latter stimulating cAMP–PKA signaling through direct coupling to $G\alpha_{olf/s}$, and participate in the so-called direct or striatonigral output pathway. Activation of MSNs belonging to the striatonigral output pathway is generally thought to promote the initiation of relevant motor or cognitive patterns, whereas activation of MSNs belonging to the striatopallidal output pathway is generally thought to suppress irrelevant motor or cognitive patterns. PDE10A inhibition is postulated to increase the excitability of both direct, striatonigral, and indirect, striatopallidal, MSNs, but whether it does so to an equivalent extent is presently debated. *DA*, dopamine; *Dyn/SubP*, dynorphin–substance P; *Enk*, enkephalin; *Glu*, glutamate; *GPe*, globus pallidus externa; *GPi*, globus pallidus interna; *SNc*, substantia nigra pars compacta; *SNr*, substantia nigra pars reticulata; *STN*, subthalamic nucleus; *VTA*, ventral tegmental area. Adapted and modified from Refs [77,78]. (See the color version of the figure in Color Plates section.)

Subcellular Localization of PDE10A

PDE10A1 and PDE10A2 are thought to be the major PDE10A mRNA transcripts expressed in human, whereas PDE10A2 and PDE10A3 are thought to be the major PDE10A mRNA transcripts expressed in rodent [29]. To compare the subcellular distribution of the products of these PDE10A-specific transcript variants, Kotera et al. generated isoform-specific antibodies to the unique N-terminal peptide sequences that distinguish these isoforms from one another (Figure 10.1). Using these isoform-specific antibodies, Western blot analysis revealed that recombinant PDE10A1 and PDE10A3 were both enriched in a cytosolic fraction, whereas recombinant PDE10A2 was enriched in a membrane fraction derived from transfected PC12h cells. When these isoform-specific antibodies were tested against similar membrane and cytosolic fractions derived from rat striatum, only native PDE10A2 was detected by Western blot analysis. Furthermore, PDE10A was found to be enriched in a membrane compared to a cytosolic fraction, the latter observation proved identical to that for PC12h cells expressing recombinant PDE10A2. Another study later confirmed and expanded upon these results, showing that rat PDE10A was enriched not only in a membrane fraction of rat striatum but also in a striatal synaptosomal fraction, and was partially solubilized by treatment with 0.4% Triton X-100 [3]. Consistent with these observations, EM immunogold labeling showed that PDE10A-specific immunogold particles were largely localized to the plasma membrane of MSN dendritic shafts, as well as to that of MSN dendritic spines, with a portion of immunogold particles also localized to the periphery of asymmetric Gray's type I postsynaptic densities (PSDs) [3]. Although it was not confirmed by double-label immunogold experiments using antibodies to specific presynaptic markers, it can be inferred that these asymmetric PSDs are apposed to presynaptic glutamatergic terminals. These findings, combined with others discussed in the following sections, suggest that PDE10A, through its close association with Gray's type I PSDs in MSN dendritic spines, is positioned to participate in the integration of cortical synaptic activity with midbrain modulatory activity, as these synapses represent a key point of convergence between cortical and limbic glutamatergic terminals and midbrain dopaminergic terminals on striatal MSNs. Moreover, even though the latter EM and subcellular fractionation studies were carried out using an antibody that did not distinguish between specific PDE10A isoforms, it can be inferred that the form of PDE10A associated with these synapses corresponds to PDE10A2, because (i) this isoform represents a major PDE10A transcript in both rodent and human brain, (ii) using specific antibodies to PDE10A1, PDE10A2, or PDE10A3, only PDE10A2, protein was adequately detected in rodent striata, and (iii) PDE10A2 is the only isoform that possesses the ability to bind membranes in striatum and in recombinant systems [29,36,49].

Regulation of PDE10A Subcellular Localization

Given the evidence that PDE10A2 is the isoform localized to MSN dendritic spines and associated with PSDs, and given the critical importance of these synapses in the regulation of information flow through the corticobasal ganglia–thalamocortical loop, it is not surprising that PDE10A2 has been found to be uniquely subject to posttranslational modifications that regulate its trafficking and localization (see Figure 10.1) [29,36,50]. As alluded to earlier, it was shown that recombinant PDE10A2, but not PDE10A1, can be phosphorylated by PKA on threonine 16 (Thr16) [29]. This event resulted in an apparent redistribution of membrane-bound PDE10A2 to the cytosolic compartment of transfected PC12h cells [36]. These initial findings were later confirmed and expanded upon using an antibody that specifically recognizes PDE10A2 only when it is phosphorylated on Thr16 [50]. This study demonstrated that a distinct pool of PDE10A2, phosphorylated on Thr16, is enriched in a striatal cytosolic fraction, while total PDE10A is enriched in a striatal membrane fraction. These results lend strong support to the hypothesis that membrane localization of PDE10A2 is dictated by Thr16 phosphorylation state *in vivo*. Furthermore, it was shown that PDE10A2 membrane attachment is controlled by palmitoylation at cysteine 11 (Cys11). Palmitoylation is a common posttranslational modification of soluble proteins that is known to facilitate membrane attachment through addition of a long-chain fatty acid to a cysteine residue by thioester linkage [51,52]. The hypothesis was strongly supported experimentally, because (i) a PDE10A2 site-specific mutant containing a serine in place of Cys11 (PDE10A2^{Cys11S}) was enriched in a cytosolic fraction of HEK293 cells and in cytosolic compartments of cultured striatal neurons; (ii) treatment of HEK293 cells or cultured striatal neurons with 2-bromopalmitate, a nonspecific inhibitor of protein acyltransferase activity, significantly reversed the localization of PDE10A2^{WT} from the membrane to the cytosolic compartments of these cells; and (iii) [^{3}H]palmitic acid was incorporated by PDE10A2^{WT} but not by PDE10A2^{Cys11S}. In addition, the mechanism underlying the regulation of PDE10A2 membrane localization by PKA-mediated phosphorylation was found likely to be a result of blockade of Cys11 palmitoylation by Thr16 phosphorylation, possibly through steric or electrostatic interference, as a phosphomimetic mutant (PDE10A2^{Thr16E}) could not incorporate [^{3}H]palmitic acid, whereas a PDE10A2^{T16A} mutant incorporated [^{3}H]palmitic acid as efficiently as PDE10A2^{WT} [50]. Because posttranslational palmitoylation can also facilitate the attachment of soluble proteins to transport vesicles or tubulovesicular structures [51], we investigated the possibility that palmitoylation of PDE10A2 on Cys11 might also be involved in the long-distance transport of PDE10A2 protein throughout the neuropil of relevant basal ganglia structures, as discussed earlier. Indeed, in cultured striatal neurons, the effective concentration of PDE10A2-specific immunofluorescence associated with a palmitoylated form of PDE10A2 remained constant for 100 μm of dendritic length, beginning at the soma boundary. In contrast, the concentration of PDE10A2-specific immunofluorescence associated with a phosphomimetic Thr16E mutation or a Cys11S mutant was greatest in the cell bodies, and rapidly decayed with dendritic length, suggesting that Cys11 palmitoylation was required for the efficient trafficking of PDE10A2 throughout the dendritic tree, and that this process can be prevented by phosphorylation of Thr16. Taken together, these findings support the idea that striatal PDE10A is primarily membrane bound and that PDE10A is transported along axons and throughout the dendritic tree of these neurons, consistent with the observation that PDE10A protein, but not mRNA, is localized throughout the neuropil of caudate, putamen, nucleus accumbens, and the olfactory tubercle, as well as the globus pallidus and the substantia nigra [1–3]. Furthermore, it is not difficult to imagine that Cys11 palmitoylation, by facilitating the association of PDE10A2 with trafficking vesicles and the plasma membrane, might at least serve as a prerequisite for the localization of PDE10A2 to MSN dendritic spines. However, additional mechanisms might be required for specific localization of PDE10A2 to these structures, as well as to the outer boundary of MSN asymmetric PSDs or other signaling microdomains.

The results from these studies have led to a model for understanding how PDE10A localization might be regulated in response to local fluctuations in cAMP levels at the site of PDE10A2 synthesis (Figure 10.2). By this model, if local cAMP concentrations rise above a given threshold level, possibly one which is set by the limits of local PDE enzyme kinetics and expression levels, PDE10A2 is likely to become phosphorylated at Thr16 as a result of increased cAMP-mediated activation of PKA. This, in turn, would be expected to interfere with palmitoylation at Cys11, resulting in the local, cytosolic accumulation of PDE10A2 and normalization of the high cAMP levels through its catalytic activity (Figure 10.2, right). Under conditions of low local cAMP at the site of PDE10A2 synthesis, PDE10A2 might be more likely to become palmitoylated. This would facilitate the association with intracellular transport vesicles, permitting distal transport and plasma membrane targeting of the enzyme, and enabling it to regulate intracellular signaling cascades associated with dopaminergic and glutamatergic synapses (Figure 10.2, left).

Selective PDE10A Inhibitors as Tools for Investigating the Role of PDE10A in Intracellular Signaling

It has long been suspected that antipsychotic medications modulate dopamine signaling, in particular through antagonism of dopamine D_2 receptors, by regulating cyclic nucleotide levels and that this activity may be a primary therapeutic mechanism of action for these drugs [53]. In addition, specific PDE inhibitors have been recognized as having therapeutic potential for a broad range of diseases by virtue of their targeted modulation of cyclic nucleotide levels in specific tissues, cell types, and even subcellular compartments [54]. Given the highly restricted expression pattern of PDE10A to the striatal MSNs, early hypotheses predicted that PDE10A inhibition would serve to broadly increase striatal MSN output in response to cortical stimulation.

Papaverine, one of several naturally occurring alkaloids isolated from the opium poppy, and which had been previously shown to inhibit PDE activity [55], was found to preferentially inhibit PDE10A, and was therefore the first tool compound used to test the above hypotheses regarding the role of PDE10A in the corticobasal ganglia–thalamocortical loop [48,56]. Papaverine exhibits an IC_{50} for PDE10A of 36 nM, and selectivity over other PDEs ranging from 9- to >277-fold [48]. Initial studies, examining the PDE10A-related neurochemical consequences of papaverine administration *in vivo*, found that treatment of CD-1 mice with papaverine resulted in significant elevations of both cAMP and cGMP in striatal microdialysates [48], consistent with the *in vitro* findings that PDE10A may function as a dual-substrate

PDE. Furthermore, these initial studies reported that administration of papaverine led to significant elevations in PDE10A-dependant phosphorylation of extracellular signal-regulated kinases (ERKs) and cAMP response element binding protein (CREB) in a cAMP-dependent manner. These are two key downstream effectors of dopamine-mediated signaling involved in the initiation of enduring neural adaptations through modulation of MSN gene expression [48,57]. Building on these initial results, Nishi et al. used D_1-DARPP-32-Flag/D_2-DARPP-32-Myc transgenic mice, which express DARPP-32-Flag, in dopamine D_1 receptor-expressing MSNs (direct, striatonigral output pathway) and DARPP-32-Myc in dopamine D_2 receptor-expressing MSNs (indirect, striatopallidal output pathway) to further examine the effects of PDE10A inhibition by papaverine on intracellular signaling in these two neuron populations [58]. DARPP-32 (dopamine- and cAMP-regulated phospho-protein of 32 kDa), which is highly enriched in neuronal populations that receive extensive dopaminergic input, including striatonigral and striatopallidal MSNs, is a major substrate for cAMP–PKA- and cGMP–protein kinase G (PKG)-mediated phosphorylation. DARPP-32 is also a major point of convergence and integration for multiple intracellular signaling pathways and is capable of mediating the precise modulation of many physiological effectors, including neurotransmitter- and voltage-gated ion channels [43]. Moreover, it has been previously shown that dopamine D_1 and D_2 receptor signaling events have opposing effects on the state of DARPP-32 phosphorylation [59]. Thus, using the phosphorylation state of threonine 34 (Thr34) of DARPP-32 as a proxy for dopamine receptor downstream signaling, the authors showed that PDE10A inhibition by papaverine activated (i) cAMP–PKA signaling, potentiating dopamine D_1 receptor signaling in striatonigral neurons, and (ii) cAMP–PKA signaling, potentiating adenosine A2A receptor signaling and attenuating dopamine D_2 receptor signaling in striatopallidal neurons. In addition, treatment with papaverine was shown to enhance the cAMP–PKA-mediated phosphorylation of the AMPA receptor subunit GluR1 at serine 845 (Ser845), an event that is also enhanced by pThr34-stimulated DARPP-32 protein phosphatase-1 (PP-1) inhibitory activity [60], and which has been linked to enhanced membrane insertion of GluR1-containing AMPA receptors, as well as to increases in glutamate-stimulated peak current amplitude and probability of channel opening [61–63]. These results are consistent with the hypothesis that PDE10A inhibition enhances striatal MSN output by potentiating cAMP–PKA-mediated signaling in both striatonigral and striatopallidal MSNs. However, by combining immunoprecipitation using flag or myc antibodies from striatal extracts derived from D_1-DARPP-32-Flag/D_2-DARPP-32-Myc transgenic mice with quantitative Western blotting using pThr34 DARPP-32 antibodies, the authors showed that PDE10A inhibition by papaverine exerted a preferential effect on striatopallidal, compared with striatonigral, MSN signaling [58].

The above results are consistent with an electrophysiological study by Threlfell et al. [64], who examined the effects of PDE10A inhibition on identified striatal MSNs. In addition to papaverine, the authors used TP-10, one of a novel class of PDE10A inhibitors with significantly improved potency and selectivity [32]. Outcomes from these studies showed that (i) PDE10A inhibition increased cortically evoked spike activity of striatal MSNs *in vivo*, and (ii) this effect was preferentially exerted in MSNs that could not be antidromically activated from the substantia nigra pars reticulata (SNr–). This suggests that the potentiating effect of TP-10 on corticostriatal transmission was only exerted on MSNs of the indirect, striatopallidal output pathway [64]. However, TP-10 administration was also shown to robustly increase the incidence of antidromic responses observed in identified striatonigral MSNs [64]. Thus, it is likely that while PDE10A inhibition by TP-10 does not affect cortically evoked spike activity in striatonigral MSNs, it does increase the axonal–terminal excitability of these direct pathway neurons (see Chapter 11 for more details).

Although the mechanism underlying the preferential effect of PDE10A inhibition on cortically evoked spike activity in the striatopallidal pathway in these studies is unknown, an intriguing hypothesis might involve subtle differences in PDE10A subcellular localization because of differences in the balance of PDE10A phospho-rylation/palmitoylation as already described [50]. For example, modeling suggests that dopamine D_2 receptors of striatopallidal MSNs are ~80% occupied during tonic activity and >95% occupied during burst activity [65]. Given this, combined with the fact that dopamine D_2 receptors are negatively coupled to cAMP–PKA activity [47], PDE10A phosphorylation might occur to a lesser degree in striatopallidal MSNs compared to striatonigral MSNs, and would, therefore, be more likely to be palmitoylated in striatopallidal neurons, possibly facilitating the localization of PDE10A to specific signal transduction microdomains associated with dopaminergic and glutamatergic synapses in these MSNs (see Figure 10.2). However, other explanations may equally account for the observed preferential effect of PDE10A inhibition on striatopallidal MSNs in these studies. For example, striatopallidal MSNs are thought to be more excitable than striatonigral MSNs, possibly as a result of a more compact dendritic arbor and greater membrane excitability and responsiveness to corticostriatal input [66]. It would be interesting to determine, for example, whether subtle differences in PDE10A subcellular localization, phosphorylation, and palmi-toylation in striatonigral versus striatopallidal MSNs might be detected using D_1-DARPP-32-Flag/D_2-DARPP-32-Myc transgenic mice, in a manner consistent with the preferential effect of PDE10A inhibition on striatopallidal MSNs in these studies [58,64].

The effect of PDE10A inhibition on major biochemical readouts, such as striatal elevations in cAMP and cGMP levels, as well as striatum-specific phosphorylation of key downstream substrates of cAMP- or cGMP-mediated signaling events (CREB, ERK, DARPP-32, GluR1), have been confirmed and expanded upon using a novel class of PDE10A inhibitor compounds, typified by MP-10 and TP-10 [32,33,67,68], with superior potency and selectivity compared to papaverine. For example, PDE10A inhibitor-induced phosphorylation of CREB is expected to drive significant and specific changes in striatal gene expression through cAMP response element (CRE)-mediated gene transcription. Indeed, lentiviral delivery of a CRE-luciferase reporter into mouse striatum confirmed that acute inhibition of PDE10A with TP-10 drives CRE-mediated gene transcription *in vivo* [68]. Moreover, the specific pattern of gene expression in response to chronic PDE10A inhibition by TP-10, determined by transcriptional profiling, predicts the induction of persistent changes in both striatopallidal and striatonigral MSNs, leading to an increase in excitability of these neurons. These observations also suggest a potential neuroprotective role for PDE10A inhibition in striatal MSNs, consistent with recent reports alluding to the efficacy of PDE10A inhibitors in animal models of Huntington's disease [69–71].

As with the above transcriptional profiling studies, other recent studies also call into question the notion that PDE10A inhibition, as already discussed, exerts a preferential effect on striatopallidal MSNs compared to striatonigral MSNs. Studies using PDE10A inhibitor-induced alterations in striatal gene expression as a measure of activation of specific MSN populations have found that MSNs belonging to both output pathways are similarly activated by PDE10A inhibition [33,67]. Thus, in addition to differential expression of dopamine receptor subtypes, striatal MSNs belonging to the two output pathways also differentially express distinct neuropeptides, such that striatopallidal MSNs preferentially express enkephalin while striatonigral MSNs preferentially express substance P/dynorphin, and the expression of these neuropeptides can be selectively activated by dopamine D_2 antagonists and dopamine D_1 agonists, respectively [72,73]. When striatal gene expression of these neuropeptides, detected by real-time PCR, was used as a proxy for specific striatopallidal and striatonigral MSN activation in response to PDE10A inhibition *in vivo*, the authors of this study found that (i) papaverine or the more potent and selective PDE10A inhibitors PQ-10 and MP-10 induced a significant increase in striatal enkephalin mRNA, similar to that elicited by the dopamine D_2 antagonist haloperidol, and (ii) these compounds induced a significant increase in striatal substance P mRNA, similar to that elicited by the dopamine D_1 agonist SKF81297 [67]. These results were confirmed in an independent study, showing that MP-10 increased striatal enkephalin and substance P mRNA expression to a similar extent *in vivo* [33]. As these data suggest, PDE10A inhibition may evenly activate striatal MSNs as measured by gene expression assays, but this still translates into an increase in cortically evoked responses preferentially in the indirect pathway. More studies in identified MSN subpopulations are needed to fully characterize the potential differential effects of PDE10A inhibition on striatal output pathways.

CURRENT PHARMACEUTICAL LANDSCAPE

Interest in PDE10A as a novel target for CNS disorders, principally schizophrenia and more recently HD, has rapidly increased over the last decade, with the majority of PDE10A inhibitor-related patents published between 2007 and 2013 [74,75]. Beginning with Pfizer in 2002, at least 17 other organizations (this is likely an underestimate) have shown an interest in discovering novel PDE10A inhibitor compounds, as indicated by published patents describing compound structures, screening assays, or crystal structures with bound inhibitor compounds. Of these discovery efforts, only Pfizer has progressed into phase II studies and reported a readout. They conducted a phase II trial of the compound MP-10 in schizophrenics as a monotherapy. Unfortunately, it was reported that MP-10 did not show a significant effect over placebo, whereas the comparator risperidone showed a significant effect [76]. It will be important for the field to learn from this and to try to understand why this experiment in human patients did not work out as predicted from the preclinical rationale and take this knowledge into account in future studies.

CONCLUSIONS

Detailed preclinical biological studies have fueled increasing interest and excitement for PDE10A inhibition as a novel mechanism to treat disorders marked by dysfunction in basal ganglia circuit activity, such as schizophrenia. Moreover, recent behavioral studies using highly potent and selective PDE10A inhibitors suggest that PDE10A inhibition might not only provide antipsychotic efficacy but also efficacy in the negative and cognitive symptom domains of schizophrenia. One of these PDE10A inhibitors, MP-10, recently failed in a phase II clinical trial for the treatment of schizophrenia. This outcome will be crucial for the field to understand and for driving modification of the hypotheses generated through extensive preclinical evaluation. Nevertheless, in addition to its perhaps thwarted therapeutic potential, the widening interest in and understanding of PDE10A biology at the molecular, cellular, and systems levels, combined with detailed explorations into the effects of selective PDE10A inhibition, including current experiments conducted in human subjects, holds the promise of further advancing our understanding of the underlying biology of this interesting biological system.

REFERENCES

1. Coskran, T.M., et al. (2006) Immunohistochemical localization of phosphodiesterase 10A in multiple mammalian species. *J. Histochem. Cytochem.*, **54**(11):1205–1213.

2. Seeger, T.F., et al. (2003) Immunohistochemical localization of PDE10A in the rat brain. *Brain Res.*, **985**(2):113–126.

3. Xie, Z., et al. (2006) Cellular and subcellular localization of PDE10A, a striatum-enriched phosphodiesterase. *Neuroscience*, **139**(2):597–607.

4. Graybiel, A.M. (2000) The basal ganglia. *Curr. Biol.*, **10**(14):R509–R511.

5. Conti, M. and Beavo, J. (2007) Biochemistry and physiology of cyclic nucleotide phosphodiesterases: essential components in cyclic nucleotide signaling. *Annu. Rev. Biochem.*, **76**:481–511.

6. Bender, A.T. and Beavo, J.A. (2006) Cyclic nucleotide phosphodiesterases: molecular regulation to clinical use. *Pharmacol. Rev.*, **58**(3):488–520.

7. Omori, K. and Kotera, J. (2007) Overview of PDEs and their regulation. *Circ. Res.*, **100**(3):309–327.

8. Fujishige, K., et al. (1999) Cloning and characterization of a novel human phosphodiesterase that hydrolyzes both cAMP and cGMP (PDE10A). *J. Biol. Chem.*, **274**(26):18438–18445.

9. Loughney, K., et al. (1999) Isolation and characterization of PDE10A, a novel human 3′,5′-cyclic nucleotide phosphodiesterase. *Gene*, **234**(1):109–117.

10. Soderling, S.H., Bayuga, S.J., and Beavo, J.A. (1999) Isolation and characterization of a dual-substrate phosphodiesterase gene family: PDE10A. *Proc. Natl. Acad. Sci. USA*, **96**(12):7071–7076.

11. Aravind, L. and Ponting, C.P. (1997) The GAF domain: an evolutionary link between diverse phototransducing proteins. *Trends Biochem. Sci.*, **22**(12):458–459.

12. Heikaus, C.C., Pandit, J., and Klevit, R.E. (2009) Cyclic nucleotide binding GAF domains from phosphodiesterases: structural and mechanistic insights. *Structure*, **17**(12):1551–1557.

13. Zoraghi, R., Bessay, E.P., Corbin, J.D., and Francis, S.H. (2005) Structural and functional features in human PDE5A1 regulatory domain that provide for allosteric cGMP binding, dimerization, and regulation. *J. Biol. Chem.*, **280**(12):12051–12063.

14. Rybalkin, S.D., Rybalkina, I.G., Shimizu-Albergine, M., Tang, X.B., and Beavo, J.A. (2003) PDE5 is converted to an activated state upon cGMP binding to the GAF A domain. *EMBO J.*, **22**(3):469–478.

15. Erneux, C., et al. (1981) Specificity of cyclic GMP activation of a multi-substrate cyclic nucleotide phosphodiesterase from rat liver. *Eur. J. Biochem.*, **115**(3):503–510.

16. Gross-Langenhoff, M., Hofbauer, K., Weber, J., Schultz, A., and Schultz, J.E. (2006) cAMP is a ligand for the tandem GAF domain of human phosphodiesterase 10 and cGMP for the tandem GAF domain of phosphodiesterase 11. *J. Biol. Chem.*, **281**(5):2841–2846.

17. Martinez, S.E., et al. (2002) The two GAF domains in phosphodiesterase 2A have distinct roles in dimerization and in cGMP binding. *Proc. Natl. Acad. Sci. USA*, **99**(20):13260–13265.

18. Martins, T.J., Mumby, M.C., and Beavo, J.A. (1982) Purification and characterization of a cyclic GMP-stimulated cyclic nucleotide phosphodiesterase from bovine tissues. *J. Biol. Chem.*, **257**(4):1973–1979.

19. Mou, H. and Cote, R.H. (2001) The catalytic and GAF domains of the rod cGMP phosphodiesterase (PDE6) heterodimer are regulated by distinct regions of its inhibitory gamma subunit. *J. Biol. Chem.*, **276**(29):27527–27534.

20. Mullershausen, F., et al. (2003) Direct activation of PDE5 by cGMP: long-term effects within NO/cGMP signaling. *J. Cell. Biol.*, **160**(5):719–727.

21. Handa, N., et al. (2008) Crystal structure of the GAF-B domain from human phosphodiesterase 10A complexed with its ligand, cAMP. *J. Biol. Chem.*, **283**(28):19657–19664.

22. Hofbauer, K., Schultz, A., and Schultz, J.E. (2008) Functional chimeras of the phosphodiesterase 5 and 10 tandem GAF domains. *J. Biol. Chem.*, **283**(37):25164–25170.

23. Matthiesen, K. and Nielsen, J. (2009) Binding of cyclic nucleotides to phosphodiesterase 10A and 11A GAF domains does not stimulate catalytic activity. *Biochem. J.*, **423**(3):401–409.

24. Kanacher, T., Schultz, A., Linder, J.U., and Schultz, J.E. (2002) A GAF-domain-regulated adenylyl cyclase from *Anabaena* is a self-activating cAMP switch. *EMBO J.*, **21**(14):3672–3680.

25. Bruder, S., et al. (2005) The cyanobacterial tandem GAF domains from the cyaB2 adenylyl cyclase signal via both cAMP-binding sites. *Proc. Natl. Acad. Sci. USA*, **102**(8):3088–3092.

26. Martinez, S.E., et al. (2005) Crystal structure of the tandem GAF domains from a cyanobacterial adenylyl cyclase: modes of ligand binding and dimerization. *Proc. Natl. Acad. Sci. USA*, **102**(8):3082–3087.

27. Wu, A.Y., Tang, X.B., Martinez, S.E., Ikeda, K., and Beavo, J.A. (2004) Molecular determinants for cyclic nucleotide binding to the regulatory domains of phosphodiesterase 2A. *J. Biol. Chem.*, **279**(36):37928–37938.

28. Fawcett, L., et al. (2000) Molecular cloning and characterization of a distinct human phosphodiesterase gene family: PDE11A. *Proc. Natl. Acad. Sci. USA*, **97**(7):3702–3707.

29. Kotera, J., Fujishige, K., Yuasa, K., and Omori, K. (1999) Characterization and phosphorylation of PDE10A2, a novel alternative splice variant of human phosphodiesterase that hydrolyzes cAMP and cGMP. *Biochem. Biophys. Res. Commun.*, **261**(3):551–557.

30. Wang, H., et al. (2007) Structural insight into substrate specificity of phosphodiesterase 10. *Proc. Natl. Acad. Sci. USA*, **104**(14):5782–5787.

31. Zhang, K.Y., et al. (2004) A glutamine switch mechanism for nucleotide selectivity by phosphodiesterases. *Mol. Cell.*, **15**(2):279–286.

32. Schmidt, C.J., et al. (2008) Preclinical characterization of selective phosphodiesterase 10A inhibitors: a new therapeutic approach to the treatment of schizophrenia. *J. Pharmacol. Exp. Ther.*, **325**(2):681–690.

33. Grauer, S.M., et al. (2009) Phosphodiesterase 10A inhibitor activity in preclinical models of the positive, cognitive, and negative symptoms of schizophrenia. *J. Pharmacol. Exp. Ther.*, **331**(2):574–590.

34. Fujishige, K., Kotera, J., Yuasa, K., and Omori, K. (2000) The human phosphodiesterase PDE10A gene genomic organization and evolutionary relatedness with other PDEs containing GAF domains. *Eur. J. Biochem.*, **267**(19):5943–5951.

35. Fujishige, K., Kotera, J., and Omori, K. (1999) Striatum- and testis-specific phosphodiesterase PDE10A isolation and characterization of a rat PDE10A. *Eur. J. Biochem.*, **266**(3):1118–1127.

36. Kotera, J., et al. (2004) Subcellular localization of cyclic nucleotide phosphodiesterase type 10A variants, and alteration of the localization by cAMP-dependent protein kinase-dependent phosphorylation. *J. Biol. Chem.*, **279**(6):4366–4375.

37. O'Connor, V., et al. (2004) Differential amplification of intron-containing transcripts reveals long term potentiation-associated up-regulation of specific PDE10A phosphodiesterase splice variants. *J. Biol. Chem.*, **279**(16):15841–15849.

38. Genin, A., et al. (2003) LTP but not seizure is associated with up-regulation of AKAP-150. *Eur. J. Neurosci.*, **17**(2):331–340.

39. Jacox, E., Gotea, V., Ovcharenko, I., and Elnitski, L. (2010) Tissue-specific and ubiquitous expression patterns from alternative promoters of human genes. *PLoS One*, **5**(8):e12274.

40. Xin, D., Hu, L., and Kong, X. (2008) Alternative promoters influence alternative splicing at the genomic level. *PLoS One*, **3**(6):e2377.

41. Hu, H., McCaw, E.A., Hebb, A.L., Gomez, G.T., and Denovan-Wright, E.M. (2004) Mutant huntingtin affects the rate of transcription of striatum-specific isoforms of phosphodiesterase 10A. *Eur. J. Neurosci.*, **20**(12):3351–3363.

42. Strick, C., Schmidt, C., and Menniti, F. (Eds) (2006) *PDE10A: A Striatum Enriched, Dual-Substrate Phosphodiesterase*, Boca Raton, FL: CRC Press, pp. 237–254.

43. Svenningsson, P., et al. (2004) DARPP-32: an integrator of neurotransmission. *Annu. Rev. Pharmacol. Toxicol.*, **44**:269–296.

44. Simpson, E.H., Kellendonk, C., and Kandel, E. (2010) A possible role for the striatum in the pathogenesis of the cognitive symptoms of schizophrenia. *Neuron*, **65**(5):585–596.

45. Kehler, J. (2011) PDE10A inhibitors: novel therapeutic drugs for schizophrenia. *Curr. Pharm. Des.*, **17**(2):137–150.

46. Herve, D., et al. (2001) Galpha(olf) levels are regulated by receptor usage and control dopamine and adenosine action in the striatum. *J. Neurosci.*, **21**(12):4390–4399.

47. Stoof, J.C. and Kebabian, J.W. (1981) Opposing roles for D-1 and D-2 dopamine receptors in efflux of cyclic AMP from rat neostriatum. *Nature*, **294**(5839):366–368.

48. Siuciak, J.A., et al. (2006) Inhibition of the striatum-enriched phosphodiesterase PDE10A: a novel approach to the treatment of psychosis. *Neuropharmacology*, **51**(2): 386–396.

49. Dlaboga, D., Hajjhussein, H., and O'Donnell, J.M. (2008) Chronic haloperidol and clozapine produce different patterns of effects on phosphodiesterase-1B, -4B, and -10A expression in rat striatum. *Neuropharmacology*, **54**(4):745–754.

50. Charych, E.I., Jiang, L.X., Lo, F., Sullivan, K., and Brandon, N.J. (2010) Interplay of palmitoylation and phosphorylation in the trafficking and localization of phosphodiesterase 10A: implications for the treatment of schizophrenia. *J. Neurosci.*, **30**(27):9027–9037.

51. Huang, K. and El-Husseini, A. (2005) Modulation of neuronal protein trafficking and function by palmitoylation. *Curr. Opin. Neurobiol.*, **15**(5):527–535.

52. Linder, M.E. and Deschenes, R.J. (2007) Palmitoylation: policing protein stability and traffic. *Nat. Rev. Mol. Cell. Biol.*, **8**(1):74–84.

53. Kebabian, J.W., Petzold, G.L., and Greengard, P. (1972) Dopamine-sensitive adenylate cyclase in caudate nucleus of rat brain, and its similarity to the "dopamine receptor." *Proc. Natl. Acad. Sci. USA*, **69**(8):2145–2149.

54. Soderling, S.H. and Beavo, J.A. (2000) Regulation of cAMP and cGMP signaling: new phosphodiesterases and new functions. *Curr. Opin. Cell. Biol.*, **12**(2):174–179.

55. Kukovetz, W.R. and Poch, G. (1970) Inhibition of cyclic-3′,5′-nucleotide-phosphodies-terase as a possible mode of action of papaverine and similarly acting drugs. *Naunyn. Schmiedebergs Arch. Pharmakol.*, **267**(2):189–194.

56. Schmidt, C., et al. (2002) Effect of PDE10 inhibition on striatal cyclic nucleotide concentration. Society for Neuroscience: Abstract, Society for Neuroscience Annual Meeting, Washington, DC, p. 43.

57. Girault, J.A. and Greengard, P. (2004) The neurobiology of dopamine signaling. *Arch. Neurol.*, **61**(5):641–644.

58. Nishi, A., et al. (2008) Distinct roles of PDE4 and PDE10A in the regulation of cAMP/ PKA signaling in the striatum. *J. Neurosci.*, **28**(42):10460–10471.

59. Nishi, A., Snyder, G.L., and Greengard, P. (1997) Bidirectional regulation of DARPP-32 phosphorylation by dopamine. *J. Neurosci.*, **17**(21):8147–8155.

60. Greengard, P., Allen, P.B., and Nairn, A.C. (1999) Beyond the dopamine receptor: the DARPP-32/protein phosphatase-1 cascade. *Neuron*, **23**(3):435–447.

61. Roche, K.W., O'Brien, R.J., Mammen, A.L., Bernhardt, J., and Huganir, R.L. (1996) Characterization of multiple phosphorylation sites on the AMPA receptor GluR1 subunit. *Neuron*, **16**(6):1179–1188.

62. Shepherd, J.D. and Huganir, R.L. (2007) The cell biology of synaptic plasticity: AMPA receptor trafficking. *Annu. Rev. Cell. Dev. Biol.*, **23**:613–643.

63. Citri, A. and Malenka, R.C. (2008) Synaptic plasticity: multiple forms, functions, and mechanisms. *Neuropsychopharmacology*, **33**(1):18–41.

64. Threlfell, S., Sammut, S., Menniti, F.S., Schmidt, C.J., and West, A.R. (2009) Inhibition of phosphodiesterase 10A increases the responsiveness of striatal projection neurons to cortical stimulation. *J. Pharmacol. Exp. Ther.*, **328**(3):785–795.

65. Dreyer, J.K., Herrik, K.F., Berg, R.W., and Hounsgaard, J.D. (2010) Influence of phasic and tonic dopamine release on receptor activation. *J. Neurosci.*, **30**(42):14273–14283.

66. Surmeier, D.J., et al. (2010) The role of dopamine in modulating the structure and function of striatal circuits. *Prog. Brain Res.*, **183**:149–167.

67. Strick, C.A., et al. (2010) Alterations in gene regulation following inhibition of the striatum-enriched phosphodiesterase, PDE10A. *Neuropharmacology*, **58**(2):444–451.

68. Kleiman, R.J., et al. (2011) Chronic suppression of phosphodiesterase 10A alters striatal expression of genes responsible for neurotransmitter synthesis, neurotransmission, and signaling pathways implicated in Huntington's disease. *J. Pharmacol. Exp. Ther.*, **336**(1):64–76.

69. Giampa, C., et al. (2010) Inhibition of the striatal specific phosphodiesterase PDE10A ameliorates striatal and cortical pathology in R6/2 mouse model of Huntington's disease. *PLoS One*, **5**(10):e13417.

70. Giampa, C., et al. (2009) Phosphodiesterase 10 inhibition reduces striatal excitotoxicity in the quinolinic acid model of Huntington's disease. *Neurobiol. Dis.*, **34**(3):450–456.

71. Leuti, A., et al. (2013) Phosphodiesterase 10A (PDE10A) localization in the R6/2 mouse model of Huntington's disease. *Neurobiol. Dis.*, **52**:104–116.

72. Wang, J.Q. and McGinty, J.F. (1997) The full D_1 dopamine receptor agonist SKF-82958 induces neuropeptide mRNA in the normosensitive striatum of rats: regulation of D_1/D_2 interactions by muscarinic receptors. *J. Pharmacol. Exp. Ther.*, **281**(2):972–982.

73. Angulo, J.A. (1992) Involvement of dopamine D_1 and D_2 receptors in the regulation of proenkephalin mRNA abundance in the striatum and accumbens of the rat brain. *J. Neurochem.*, **58**(3):1104–1109.

74. Kehler, J. (2013) Phosphodiesterase 10A inhibitors: a 2009–2012 patent update. *Expert Opin. Ther. Pat.*, **23**(1):31–45.

75. Kehler, J. and Kilburn, J.P. (2009) Patented PDE10A inhibitors: novel compounds since 2007. *Expert Opin. Ther. Pat.*, **19**(12):1715–1725.

76. de Martinis, N. (2012) SIRS Conference Abstract.

77. Alexander, G.E. and Crutcher, M.D. (1990) Functional architecture of basal ganglia circuits: neural substrates of parallel processing. *Trends Neurosci.*, **13**(7):266–271.

78. DeLong, M.R. and Wichmann, T. (2007) Circuits and circuit disorders of the basal ganglia. *Arch. Neurol.*, **64**(1):20–24.

CHAPTER 11

ROLE OF CYCLIC NUCLEOTIDE SIGNALING AND PHOSPHODIESTERASE ACTIVATION IN THE MODULATION OF ELECTROPHYSIOLOGICAL ACTIVITY OF CENTRAL NEURONS

SARAH THRELFELL
Department of Physiology, Anatomy and Genetics, University of Oxford, Oxford, UK

ANTHONY R. WEST
Department of Neuroscience, Rosalind Franklin University of Medicine and Science, North Chicago, IL, USA

INTRODUCTION

Cyclic Nucleotide Synthesis

Numerous central nervous system (CNS) neurotransmitters exert their actions via activation of membrane-bound G-protein coupled receptors (Gpcrs). Activation of GPCRs initiates intracellular signaling cascades that are linked to the regulation of second-messenger synthesis and degradation. The most well studied of these ubiquitous second messengers is adenosine $3'$-$5'$-cyclic monophosphate (cAMP), which is synthesized from adenosine triphosphate (ATP) following activation of adenylyl cyclase (AC) (Figure 11.1, left). A wide range of neurotransmitters are functionally coupled to AC via stimulatory G-protein (G_{olf}, G_s) coupled receptors (e.g., norepinephrine (NE) β-adrenergic receptors, dopamine (DA) $D_1/_5$ receptors, adenosine A2 receptors, serotonin (5-HT) receptors), and inhibitory G-protein (G_i) coupled receptors (e.g., DA $D_2/_3/_4$ receptors). Furthermore, the resulting rise of intracellular calcium levels following activation of N-methyl-D-aspartate (NMDA) receptors or

Cyclic-Nucleotide Phosphodiesterases in the Central Nervous System: From Biology to Drug Discovery,
First Edition. Edited by Nicholas J. Brandon and Anthony R. West.
© 2014 John Wiley & Sons, Inc. Published 2014 by John Wiley & Sons, Inc.

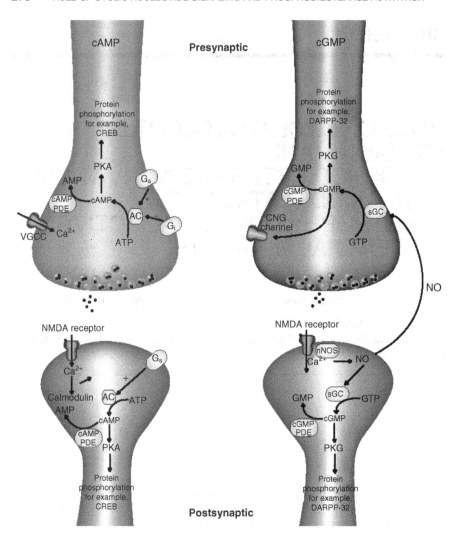

FIGURE 11.1   Summary of the roles of AC-cAMP-PKA and NO-sGC-cGMP-PKG signaling and PDE function in the regulation of neuronal excitability and synaptic plasticity. *Left:* Most studies in cortex, hippocampus, and striatum show that regulation of membrane excitability and synaptic plasticity depends on a signaling cascade triggered by an increase in AC activity and cAMP production, with subsequent activation of PKA. Long-term changes in synaptic plasticity usually require phosphorylation of CREB and related changes in gene expression in both pre- and postsynaptic elements. *Right:* At many synapses in these same brain regions, increases in membrane excitability and synaptic plasticity can also be induced by NO-dependent activation of sGC activity and the resulting increase in cGMP production. The subsequent activation of PKG and phosphorylation of DARPP-32 is thought to lead to long-term changes in synaptic plasticity in both pre- and postsynaptic elements. Activation of subtype-specific or dual-substrate PDEs regulates these processes by metabolizing cAMP and cGMP to 5′-AMP or 5′-GMP, respectively. Thus, inhibition of specific PDE subtypes can be expected to augment membrane excitability, and in some cases various forms of synaptic plasticity via increasing the gain of cAMP and/or cGMP signaling.

opening of voltage-gated calcium channels can facilitate a calmodulin-dependent activation of some isoforms of AC, whereas other isoforms are inhibited by calcium transients [1]. Given the above, it is likely that cAMP synthesis is controlled via numerous transmitter interactions targeting both ligand- and voltage-gated ion channels and GPCRs. The downstream effector pathways targeted by AC–cAMP signaling include protein kinase A (PKA), guanosine triphosphate (GTP) exchange protein activated by cAMP (EPAC), and cyclic nucleotide-gated (CNG) channels, all of which are discussed in more detail below.

Although less well studied, guanosine $3'$-$5'$-cyclic monophosphate (cGMP) has been shown to play a variety of critical roles in key regions of the CNS (see below). cGMP is synthesized via activation of either membrane-bound particulate guanylyl cyclase or soluble guanylyl cyclase (sGC) by natriuretic peptides and the gaseous neuromodulator nitric oxide (NO), respectively [2–6]. In the CNS, the NO-sensitive sGC isoform is the major enzyme responsible for cGMP synthesis [7]. Indeed, Garthwaite [5]was the first to demonstrate that cGMP is produced in abundance in the cerebellum following NMDA receptor-dependent activation of neuronal NO synthase (nNOS) (see Figure 11.1, right). A series of subsequent studies demonstrated that, like sGC, nNOS is expressed throughout the CNS [5–7]. High concentrations of cGMP produced by NO-dependent activation of sGC modulate the function of numerous enzymes, such as cGMP-gated ion channels, cGMP-dependent protein kinases (protein kinase Gs (PKGs)), and various cyclic nucleotide phosphodiesterases (PDEs) [2].

Cyclic Nucleotide Degradation

PDEs regulate both spatial and temporal components of cAMP and cGMP signaling within distinct subcellular compartments of central neurons [8,9]. Both cAMP and cGMP are metabolized by subtype-specific or dual-substrate PDEs to $5'$-AMP or $5'$-GMP, respectively (see Figure 11.1). The PDE gene superfamily is organized into 11 distinct families based on sequence homology, substrate specificity, and regulation [8,10,11]. Splice variants containing unique N-terminal regulatory domains exist within each PDE family, resulting in close to 100 functionally unique isoforms [10]. PDE families can be divided into three groups according to substrate specificity: cAMP PDEs (PDE4, PDE7, and PDE8), cGMP PDEs (PDE5, PDE6, and PDE9), and dual-substrate PDEs (PDE1–PDE3, PDE10, and PDE11) (see Chapter 1 for more details). The vast majority of the above PDE isoforms are abundant in the CNS. For example, a wide variety of PDE subtypes are expressed within the hippocampus (PDE1–PDE4 and PDE7–PDE10), striatum (PDE1–PDE4 and PDE8–PDE10), and cortex (PDE1, PDE2, PDE4, PDE8, and PDE9) [10,12]. Interestingly, each isoform is differentially distributed and compartmentalized within individual neurons, and presumably this differential subcellular targeting plays a key role in shaping cyclic nucleotide signaling (see Chapter 3) and, consequently, neuronal membrane excitability and synaptic plasticity. Thus, PDE inhibitors have the potential to elevate cyclic nucleotide concentrations to enhance cyclic nucleotide-mediated signaling within specific subcompartments of the cell. Indeed, for many years nonselective PDE inhibitors such as IBMX have been used to investigate physiological effects of cyclic nucleotides. The increasing accessibility of isoform-specific PDE inhibitors will facilitate more selective enhancement of cyclic nucleotide availability within given cell populations and even within cellular compartments. Increasing numbers of studies have begun to evaluate the role of a variety of subtype specific PDEs either pharmacologically using isoform-specific PDE inhibitors or genetically using mice lacking individual PDE isoforms (see Chapters 4 and 5). Many of these studies have examined the ability of PDEs to affect measures of excitability of neurons and plasticity of neuronal pathways, with strong emphasis on their actions in brain regions such as the hippocampus, cortex, and striatum. This information will be helpful in assessing the potential of PDE inhibitors as therapeutic agents in a variety of neurological disorders characterized by impaired neuronal function and synaptic plasticity. An extensive review of PDE inhibitors and their effects on cognition and behavioral tasks has been covered elsewhere [13]. Consequently, the present work will focus on summarizing the effects of PDE inhibitors on neuronal activity and synaptic plasticity.

MODULATION OF NEURONAL EXCITABILITY AND SYNAPTIC PLASTICITY

Action potential generation in neurons of the CNS arises as a result of integration of thousands of excitatory, inhibitory and modulatory afferent signals, and the interplay of this synaptic activity with the intrinsic membrane properties of the given neuron (reviewed in Ref. [14]). These highly integrative neurons must be capable of changing their output readily in response to strong afferent drive. Importantly, the responsiveness to inhibitory or excitatory afferent drive [measured as excitatory postsynaptic potentials (EPSPs) or inhibitory postsynaptic potentials (IPSPs) or as excitatory postsynaptic currents (EPSCs) or inhibitory postsynaptic currents (IPSCs)] is often facilitated or depressed by the actions of numerous neuromodulators. Considerable evidence indicates that the action of many neuromodulators is mediated by up- or downregulation of the ubiquitous second messengers cAMP and cGMP [15]. Indeed, cyclic nucleotides are produced in both presynaptic and postsynaptic elements and are, therefore, well positioned to play a pivotal role in the modulation of synaptic transmission.

Synaptic plasticity involves short- or long-acting increases or decreases in strength of synaptic connections. The most well characterized phenomena associated with increased and decreased synaptic efficacy are long-term potentiation (LTP) and long-term depression (LTD), respectively. LTP represents a long-lasting usage-dependent increase in synaptic efficacy and neurotransmitter release that may provide a useful electrophysiological model for synaptic changes occurring during memory formation. In contrast, LTD represents a long-lasting decrease in presynaptic neurotransmitter release, resulting in decreased synaptic strength. Both LTP and LTD are considered neurophysiological correlates of learning and memory [16,17] and are often associated with changes in gene expression, protein synthesis, and formation of new synaptic connections. Interestingly, LTP exists in different forms: early-phase LTP

(E-LTP) and late-phase LTP (L-LTP). Presynaptic cGMP is thought to be involved in E-LTP, which lasts less than 3 h and is not dependent on gene expression/protein synthesis, and, therefore, perhaps contributes to short-term memory. Postsynaptic cAMP and cGMP pathways are thought to be involved in L-LTP, which is likely to encode longer-term memory that lasts more than 3 h [16,17].

The precise mechanisms underlying synaptic plasticity at identified synapses have been intensely investigated for several decades. The vast majority of studies have focused on the hippocampus, as this nucleus is pivotal in both the normal consolidation of memories and in disorders characterized by memory impairment, such as Alzheimer's disease (AD). However, recent studies of synaptic plasticity have begun to focus more frequently on other brain nuclei where learning and memory are thought to be critical for normal function (e.g., cortex and striatum). Plasticity at synapses in all of these brains regions can be regulated presynaptically by altering neurotransmitter release or postsynaptically by altering the number, type, or properties of neurotransmitter receptors. Postsynaptic generation of cyclic nucleotides is thought to modify plasticity via the regulation of AMPA receptor trafficking. AMPA receptors can be inserted or removed from the synapse following phosphorylation by PKA and PKG [18]. Presynaptic generation of cyclic nucleotides can modify ion channel activity via phosphorylation, and in turn, alter presynaptic release mechanisms and synaptic plasticity. In general, it is thought that increased levels of cAMP and/or cGMP enhance processes that maintain LTP, thereby augmenting synaptic plasticity. The induction and regulation of these processes in specific cell types in different brain regions are discussed in more detail in the following sections.

Cyclic Nucleotide Signaling

As mentioned above, cAMP and cGMP can affect a multitude of downstream signaling cascades via the activation of PKA and PKG, respectively. Once activated, these kinases can subsequently phosphorylate numerous other proteins, including ion channels and phosphatases, to alter their activity and modify cell excitability. For example, in the cortex and striatum, activation of DA- and cAMP-regulated phosphoprotein molecular weight 32 kDa (DARPP-32) is thought to play a key role in mediating many of the physiological effects of DA and the interactions between DA and fast-acting neurotransmitters such as γ-aminobutyric acid (GABA) and glutamate that control cortical and striatal output [15]. The PKA-mediated phosphorylation of DARPP-32 (on Thr34) activates the inhibitory phosphatase action of DARPP-32 and results in the dephosphorylation and inhibition of protein phosphatase-1 (PP1). This enables PKA to drive the phosphorylation of other substrates such as cAMP response element binding protein (CREB). Activation of CREB initiates gene transcription thought to be involved in long-term memory formation [19–21].

PKG is also a potent activator of DARPP-32 [15]. Indeed, studies performed in brain slices from rodents have shown that the NO-sGC-cGMP mediated stimulation of PKG results in increased DARPP-32 (Thr34) phosphorylation in striatal medium-sized spiny projection neurons (MSNs) [22,23]. The PKG-dependent facilitation of DARPP-32 phosphorylation (also on Thr34) is rapid, transient, and requires a concurrent increase in intracellular calcium levels induced following glutamatergic stimulation of NMDA, AMPA, and metabotropic glutamate subtype 5 receptors [22,23]. Together, the PKA–PKG effector pathways have been shown to play a key role in regulating long-term changes in striatal synaptic efficacy [24], and as such are likely to be critically involved in the planning and execution of motor behavior [25]. This is supported further by studies showing that local infusion of cAMP/cGMP analogs and PDE inhibitors potently modulates the membrane excitability of central neurons [26,27].

The remainder of this chapter summarizes how regulation of cAMP and cGMP transients modify neuronal excitability and synaptic plasticity in several key regions of the CNS. The role of specific PDE subtypes in regulating these effects is also reviewed. Lastly, the potential utility of subtype specific PDE inhibitors/activators for reversing pathophysiological changes in membrane excitability and synaptic plasticity are discussed. Given the massive body of literature pertaining to this topic (there are probably thousands of studies that manipulate cAMP and cGMP signaling and downstream effectors in the hippocampus alone), our attention is focused largely on recent studies performed in a handful of major brain regions that have significantly advanced our understanding of cyclic nucleotide–PDE function in the CNS.

MODULATION OF CORTICAL NEURONAL EXCITABILITY BY CYCLIC NUCLEOTIDES AND PDES

Both cAMP and cGMP potently modulate the neuronal excitability and synaptic plasticity of cortical pyramidal cells. Pioneering studies by Woody and Gruen using *in vivo* intracellular recordings in awake cats showed that intracellular injection of cAMP into identified pyramidal cells recorded in layer 5 of the motor cortex produced increased rates of spike activity [28]. Later studies in reduced preparations showed that bath application of the cAMP analog dibutyryl cAMP increased the membrane excitability of neurons recorded in cortical slices [29]. Other studies performed in pyramidal neurons recorded in the prefrontal cortex (PFC) have found that alpha2A adrenoreceptor activation by NE inhibits cAMP production, which leads to the closing of hyperpolarization-activated cation (HCN) channels [30]. This, in turn, induced an enhancement of spatially tuned delay-related firing [30]. Additional recent work performed on pyramidal cells in layer 5 of the parietal cortex shows that bath application of the AC activator forskolin increases membrane excitability by suppressing a tonic outward K^+ current via a cAMP- and PKA-dependent mechanism [31]. This suppression was potentiated by the β_1-agonist isoproterenol and the PDE4 inhibitor rolipram. The effect of rolipram was also found to translate into increased spike generation in response to depolarizing current application [31]. These studies suggest that cAMP production, stimulated by NE activation of β_1 receptors, may act to increase the intrinsic excitability and spontaneous spiking of pyramidal cells by suppressing outward K^+ currents that normally act to hyperpolarize the plasma membrane. Furthermore, in a PFC nerve terminal preparation, rolipram potentiates calcium-dependent release of glutamate evoked by 4-aminopyridine

(4-AP) by increasing terminal excitability [32]. These findings are consistent with studies showing that PDE4 activity is necessary for proper integration of the neuro-modulatory signals that regulate cAMP signaling and excitatory transmission in the cortex [33].

As suggested by the above studies, cAMP production in diverse cortical regions is likely to be stimulated by the activation of GPCRs by neuromodulators such as NE, DA, and 5-HT. In particular, studies examining the impact of DA receptor activation on neuronal activity of cortical pyramidal neurons have provided tre-mendous insight as to how cyclic nucleotides regulate cortical networks. For example, recent studies have shown that DA decreases the excitability of layer 5 pyramidal neurons of the entorhinal cortex by a DA D_1 receptor-dependent mechanism involving activation of AC and subsequent non–PKA-related actions of cAMP on hyperpolarization activated channels (h channels) [34]. In the PFC, DA reduces inwardly rectifying K^+ currents (IRKCs) by two mechanisms: (i) D_1 receptor activation of cAMP and direct action of cAMP on the IRK channels, and (ii) D_2 receptor-mediated dephosphorylation of IRK channels [35]. DA has also been reported to increase the excitability of PFC neurons via a D_1 receptor–cAMP–PKA-dependent mechanism by decreasing the rapidly inactivating component of a Na^+ current arising from $Na_v1.1$ and 1.2 channels [36]. D_1 receptor activation also facilitates NMDA-induced increases in the membrane excitability of PFC pyramidal neurons in a calcium and PKA-dependent manner [37]. Together, these studies indicate that DA-receptor activation produces complex regionally specific effects on pyramidal neuron excitability, which depend on the specific downstream signaling mechanisms targeted by cAMP- and PKA-dependent processes.

Studies examining the role of NO-sGC-cGMP and PKG have shown that this cascade also produces complex modulatory changes in cortical neuron activity. Early studies using *in vivo* intracellular recordings in awake cats showed that intracellular injection of cGMP into identified pyramidal cells recorded in layer 5 of the motor cortex did not affect firing rate [38]. However, a large increase in the input resistance of recorded pyramidal cells was observed, suggesting that cGMP may act to increase the responsiveness of these cells to excitatory afferent input. In the rodent visual cortex, production of cGMP and activation of PKG exert depressive effects via presynaptic mechanisms at excitatory synapses, and facilita-tory effects via postsynaptic changes [39]. Thus, application of cGMP or a PKG activator induced a decrease in stimulus-evoked EPSPs in slices of rodent visual cortex when applied extracellularly (but not intracellularly), which is indicative of a presynaptic mechanism. In contrast to this, cGMP enhanced NMDA-evoked responses that were prevented by intracellular application of a PKG inhibitory peptide in the postsynaptic cell [39]. This is consistent with work performed in the intact cat primary visual cortex, which showed that iontophoretic application of a nNOS inhibitor often resulted in an inhibition of visually evoked activity, an effect that could be reversed by either the NO precursor L-ARG or the cGMP analog 8-bromo-cGMP [40]. However, the authors also reported evidence for opposite effects in a subpopulation of neurons [40]. Thus, an elevation in cGMP production in cortical pyramidal neurons is likely to have complex effects as a result of a combination of presynaptic and postsynaptic mechanisms involved in controlling neuronal excitability.

MODULATION OF CORTICAL SYNAPTIC PLASTICITY BY CYCLIC NUCLEOTIDES AND PDES

Work examining mechanisms controlling the LTP of synaptic transmission in the cortex has focused on plasticity of inputs to pyramidal neurons. For example, LTP in the anterior cingulate cortex is NMDA-receptor dependent, but also requires activation of L-type voltage-gated calcium channels and the calcium-stimulated AC subtype 1 [41]. Consequently, calcium-stimulated cAMP-dependent signaling pathways play a crucial role in LTP in the anterior cingulate cortex. The perirhinal cortex, which is involved in visual perception, also exhibits LTP, which is dependent on phosphorylation of CREB [42], indicating a role for cAMP signaling cascades in LTP. Expression of a dominant negative inhibitor of CREB impaired LTP and induced several behavioral deficits, including decreased object recognition [42]. Interestingly, the PDE1 inhibitor vinpocetine was shown to enhance LTP and reverse deficits in ocular dominance plasticity in the visual cortex of ferrets in a model of fetal alcohol exposure [43].

Although less well studied, activation of the NO-sGC-cGMP signaling pathway is required for LTP in the rat PFC and mouse visual cortex [44,45]. In the visual cortex, both isoforms of sGC (designated NO-sGC1 and NO-sGC2) are involved in LTP as knockout of either isoform prevents LTP, in a manner that is restored by administration of a cGMP analog [44].

MODULATION OF HIPPOCAMPAL NEURONAL EXCITABILITY BY CYCLIC NUCLEOTIDES AND PDES

Various neuromodulators such as 5-HT, DA, and NE are thought to play a complex role in regulating cAMP production and the excitability of hippocampal pyramidal neurons. For example, 5-HT activation of 5-HT_4 receptors in the CA1 region of the hippocampus increases membrane excitability by reducing the calcium-activated K^+ current responsible for the slow afterhyperpolarization (sIAHP) in a manner that is dependent on cAMP and PKA [46]. The cAMP-dependent reduction of the sIAHP of hippocampal neurons by 5-HT_4 receptors is the result of reduced calcium-induced calcium release [47]. Similarly, elevations in cAMP and PKA activity induced by DA increases CA1 hippocampal pyramidal neuron excitability by suppressing the sIAHP and spike frequency adaptation [48,49]. Interestingly, the nonselective PDE inhibitor IBMX enhances the ability of 5-HT_4 receptors to modulate this sIAHP in CA1 pyramidal cells [46]. Given that this effect is reproduced by application of the selective PDE4 inhibitors rolipram and Ro 20-1724, it is likely that PDE4 plays a critical role in the modulation of the sIAHP in CA1 pyramidal neurons [48].

Conversely, cGMP negatively regulates the basal membrane excitability of hippocampal pyramidal neurons. For example, the secreted form of the Alzheimer β-amyloid precursor protein (sAPP) suppresses action potential generation and hyperpolarizes hippocampal neurons by activating high conductance, charybdo-toxin-sensitive K^+ channels via a cGMP-dependent mechanism [50]. Another study examining effects of cGMP pathways on rat hippocampal CA1 pyramidal neurons revealed that the PDE3 inhibitor olprinone decreases glutamate release and mini excitatory postsynaptic currents (mEPSCs) induced via AMPA receptor activation in a manner dependent on cGMP [51]. These findings indicate that cGMP plays a role in suppressing the activity of hippocampal pyramidal neurons via both presynaptic and postsynaptic mechanisms.

MODULATION OF HIPPOCAMPAL SYNAPTIC PLASTICITY BY CYCLIC NUCLEOTIDES AND PDES

The most widely studied form of hippocampal synaptic plasticity is LTP of excitatory synaptic transmission. LTP is often induced by high-frequency stimulation (HFS) of afferent fibers, but more recently it has been studied with spike timing-dependent plasticity protocols [52]. LTP can be detected in various areas of the hippocampus, but is most frequently studied in the CA1 pyramidal cells and presynaptic fibers forming the Schaffer collateral. Various studies show that hippocampal LTP depends on a signaling cascade triggered by an increase in cAMP, with subsequent activation of PKA and phosphorylation of CREB [53–58]. The GTPase Rap1 couples cAMP signaling to the membrane-associated pool of p42/MAPK to control the excitability of pyramidal cells, the early and late phases of LTP, and storage of spatial memory [59]. Long-lasting forms of LTP such as L-LTP typically require an elevation of both cAMP and cGMP, recruitment of PKA, and, ultimately, activation of transcription and translation. The rise in cAMP observed during L-LTP in CA1 hippocampal Schaffer collateral synapses activates brain-derived neurotrophic factor (BDNF) receptor TrkB [60]. During LTP, back propagating action potentials in apical dendrites increase in amplitude in a manner that is dependent on PKA activation of mitogen-activated protein kinase (MAPK) [61]. Inhibition of PKA prevents both LTP and enhanced dendritic excitability, whereas inhibition of MAPK pathway only prevents the enhancement of dendritic excitability [61].

Some forms of LTP at CA1 hippocampal synapses also require presynaptic activation of cAMP and PKA [62]. Genetic and pharmacological studies have revealed that cAMP primarily regulates L-LTP at CA1 hippocampal terminals [53,62–65], and therefore may play a crucial role in consolidation of long-term memories.

Generation of mutant mice lacking specific PDE4 isoforms has recently been used as an approach to assess the role cAMP and PDEs in synaptic plasticity. For example, mice lacking PDE4D showed similar paired-pulse facilitation in the CA1 hippocampal region compared to wild-type animals, suggesting normal presynaptic glutamate release in these animals [66]. However, L-LTP was significantly increased in the CA1 region of mice lacking PDE4D [66]. These findings in mutant mice are consistent with pharmacological studies revealing that rolipram enhances L-LTP [67–69]. Rolipram is also able to restore deficits in CA1 hippocampal LTP in (i) sleep-deprived mice [70], (ii) the APP/PS1 double-transgenic mouse model of AD [71], and (iii) hippocampal slices acutely treated with amyloid beta (Aβ) [72]. Rolipram is thought to achieve this enhancement of LTP via increased cAMP-mediated basal CREB expression and phosphorylation [73]. Consistent with this, mutant mice overexpressing AC subtype 1 exhibit enhanced CA1 LTP [64]. The PDE1 inhibitor vinpocetine is also able to enhance LTP in hippocampus of guinea pigs [74,75].

NMDA-independent LTP at the mossy fiber synapse is also dependent on cAMP [76]. Mice expressing a mutant form of AC subtype 1 show impaired LTP that can be rescued by high concentrations of the AC activator forskolin [77]. In addition, genetic disruption of PKA produces a selective defect in mossy fiber LTP; however, this did not disrupt spatial or contextual learning [78].

The medial perforant pathway (MPP) forms another major input to the hippocampus. Short-lasting LTP in the MPP is NMDA receptor dependent but independent of PKA activation, whereas L-LTP in the MPP is PKA dependent [79]. L-LTP in all three pathways (Schaffer, mossy fiber, and MPP) within the hippocampus appears to require cAMP-mediated transcription of new proteins.

As already mentioned, in addition to cAMP, a critical role is also played by NO-sGC-cGMP-PKG signaling in some circumstances that may act in parallel with the cAMP-PKA-CREB pathway. NO is required for formation of hippocampal LTP at some synapses, as blockade of nNOS with L-NAME markedly reduces the potentiation induced by HFS. This reduction in potentiation is partially blocked by cGMP analogs or the NO precursor L-arginine [80]. Furthermore, LTP of cultured hippocampal neurons is prevented by prior application of inhibitors of either sGC or PKG and induced by application of cGMP. cGMP-induced LTP also involves an increase in transmitter release, suggesting a presynaptic role of cGMP in this form of synaptic plasticity [81]. Lastly, NO contributes to L-LTP in the hippocampus via a sGC-PKG pathway that acts in parallel with cAMP to increase CREB activity [82].

Mechanisms by which cGMP is thought to contribute to LTP include activation of cGMP-specific PDEs (see Chapter 1 for more information on PDE activation). As detailed above, this activation results in long-lasting decreases in cGMP levels and increases in GMP metabolites [83]. Transient elevations in cGMP can also activate PDE2A. This isoform is a dual-substrate PDE that is stimulated by cGMP, but capable of regulating both cAMP and cGMP levels. Inhibition of PDE2A with the novel, potent, and selective inhibitor BAY 60-7550 enhances hippocampal LTP without affecting basal synaptic transmission [84]. Interestingly, PDE9 appears to regulate intracellular levels of cGMP and hippocampal synaptic plasticity in a similar manner. Thus, the novel PDE9 inhibitor BAY 73-6691 has been shown to enhance hippocampal LTP in rats without affecting basal synaptic transmission [85].

Although not strongly expressed in the brain in general, PDE5 also regulates cGMP levels and synaptic plasticity in the hippocampus (see Chapters 2 and 9). PDE5 inhibitors are best known for their use in treatment of erectile dysfunction and pulmonary hypertension; however their role within the CNS is just beginning to be unveiled. For example, a recent study shows that the PDE5 inhibitors sildenafil and tadalafil can restore deficits in hippocampal CA1-LTP in a double-transgenic mouse model (i.e., APP/PS1 mice) of AD [86], in a manner similar to that observed with the PDE4 inhibitor rolipram. Reduction of the CA1-LTP impairment with PDE5 inhibitors occurred without affecting basal hippocampal synaptic transmission. Aβ has also been found to markedly impair hippocampal LTP [87,88] by reducing activity in the nNOS-cGMP-CREB pathway [89]. Thus, elevation of cGMP tone with cGMP analogs or PDE inhibitors effectively alleviates impaired CA1-LTP following exposure to Aβ [89].

Interestingly, α-synuclein, a major component of the Lewy bodies found in brains of Parkinson's disease (PD) patients, also seems to be involved in NO-cGMP-dependent changes in plasticity in the hippocampus. Thus, the long-lasting enhancement of evoked and spontaneous miniature transmitter release by NO or cGMP analogs is abolished in cultured hippocampal neurons from mice lacking α-synuclein, suggesting that the nNOS-cGMP-PKG pathway plays a key role in the redistribution of α-synuclein during plasticity [90].

Although less well studied, cGMP has also been reported to induce both short- and long-term depression *in vivo* in the hippocampus. Increasing the concentration of cGMP *in vivo* induces short-term depression (STD) in the hippocampal CA1 region in the freely moving rat, and this chemical LTD was found to occlude electrically induced STD [91]. LTD induced in the CA1 region by low-frequency stimulation was prevented by a sGC inhibitor, suggesting the involvement of cGMP in these effects [91]. Furthermore, simultaneous inhibition of PKA and increased availability of cGMP is sufficient to induce LTD of the Schaffer collateral–CA1 hippocampal synapse in the absence of afferent stimulation [92]. Repeated administration of zaprinast, a cGMP PDE-selective inhibitor, has been shown to induce persistent LTD in CA1 of freely behaving rats [91].

MODULATION OF STRIATAL NEURONAL EXCITABILITY BY CYCLIC NUCLEOTIDES AND PDES

Striatal projections cells (i.e., MSNs) contain very high levels of key enzymes involved in the synthesis and metabolism of cyclic nucleotides, including ACs, sGCs, and various PDEs. Recent studies have shown that numerous PDEs play a complex role in the regulation of both cGMP and cAMP signaling cascades in striatal MSNs [9]. Indeed, several PDE isoforms/families are highly expressed in the striatum (see Chapter 2), including the dual-substrate enzymes PDE1B, PDE2A, and PDE10A. The cAMP-specific enzymes PDE3A, PDE3B, PDE4D, PDE7B, and PDE8B are also prominently expressed in striatum. The cGMP-specific enzyme PDE9A is moderately expressed in the striatum and appears to regulate tonic levels of cGMP [10].

Similar to cortical and hippocampal regions (see above), the activation of AC and sGC in the striatum is controlled by a variety of neurotransmitters (e.g., DA, 5-HT, adenosine, glutamate) that act through GPCRs and stimulation of NO synthesis, respectively [15]. Our previous studies and the work of others show that under physiological conditions, drugs that augment cAMP or cGMP levels in MSNs

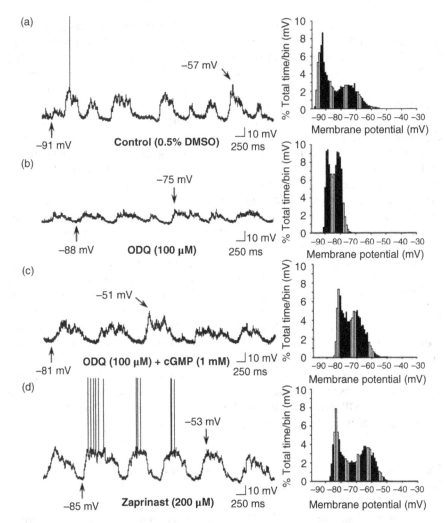

FIGURE 11.2   Opposite effects of sGC inhibition and PDE inhibition on the membrane properties of striatal MSNs. Striatal MSNs were recorded after intracellular application (~5 min) of either vehicle, the drug ODQ, which blocks cGMP synthesis by inhibiting the synthetic enzyme guanylyl cyclase, ODQ plus cGMP, or the drug zaprinast, which inhibits PDE enzymes responsible for degrading cGMP. (a) *Left*: Following vehicle injection, striatal MSNs exhibited typical rapid spontaneous shifts in steady-state membrane potential and irregular spontaneous spike discharge. *Right*: Time interval plot of membrane potential activity recorded from control MSNs. (b) *Left*: Striatal MSNs recorded following ODQ injection exhibited significantly lower amplitude depolarizing events compared to vehicle-injected controls and rarely fired action potentials. *Right*: The depolarized portion of the membrane potential distribution of MSNs recorded following ODQ injection was typically shifted leftward (hyperpolarized) compared to controls. (c) *Left*: Striatal MSNs recorded following ODQ and cGMP coinjection rarely fired action potentials, but exhibited high amplitude depolarizing events with extraordinarily long durations. *Right*: The membrane potential distribution of MSNs recorded following ODQ and cGMP injection was similar to controls, indicating that

(e.g., PDE inhibitors or cyclase activators) generally facilitate spontaneous and evoked corticostriatal transmission [26,27,93], whereas drugs that decrease cyclic nucleotide production (e.g., cyclase inhibitors) or block kinase activity have the opposite effect [26,27,94]. For example, studies by Colwell and Levine show that bath application of the AC activator forskolin enhanced the amplitude and duration of cortically evoked EPSPs in a dose-dependent manner [26]. This effect was blocked by local application of PKA inhibitors and mimicked by similar application of a PKA activator. Forskolin was also shown to enhance the depolarizing effects of bath application of glutamate receptor agonists [26]. Furthermore, treatment with a PKA inhibitor alone decreased the amplitude of cortically evoked EPSPs [26]. These findings indicate that activation of AC-cAMP-PKA signaling in striatal MSNs facilitates corticostriatal transmission and the excitatory effects of AMPA and NMDA receptor activation. Additionally, it is now clear that NMDA-mediated excitatory responses are potentiated by D_1 receptor and attenuated by D_2 receptor stimulation [95–100]. Given that D_1-like receptors are positively coupled to AC and D_2 receptors are negatively coupled to AC [15], it is likely that activation of AC-cAMP-PKA signaling by D_1-like receptors is a primary mechanism by which DA facilitates corticostriatal transmission. Conversely, D_2 receptor activation produces the opposite effect in part by suppressing AC-cAMP-PKA signaling.

Studies from our laboratory have focused on how striatal NO-sGC-cGMP signaling modulates the membrane properties of MSNs and their responsiveness to excitatory synaptic transmission. In the striatum, NO is synthesized primarily in relatively aspiny interneurons that express nNOS [101]. These NO interneurons make synaptic connections with the more distal dendrites of sGC-expressing MSNs of both the direct and indirect projection pathways [102]. A variety of studies found that striatal NO-sGC-cGMP signaling acts to enhance the membrane excitability of striatal MSNs and facilitate corticostriatal transmission [27,93,94,101,103,104]. Our early work showed that reverse dialysis of the NO-generating compound *S*-nitroso-*N*-acetylpenicillamine (SNAP) increased the basal firing rate and bursting of striatal neurons recorded in anesthetized rats [101]. Conversely, drugs limiting the availability of NO, such as nNOS inhibitors or NO scavengers, decrease the amplitude of spontaneously occurring, glutamate-driven depolarized plateau potentials (i.e., up events) in striatal MSNs recorded *in vivo* [27] and suppress spontaneous firing [103]. NO scavengers also decrease the responsiveness of striatal MSNs to both depolarizing current pulses [101] and electrical stimulation of the orbital PFC [27]. Similar effects are observed with systemic or intracellular administration of sGC inhibitors (Figure 11.2a–c) [27,94,104]. The inhibitory effects of nNOS and sGC inhibitors

◀————————————————————————————————————

cGMP partially reversed some of the effects of ODQ. (d) *Left*: Striatal MSNs recorded following intracellular injection of zaprinast exhibited high amplitude depolarizing events with extraordinarily long durations. Additionally, all of the MSNs fired action potentials at relatively high rates (0.4–2.2 Hz). *Right*: The membrane potential distribution of these MSNs was typically shifted rightward (depolarized) compared to controls. Because zaprinast is blocking the degradation of endogenous cGMP, we can conclude that basal levels of cGMP depolarize the membrane potential of striatal MSNs and facilitate spontaneous post-synaptic potentials (modified from Ref. [27]).

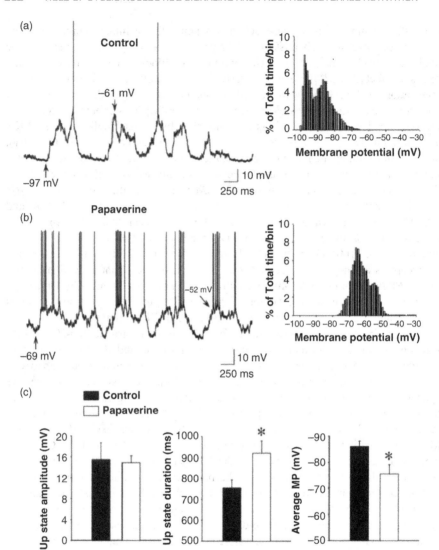

FIGURE 11.3   Inhibition of intracellular cyclic nucleotide metabolism modulates the membrane activity of striatal neurons. Striatal neurons were recorded after intracellular application (~5 min) of either vehicle (3 M potassium acetate) or the PDE inhibitor papaverine (200 μM). (a) *Left:* Following vehicle injection, striatal neurons exhibited typical rapid spontaneous shifts in steady-state membrane potential and irregular and infrequent spontaneous spike discharge. *Right:* Time interval histogram of membrane potential activity (30 s sampled at 20 kHz) recorded from the same control neuron. (b) *Left:* Striatal neurons recorded following intracellular injection of papaverine exhibited robust up events with extraordinarily long durations. Some cells fired action potentials at relatively high rates (>2 Hz). *Right:* The membrane potential distribution of neurons injected with papaverine was typically shifted rightward (depolarized) compared to controls. Arrows indicate the membrane potential at its maximal depolarized and hyperpolarized levels. (c) *Left:* The mean ± SEM up state amplitude was not

on the activity of striatal MSNs are partially reversed by cGMP (Figure 11.2c), indicating that NO and cGMP act to enhance the excitability of MSNs and increase their responsiveness to corticostriatal input [27].

Our recent work shows that many of the effects observed with NO generators and cGMP are reproduced following systemic, intrastriatal, or intracellular administration of various selective and nonselective PDE inhibitors (Figures 11.2–11.5). Work depicted in Figure 11.2d demonstrates that intracellular application of the PDE inhibitor zaprinast depolarized the steady-state membrane potential and increased the responsiveness of MSNs to cortical input, as observed by the increased duration of the depolarized up states [27]. Similarly, intracellular injection of the PDE10A inhibitor papaverine depolarized the average membrane potential and increased the responsiveness of MSNs to cortical input (Figure 11.3). As shown in Figure 11.4, further studies revealed that intrastriatal infusion of the PDE10A inhibitors papaverine or TP-10 during cortical stimulation increased the spike probability of MSNs and total number of spikes per stimulation [93]. Infusion of TP-10 also decreased the onset latency of cortically evoked spikes and reduced the variance in latency to spike, suggesting that when activated, PDE10A may filter asynchronous or weak cortical input to MSNs by suppressing cyclic nucleotide production [93]. Taken together, these findings indicate that inhibition of striatal PDE10A activity increases the responsiveness of MSNs to glutamatergic drive (i.e., corticostriatal transmission).

As discussed in Chapter 12, striatal output is relayed by MSNs making up one of two pathways: (i) the "direct" pathway (striatonigral MSNs) or (ii) the "indirect" pathway (striatopallidal MSNs). There is no evidence for significant differential expression of PDE10A mRNA or protein in striatopallidal or striatonigral MSNs [105–108]. Therefore, we predicted that PDE10A inhibitors would have similar effects on identified striatopallidal and striatonigral MSNs. However, as shown in Figure 11.5, we found that a robust increase in cortically evoked activity was apparent in striatopallidal MSNs. Interestingly, TP-10 administration did not affect cortically evoked activity in identified striatonigral MSNs. Given these outcomes, we believe that PDE10A inhibition may have a greater facilitatory effect on corticostriatal synaptic activity in striatopallidal MSNs [93]. This interpretation is strongly supported by recent studies by Snyder and coworkers showing that papaverine increases DARPP-32 and GluA1 (AMPA) receptor phosphorylation much more robustly in identified striatopallidal as compared to striatonigral MSNs [109].

Interestingly, TP-10 administration was also found to increase the incidence of antidromic responses observed in identified striatonigral MSNs [93]. Thus, it is likely that PDE10A inhibition increases the axonal/terminal excitability of striatonigral MSNs. Indeed, PDE10A is transported to the axons and terminals of MSNs and, based on Western blot analysis, the terminal concentration of the protein appears to be at

◄───

changed following intracellular injection of papaverine ($p > 0.05$, $n = 7–8$ cells/group). *Middle:* The mean ± SEM up state duration was significantly increased following intracellular injection of papaverine (*$p < 0.05$, t-test, $n = 7–8$ cells/group). *Right:* The mean ± SEM membrane potential was significantly depolarized following intracellular injection of papaverine (*$p < 0.05$, t-test, $n = 7–9$ cells/group).

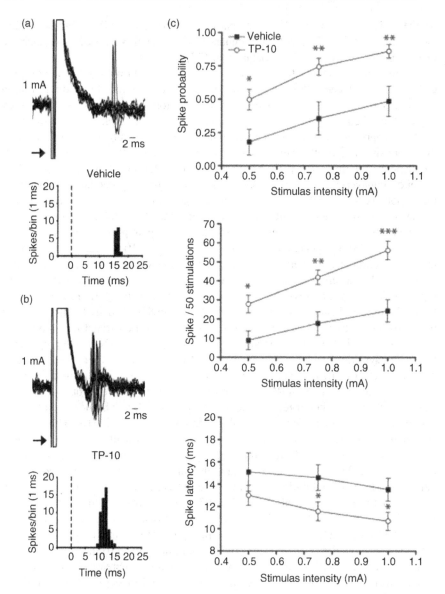

FIGURE 11.4 Intrastriatal infusion of the PDE10A inhibitor TP-10 increases the responsiveness of striatal neurons to electrical stimulation of the frontal cortex. (a) *Top:* Representative traces of cortically evoked spike activity of a single unit recorded following intrastriatal vehicle infusion. *Bottom:* Corresponding peristimulus time interval histogram showing the response of the same striatal neuron to cortex stimulation delivered over 50 stimulation trials. (b) *Top:* Representative traces of cortically evoked spike activity of a single unit recorded following local TP-10 (2 μM) infusion. *Bottom:* Corresponding peristimulus time interval histogram showing the response of the same striatal neuron to cortex stimulation delivered over 50 stimulation trials. For both parts (a) and (b), 10 superimposed traces of cortically evoked

least as high as in striatum [107]. Taken together, these findings further support the hypothesis that PDE10A inhibition increases striatal output, in this case potentially via the direct facilitation of axonal excitability.

MODULATION OF STRIATAL SYNAPTIC PLASTICITY BY CYCLIC NUCLEOTIDES AND PDES

In contrast to hippocampal plasticity studies, LTD is the best characterized form of striatal synaptic plasticity [110,111]. HFS of striatal afferents leads to LTD at corticostriatal synapses onto MSNs that is initiated postsynaptically, yet expressed presynaptically via a reduction in neurotransmitter release [112,113]. This LTD requires D_2 receptors, G_q-coupled group 1 mGlu receptors, L-type calcium channels, and CB_1 receptor activation, but not NMDA or muscarinic receptor stimulation [102,112,114–117]. Postsynaptic D_2 receptors and mGlu receptor activation stimulates the synthesis of endocannabinoids (eCBs) in MSNs [118], which, when released, subsequently activate CB_1 receptors on presynaptic glutamatergic inputs [119], thereby inducing LTD. Presynaptic activation of CB_1 receptors decreases presynaptic release probability, likely via inhibition of AC-cAMP-PKA signaling and $G\beta\gamma$ inhibition of voltage-gated calcium channels. eCB-LTD at MSNs in the ventral striatum requires protracted inhibition of AC-cAMP-PKA signaling and also of P/Q-type calcium channels [120].

Corticostriatal synaptic plasticity is thought to be bidirectional and both LTP and LTD can be observed in direct and indirect MSNs under appropriate conditions [121]. mGlu receptor–eCB-LTD in dorsal striatum is present predominantly at synapses onto indirect pathway MSNs that express D_2 receptors [122]. mGlu receptor–eCB-LTD is only observed in direct pathway MSNs that express D_1 receptors under conditions whereby D_1 receptors are inhibited [121]. D_1 receptor blockade may be necessary to reveal LTD in direct pathway MSNs as D_1 receptors are thought to

◄───

spike responses from a single striatal neuron are shown for trials delivered at the 1.0 mA stimulus intensity. Arrows indicate the location of the stimulus artifact. (c) *Top*: TP-10 (2 μM) infusion significantly increased the mean ± SEM probability of eliciting spike activity during cortical stimulation as compared to cells recorded during a CSF infusion ($p < 0.001$; two-way ANOVA with Bonferroni t-test; $*p < 0.05$ for trials using 0.5 mA stimulus intensities; $**p$ <0.01 for trials using 0.75 mA and 1.0 mA stimulus intensities; $n = 12$–19 cells, $N = 6$–8 rats). *Middle*: TP-10 infusion significantly increased the mean ± SEM number of spikes evoked per trial (50 stimulus pulses) of cortical stimulation as compared to cells recorded during a CSF infusion ($p < 0.001$; two-way ANOVA with Bonferroni t-test; $*p < 0.05$ for trials using 0.5 mA stimulus intensities; $**p < 0.01$ for trials using 0.75 mA stimulus intensities; $***p < 0.001$ for trials using 1.0 mA stimulus intensities; $n = 12$–19 cells, $N = 6$–8 rats). *Bottom*: TP-10 infusion significantly increased the mean ± SEM onset latency of spike responses evoked during single-pulse electrical stimulation of the frontal cortex compared to cells recorded during a CSF infusion ($p < 0.001$; two-way ANOVA with Bonferroni t-test; $*p < 0.05$ for trials using 0.75 and 1.0 mA stimulus intensities; $n = 6$–19 cells, $N = 6$–8 rats) (modified from Ref. [93]).

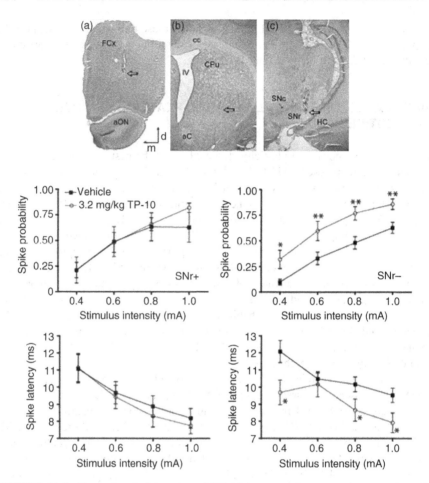

FIGURE 11.5 Systemic administration of TP-10 increases the responsiveness of a subpopulation of MSNs to cortical stimulation. *Top:* Position of implants in the frontal cortex, dorsal striatum, and SNr. (a) Stimulating electrodes were implanted into the frontal cortex. (b) Electrophysiological recording electrodes were implanted into the dorsocentral striatum. (c) In studies involving antidromic stimulation of the SNr, stimulating electrodes were implanted into the SNr. The large unfilled arrows indicate the termination sites of implants in the frontal cortex, dorsal striatum, and SNr. aC, anterior commissure; aON, anterior olfactory nucleus; CC, corpus callosum; CPu, caudate putamen; d, dorsal; FCx, frontal cortex; HC, hippocampus; IV, lateral ventricle; m, medial; SNc, substantia nigra pars compacta; SNr, substantia nigra pars reticulata. *Bottom*: Effect of systemic TP-10 administration on cortically evoked activity of identified striatonigral (SNr+) MSNs. *Left:* The mean ± SEM spike probability and spike onset latency observed in SNr+ MSNs were not changed following TP-10 administration compared to vehicle controls ($p > 0.05$; two-way ANOVA). *Right:* A significant increase in the mean ± SEM spike probability was observed across all stimulus intensities in SNr− MSNs following TP-10 administration compared to vehicle controls ($p < 0.001$; two-way ANOVA with Bonferroni *t*-test; *$p < 0.05$ for trials using 0.4 mA stimulus intensities; **$p < 0.01$ for trials using 0.6–1.0 mA stimulus intensities). The mean ± SEM onset latency of cortically evoked spikes observed in SNr+ MSNs was not changed following TP-10

contribute to striatal LTP [123]. Consistent with this, LTP in striatal slices from DA-depleted mice can be restored by bath application of a D_1 receptor agonist [124]. CREB is essential for bidirectional long-term synaptic plasticity in the striatum. Corticostriatal LTD and LTP are abolished in mice lacking striatal CREB [125]. AC subtype 5 also contributes to corticostriatal plasticity [126]. Corticostriatal HFS-LTD is reduced in mice lacking AC subtype 5, and rescued by CB_1 receptor or mGlu receptor activation. Coactivation of D_2 receptors and mGlu receptors induced LTD in wild-type mice, but not in AC subtype 5-null mice [126]. Furthermore, elevation of cAMP in striatal neurons from wild-type mice eliminated corticostriatal LTD [126].

Numerous findings also indicate that interactions between striatal DA and NO signaling play a key role in the regulation of short and long-term synaptic plasticity [94,103,127–129] via cyclic nucleotide-dependent actions on protein kinase and phosphatase activities [22,130]. The NO-cGMP pathway is also thought to be involved in corticostriatal LTD as HFS of the corticostriatal pathway induces LTD, which is blocked by sGC inhibitor ODQ and by nNOS inhibitors [128]. LTD elicited by HFS can also be induced by the cGMP PDE inhibitor zaprinast [128].

In the intact animal, NO release can be evoked by HFS of corticostriatal pathway in a manner that depends on NMDA receptor activation [104], and may retrogradely potentiate presynaptic glutamate release [103,131]. Interestingly, inhibition of nNOS activity increases the magnitude of D_2 receptor-mediated STD of cortically evoked spike activity induced during phasic stimulation of frontal cortical afferents [103]. Thus, under physiological conditions, corticostriatal transmission may be preferentially detected and amplified by nNOS interneurons in a feed-forward manner that may initiate the synchronization of local network activity with glutamate-driven events. Furthermore, interruption of NO neuromodulation would be expected to disrupt the integration of corticostriatal transmission and short-term plasticity within striatal networks [27,103,104]. This is supported by observations that pharmacological or genetic downregulation of striatal nNOS activity has profound effects on striatal output as measured in electrophysiological [101,132] and behavioral studies [25].

MODULATION OF NEURONAL EXCITABILITY AND SYNAPTIC PLASTICITY BY CYCLIC NUCLEOTIDES AND PDES: MIDBRAIN AND BRAIN STEM

A complete survey of the effects of cyclic nucleotides and PDEs on midbrain, brain stem, and spinal cord is beyond the scope of this chapter. However, representative studies show that increased availability of cAMP within the midbrain and brain stem

◄————————————————————————————————

administration compared to vehicle controls ($p > 0.05$; two-way ANOVA). A significant stimulus intensity-dependent decrease in the mean $\pm$ SEM onset latency of cortically evoked spikes was observed in SNr– MSNs following TP-10 administration compared to vehicle controls ($p < 0.001$; two-way ANOVA with Bonferroni t-test; $*p < 0.05$ for trials using 0.4, 0.8, and 1.0 mA stimulus intensities). Data are derived from $n = 15–37$ cells per group; $N = 7–9$ rats per group (modified from Ref. [93]).

appears to enhance the excitability of local neurons. In the locus ceruleus (LC), a brain stem center that gives rise to norepinephrine projections, the transcription factor CREB plays an important role in regulating the excitability of LC neurons [133]. Increases in either calcium or cAMP can trigger the phosphorylation of CREB by PKA [133]. Expression of a constitutively active form of CREB in LC neurons results in more depolarized neurons firing at higher rates, whereas expression of a dominant negative form of CREB hyperpolarizes LC neurons and decreases their firing rate [133].

The ventral tegmental area (VTA) is a dopaminergic midbrain center involved in motivated behavior and addiction to psychostimulants such as cocaine [110]. Cocaine increases VTA DA neuron activity via a DA D_1/D_5 receptor-dependent mechanism [134]. This mechanism requires a cAMP-mediated redistribution of NMDA receptors to the synaptic membrane of DA neurons [134]. Similar to the striatum, D_2 receptors in the VTA cooperate with mGlu receptors to induce eCB-mediated LTD of GABAergic inhibitory synaptic transmission onto DA neurons [135]. In the VTA, mGlu receptor activation contributes to LTD of GABAergic synapses by enhancing eCB release and presynaptic CB_1 receptor activation, whereas D_2 receptor-mediated LTD contributes via direct inhibition of PKA signaling [136]. Excitatory synapses onto DA neurons in the VTA also exhibit LTD, resulting from a decrease in AMPA receptor function associated with decreased cell surface expression of AMPA receptors [136]. Importantly, activation of PKA is necessary and sufficient to trigger LTD in the VTA [137].

LTP of GABAergic synaptic transmission onto DA neurons is mediated via the NO-sGC-cGMP signaling pathway and blocked by opioids [138,139]. Given the role of VTA DA neurons in motivated behavior and drug abuse (see above), these observations indicate that aberrant plasticity in NO-sGC-cGMP signaling pathways may be involved in the development of some forms of addiction. Furthermore, studies by Kauer and coworkers show that cGMP-specific PDE inhibitors may have potential as novel agents for the treatment of heroin addiction [138,139].

The above studies indicate that augmentation of cyclic nucleotide signaling occurring during AC or sGC activation, or following PDE inhibition has potent actions on the membrane properties and synaptic plasticity of central neurons. Given that disrupted neuron activity and synaptic plasticity are believed to play a pathophysiological role in many neurological disorders, it is likely that treatment with selective PDE inhibitors will be of use for restoring normal function in the degenerating brain. The remainder of this chapter focuses on recent studies that have assessed the impact of PDE inhibitor treatment on neuronal dysfunction in some of the most prevalent neurodegenerative disorders.

IMPLICATIONS FOR THE TREATMENT OF NEUROLOGICAL DISORDERS

The ability of a multitude of PDE isoforms to tightly regulate the intracellular availability of cyclic nucleotides within specific CNS nuclei, and thereby control activity patterns of neurons, has tremendous ramifications for the treatment of neurological disorders. Aberrant forms of synaptic plasticity within the brain are thought to exist in a number of prevalent movement disorders. Functional synaptic plasticity is likely to be essential for normal learning and memory processing and other neural activities. Therefore, novel treatments for diseases such as AD that are characterized by memory loss or PD whereby normal planning and initiation of motor behaviors are impaired will have to manipulate therapeutic targets in a manner that is able to rectify abnormal neuronal excitability and synaptic plasticity. Advances toward this goal are detailed in the following sections.

Alzheimer's Disease

AD is a progressive neurodegenerative disorder characterized by deterioration of all forms of memory and cognitive problems leading to dementia [140]. Although there is eventual loss of synapses in both AD and animal models of AD, deficits in both spatial memory and disruption of LTP precede neuronal death in animal models, implying earlier biochemical changes in the disease [141]. A wide variety of neuromodulators and neurotransmitters have been implicated in processes underlying learning and memory. Perhaps a common feature of all of the neurotransmitter systems implicated in learning and memory lies downstream of the receptors at which these neurotransmitters act. Indeed, as described above, the AC-cAMP-PKA and NO-sGC-cGMP-PKG signaling pathways may represent good therapeutic targets that could be regulated in a specific manner using selective PDE inhibitors.

In support of the above tenet, recent studies show that increasing the availability of cyclic nucleotides using selective PDE4 inhibitors (e.g., rolipram) enhances synaptic plasticity in critical brain regions such as the hippocampus, and improves cognitive performance in animal models of AD. The cloning of the mammalian PDE4 gene arose from work in *Drosophila* focusing on the identification of *dunce* as a PDE4 homolog. As the name implies, *dunce* was originally identified because of its detrimental effects on learning in flies [142]. Consistent with these pioneering studies, PDE4 inhibition enhances hippocampal LTP and improves cognitive performance in multiple animal models [67]. PDE4 inhibitors have also been shown to reverse memory impairments in genetic mouse models of AD [71,143]. Furthermore, rolipram has been shown to reverse decreases in dendritic spine morphology in the hippocampus of mouse models of AD [144]. Transcription of a particular isoform of PDE4, PDE4B, has also been shown to change across different stages of LTP. PDE4B expression is transiently upregulated 2 h after LTP initiation, peaking 6 h after induction [145]. PDE4B expression then rapidly declines over a subsequent 8 h period following LTP induction [145]. Unfortunately, the development of PDE4 inhibitors for AD has been hampered by the considerable side effects of emesis, nausea, and headache, which severely restrict their therapeutic use.

Selective PDE5 inhibitors (e.g., Sildenafil) act to elevate hippocampal cGMP levels, and as a result, improve memory processes [86]. This is thought to be driven by enhanced CREB phosphorylation and LTP via reduced Aβ peptide levels [86]. These outcomes are similar to those previously observed following elevations in NO-cGMP signaling using NO donors and/or cGMP analogs [89].

Huntington's Disease

Huntington's disease (HD) is a genetic neurodegenerative movement disorder characterized by the presence of excessive trinucleotide repeats in the Huntingtin gene [146]. Hallmark pathology of HD includes loss of striatal MSNs, particularly those encompassing the indirect pathway through the basal ganglia. Striatal MSNs are thought to degenerate as a result of overexpression of huntingtin protein produced by the HD gene [146]. Preclinical studies of HD mouse models show that the HD mutation results in age-dependent changes in the modulation of excitatory corticostriatal transmission by DA [147,148]. These abnormalities are characterized by early symptomatic increases in glutamate release and decreases in glutamate release in late stages of the disease [149]. Agents that decrease DA transmission (e.g., tetrabenazine, DA antagonists) have been shown to reverse these pathophysiological changes in HD models [149] and to reduce chorea and motor symptoms in patients with HD [150]. Interestingly, in animal models of HD, steady-state levels of cAMP decrease as the disease progresses from presymptomatic to symptomatic stages [151]. Additional work showed that the mutant huntingtin protein impairs cAMP signaling and CREB function [151]. Moreover, cAMP levels are reduced in the cerebrospinal fluid (CSF) of HD patients compared to controls [152,153]. Furthermore, nNOS mRNA is reduced in the striatum of HD patients [154], indicating that NO synthesis is likely to be depressed, an outcome that would lead to reduced cGMP tone in sGC containing MSNs.

It is now clear that drugs such as PDE inhibitors, which augment cAMP and cGMP levels in striatal MSNs, increase measures of spontaneous and evoked corticostriatal transmission in normal animals (see above) (Figures 11.1–11.4). Thus, it is not immediately apparent how huntingtin-induced decreases in cyclic nucleotide levels could lead to the increased glutamatergic transmission proposed in early-stage HD. However, other studies report that PDE10A and PDE1B mRNA and protein are reduced in the striatum in mouse models of HD [155,156]. Similar results have been reported in postmortem studies comparing time-matched brain samples from controls and HD patients [155]. Furthermore, the PDE10A inhibitor TP-10 has been shown to have potential as a neuroprotective agent in HD [157]. Thus, quinolinic acid lesions of the striatum and pathology in the cortex and striatum observed in R6/2 mice were all markedly reduced by treatment with TP-10 [157].

Given the above, it is possible that persistent abnormalities in the synthesis and catabolism of cyclic nucleotides may contribute to the enduring dysregulation of corticostriatal transmission observed in preclinical models and patients with HD. It is also plausible that treatment with PDE inhibitors may normalize cyclic nucleotide tone and reverse the potential homeostatic imbalance between various PDE isoforms that may be up- or downregulated as a result of pathophysiological changes (e.g., hyperdopaminergia) in the HD striatum. To understand how these complex phenomena may contribute to the HD phenotype, future studies will have to characterize the role of various PDE isoforms in regulating MSN neuron activity and corticostriatal transmission in identified projection pathways. The precise role of cAMP and cGMP in this PDE regulation will also need to be determined in presymptomatic and symptomatic HD models.

Parkinson's Disease and Related Disorders

PD is a neurodegenerative movement disorder initially described by James Parkinson almost two centuries ago [158] and characterized by the loss of DAergic neurons in the substantia nigra pars compacta [159]. The DAergic loss in PD results in dysregulation of activity within the basal ganglia, in particular an increase in activity in striatopallidal pathway neurons and a decrease in activity in the striatonigral pathway [160]. Furthermore, striatal mGlu receptor–eCB-LTD is absent in rodent models of PD [121,122]. Current pharmacotherapy for PD involves treatment with the DAergic precursor L-DOPA; however, prolonged treatment with L-DOPA can result in the development of dyskinesias and other motor fluctuations [161].

As already discussed, PDE10A inhibition has been shown to increase the responsiveness of indirect pathway MSNs to cortical inputs [93], and as such would be expected to exacerbate motor deficits observed in PD. However, recent studies indicate that PDE inhibition may be useful in the treatment of dyskinesias observed in patients with PD who have been chronically treated with L-DOPA. In support of this, studies in animal models of L-DOPA-induced dyskinesias show that the nonselective PDE inhibitor zaprinast reverses decreases in striatal cyclic nucleotide levels and abnormal involuntary movements induced by L-DOPA administration [162,163]. Dyskinetic animals lack the ability to reverse previously induced LTP (i.e., there is a lack of bidirectional plasticity) within the corticostriatal pathway [164]. Because this is reversed by PDE inhibitors, it is likely that deficits in cyclic nucleotide signaling underlie abnormalities in corticostriatal LTP observed in models of L-DOPA-induced dyskinesias.

Consistent with the above studies, the PDE10A inhibitor papaverine has been shown to reduce stereotypies observed following apomorphine administration [165], and hyperactivity induced by the psychostimulants phencyclidine and amphetamine [166]. Given these reports, it is plausible that selective PDE inhibitors will be useful therapeutic agents for reversing pathophysiological correlates of hyperdopaminergia because these agents are able to enhance the bidirectionality of corticostriatal plasticity.

A related neurological disorder, autosomal-dominant striatal degeneration (ADSD) is a movement disorder affecting the striatum, which is characterized by bradykinesia, dysarthria, and muscle rigidity, but lacking the presence of tremors associated with PD. ADSD is caused by a complex frameshift mutation in the PDE8B gene [167]. Together with the above studies, the involvement of PDE8B in the etiology of ADSD highlights the importance of PDEs in regulation of motor function.

CONCLUSIONS

Currently, little is known about the function of many of the PDE families, isoforms, and splice variants as it pertains to the regulation of neuronal excitability and synaptic transmission. Given the strong and diverse expression of numerous PDEs in central neurons, it is likely that future studies examining the impact of selective PDE inhibitors on the regulation of neural excitability and synaptic plasticity in key regions of the CNS will be invaluable. Several examples highlighted above support this contention: (i) the enhancing and restorative effects of PDE4 inhibitors such as rolipram on synaptic plasticity and memory formation in the hippocampus, and (ii) the facilitatory influence of PDE10A inhibitors such as TP-10 on corticostriatal transmission and striatal output. Indeed, a better understanding of the complexities of PDE signaling in identified central neurons in normal and pathophysiological states holds considerable promise for the development of novel therapeutic strategies for restoring motor function in patients with neurodegenerative disorders.

ACKNOWLEDGMENTS

Some of the work reviewed in this chapter was supported by NIH grant R01 047452 (ARW) and grants from Pfizer Inc. (ARW), the Parkinson's Disease Foundation (ARW), and the Brain Research Foundation (ARW).

REFERENCES

1. Mons, N., Decorte, L., Jaffard, R., and Cooper, D.M. (1998) Ca^{2+}-sensitive adenylyl cyclases, key integrators of cellular signalling. *Life Sci.*, **62**(17–18):1647–1652.
2. Murad, F. (2006) Shattuck Lecture. Nitric oxide and cyclic GMP in cell signaling and drug development. *N. Engl. J. Med.*, **355**(19):2003–2011.
3. Boehning, D. and Snyder, S.H. (2003) Novel neural modulators. *Annu. Rev. Neurosci.*, **26**:105–131.
4. Bredt, D.S. (2003) Nitric oxide signaling specificity: the heart of the problem. *J. Cell Sci.*, **116**(Part 1):9–15.
5. Garthwaite, J. (2008) Concepts of neural nitric oxide-mediated transmission. *Eur. J. Neurosci.*, **27**(11):2783–2802.
6. Bredt, D.S., Hwang, P.M., and Snyder, S.H. (1990) Localization of nitric oxide synthase indicating a neural role for nitric oxide. *Nature*, **347**(6295):768–770.
7. Bredt, D.S. and Snyder, S.H. (1990) Isolation of nitric oxide synthetase, a calmodulin-requiring enzyme. *Proc. Natl. Acad. Sci. USA*, **87**(2):682–685.
8. Lugnier, C. (2006) Cyclic nucleotide phosphodiesterase (PDE) superfamily: a new target for the development of specific therapeutic agents. *Pharmacol. Ther.*, **109**(3):366–398.
9. Menniti, F.S., Faraci, W.S., and Schmidt, C.J. (2006) Phosphodiesterases in the CNS: targets for drug development. *Nat. Rev. Drug Discov.*, **5**(8):660–670.

10. Bender, A.T. and Beavo, J.A. (2006) Cyclic nucleotide phosphodiesterases: molecular regulation to clinical use. *Pharmacol. Rev.*, **58**(3):488–520.

11. Omori, K. and Kotera, J. (2007) Overview of PDEs and their regulation. *Circ. Res.*, **100**(3):309–327.

12. Kleppisch, T. (2009) Phosphodiesterases in the central nervous system. *Handb. Exp. Pharmacol.*, **(191)**:71–92.

13. Reneerkens, O.A., Rutten, K., Steinbusch, H.W., Blokland, A., and Prickaerts, J. (2009) Selective phosphodiesterase inhibitors: a promising target for cognition enhancement. *Psychopharmacology (Berl.)*, **202**(1–3):419–443.

14. Shepherd, G.M. (Ed.) (2004) *The Synaptic Organisation of the Brain*, Oxford University Press, p. 736.

15. Greengard, P. (2001) The neurobiology of slow synaptic transmission. *Science*, **294**(5544):1024–1030.

16. Bliss, T.V. and Collingridge, G.L. (1993) A synaptic model of memory: long-term potentiation in the hippocampus. *Nature*, **361**(6407):31–39.

17. Lynch, M.A. (2004) Long-term potentiation and memory. *Physiol. Rev.*, **84**(1): 87–136.

18. Kessels, H.W., Kopec, C.D., Klein, M.E., and Malinow, R. (2009) Roles of stargazin and phosphorylation in the control of AMPA receptor subcellular distribution. *Nat. Neurosci.*, **12**(7):888–896.

19. Deisseroth, K., Bito, H., and Tsien, R.W. (1996) Signaling from synapse to nucleus: postsynaptic CREB phosphorylation during multiple forms of hippocampal synaptic plasticity. *Neuron*, **16**(1):89–101.

20. Deisseroth, K. and Tsien, R.W. (2002) Dynamic multiphosphorylation passwords for activity-dependent gene expression. *Neuron*, **34**(2):179–182.

21. Frank, D.A. and Greenberg, M.E. (1994) CREB: a mediator of long-term memory from mollusks to mammals. *Cell*, **79**(1):5–8.

22. Nishi, A., et al. (2005) Glutamate regulation of DARPP-32 phosphorylation in neostriatal neurons involves activation of multiple signaling cascades. *Proc. Natl. Acad. Sci. USA*, **102**(4):1199–1204.

23. Tsou, K., Snyder, G.L., and Greengard, P. (1993) Nitric oxide/cGMP pathway stimulates phosphorylation of DARPP-32, a dopamine- and cAMP-regulated phosphoprotein, in the substantia nigra. *Proc. Natl. Acad. Sci. USA*, **90**(8):3462–3465.

24. Calabresi, P., Picconi, B., Tozzi, A., and Di Filippo, M. (2007) Dopamine-mediated regulation of corticostriatal synaptic plasticity. *Trends Neurosci.*, **30**(5):211–219.

25. Del Bel, E.A., et al. (2005) Role of nitric oxide on motor behavior. *Cell Mol. Neurobiol.*, **25**(2):371–392.

26. Colwell, C.S. and Levine, M.S. (1995) Excitatory synaptic transmission in neostriatal neurons: regulation by cyclic AMP-dependent mechanisms. *J. Neurosci.*, **15**(3 Part 1): 1704–1713.

27. West, A.R. and Grace, A.A. (2004) The nitric oxide-guanylyl cyclase signaling pathway modulates membrane activity states and electrophysiological properties of striatal medium spiny neurons recorded *in vivo*. *J. Neurosci.*, **24**(8):1924–1935.

28. Woody, C. and Gruen, E. (1986) Responses of morphologically identified cortical neurons to intracellularly injected cyclic AMP. *Exp. Neurol.*, **91**(3):596–612.

29. Boulton, C.L., McCrohan, C.R., and O'Shaughnessy, C.T. (1993) Cyclic AMP analogues increase excitability and enhance epileptiform activity in rat neocortex *in vitro*. *Eur. J. Pharmacol.*, **236**(1):131–136.

30. Huang, C.C. and Hsu, K.S. (2006) Presynaptic mechanism underlying cAMP-induced synaptic potentiation in medial prefrontal cortex pyramidal neurons. *Mol. Pharmacol.*, **69**(3):846–856.

31. Castro, L.R., et al. (2010) Type 4 phosphodiesterase plays different integrating roles in different cellular domains in pyramidal cortical neurons. *J. Neurosci.*, **30**(17):6143–6151.

32. Wang, S.J. (2006) An investigation into the effect of the type IV phosphodiesterase inhibitor rolipram in the modulation of glutamate release from rat prefrontocortical nerve terminals. *Synapse*, **59**(1):41–50.

33. Ye, Y. and O'Donnell, J.M. (1996) Diminished noradrenergic stimulation reduces the activity of rolipram-sensitive, high-affinity cyclic AMP phosphodiesterase in rat cerebral cortex. *J. Neurochem.*, **66**(5):1894–1902.

34. Rosenkranz, J.A. and Johnston, D. (2006) Dopaminergic regulation of neuronal excitability through modulation of Ih in layer V entorhinal cortex. *J. Neurosci.*, **26**(12):3229–3244.

35. Dong, Y., Cooper, D., Nasif, F., Hu, X.T., and White, F.J. (2004) Dopamine modulates inwardly rectifying potassium currents in medial prefrontal cortex pyramidal neurons. *J. Neurosci.*, **24**(12):3077–3085.

36. Maurice, N., Tkatch, T., Meisler, M., Sprunger, L.K., and Surmeier, D.J. (2001) D_1/D_5 dopamine receptor activation differentially modulates rapidly inactivating and persistent sodium currents in prefrontal cortex pyramidal neurons. *J. Neurosci.*, **21**(7):2268–2277.

37. Tseng, K.Y. and O'Donnell, P. (2004) Dopamine-glutamate interactions controlling prefrontal cortical pyramidal cell excitability involve multiple signaling mechanisms. *J. Neurosci.*, **24**(22):5131–5139.

38. Woody, C., Gruen, E., Sakai, H., Sakai, M., and Swartz, B. (1986) Responses of morphologically identified cortical neurons to intracellularly injected cyclic GMP. *Exp. Neurol.*, **91**(3):580–595.

39. Wei, J.Y., Jin, X., Cohen, E.D., Daw, N.W., and Barnstable, C.J. (2002) cGMP-induced presynaptic depression and postsynaptic facilitation at glutamatergic synapses in visual cortex. *Brain Res.*, **927**(1):42–54.

40. Cudeiro, J., et al. (1997) Actions of compounds manipulating the nitric oxide system in the cat primary visual cortex. *J. Physiol.* **504**(Part 2):467–478.

41. Liauw, J., Wu, L.J., and Zhuo, M. (2005) Calcium-stimulated adenylyl cyclases required for long-term potentiation in the anterior cingulate cortex. *J. Neurophysiol.*, **94**(1):878–882.

42. Warburton, E.C., et al. (2005) cAMP responsive element-binding protein phosphorylation is necessary for perirhinal long-term potentiation and recognition memory. *J. Neurosci.*, **25**(27):6296–6303.

43. Medina, A.E., Krahe, T.E., and Ramoa, A.S. (2006) Restoration of neuronal plasticity by a phosphodiesterase type 1 inhibitor in a model of fetal alcohol exposure. *J. Neurosci.*, **26**(3):1057–1060.

44. Haghikia, A., et al. (2007) Long-term potentiation in the visual cortex requires both nitric oxide receptor guanylyl cyclases. *J. Neurosci.*, **27**(4):818–823.

45. Nowicky, A.V. and Bindman, L.J. (1993) The nitric oxide synthase inhibitor, *N*-monomethyl-L-arginine blocks induction of a long-term potentiation-like phenomenon in rat medial frontal cortical neurons *in vitro*. *J. Neurophysiol.*, **70**(3):1255–1259.

46. Torres, G.E., Chaput, Y., and Andrade, R. (1995) Cyclic AMP and protein kinase A mediate 5-hydroxytryptamine type 4 receptor regulation of calcium-activated potassium current in adult hippocampal neurons. *Mol. Pharmacol.*, **47**(1):191–197.

47. Torres, G.E., Arfken, C.L., and Andrade, R. (1996) 5-Hydroxytryptamine4 receptors reduce afterhyperpolarization in hippocampus by inhibiting calcium-induced calcium release. *Mol. Pharmacol.*, **50**(5):1316–1322.

48. Pedarzani, P., Krause, M., Haug, T., Storm, J.F., and Stuhmer, W. (1998) Modulation of the Ca^{2+}-activated K^+ current sIAHP by a phosphatase-kinase balance under basal conditions in rat CA1 pyramidal neurons. *J. Neurophysiol.*, **79**(6):3252–3256.

49. Pedarzani, P. and Storm, J.F. (1995) Dopamine modulates the slow Ca^{2+}-activated K^+ current IAHP via cyclic AMP-dependent protein kinase in hippocampal neurons. *J. Neurophysiol.*, **74**(6):2749–2753.

50. Furukawa, K., Barger, S.W., Blalock, E.M., and Mattson, M.P. (1996) Activation of K^+ channels and suppression of neuronal activity by secreted beta-amyloid-precursor protein. *Nature*, **379**(6560):74–78.

51. Fujikawa, H., Kanno, T., Nagata, T., and Nishizaki, T. (2008) The phosphodiesterase III inhibitor olprinone inhibits hippocampal glutamate release via a cGMP/PKG pathway. *Neurosci. Lett.*, **448**(2):208–211.

52. Caporale, N. and Dan, Y. (2008) Spike timing-dependent plasticity: a Hebbian learning rule. *Annu. Rev. Neurosci.*, **31**:25–46.

53. Abel, T., et al. (1997) Genetic demonstration of a role for PKA in the late phase of LTP and in hippocampus-based long-term memory. *Cell*, **88**(5):615–626.

54. Huang, Y.Y. and Kandel, E.R. (1994) Recruitment of long-lasting and protein kinase A-dependent long-term potentiation in the CA1 region of hippocampus requires repeated tetanization. *Learn. Mem.*, **1**(1):74–82.

55. Montminy, M. (1997) Transcriptional regulation by cyclic AMP. *Annu. Rev. Biochem.*, **66**:807–822.

56. Murphy, D.D. and Segal, M. (1997) Morphological plasticity of dendritic spines in central neurons is mediated by activation of cAMP response element binding protein. *Proc. Natl. Acad. Sci. USA*, **94**(4):1482–1487.

57. Otmakhova, N.A., Otmakhov, N., Mortenson, L.H., and Lisman, J.E. (2000) Inhibition of the cAMP pathway decreases early long-term potentiation at CA1 hippocampal synapses. *J. Neurosci.*, **20**(12):4446–4451.

58. Roberson, E.D. and Sweatt, J.D. (1996) Transient activation of cyclic AMP-dependent protein kinase during hippocampal long-term potentiation. *J. Biol. Chem.*, **271**(48):30436–30441.

59. Morozov, A., et al. (2003) Rap1 couples cAMP signaling to a distinct pool of p42/44MAPK regulating excitability, synaptic plasticity, learning, and memory. *Neuron*, **39**(2):309–325.

60. Patterson, S.L., et al. (2001) Some forms of cAMP-mediated long-lasting potentiation are associated with release of BDNF and nuclear translocation of phospho-MAP kinase. *Neuron*, **32**(1):123–140.

61. Rosenkranz, J.A., Frick, A., and Johnston, D. (2009) Kinase-dependent modification of dendritic excitability after long-term potentiation. *J. Physiol.*, **587**(Part 1):115–125.

62. Wu, Z.L., et al. (1995) Altered behavior and long-term potentiation in type I adenylyl cyclase mutant mice. *Proc. Natl. Acad. Sci. USA*, **92**(1):220–224.

63. Frey, U., Huang, Y.Y., and Kandel, E.R. (1993) Effects of cAMP simulate a late stage of LTP in hippocampal CA1 neurons. *Science*, **260**(5114):1661–1664.

64. Wang, H., Ferguson, G.D., Pineda, V.V., Cundiff, P.E., and Storm, D.R. (2004) Over-expression of type-1 adenylyl cyclase in mouse forebrain enhances recognition memory and LTP. *Nat. Neurosci.*, **7**(6):635–642.

65. Wong, S.T., et al. (1999) Calcium-stimulated adenylyl cyclase activity is critical for hippocampus-dependent long-term memory and late phase LTP. *Neuron*, **23**(4):787–798.

66. Rutten, K., et al. (2008) Enhanced long-term potentiation and impaired learning in phosphodiesterase 4D-knockout (PDE4D) mice. *Eur. J. Neurosci.*, **28**(3):625–632.

67. Barad, M., Bourtchouladze, R., Winder, D.G., Golan, H., and Kandel, E. (1998) Rolipram, a type IV-specific phosphodiesterase inhibitor, facilitates the establishment of long-lasting long-term potentiation and improves memory. *Proc. Natl. Acad. Sci. USA*, **95**(25):15020–15025.

68. Marchetti, C., et al. (2010) Synaptic adaptations of CA1 pyramidal neurons induced by a highly effective combinational antidepressant therapy. *Biol. Psychiatry*, **67**(2):146–154.

69. Navakkode, S., Sajikumar, S., and Frey, J.U. (2005) Mitogen-activated protein kinase-mediated reinforcement of hippocampal early long-term depression by the type IV-specific phosphodiesterase inhibitor rolipram and its effect on synaptic tagging. *J. Neurosci.*, **25**(46):10664–10670.

70. Vecsey, C.G., et al. (2009) Sleep deprivation impairs cAMP signalling in the hippocampus. *Nature*, **461**(7267):1122–1125.

71. Gong, B., et al. (2004) Persistent improvement in synaptic and cognitive functions in an Alzheimer mouse model after rolipram treatment. *J. Clin. Invest.*, **114**(11):1624–1634.

72. Vitolo, O.V., et al. (2002) Amyloid beta-peptide inhibition of the PKA/CREB pathway and long-term potentiation: reversibility by drugs that enhance cAMP signaling. *Proc. Natl. Acad. Sci. USA*, **99**(20):13217–13221.

73. Monti, B., Berteotti, C., and Contestabile, A. (2006) Subchronic rolipram delivery activates hippocampal CREB and arc, enhances retention and slows down extinction of conditioned fear. *Neuropsychopharmacology*, **31**(2):278–286.

74. Ishihara, K., Katsuki, H., Sugimura, M., and Satoh, M. (1989) Idebenone and vinpocetine augment long-term potentiation in hippocampal slices in the guinea pig. *Neuropharmacology*, **28**(6):569–573.

75. Molnar, P. and Gaal, L. (1992) Effect of different subtypes of cognition enhancers on long-term potentiation in the rat dentate gyrus *in vivo*. *Eur. J. Pharmacol.*, **215**(1):17–22.

76. Weisskopf, M.G., Castillo, P.E., Zalutsky, R.A., and Nicoll, R.A. (1994) Mediation of hippocampal mossy fiber long-term potentiation by cyclic AMP. *Science*, **265**(5180):1878–1882.

77. Villacres, E.C., Wong, S.T., Chavkin, C., and Storm, D.R. (1998) Type I adenylyl cyclase mutant mice have impaired mossy fiber long-term potentiation. *J. Neurosci.*, **18**(9):3186–3194.

78. Huang, Y.Y., et al. (1995) A genetic test of the effects of mutations in PKA on mossy fiber LTP and its relation to spatial and contextual learning. *Cell*, **83**(7):1211–1222.

79. Nguyen, P.V. and Kandel, E.R. (1996) A macromolecular synthesis-dependent late phase of long-term potentiation requiring cAMP in the medial perforant pathway of rat hippocampal slices. *J. Neurosci.*, **16**(10):3189–3198.

80. Haley, J.E., Wilcox, G.L., and Chapman, P.F. (1992) The role of nitric oxide in hippocampal long-term potentiation. *Neuron*, **8**(2):211–216.

81. Arancio, O., Kandel, E.R., and Hawkins, R.D. (1995) Activity-dependent long-term enhancement of transmitter release by presynaptic 3′,5′-cyclic GMP in cultured hippocampal neurons. *Nature*, **376**(6535):74–80.

82. Lu, Y.F., Kandel, E.R., and Hawkins, R.D. (1999) Nitric oxide signaling contributes to late-phase LTP and CREB phosphorylation in the hippocampus. *J. Neurosci.*, **19**(23):10250–10261.

83. Monfort, P., Munoz, M.D., Kosenko, E., and Felipo, V. (2002) Long-term potentiation in hippocampus involves sequential activation of soluble guanylate cyclase, cGMP-dependent protein kinase, and cGMP-degrading phosphodiesterase. *J. Neurosci.*, **22**(23):10116–10122.

84. Boess, F.G., et al. (2004) Inhibition of phosphodiesterase 2 increases neuronal cGMP, synaptic plasticity and memory performance. *Neuropharmacology*, **47**(7):1081–1092.

85. van der Staay, F.J., et al. (2008) The novel selective PDE9 inhibitor BAY 73-6691 improves learning and memory in rodents. *Neuropharmacology*, **55**(5):908–918.

86. Puzzo, D., et al. (2009) Phosphodiesterase 5 inhibition improves synaptic function, memory, and amyloid-beta load in an Alzheimer's disease mouse model. *J. Neurosci.*, **29**(25):8075–8086.

87. Cullen, W.K., Suh, Y.H., Anwyl, R., and Rowan, M.J. (1997) Block of LTP in rat hippocampus *in vivo* by beta-amyloid precursor protein fragments. *Neuroreport*, **8**(15):3213–3217.

88. Itoh, A., et al. (1999) Impairments of long-term potentiation in hippocampal slices of beta-amyloid-infused rats. *Eur. J. Pharmacol.*, **382**(3):167–175.

89. Puzzo, D., et al. (2005) Amyloid-beta peptide inhibits activation of the nitric oxide/cGMP/cAMP-responsive element-binding protein pathway during hippocampal synaptic plasticity. *J. Neurosci.*, **25**(29):6887–6897.

90. Liu, S., et al. (2007) Alpha-synuclein involvement in hippocampal synaptic plasticity: role of NO, cGMP, cGK and CaMKII. *Eur. J. Neurosci.*, **25**(12):3583–3596.

91. Stricker, S. and Manahan-Vaughan, D. (2009) Regulation of long-term depression by increases in [guanosine 3′,5′-cyclic monophosphate] in the hippocampal CA1 region of freely behaving rats. *Neuroscience*, **158**(1):159–166.

92. Santschi, L., Reyes-Harde, M., and Stanton, P.K. (1999) Chemically induced, activity-independent LTD elicited by simultaneous activation of PKG and inhibition of PKA. *J. Neurophysiol.*, **82**(3):1577–1589.

93. Threlfell, S., Sammut, S., Menniti, F.S., Schmidt, C.J., and West, A.R. (2009) Inhibition of phosphodiesterase 10A increases the responsiveness of striatal projection neurons to cortical stimulation. *J. Pharmacol. Exp. Ther.*, **328**(3):785–795.

94. Sammut, S., Threlfell, S., and West, A.R. (2010) Nitric oxide-soluble guanylyl cyclase signaling regulates corticostriatal transmission and short-term synaptic plasticity of striatal projection neurons recorded *in vivo*. *Neuropharmacology*, **58**(3):624–631.

95. Brady, A.M. and O'Donnell, P. (2004) Dopaminergic modulation of prefrontal cortical input to nucleus accumbens neurons *in vivo*. *J. Neurosci.*, **24**(5):1040–1049.

96. Cepeda, C., Buchwald, N.A., and Levine, M.S. (1993) Neuromodulatory actions of dopamine in the neostriatum are dependent upon the excitatory amino acid receptor subtypes activated. *Proc. Natl. Acad. Sci. USA*, **90**(20):9576–9580.

97. Chergui, K. and Lacey, M.G. (1999) Modulation by dopamine D_1-like receptors of synaptic transmission and NMDA receptors in rat nucleus accumbens is attenuated by the protein kinase C inhibitor Ro 32-0432. *Neuropharmacology*, **38**(2):223–231.

98. Levine, M.S., Li, Z., Cepeda, C., Cromwell, H.C., and Altemus, K.L. (1996) Neuro-modulatory actions of dopamine on synaptically-evoked neostriatal responses in slices. *Synapse*, **24**(1):65–78.

99. Snyder, G.L., Fienberg, A.A., Huganir, R.L., and Greengard, P. (1998) A dopamine/D_1 receptor/protein kinase A/dopamine- and cAMP-regulated phosphoprotein (Mr 32 kDa)/ protein phosphatase-1 pathway regulates dephosphorylation of the NMDA receptor. *J. Neurosci.*, **18**(24):10297–10303.

100. Jocoy, E.L., et al. (2011) Dissecting the contribution of individual receptor subunits to the enhancement of *N*-methyl-D-aspartate currents by dopamine D_1 receptor activation in striatum. *Front. Syst. Neurosci.*, **5**:28.

101. West, A.R., Galloway, M.P., and Grace, A.A. (2002) Regulation of striatal dopamine neurotransmission by nitric oxide: effector pathways and signaling mechanisms. *Synapse*, **44**(4):227–245.

102. Calabresi, P., et al. (1997) Abnormal synaptic plasticity in the striatum of mice lacking dopamine D_2 receptors. *J. Neurosci.*, **17**(12):4536–4544.

103. Ondracek, J.M., et al. (2008) Feed-forward excitation of striatal neuron activity by frontal cortical activation of nitric oxide signaling *in vivo*. *Eur. J. Neurosci.*, **27**(7):1739–1754.

104. Sammut, S., Park, D.J., and West, A.R. (2007) Frontal cortical afferents facilitate striatal nitric oxide transmission *in vivo* via a NMDA receptor and neuronal NOS-dependent mechanism. *J. Neurochem.*, **103**(3):1145–1156.

105. Coskran, T.M., et al. (2006) Immunohistochemical localization of phosphodiesterase 10A in multiple mammalian species. *J. Histochem. Cytochem.*, **54**(11):1205–1213.

106. Sano, H., Nagai, Y., Miyakawa, T., Shigemoto, R., and Yokoi, M. (2008) Increased social interaction in mice deficient of the striatal medium spiny neuron-specific phosphodies-terase 10A2. *J. Neurochem.*, **105**(2):546–556.

107. Seeger, T.F., et al. (2003) Immunohistochemical localization of PDE10A in the rat brain. *Brain Res.*, **985**(2):113–126.

108. Xie, Z., et al. (2006) Cellular and subcellular localization of PDE10A, a striatum-enriched phosphodiesterase. *Neuroscience*, **139**(2):597–607.

109. Nishi, A., et al. (2008) Distinct roles of PDE4 and PDE10A in the regulation of cAMP/ PKA signaling in the striatum. *J. Neurosci.*, **28**(42):10460–10471.

110. Kreitzer, A.C. and Malenka, R.C. (2008) Striatal plasticity and basal ganglia circuit function. *Neuron*, **60**(4):543–554.

111. Luscher, C. and Huber, K.M. (2010) Group 1 mGluR-dependent synaptic long-term depression: mechanisms and implications for circuitry and disease. *Neuron*, **65**(4):445–459.

112. Choi, S. and Lovinger, D.M. (1997) Decreased probability of neurotransmitter release underlies striatal long-term depression and postnatal development of corticostriatal synapses. *Proc. Natl. Acad. Sci. USA*, **94**(6):2665–2670.

113. Robbe, D., Kopf, M., Remaury, A., Bockaert, J., and Manzoni, O.J. (2002) Endogenous cannabinoids mediate long-term synaptic depression in the nucleus accumbens. *Proc. Natl. Acad. Sci. USA*, **99**(12):8384–8388.

114. Calabresi, P., Maj, R., Pisani, A., Mercuri, N.B., and Bernardi, G. (1992) Long-term synaptic depression in the striatum: physiological and pharmacological characterization. *J. Neurosci.*, **12**(11):4224–4233.

115. Gerdeman, G.L., Ronesi, J., and Lovinger, D.M. (2002) Postsynaptic endocannabinoid release is critical to long-term depression in the striatum. *Nat. Neurosci.*, **5**(5):446–451.

116. Kreitzer, A.C. and Malenka, R.C. (2005) Dopamine modulation of state-dependent endocannabinoid release and long-term depression in the striatum. *J. Neurosci.*, **25**(45):10537–10545.

117. Sung, K.W., Choi, S., and Lovinger, D.M. (2001) Activation of group I mGluRs is necessary for induction of long-term depression at striatal synapses. *J. Neurophysiol.*, **86**(5):2405–2412.

118. Giuffrida, A., et al. (1999) Dopamine activation of endogenous cannabinoid signaling in dorsal striatum. *Nat. Neurosci.*, **2**(4):358–363.

119. Ronesi, J., Gerdeman, G.L., and Lovinger, D.M. (2004) Disruption of endocannabinoid release and striatal long-term depression by postsynaptic blockade of endocannabinoid membrane transport. *J. Neurosci.*, **24**(7):1673–1679.

120. Mato, S., Lafourcade, M., Robbe, D., Bakiri, Y., and Manzoni, O.J. (2008) Role of the cyclic-AMP/PKA cascade and of P/Q-type Ca++ channels in endocannabinoid-mediated long-term depression in the nucleus accumbens. *Neuropharmacology*, **54**(1):87–94.

121. Shen, W., Flajolet, M., Greengard, P., and Surmeier, D.J. (2008) Dichotomous dopaminergic control of striatal synaptic plasticity. *Science*, **321**(5890):848–851.

122. Kreitzer, A.C. and Malenka, R.C. (2007) Endocannabinoid-mediated rescue of striatal LTD and motor deficits in Parkinson's disease models. *Nature*, **445**(7128):643–647.

123. Reynolds, J.N., Hyland, B.I., and Wickens, J.R. (2001) A cellular mechanism of reward-related learning. *Nature*, **413**(6851):67–70.

124. Kerr, J.N. and Wickens, J.R. (2001) Dopamine D-1/D-5 receptor activation is required for long-term potentiation in the rat neostriatum *in vitro*. *J. Neurophysiol.*, **85**(1):117–124.

125. Pittenger, C., et al. (2006) Impaired bidirectional synaptic plasticity and procedural memory formation in striatum-specific cAMP response element-binding protein-deficient mice. *J. Neurosci.*, **26**(10):2808–2813.

126. Kheirbek, M.A., et al. (2009) Adenylyl cyclase type 5 contributes to corticostriatal plasticity and striatum-dependent learning. *J. Neurosci.*, **29**(39):12115–12124.

127. Calabresi, P., Centonze, D., Gubellini, P., Marfia, G.A., and Bernardi, G. (1999) Glutamate-triggered events inducing corticostriatal long-term depression. *J. Neurosci.*, **19**(14):6102–6110.

128. Calabresi, P., et al. (1999) A critical role of the nitric oxide/cGMP pathway in cortico-striatal long-term depression. *J. Neurosci.*, **19**(7):2489–2499.

129. Doreulee, N., et al. (2003) Cortico-striatal synaptic plasticity in endothelial nitric oxide synthase deficient mice. *Brain Res.*, **964**(1):159–163.

130. Calabresi, P., et al. (2000) Dopamine and cAMP-regulated phosphoprotein 32 kDa controls both striatal long-term depression and long-term potentiation, opposing forms of synaptic plasticity. *J. Neurosci.*, **20**(22):8443–8451.

131. West, A.R. and Galloway, M.P. (1997) Endogenous nitric oxide facilitates striatal dopamine and glutamate efflux *in vivo*: role of ionotropic glutamate receptor-dependent mechanisms. *Neuropharmacology*, **36**(11–12):1571–1581.

132. West, A.R. and Grace, A.A. (2000) Striatal nitric oxide signaling regulates the neuronal activity of midbrain dopamine neurons *in vivo*. *J. Neurophysiol.*, **83**(4):1796–1808.

133. Han, M.H., et al. (2006) Role of cAMP response element-binding protein in the rat locus ceruleus: regulation of neuronal activity and opiate withdrawal behaviors. *J. Neurosci.*, **26**(17):4624–4629.

134. Schilstrom, B., et al. (2006) Cocaine enhances NMDA receptor-mediated currents in ventral tegmental area cells via dopamine D5 receptor-dependent redistribution of NMDA receptors. *J. Neurosci.*, **26**(33):8549–8558.

135. Pan, B., Hillard, C.J., and Liu, Q.S. (2008) Endocannabinoid signaling mediates cocaine-induced inhibitory synaptic plasticity in midbrain dopamine neurons. *J. Neurosci.*, **28**(6):1385–1397.

136. Pan, B., Hillard, C.J., and Liu, Q.S. (2008) D_2 dopamine receptor activation facilitates endocannabinoid-mediated long-term synaptic depression of GABAergic synaptic transmission in midbrain dopamine neurons via cAMP-protein kinase A signaling. *J. Neurosci.*, **28**(52):14018–14030.

137. Gutlerner, J.L., Penick, E.C., Snyder, E.M., and Kauer, J.A. (2002) Novel protein kinase A-dependent long-term depression of excitatory synapses. *Neuron*, **36**(5):921–931.

138. Nugent, F.S., Niehaus, J.L., and Kauer, J.A. (2009) PKG and PKA signaling in LTP at GABAergic synapses. *Neuropsychopharmacology*, **34**(7):1829–1842.

139. Nugent, F.S., Penick, E.C., and Kauer, J.A. (2007) Opioids block long-term potentiation of inhibitory synapses. *Nature*, **446**(7139):1086–1090.

140. Carlesimo, G.A. and Oscar-Berman, M. (1992) Memory deficits in Alzheimer's patients: a comprehensive review. *Neuropsychol. Rev.*, **3**(2):119–169.

141. Selkoe, D.J. (2002) Alzheimer's disease is a synaptic failure. *Science*, **298**(5594):789–791.

142. Dudai, Y., Jan, Y.N., Byers, D., Quinn, W.G., and Benzer, S. (1976) *dunce*, a mutant of *Drosophila* deficient in learning. *Proc. Natl. Acad. Sci. USA*, **73**(5):1684–1688.

143. Comery, T.A., et al. (2005) Acute gamma-secretase inhibition improves contextual fear conditioning in the Tg2576 mouse model of Alzheimer's disease. *J. Neurosci.*, **25**(39):8898–8902.

144. Smith, D.L., Pozueta, J., Gong, B., Arancio, O., and Shelanski, M. (2009) Reversal of long-term dendritic spine alterations in Alzheimer disease models. *Proc. Natl. Acad. Sci. USA*, **106**(39):16877–16882.

145. Ahmed, T. and Frey, J.U. (2003) Expression of the specific type IV phosphodiesterase gene PDE4B3 during different phases of long-term potentiation in single hippocampal slices of rats *in vitro*. *Neuroscience*, **117**(3):627–638.

146. MacDonald, M.E., et al. (1993) A novel gene containing a trinucleotide repeat that is expanded and unstable on Huntington's disease chromosomes. *Cell*, **72**(6):971–983.

147. Andre, V.M., Cepeda, C., and Levine, M.S. (2010) Dopamine and glutamate in Huntington's disease: a balancing act. *CNS Neurosci. Ther.*, **16**(3):163–178.

148. Cepeda, C., Bamford, N.S., Andre, V.M., and Levine, M.S. (2010) Alterations in corticostriatal synaptic function in Huntington's and Parkinson's diseases. In: Steiner, H. and Tseng, K.Y. (Eds), *Handbook of Basal Ganglia Structure and Function, a Decade of Progress*, London: Elsevier Inc., pp. 607–623.

149. Andre, V.M., et al. (2011) Differential electrophysiological changes in striatal output neurons in Huntington's disease. *J. Neurosci.*, **31**(4):1170–1182.

150. Huntington Study Group (2006) Tetrabenazine as antichorea therapy in Huntington disease: a randomized controlled trial. *Neurology*, **66**(3):366–372.

151. Gines, S., et al. (2003) Specific progressive cAMP reduction implicates energy deficit in presymptomatic Huntington's disease knock-in mice. *Hum. Mol. Genet.*, **12**(5):497–508.

152. Cramer, H., Kohler, J., Oepen, G., Schomburg, G., and Schroter, E. (1981) Huntington's chorea: measurements of somatostatin, substance P and cyclic nucleotides in the cerebrospinal fluid. *J. Neurol.*, **225**(3):183–187.

153. Cramer, H., Warter, J.M., and Renaud, B. (1984) Analysis of neurotransmitter metabolites and adenosine 3′,5′-monophosphate in the CSF of patients with extrapyramidal motor disorders. *Adv. Neurol.*, **40**:431–435.

154. Norris, P.J., Waldvogel, H.J., Faull, R.L., Love, D.R., and Emson, P.C. (1996) Decreased neuronal nitric oxide synthase messenger RNA and somatostatin messenger RNA in the striatum of Huntington's disease. *Neuroscience*, **72**(4):1037–1047.

155. Hu, H., McCaw, E.A., Hebb, A.L., Gomez, G.T., and Denovan-Wright, E.M. (2004) Mutant huntingtin affects the rate of transcription of striatum-specific isoforms of phosphodiesterase 10A. *Eur. J. Neurosci.*, **20**(12):3351–3363.

156. Hebb, A.L.O., Robertson, H.A., and Denovan-Wright, E.M. (2004) Striatal phosphodiesterase mRNA and protein levels are reduced in Huntington's disease transgenic mice prior to the onset of motor symptoms. *Neuroscience*, **123**(4):967–981.

157. Giampa, C., et al. (2009) Phosphodiesterase 10 inhibition reduces striatal excitotoxicity in the quinolinic acid model of Huntington's disease. *Neurobiol. Dis.*, **34**(3):450–456.

158. Parkinson, J. (2002) An essay on the shaking palsy. 1817. *J. Neuropsychiatry Clin. Neurosci.*, **14**(2):223–236; discussion 222.

159. Rodriguez-Oroz, M.C., et al. (2009) Initial clinical manifestations of Parkinson's disease: features and pathophysiological mechanisms. *Lancet Neurol.*, **8**(12):1128–1139.

160. Albin, R.L., Young, A.B., and Penney, J.B. (1989) The functional anatomy of basal ganglia disorders. *Trends Neurosci.*, **12**(10):366–375.

161. Obeso, J.A., et al. (1989) Motor complications associated with chronic levodopa therapy in Parkinson's disease. *Neurology*, **39**(11 Suppl. 2):11–19.

162. Giorgi, M., et al. (2008) Lowered cAMP and cGMP signalling in the brain during levodopa-induced dyskinesias in hemiparkinsonian rats: new aspects in the pathogenetic mechanisms. *Eur. J. Neurosci.*, **28**(5):941–950.

163. Picconi, B., et al. (2011) Inhibition of phosphodiesterases rescues striatal long-term depression and reduces levodopa-induced dyskinesia. *Brain*, **134**(Part 2):375–387.

164. Picconi, B., et al. (2003) Loss of bidirectional striatal synaptic plasticity in L-DOPA-induced dyskinesia. *Nat. Neurosci.*, **6**(5):501–506.

165. Kostowski, W., Gajewska, S., Bidzinski, A., and Hauptman, M. (1976) Papaverine, drug-induced stereotypy and catalepsy and biogenic amines in the brain of the rat. *Pharmacol. Biochem. Behav.*, **5**(1):15–17.

166. Siuciak, J.A., et al. (2006) Inhibition of the striatum-enriched phosphodiesterase PDE10A: a novel approach to the treatment of psychosis. *Neuropharmacology*, **51**(2):386–396.

167. Appenzeller, S., et al. (2010) Autosomal-dominant striatal degeneration is caused by a mutation in the phosphodiesterase 8B gene. *Am. J. Hum. Genet.*, **86**(1):83–87.

CHAPTER 12

THE ROLE OF PHOSPHODIESTERASES IN DOPAMINE SYSTEMS GOVERNING MOTIVATED BEHAVIOR

GRETCHEN L. SNYDER and JOSEPH P. HENDRICK
Intra-Cellular Therapies Inc., New York, NY, USA

AKINORI NISHI
Department of Pharmacology, Kurume University School of Medicine, Fukuoka, Japan

DOPAMINE: A CENTRAL REGULATOR OF MOTIVATION AND VOLITIONAL BEHAVIOR

The neurotransmitter dopamine plays a pivotal role in the regulation of motivated behaviors. Dopamine activity in the central nervous system is essential for initiation and control of the fine motor movement, behavioral flexibility necessary for learning and memory, and maintaining motivational states that drive the acquisition of food, water, and sex [1,2]. Dopamine activity in the brain is finely tuned. Drugs and disease processes that either interrupt or decrease brain dopamine activity [e.g., Parkinson's disease (PD)] or abnormally elevate dopamine's effects (e.g., psychosis, psychostimulant drugs) result in neurological and psychiatric dysfunction [2–6].

The broad nature of dopamine's effects on diverse behaviors likely owes to the neuroanatomical organization of brain dopamine systems [7–9]. Dopamine-containing cell bodies located in discrete midbrain nuclei project to distinct midbrain and forebrain regions. Neurons in the A9 dopamine cell group, located in the substantial nigra (pars compacta), innervate the basal ganglia nuclei, caudate nucleus, and putamen in primates (or the fused structure known as the neostriatum in other mammals). This nigrostriatal pathway mediates actions of dopamine on motor movement. Selective depletion of dopamine in the nigrostriatal pathway recapitulates akinesia and rigidity

Cyclic-Nucleotide Phosphodiesterases in the Central Nervous System: From Biology to Drug Discovery, First Edition. Edited by Nicholas J. Brandon and Anthony R. West.

characteristic of motor deficits in Parkinson's disease [10]. Dopamine neurons in the A10 dopamine cell group, located in the ventral tegmental area (VTA), send out projections to limbic system targets, including the nucleus accumbens and prefrontal cortex. There, mesolimbic and mesocortical projection pathways are implicated in dopamine's influence over appetitive behaviors, drug craving and addiction, and working memory [2,6]. Selective destruction of mesolimbic dopamine projections, using microinjection of the neurotoxin 6-OHDA into the VTA, attenuates the behavioral effects of psychostimulant drugs [11]. Drugs of abuse, such as the psychostimulant cocaine, preferentially elicit dopamine release in mesolimbic compared with nigrostriatal pathways [6]. Another elegant demonstration of the distinct pathway-specific effects of dopamine is the dopamine knockout mouse. Mice lacking dopamine display deficits in motor movement, learning and memory, and appetitive behaviors anticipated after a global compromise of nigrostriatal, mesolimbic, and mesocortical pathways [12]. Thus, specific dopamine-related behaviors can be rescued by selective stimulation of either nigrostriatal or mesolimbic pathways [1].

ANATOMICAL AND CHEMICAL ORGANIZATION OF STRIATUM

The mammalian striatum contains the highest concentration of dopamine in the central nervous system (CNS), by virtue of its dense network of dopamine-containing nigrostriatal terminals [8,9]. Striatum also contains the highest concentration of dopamine receptors in the brain. This brain region has been used to identify and characterize the known dopamine receptor subclasses. Striatum has been actively studied with regard to the biochemical mechanisms for dopamine signal transduction and has served as an ideal model for unraveling the receptor proteins and signaling molecules that translate dopamine's actions. A striking advantage of striatum for biochemical analysis is the relative anatomical homogeneity of the striatal neuronal populations. The principal target neurons for nigrostriatal dopamine terminals are γ-aminobutyric acid (GABA)-containing medium spiny-type neurons (MSNs) that comprise >95% of all striatal neurons [13,14]. These MSNs form two major output pathways of striatum that modulate basal ganglia output to the thalamus and cortex to affect voluntary motor movement, as shown in Figure 12.1. MSNs comprising a "direct pathway" project to the substantia nigra pars reticulata (SNr). These "striatonigral" neurons also provide a feedback loop to the dopaminergic nigrostriatal neurons. A second population of MSNs forms an "indirect pathway," synapsing to second-order neurons in the globus pallidus (GP) that control glutamatergic neurons within the subthalamic nucleus (STN). This "striatopallidal" pathway also provides feedback to substantia nigra dopamine neurons. However, the segregation of striatal feedback to the midbrain in "direct" and "indirect" pathways confers opposing activities to these pathways as "go" and "no go" controls on neuronal output from the basal ganglia to motor areas of the thalamus and cortex. Thus, the apparent anatomical homogeneity of striatal MSNs contrasts with the more complicated organization of these neurons into systems of on–off controls for motor output.

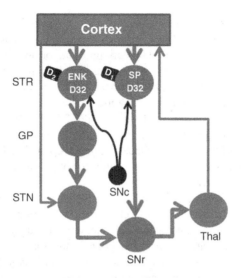

FIGURE 12.1 Schematic showing neuronal pathways that connect nuclei of the basal ganglia. Excitatory pathways are depicted in blue, inhibitory pathways are depicted in red. Dopamine-containing neurons and pathways are depicted in black. D_1, D_1 receptors; D_2, D_2 receptors; D-32, DARPP-32; ENK, enkephalin; GP, globus pallidus; SNc, substantia nigra pars compacta; SNr, substantia nigra pars reticulata; SP, substance P; STN, subthalamic nucleus; STR, striatum; Thal, thalamus. (See the color version of the figure in Color Plates section.)

The biochemical composition of MSNs parallels the organization of direct and indirect striatal output pathways. Though all MSNs use GABA as a neurotransmitter, striatonigral and striatopallidal neurons express separate complements of cotransmitters and signaling proteins [15,16]. Striatonigral MSNs express immunoreactivity for the peptides, dynorphin and substance P. In contrast, striatopallidal pathway neurons express the opioid peptide, enkephalin. Expression levels for both dynorphin and enkephalin are regulated by dopaminergic activity [15,17]. As revealed by these and other studies, nigral and pallidal output neurons are chemically unique [18]. The data also support the concept that dopamine actions on the two subclasses of MSNs are independently controlled by different subclasses of dopamine receptor proteins.

Dopamine receptors are known to exist as five distinct receptor types, D_1–D_5, grouped into two functionally related receptor subclasses [19,20]. Receptors of the D_1 subclass (D_1, D_5) are enriched in MSNs of the striatonigral pathway, whereas receptors of the D_2 subclass (D_2–D_4) are expressed in MSNs of the striatopallidal pathway [21]. D_1- and D_2-type receptors couple to separate classes of G proteins that exert opposing effects on signal transduction pathways governed by second messenger molecules known as cyclic nucleotides [22,23]. For example, D_1-type receptors expressed in striatonigral neurons couple to specific G proteins (i.e., G[olf] and Gs) and the enzyme adenylyl cyclase to stimulate the formation of the cyclic nucleotide, adenosine 3',5'- monophosphate, also known as cyclic adenosine monophosphate) (cAMP) [24]. D_2-type dopamine receptors, predominating in striatopallidal MSNs, couple to different classes of G proteins (i.e., Gi) to inhibit the formation of cAMP. Levels of cAMP in striatal MSNs are also controlled, in part, by other neurotransmitters that couple to G proteins to promote or inhibit the formation of cAMP. For instance, striatonigral neurons are enriched in μ-opioid receptors that respond to enkephalin and suppress cAMP formation through a Gi-dependent process [25]. Adenosine A2A-type receptors, which are enriched in striatopallidal neurons, stimulate cAMP levels in these cells [26,27]. cAMP accumulation in the striatum, then, is determined by multiple G-protein processes, controlled by the opposing influences of D_1- and D_2-type dopamine receptors in addition to the actions of membrane-bound receptors that respond to other striatal neurotransmitters impacting striatal MSNs [28,29].

Although cyclic nucleotide signaling involving cAMP has been more thoroughly studied, striatum is also a rich source of a closely related second messenger, guanosine $3',5'$-monophosphate, or cyclic guanosine monophosphate (cGMP) [30,31]. The guanylate cyclases generate cGMP (analogous to the formation of cAMP by the adenylyl cyclases). These enzymes exist as distinct particulate (membrane-bound) and soluble forms. Both enzymes are highly expressed in rodent striatum [30–35]. Striatal MSNs exhibit the highest concentrations of the soluble guanylate cyclase in the brain, as identified by immunohistochemical staining [35]. The soluble form of guanylate cyclase generates cGMP in response to one of the major regulators of the enzyme in brain, the inter-cellular signaling molecule nitric oxide (NO) [36–38]. NO is abundant in striatum [39], where it is synthesized by nitric oxide synthases (NOSs) that are enriched in specific populations of striatal interneurons [36–40]. Activation of NOS activity in striatum is a calcium/calmodulin-dependent process [40]. NO synthesis by NOS is mediated by glutamate stimulation of *N*-methyl-D-aspartate (NMDA)-type receptors, although a recent study demonstrates that NOS activation by NMDA receptors also requires coactivation of dopamine D_1 receptors [41]. NO, synthesized under the control of neuronal receptors, then diffuses freely among striatal MSNs where it activates soluble guanylate cyclase to form cGMP [40]. The generation of cGMP under the dual control of glutamate and dopamine activity provides an interesting counterpoint to the generation of cAMP under the predominant control of dopamine. The actions of the two major striatal neurotransmitters, dopamine and glutamate, may be integrated, in part, by the relative balance between cAMP and cGMP.

PHOSPHODIESTERASES AND DOPAMINE SYSTEMS: OVERLAPPING TISSUE DISTRIBUTIONS

Dopamine neurotransmission in the striatum is transduced in MSNs through regulation of intracellular levels of second-messenger molecules, such as the cyclic nucleotides. Intracellular cyclic nucleotide pools, furthermore, reflect the influences of other neurotransmitter inputs to MSNs through membrane-bound (e.g., adenosine, enkephalin) and intracellular (e.g., NO) receptor proteins. The levels of cAMP and cGMP are determined by interplay between the adenylyl and guanylyl cyclases that catalyze their formation, under the direction of these receptors, and several classes of phosphodiesterase (PDE) enzymes, which catalyze their hydrolysis and inactivation. PDEs convert active cyclic nucleotide molecules to inactive 5′-nucleotide forms, thereby terminating their ability to activate their respective target proteins, the cAMP- and cGMP-dependent protein kinases. To date, 11 families of PDE enzymes have been identified, encoded by 21 genes (see Chapter 1) [42–45]. Individual PDEs have been classified based on substrate specificities (e.g., cAMP only, cGMP only, dual-substrate or cAMP–cGMP, and either cAMP- or cGMP-preferring) and regulatory factors (e.g., activation or inhibition by cAMP or cGMP, activation by calcium/ calmodulin, and regulation by phosphorylation). They also display unique tissue distributions that distinguish individual PDE families and sometimes structurally related isoforms within the same enzyme family. Many of these enzymes are expressed in the CNS (Table 12.1), several in striatum [45]. The diversity of PDE enzyme families and isoforms present in the brain, the complexity of their regulation, and the high degree of overlap in their brain distribution indicate multiple opportunities for interaction with dopamine systems and modulation of dopamine-mediated motivated behaviors.

PDEs Enriched in Striatum

Several PDE isoforms are highly enriched in the mammalian striatum, based on mRNA abundance and protein detection using immunohistochemistry (this is nicely exemplified in Chapter 2). The 1B isoform of the PDE1 family of enzymes is expressed at high levels in striatal MSNs [49,50]. Three PDE1 family isoforms have been identified, PDE1A, PDE1B, and PDE1C, sharing highly homologous structure and regulation by calcium/calmodulin [69,70]. PDE1B is a dual-substrate, cGMP-preferring enzyme. Unlike the other PDE enzyme families, the activity of PDE1 isoforms is controlled by cellular calcium/calmodulin availability, indicating that the activity of the enzyme is recruited under conditions of neuronal activation. The PDE1B isoform is highly abundant in basal ganglia structures, including striatum, substantia nigra and olfactory tubercle [50]. Although enriched in striatum, PDE1B is also found in other brain regions involved in learning and memory [49,51]. PDE1B immunoreactivity is prominent in the hippocampus, with high levels expressed in dentate granule cells, and in pyramidal cells within layers 5 and 6 of the cerebral cortex [49,51]. PDE1A and PDE1C mRNA are found in regions of the CNS involved in learning and memory and volitional behavior. PDE1A mRNA is prominent in deep layers of the cerebral cortex and dentate gyrus of the hippocampus, with low levels apparent throughout the mouse brain [51]. PDE1C mRNA is expressed at highest level in mouse cerebellum, with low diffuse staining present in cortex and hippocampus [51].

Although PDE1B is predominantly expressed in brain, PDE1A and PDE1C isoforms are also expressed in the peripheral vasculature, specifically in vascular smooth muscle cells and cardiac myocytes. A review by Rybalkin [71] notes the presence of phosphodiesterases in arterial smooth muscle, including PDE1A and PDE1C, PDE3A and PDE3B, and PDE5. Under basal conditions (i.e., low calcium

TABLE 12.1 Summary of Properties of PDE Isoforms with Significant CNS Distribution

PDE	Substrate	Modulated by	Inhibitors	CNS Distribution	References
			PDEs enriched in striatum		
PDE10A	cGMP/cAMP	cAMP/phosphorylation	Papaverine, MP-10, TP-10	Striatum	[46–48]
PDE1B	cGMP > cAMP	Ca^{2+}/calmodulin	IC2224, IC86340	Striatum, hippocampus	[49–51]
PDE7B	cAMP	—	—	Striatum, hippocampus, cerebellum	[52–56]
			PDEs widely distributed in striatum and forebrain		
PDE4B	cAMP	Phosphorylation	Rolipram	Nucleus accumbens, cerebellum, hypothalamus, cortex	[57–59]
PDE4D	cAMP	Phosphorylation	Rolipram	Hippocampus cerebellum, thalamus	[57–59]
PDE4A	cAMP	Phosphorylation	Rolipram	Olfactory tubercle, olfactory bulb, layer V cortical cells	[57–59]
PDE2A	cAMP/cGMP	cGMP stimulated	BAY 60-7550	Broadly throughout the brain, highest in hippocampus	[60]
			PDEs with unknown or low-level striatal expression		
PDE3A	cAMP	cGMP inhibited	Milrinone, cilostamide	Widespread during development; large striatal neurons in adult	[61]
PDE5A	cGMP	cGMP/phosphorylation	Sildenafil, vardenafil, tadalafil	Cerebellum	[62,63]
PDE9A	cGMP	—	BAY 73-6991	Broadly throughout the brain	[64,65]
PDE11A	cAMP/cGMP	cGMP	Tadalafil	Hippocampus	[66–68]

levels), it is thought that the most active cGMP hydrolyzing PDE in smooth muscle is the cGMP-specific, cGMP-activated PDE, PDE5. Under higher calcium conditions (e.g., during muscle contraction and possibly in cells being stimulated to divide), one or more of the PDE1 variants may become predominant [71]. Recently, it was found that PDE1C1 splice variant is expressed at high levels in human cardiac myocytes with an intracellular distribution distinct from that of PDE3A, supporting a role in the integration of cGMP-, cAMP-, and Ca^{2+}-mediated signaling in these cells [72].

The PDE with a brain expression most highly restricted to striatum is PDE10A, a cGMP-preferring, dual-substrate PDE [73,74] (see also Chapter 10). PDE10A was first detected as a highly abundant PDE in the rat striatum and testis [75]. Subsequent neuroanatomical analysis confirmed the presence of PDE10A mRNA and protein within MSNs of rat striatum and nucleus accumbens, in addition to other structures with connections to the basal ganglia, including the olfactory tubercle [46]. This distribution is strikingly similar to the brain distribution of dopamine innervation and the composite expression pattern for D_1- and D_2-type dopamine receptors. No PDE10A immunoreactivity is detected in striatal interneurons, suggesting a selective expression of this PDE in dopaminoceptive neurons, specifically GABAergic MSNs expressing dopamine receptors [47]. Subcellular fractionation studies further reveal that PDE10A, specifically the PDE10A2 splice variant, associates with membranes, being present in striatal synaptosomal fractions [48]. Supporting this conclusion with electron microscopy, PDE10A immunoreactivity has been localized to membrane surfaces of striatal dendrites and dendritic spines [48]. Interestingly, this PDE is absent in the postsynaptic density. Recent studies indicate that subcellular localization of the PDE10A2 splice variant of PDE10A between plasma membrane and cytosol is controlled by posttranslational modification of the enzyme mediated by palmitoylation and phosphorylation [76,77].

PDE7B has been identified as a cAMP-specific PDE with high expression in the striatum [52,53]. The PDE7 family of enzymes is comprised of PDE7A and PDE7B isoforms. PDE7A expression is predominantly expressed in T-cells and airway and vascular smooth muscle cells [54,55]. The PDE7B isoform is present in three splice variants, PDE7B1–PDE7B3, with distinct tissue distribution. The B2 isoform is expressed in testes, whereas mRNA for the B3 variant is found in heart, lung, and skeletal muscle [56]. Only the B1 splice variant of PDE7B is expressed at appreciable level in brain. Within the brain, mRNA for this variant was most abundant in striatum, with lower levels detected in olfactory tubercle and dorsal thalamus [56]. A follow-up study using embryonic striatal cultures reported the detection of PDE7B1 in striatal neurons, indicating that striatal expression of the PDE is likely in MSNs [78]. However, little else is known at present regarding the cellular and subcellular distribution of PDE7B1, both within striatum and in other brain regions. In striatum, the cAMP selectivity of PDE7B contrasts with the dual-substrate specificity of PDE10A and PDE1B enzymes. In addition, the high affinity of PDE7B1 for its substrate, cAMP, compared with other striatal-expressed PDEs, including PDE1B, PDE2A, PDE4, and PDE10A [53,78], suggests a major role for this enzyme in the regulation of cAMP signaling.

In summary, several PDE isoforms, including PDE1B, PDE10A, and PDE7B, are enriched in dopamine-receptive neurons in mammalian striatum. Unfortunately, the relative roles of these various PDEs in striatal function have yet to be explored. Principally, the relative contributions of single PDE isoforms to total striatal PDE activity have not been well studied. Factors such as subcellular distribution and compartmentalization will likely contribute to the role(s) that each PDE isoform plays in striatal activity.

PDEs Widely Distributed in Striatum and Forebrain

The PDE4 family of enzymes is perhaps the best studied of the PDE superfamilies, with regard to anatomical localization and function [79]. Unlike PDE1B, PDE10A, and PDE7B, PDE4 enzymes display a broad distribution throughout the brain and peripheral tissues. The wide-ranging distribution of PDE4 owes in part to the large number of enzymes that make up the family. Currently, at least 20 family members have been identified, including four enzyme isoforms (A–D), and numerous splice variants [80]. Certain of the PDE4 family enzymes have been found to selectively associate with an array of scaffolding proteins, including RACK1 (i.e., receptor for activated C kinase) and the AKAPs (i.e., A-kinase anchoring proteins), which mediate cell-compartment-specific aspects of intracellular signaling [81,82], and NUDEL and Lis1, which are involved in neurodevelopment [83]. The role of PDE4 in cAMP signaling is of particular interest as all of the identified PDE4 enzymes are cAMP specific and insensitive to calcium. First identified in *Drosophila* based on homology to the *dunce* gene, which is involved in learning and memory, neuroanatomical studies have focused on the presence of PDE4 isoforms in brain regions subserving memory function, including cortex and hippocampus [57]. Cherry and Davis [57] performed a thorough study of the distribution of the PDE4A, PDE4B, and PDE4D isoforms in the mouse brain, comparing the relative enrichment of these major isoforms using immunohistochemistry. Of the three isoforms studied, only PDE4B is expressed at a significant level in striatum and substantia nigra, as shown using an isoform-specific polyclonal antibody. The most striking feature of PDE4B distribution is the heavy PDE4B immunoreactivity in nucleus accumbens, which is comparatively stronger than labeling in the striatum. This feature suggests a preferential association of PDE4B with dopamine systems subserving functions related to affect, emotion, and motivation [1,6]. Modest levels of enzyme are also detected in cerebellum, hypothalamus, and cerebral cortex. By contrast, PDE4D is the PDE4 isoform most highly expressed in hippocampus, where it is localized to CA2/CA3 pyramidal cells, with significant expression in cerebellum, thalamus, and thalamic output pathways. PDE4A displays the most restricted distribution of the three isoforms in mouse brain, with major expression in olfactory tubercle, olfactory bulb, and layer 5 cells of the cerebral cortex [57]. The conspicuous presence of PDE4A in olfactory bulb overlaps with the expression pattern of PDE1B. Ultrastructural studies indicate that these two PDEs exhibit distinct cellular and subcellular distributions in olfactory tubercle, with PDE4A expressed in dendrites, axons, and cell bodies of the neuroepithelium, and PDE1B restricted to the olfactory cilia [58,59].

Functional data also indicate that PDE4 is expressed in presynaptic dopamine-containing terminals in the striatum, based on the ability of the PDE4A–PDE4D inhibitor, rolipram, to increase dopamine synthesis, release, and turnover [84–86]. No neuroanatomical data currently exists to confirm the subcellular location of PDE4 in terminals or to identify the relevant PDE4 isoform present. Little is known regarding the cellular distribution of PDE4C and its relation to other PDE isoforms. The 4C isoform is present in rodent brain with expression primarily restricted to the olfactory bulb [87].

As a cGMP-stimulated, dual-substrate PDE, PDE2A was originally identified as a component of adrenal cortex and heart [88,89] where it was found to modulate the release of atrial natriuretic factor (ANF), a peptide hormone that controls salt retention in the kidney [90]. These data stimulated an interest in the functional potential of PDE2A inhibition for the regulation of fluid volume regulation and the treatment of salt-dependent hypertension. Recently, it has become clear that PDE2A is also broadly expressed in the brain [91], suggesting it has CNS roles in addition to effects on peripheral volume regulation. In fact, a recent immunohistochemical study [60] confirms the brain expression of PDE2A in several mammalian species, including mice, rats, dogs, nonhuman primates, and humans. PDE2A immunoreactivity was found to be regionally expressed in the brain in a similar manner across the five species. Furthermore, a higher relative abundance of the enzyme was detected by Western blotting in brain compared with peripheral tissues, namely adrenal and heart [60], supporting the idea that this PDE is a significant cyclic nucleotide regulator in the brain. Analogous to PDE9A and PDE1B, PDE2A is most abundant in brain regions known to be involved in learning and memory and the control of motivated behaviors, including hippocampus, cortex, striatum, substantia nigra, olfactory bulb, and amygdala. In contrast with other PDE isoforms, hippocampal PDE2A is absent in dentate granule cells and CA2/CA3 pyramidal cells, but enriched in the hilus and molecular layer of the dentate gyrus, and in CA3 mossy fibers and the subiculum [60]. Thus, PDE2A appears to display a cellular distribution that complements the expression pattern for other PDEs in hippocampus. High levels are also present in the substantia nigra. In contrast to PDE10A and PDE1B, which are expressed relatively equally within striatonigral and striatopallidal MSNs and their fiber projections, PDE2A immunoreactivity appears strongest in nigrostriatal fibers and dopamine cell body regions in the substantia nigra pars compacta. Although no double-staining data are available to confirm the neurochemical identity of PDE2A-containing cells, the data suggest that PDE2A is present in dopamine-containing pathways innervating striatum, placing this PDE in a position to modulate presynaptic dopamine activity in a manner that is similar to PDE4 [84,85]. PDE2A is also expressed in cell bodies of subsets of neurons in the VTA, suggesting an additional role in the regulation of mesocortical and mesolimbic regions innervated by VTA [60]. Cortical immunoreactivity for PDE2A appears in several layers with prominent staining in bipolar neurons of layer 5. Significantly, strong expression of PDE2A is detected in human prefrontal cortex, a key region involved in the control of working memory in primates and rodents, as determined by Western blotting of postmortem tissue samples.

PDEs with Unknown or Low-Level Striatal Expression

Several PDEs are expressed in the brain with low abundance in striatum and other dopamine-innervated brain regions. They may, however, have significant expression in regions such as cortex or hippocampus, which interconnect with basal ganglia systems to affect volitional behavior. For example, two members of the PDE3 family of enzymes, PDE3A and PDE3B, are cGMP-inhibited, cAMP-specific PDEs, and known constituents of vascular endothelial cells [92,93]. Both isoforms are present in the brain, but with different cellular distributions [61]. *In situ* hybridization reveals PDE3A mRNA to be uniformly distributed in neurons and glial cells in the CNS throughout early development. In contrast, in adult mice, PDE3A is transiently expressed at high levels in dentate gyrus of hippocampus and in striatum. Of interest, PDE3A immunoreactivity appears in large aspiny striatal neurons that may be the cholinergic interneurons that function to regulate the excitability of the more abundant striatal MSNs [94].

PDE9A is a cGMP-specific PDE with a broad distribution throughout the brain [64,65]. This enzyme has the highest affinity for cGMP yet reported ($K_m = 170$ nM) [64,95]. PDE9A mRNA is detectable by *in situ* hybridization in rodent brain. Although this PDE is not enriched in striatum comparably to PDE1B, there are similarities in the regional expression pattern of the two PDE isoforms. For example, both PDE1B and PDE9A are prominent in cerebellar granule cells, dentate granule cells of hippocampus, olfactory tubercle, and layer 5 pyramidal cells of cerebral cortex [65]. The possible importance of these PDEs in the regulation of cGMP levels in the brain is also illustrated by the similarities between the brain expression pattern for PDE9A (and PDE1B) and that of the soluble form of guanylyl cyclase [65,96].

PDE5 and PDE11A have established emerging roles in peripheral systems and are being reevaluated for CNS roles. PDE5, a cGMP-selective PDE [97], is best known for its expression and control of cGMP signaling in vascular smooth muscle cells [98]. PDE5 inhibitors mediate vascular relaxation that is beneficial for the treatment of erectile dysfunction [99]. Recently, PDE5 has been considered for possible CNS functions, including memory [100], as highlighted in Chapter 9. PDE5 inhibitors have been found to favorably affect cognitive abilities in normal and cognitively impaired animals [101,102]. Surprisingly, PDE5 displays rather low-level expression in rat brain regions involved in cognition, including cortex and hippocampus, and is nearly undetectable in striatum [62,63]. Transient staining for PDE5 is found in rat cerebellar Purkinje cell bodies and dendrites in young but not adult rats. To date, PDE5 expression has not been confirmed in human brain [103]. Recently, human clinical evaluation of sildenafil, a potent PDE5 inhibitor, for effects on cognition failed to reveal significant memory improvement [104]. Thus, despite animal data supporting behavioral effects of PDE5 inhibitors in cognition, there is a lack of compelling evidence in humans for either brain localization of the PDE5 enzyme or cognition-enhancing effects. PDE11A has been identified and characterized as three major variants with expression in several peripheral tissues, including skeletal muscle, prostate, kidney, and testes [66]. Subsequent analysis of the regional distribution of the rat splice variants (i.e., PDE11A2–PDE11A4) has revealed different regional expression patterns, with at least one of the rat variants (A2) present in brain [67]. Recent studies confirm that PDE11A message and protein are present at low levels with a restricted enrichment in rodent hippocampus [68].

ACTIVITY-DEPENDENT REGULATION OF PDE EXPRESSION

The control of PDEs by neuronal activity imposes an additional level of control on cAMP-dependent signaling that adjusts system function under conditions of high or low activity, and in response to disease states (e.g., Parkinson's disease, Huntington's disease, or psychosis). PDE enzymes that distribute to brain regions rich in dopamine innervation can be regulated by cAMP-dependent activity, in some instances under the control of dopamine. Dopamine activity controls the expression level of several PDEs. For example, PDE7B expression is modified by dopamine receptor activation, as measured in primary striatal cells in culture [78]. D_1-type dopamine receptor activation increases PDE7B transcription, whereas D_2-type receptor agonists have no effect on levels of the enzyme. The expression levels of PDE1B and PDE10A change under disease conditions that affect dopamine system activity. Mice expressing mutant huntingtin protein, mimicking the presymptomatic stages of Huntington's disease, show decreased expression of both PDE10A and PDE1B protein [105,106]. Dopamine-depleting brain lesions that model the early stages of dopamine loss in Parkinson's disease result in increased striatal levels of PDE1B mRNA and protein [107]. Dopamine depletion also impacts other components of cAMP- and cGMP-dependent signaling in striatum. Dopamine depletion with the neurotoxin, 6-OHDA, substantially reduces striatal NOS levels and NO generating capacity, and results in an increase in mRNA levels and protein expression for PDE1B in the lesioned striatum [107]. The increase in PDE1B expression level after dopamine depletion may be explained, in part, by the loss of dopamine activity via D_2-type receptors, since chronic treatment with potent D_2 receptor antagonists such as haloperidol mimics the increase in PDE1B expression level seen after 6-OHDA lesions [108].

Therapeutic treatments that engage multiple brain systems, for example electro-convulsive shock (ECS), a rapid and effective intervention for depression, elicit changes in expression of specific PDE enzyme families. Drug treatments for depression, including the selective serotonin reuptake inhibitor (SSRI) fluoxetine, and chronic ECS treatment increase expression levels of PDE4A and PDE4B isoforms in cerebral cortex and hippocampus [109–112]. Other classes of antidepressant medications, including the tricyclic antidepressants imipramine and desipramine, also induce PDE4 isoform expression when given chronically to animals [113,114]. There is some evidence to suggest that antidepressant treatments evoke PDE4 expression by increasing synaptic availability of the neurotransmitters serotonin, norepinephrine, and histamine. For instance, pharmacological treatments that elevate these transmitters in the brain result in increased labeling of PDE4 with [11C]-rolipram, as revealed by positron emission tomography [115]. Inhibitors of serotonin, norepinephrine, and histamine receptors attenuate these increases. In contrast, agents that increase or decrease dopamine activity in the brain had no effect on [^{11}C]-rolipram binding, supporting the conclusion that the short-term effects of antidepressant treatment are a result of modulation of specific neurotransmitter systems. Furthermore, the impact of antidepressant treatment on PDE4 expression appears to be specific for certain enzyme splice variants. For example, D'Sa et al. [112] used *in situ* hybridization to demonstrate region-specific effects of antidepressants on three splice variants of PDE4A, including the brain-enriched short splice variant PDE4A1 [116] and two long splice variants, PDE4A5 [117] and PDE4A10 [118]. Chronic ECS or fluoxetine exposure selectively elevates PDE4A1 in cortex and PDE4A10 in hippocampus, suggesting that certain variants predominate in different brain regions involved in antidepressant responses [112]. Though most studies investigating links between antidepressant drug therapy and PDE function have focused upon the cAMP-specific PDE4, recent studies indicate that cGMP signaling is also affected. Reierson et al. report that hippocampal levels of cGMP, but not cAMP, are significantly elevated in rats treated chronically with either fluoxetine or amitriptyline [119]. These data suggest that the full impact of antidepressant therapies on cyclic nucleotide signaling and the function of specific PDE isoforms require further study.

DARPP-32 REGULATES CYCLIC NUCLEOTIDE-DEPENDENT DOPAMINE SIGNALING AND BEHAVIOR: A MONITOR FOR PDE ACTIVITY

DARPP-32 as a Biochemical Mediator of Cyclic Nucleotide Effects within Dopamine Systems

Biochemical tools are needed to characterize the functional impact of the multiple brain-expressed PDE isoforms that associate with dopamine systems. One of the best-studied dopamine signaling pathways in the brain is the biochemical cascade controlling the activity of the *d*opamine and cAMP-*r*egulated *p*hospho*p*rotein, MW = 32 kDa, also known as DARPP-32 [120]. DARPP-32 was identified as a substrate for the cAMP-dependent protein kinase A (PKA) in brain [121,122]. DARPP-32 has an uneven brain distribution, with highest levels of the protein expressed in areas with prominent dopamine innervation and high levels of D$_1$- and D$_2$-type dopamine receptors [122–124]. It is highly enriched in striatum, specifically within MSNs, but is undetectable in dopamine and glutamatergic striatal nerve terminals or in striatal interneurons [123,125–127]. DARPP-32 is also expressed in the nucleus accumbens, globus pallidus, cortical neurons, most prominently in pyramidal cells of layers 5 and 6 neurons that project to thalamus, in striatonigral nerve terminals in the substantia nigra pars reticulata, olfactory tubercle, and in lower levels in the bed nucleus or the stria terminalis, the amygdaloid complex, cerebellar Purkinje cells, granule cells of the dentate gyrus of the hippocampus, and the arcuate nucleus of the hypothalamus [126]. Although DARPP-32 is a cytosolic protein, significant immunoreactivity for the protein is found in nuclei of MSNs, where it may regulate cAMP-dependent gene expression [21,127,128].

DARPP-32 Mediates cAMP- and cGMP-Dependent Signaling

The restricted distribution pattern of DARPP-32 and its overlap with dopamine innervation in the brain led to the hypothesis that DARPP-32 was an effector for dopamine signaling via PKA [121,127]. DARPP-32 is phosphorylated by PKA at a single threonine (Thr) residue at position 34 (Thr34) as the result of a cascade of biochemical events involving stimulation of dopamine D_1-type receptors by dopamine, activation of adenylyl cyclase, increased formation of cAMP, and activation of PKA [121,122]. Importantly, phosphorylation of DARPP-32 at Thr34 functionally converts the protein into a potent inhibitor of a major serine–threonine protein phosphatase, protein phosphatase-1 (PP-1) [129]. DARPP-32 is, in fact, structurally homologous to another PP-1 inhibitor protein, called protein phosphatase inhibitor-1 [130–133]. This 29 kDa cytosolic protein is converted to a PP-1 inhibitor (and is designated as PP1-R1B) upon phosphorylation within an N-terminal threonine sequence in a manner that is nearly identical to that occurring in DARPP-32.

DARPP-32 (like inhibitor-1) is phosphorylated and activated in response to both cAMP- and cGMP-dependent signaling pathways. Thr34 sites on both proteins are excellent substrates for the cGMP-dependent protein kinase G (PKG) [122]. Consistent with the high level of expression of soluble and particulate forms of guanylyl cyclase, cGMP, and PKG within striatum and substantia nigra [31–33,134], it is likely that cGMP-dependent signaling via DARPP-32 and inhibitor-1 plays a significant role in integrating the effects of multiple neurotransmitter systems with those of dopamine by impacting the phosphorylation state of these two PP-1 inhibitors. A striking illustration of this concept is the phosphorylation of DARPP-32 at Thr34 seen in slices of rat substantia nigra treated with NO generators to selectively increase cGMP levels (as opposed to cAMP) [135]. Thus, DARPP-32 and inhibitor-1 represent effectors for cAMP- and cGMP-dependent signaling with overlapping regional and cellular expression patterns.

Calcium-Dependent Phosphatase Activation Opposes Cyclic Nucleotide Signaling via DARPP-32

A major role for dopamine in the striatum is to modulate glutamatergic inputs from cortex and thalamus [29]. Glutamate, released from corticostriatal nerve terminals, has complex effects on striatal signaling cascades that impinge upon cAMP and cGMP cascades and upon DARPP-32 [136]. Glutamatergic activity in striatal interneurons, mediated by activation of NMDA-type receptors, also rapidly activates NOS, leading to generation of NO that diffuses into MSNs [40]. The resulting increase in soluble guanylyl cyclase and PKG activity caused by NO is responsible for transient increases in phosphorylation of DARPP-32 at Thr34 [136], consistent with the role of this protein as a biochemical sensor for cGMP-dependent activity [135]. In contrast, cyclic nucleotide signaling pathways in the striatum, including both cAMP- and cGMP-dependent cascades, are strongly opposed by calcium-dependent signaling pathways that effect the dephosphorylation of DARPP-32 through activation of calcium-dependent phosphatase activity. Persistent activation of NMDA-type receptors on MSNs is sufficient to increase calcium influx and activate the striatal-enriched, calcium-dependent phosphatase, protein phosphatase-2B (PP-2B) or calcineurin [137]. Activation of calcineurin rapidly dephosphorylates DARPP-32 at Thr34 [136,138–140], thereby inactivating it as a PP-1 inhibitor [129]. Activation of D_2-type dopamine receptors in striatum can also result in dephosphorylation of DARPP-32 at Thr34 that is mediated by increased calcium efflux and activation of calcineurin [141]. This effect is blocked by calcium removal or by treatment of slices with the selective calcineurin inhibitor, cyclosporin A. The data support the idea that D_2 receptor activation reduces cyclic nucleotide signaling via DARPP-32 pathways by multiple distinct mechanisms, including the inhibition of cAMP formation through adenylyl cyclase, activation of Gi and PI turnover, and increased calcium influx leading to activation of calcineurin-dependent dephosphorylation of DARPP-32 [141].

SPECIFIC PDE ISOFORMS REGULATE DOPAMINE SIGNALING BEHAVIORS

Tools for Studying PDE1B Regulation of Dopamine Signaling and Behavior

PDE1B is most abundantly expressed in striatum [49,50]. The enzyme is also expressed in neurons within cerebral cortex (layers 5 and 6) and hippocampus (dentate granule cells) that make up brain circuits governing memory and learning [50]. The anatomical distribution of PDE1B suggests a prominent role for this PDE in modulating dopamine's role in memory and motor movement [49]. The biochemical regulation of PDE1 enzymes has been studied *in vitro* [142]. PDE1 enzymes, including PDE1A–PDE1C isoforms, are uniquely regulated by calcium and calmodulin, as distinct from all other PDEs [142,143]. PDE1 activity, then, is activity-dependent, responding to changes in intracellular calcium level resulting from neuronal activity. As has been argued by Sharma and coworkers, PDE1 enzymes provide a means for crosstalk between cyclic nucleotide and calcium signaling pathways [143].

The study of PDE1B effects on biochemistry and behavior has been limited because of the lack of highly specific pharmacological inhibitors of the enzyme. Previously, vinpocetine, a vascular vasodilator, was employed to block calcium-dependent PDE activity in isolated tissue preparations, including primary dopamine neuron cultures [84]. The compound has multiple pharmacological actions, including the ability to block sodium channels, in addition to its effects on PDE1. Owing to its potent vascular effects, vinpocetine has mostly been used to study the function of PDE1 isoforms enriched in cardiac myocytes and peripheral vasculature [144,145]. Relatively few studies have employed vinpocetine as a means to study PDE1 activity in brain preparations. Because of the multiple pharmacological effects of this compound, caution should be exercised in the use of vinpocetine as a tool for characterizing PDE1B function. For example, vinpocetine has been reported to deplete dopamine levels in cultured striatal neurons [84]. These results are difficult to interpret given that PDE1B is not known to be expressed in dopamine-containing mesencephalic neurons. Thus, it is possible that channel blocking effects of the compound could mediate neurochemical effects independently of actions on PDE1 activity.

PDE1B Knockout Displays Enhanced Locomotor Stimulation by Dopamine Agonists

Much of what is known about PDE1B and its regulation of dopamine signaling and behavior has been obtained by analysis of PDE1B knockout mice. PDE1B has a cytoplasmic subcellular localization in brain as determined by subcellular fractionation techniques [49,50,69,70]. In normal mice, striatal PDE1B has been localized to DARPP-32-positive MSNs using fluorescent double-labeling with antibodies specific for PDE1B and those against DARPP-32 (A. Nishi and M. Kuroiwa, unpublished observations). All DARPP-32-positive MSNs detected in striatum also express immunoreactivity for PDE1B, indicating that the phosphodiesterase is expressed in both striatopallidal and striatonigral MSNs. The effects of PDE1B knockout on various aspects of monoamine neurochemistry and dopamine-mediated behaviors have been reported in mice by two groups using the same mouse model derived on a C57BL/6N background [146,147]. The major observation made by both groups is a significant increase in the locomotor activity response of PDE1B knockout mice to low doses of dopamine agonists, including methamphetamine, compared to wild-type mice. A trend toward a sex difference was also noted in both studies, as the effect of the knockout on locomotor responses was more consistently pronounced in male, compared with female, mice. The enhancement of agonist-induced hyperlocomotion is consistent with the activity-dependent nature of PDE1B regulation coupled with the enrichment of the enzyme in dopaminoceptive neurons in striatum. PDE1B knockout does not consistently or significantly affect spontaneous locomotor activity. Siuciak et al. reported a small increase in spontaneous locomotor activity in PDE1B knockout compared with wild-type mice [147]. Reed et al., in contrast, observed clear sex differences in the effect of gene knockout on activity levels. Female, but not male, knockout mice displayed a small but significant increase in spontaneous locomotor activity that was transiently expressed when placed in a novel environment, but diminished with habituation. Elevated locomotor activity in female knockout mice decreased to normal levels after habituation to a novel test chamber, suggesting that the knockout may elicit increased attention to novelty in female mice [146].

Biochemical Changes in PDE1B Knockout Mice

Changes in striatal neurochemistry after PDE1B knockout have been measured in an effort to characterize the biochemical basis for the behavioral phenotype. Two studies have reported a detailed analysis of dopamine and serotonin (5-HT) tissue levels and turnover rates in PDE1B knockout and wild-type mice, but have yielded conflicting results. Ehrman et al. reported no significant differences in striatal levels of dopamine or 5-HT or in levels of their major metabolites, suggesting that the knockout had no effect on monoamine turnover in striatum [148]. A subsequent analysis by Siuciak et al. reported a decrease in striatal tissue levels of dopamine such that the DOPAC/DA ratio (i.e., dopamine turnover) is significantly elevated. Striatal levels of 5-HT were decreased, with no change in 5-HT turnover noted [147]. Although both studies noted a decrease in 5-HT tissue levels (especially in female mice), the differences in dopamine turnover are difficult to resolve. It is possible that procedural variations, including differences in the number of generations of backcrossing in the mice used for the respective studies, could contribute to neurochemical variation. Because of the absence of PDE1B in dopamine-containing nerve terminals, it must be assumed that the increased turnover detected in the Siuciak study indicates the potential for a change in feedback control of nigrostriatal neurons mediated via striatal output neurons normally expressing PDE1B. As Reed et al. did not perform similar neurochemical measures, it is also difficult to relate possible differences in striatal dopamine turnover to the observed differences in spontaneous activity reported in these two studies.

PDE1B knockout mice show increased protein phosphorylation in striatum after dopamine agonist treatment. Striatal slices were prepared from PDE1B knockout mice and wild-type mice and treated with dopamine agonists *in vitro*. The basal level of phosphorylation found in striatum at PKA-dependent sites—Thr34 of DARPP-32 and Ser845 of GluR1—was unaffected by the loss of PDE1B expression [146]. The lack of effect of PDE1B knockout on the basal phosphorylation state of these dopamine and cAMP-dependent substrates is consistent with the lack of effect of PDE1B knockout on spontaneous locomotor activity in the same study [146]. In contrast, treatment of slices with the dopamine D_1-type receptor agonist, SKF81297, evoked a larger increase in the state of phosphorylation of both DARPP-32 and GluR1 in slices from knockout mice, compared with slices from wild-type littermates, an observation that is in line with the greater increase in methamphetamine-induced hyperlocomotion noted in knockout mice [146,147]. A similar increase in phosphorylation is also seen when slices are treated with forskolin, a direct activator of adenylyl cyclase. Forskolin-induced increases in Thr34 DARPP-32 and Ser845 GluR1 are larger in slices from knockout mice, relative to wild-type mice, suggesting further that the loss of PDE1B is unlikely to be affecting phosphorylation via unanticipated alterations in the sensitivity of D_1 receptors. However, the hyperlocomotion observed in PDE1B knockout mice after dopamine agonist treatment is clearly dependent upon DARPP-32. The importance of the DARPP-32 signaling pathway in the motor phenotype of PDE1B-deficient mice has been studied in mice bearing a double knockout of both PDE1B and DARPP-32 [148]. This study confirms the enhanced locomotor response of PDE1B knockout mice to methamphetamine treatment and demonstrates that this effect is lost in double-knockout ($PDE1B^{-/-} \times DARPP-32^{-/-}$) mice. These data support the biochemical data indicating an enhanced DARPP-32 phosphorylation response to PDE1B knockout. Furthermore, the absence of changes in basal phosphorylation state after PDE1B knockout is consistent with the activity-dependent recruitment of PDE1B in the regulation of dopamine signaling.

PDE1B Knockout and Memory Performance

PDE1B knockout results in a stimulation-dependent enhancement in motor activity. Less is known about the role of PDE1B in memory performance. Based on the prominent expression of PDE1B in hippocampal regions that mediate dopamine effects on spatial memory [149], PDE1B mutant and wild-type mice were tested for spatial performance in the Morris water maze paradigm [148]. PDE1B null mice display no difference from wild-type mice in time spent to locate the submerged platform or in total distance traveled during the test. A small, but significant increase in latency to learn the maze position was noted in PDE1B knockout mice under both the task acquisition phase and in a reversal paradigm [148]. Factors that can influence memory performance, including anxiety, are not affected by gene mutation [147,148]; PDE1B null mice perform normally in elevated plus maze and other anxiety screens.

PDE1B null mice also perform normally in other mouse behavioral models [147]. For example, null and wild-type mice have been evaluated in the conditioned avoidance response (CAR) paradigm. Animals learn to escape a mild electrical shock in one compartment of the test chamber by attending to an auditory or visual cue that precedes shock delivery. PDE1B null mice learn this paradigm comparably to wild-type littermates. Another learning paradigm, passive avoidance, also utilizes mild shock to motivate mice to learn an escape strategy. Knockout mice also perform this test as well as genetically normal mice, indicating that the PDE1B mouse does not display generalized deficits in learning or memory.

One factor to consider in evaluating the biochemistry and behavior of the PDE1B knockout mouse is the potential effect on function of the other PDE1 family members. As reviewed above, PDE1A and PDE1C isoforms are expressed within the same brain regions, and perhaps even the same cells, as PDE1B [51]. For example, mRNA for PDE1A–PDE1C isoforms is present within hippocampal dentate gyrus and in deep layers of the cortex—areas likely to subsume prominent roles in the memory tasks in which PDE1B null mice have been tested [150,151]. Thus, gene knockout of PDE1B alone may be insufficient to reveal the biological role of PDE1 enzymes in memory performance as a result of residual effects of PDE1A and PDE1C isoforms in hippocampus and striatum. The development of high-potency inhibitors of PDE1 enzymes with selectivity over other PDE enzyme families will be of great importance in deciphering the role of this family of enzymes in learning and memory. To this end, the recent discovery of selective and potent small-molecule inhibitors of PDE1 has led to the demonstration of unique cognitive enhancement in memory models, such as the novel object recognition paradigm (J. Prickaerts, L. Wennogle, and G.L. Snyder, unpublished observations).

PDE10A

PDE10A Regulates Basal Phosphorylation in Striatal MSNs

Immunohistochemical studies have localized striatal PDE10A to GABAergic MSNs [46]. The PDE10A2 splice variant of PDE10A, which is the predominant form expressed in striatum, is localized to either plasma membrane or cytosol of striatal neurons, based on the state of posttranslational modification of the enzyme mediated by palmitoylation and phosphorylation [76,77]. PDE10A is anticipated to possess high constitutive activity, consistent with a role in regulating basal levels of cAMP and cGMP in striatal neurons [48]. Consistent with this role, papaverine, a potent and reasonably selective inhibitor for the enzyme, elicits large (i.e., several-fold) increases in both cAMP and cGMP levels in the striatum of normal mice *in vivo* [152]. Similar increases in bulk cyclic nucleotide levels are seen with novel inhibitors for PDE10A with high specificity for this PDE isoform [153].

The impact of elevated striatal cyclic nucleotides on protein phosphorylation *in vivo* has been studied in mice and rats using focused microwave irradiation of the head as a means to "fix" phosphoprotein levels and avoid postmortem modification of phosphoproteins, as can occur during brain collection and dissection [154]. When paired with the use of multiple, high- affinity, phospho-specific antibodies, this procedure enables researchers to estimate the state of phosphorylation of phospho-proteins present in living animals during systemic drug administration. By monitoring the phosphorylation state of multiple phosphoproteins, it is possible to identify the biochemical cascades that are specifically regulated by distinct PDEs, including PDE10A, in striatum. Using these techniques, PDE10A activity is observed to exert a dominant role in setting the state of cyclic nucleotide-dependent protein phosphoryl-ation in striatal MSNs. Phosphorylation of DARPP-32 at Thr34 is strongly increased by papaverine *in vivo*, as is the phosphorylation state of other neuronal effectors that are substrates for regulation by DARPP-32 and are similarly enriched in striatal MSNs [86]. For example, papaverine treatment of rats *in vivo* causes a rapid and persistent increase in phosphorylation of cAMP response element binding (CREB) [86,152,153,155] at a serine residue (Ser133) that controls transcriptional activity [156]. PDE10A inhibitors also increase phosphorylation of the AMPA receptor subunit, GluR1, at Ser845 [86,155], a PKA site that is regulated by DARPP-32, via the control of dephosphorylation mediated via PP-1 [153,157]. The increases in phosphorylation of DARPP-32 and GluR1 seen *in vivo* are likely mediated by direct effects of PDE10A in striatum as they are replicated in mouse striatal slices incubated with inhibitors *in vitro*. Papaverine and novel, isoform-selective inhibitors of PDE10A (e.g., TP-10, MP-10) cause dose-dependent increases in Thr34-phosphorylated DARPP-32 and Ser845 GluR1, when incubated with mouse striatal slices *in vitro* [86,155].

PDE10A inhibitors such as papaverine [86,152] and the more selective and potent MP-10 [153] potently increase phosphorylation of proteins that are known to be enriched in striatal MSNs, using both *in vivo* and *in vitro* techniques. The phospho-rylation state of substrates such as CREB, which are likely expressed in MSNs and in interneurons and presynaptic locations in striatum, are also increased in response to PDE10A inhibitors [153,155]. Papaverine treatment of mice *in vivo* or of striatal slices *in vitro* was found to have no significant effect on the phosphorylation state of proteins with expression patterns restricted to presynaptic terminals [86]. For instance, neither phosphorylation of the synaptic vesicle-associated protein, synapsin, nor dopamine synthetic enzyme, tyrosine hydroxylase (TH) were altered by papaverine treatment. However, Grauer and coworkers have reported that PDE10A inhibitors with longer brain half-lives, such as MP-10, induce larger effects on striatal CREB phosphorylation, compared with papaverine treatment [155]. Thus, PDE10A inhibitors strongly regulate the phosphorylation state of striatal substrates whose expression is restricted to MSNs, such as DARPP-32. The impact of PDE10A inhibition on targets expressed in striatal nerve terminals and interneurons is a topic for ongoing investigation.

PDE10A Regulates Phosphorylation Preferentially in Striatopallidal Neurons

A recent study [86] employed bacterial artificial chromosome (BAC) technology [18,158,159] to investigate the cell type-specific expression of PDE10A in striatum. In this study, mice were engineered to express flag-tagged DARPP-32 under control of the D_1-receptor promoter in neurons comprising the striatonigral pathway, and myc-tagged DARPP-32 expressed under control of the D_2-receptor promoter in neurons comprising the striatopallidal pathway (Figure 12.2a) [86]. Fluorescent tags enable visualization of individual "direct" and "indirect" pathway neurons in these D_1-flag DARPP-32/D_2-myc-DARPP-32 mice enabling the study of coexpressed factors (using double-label techniques) and the biochemical evaluation (using serial immunoprecipitation with anti-flag and anti-myc antibodies) of the phosphorylation state of DARPP-32 expressed in these distinct cell types. Immunofluorescent detection of striatum from D_1-flag DARPP-32/D_2-myc-DARPP-32 mice demonstrates separate populations of flag (green)-tagged and myc (red)-tagged MSNs (Figure 12.2a). The effect of PDE10A inhibition in regulating phosphorylation in these two cell populations was studied by sequential immunoprecipitation of flag- and myc-tagged DARPP-32 from striatal slices treated *in vitro* with papaverine. Surprisingly, despite the apparently similar level of expression of PDE10A in the two cell populations, biochemical analysis reveals neuron-specific effects of papaverine on DARPP-32 phosphorylation primarily in striatopallidal neurons [86] (Figure 12.3). PDE10A inhibition increases Thr34 DARPP-32 phosphorylation several-fold in myc-tagged neurons, with only modest effects (less than twofold increase) seen in flag-tagged cells. By enhancing the Thr34 phosphorylation of DARPP-32 in myc-tagged, D_2 receptor-expressing striatal MSNs, PDE10A inhibitors mimic the biochemical actions of D_2 receptor antagonists, including most antipsychotic drugs [159]. D_2 antagonists increase levels of T34 phosphorylated DARPP-32 in striatal slices, an effect mediated by the loss of D_2 inhibition of adenylyl cyclase and by blockade of calcineurin-mediated dephosphorylation at T34 [141]. Recently, antipsychotic drugs, including typical (haloperidol) and atypical (clozapine) medications were shown to selectively increase phospho-Thr34 DARPP-32 levels in the D_2 receptor-expressing (myc-tagged) striatal neurons [159], an effect identical to that seen with PDE10A inhibitors [86,155].

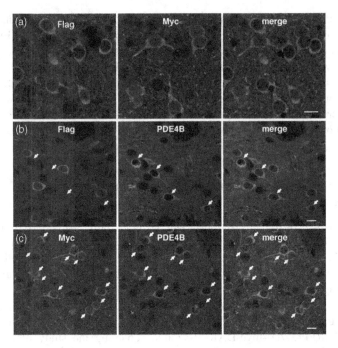

FIGURE 12.2 Photomicrograph showing high expression of PDE4B in Myc-positive, striatopallidal neurons in the striatum of D_1-DARPP-32-Flag/D_2-DARPP-32-Myc mutant mice. (a) Flag- and Myc-tagged DARPP-32 are expressed in striatonigral and striatopallidal neurons, respectively, in the striatum of D_1-DARPP-32-Flag/D_2-DARPP-32-Myc mutant mice. (b) Striatal sections from D_1-DARPP-32-Flag/D_2-DARPP-32-Myc mutant mice double labeled with antibodies against Flag and PDE4B. (c) Striatal sections from D_1-DARPP-32-Flag/D_2-DARPP-32-Myc mutant mice double labeled with antibodies against Myc and PDE4B. Striatal neurons showing strong PDE4B immunoreactivity, indicated by arrows, correspond to Myc-positive, striatopallidal neurons. Scale bars: 10 μm. (See the color version of the figure in Color Plates section.)

Relevant to the idea of selective activation of striatopallidal neurons, a recent study shows that PDE10A inhibition preferentially enhances the responsiveness of this subpopulation of MSNs to activation of cortical inputs. Threlfell et al. demonstrate that intrastriatal infusion of the PDE10A inhibitor, TP-10, increased activity of striatal neurons in response to stimulation of cortical afferents [160]. PDE10A inhibition did not alter spontaneous firing of striatal neurons, but did significantly increase the probability of firing after activation of corticostriatal inputs. Furthermore, the responsive cells were not activated by antidromic stimulation of the substantia nigra, implying that they were most likely striatopallidal neurons. These data support the idea that functional activation of striatopallidal neurons may predominate in the presence of PDE10A inhibition, owing to the special electrophysiological properties of D_2 receptor-containing neurons, and suggesting that a selective targeting of this PDE within D_2 receptor-containing neurons may be responsible.

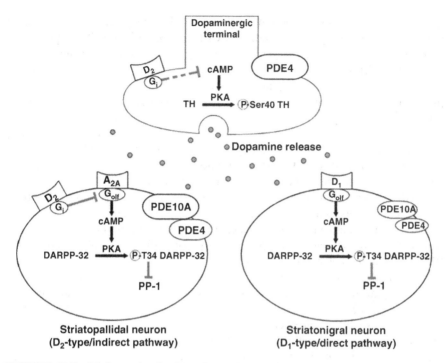

FIGURE 12.3 Distinct roles for PDE10A and PDE4 in striatal MSNs and in dopamine terminals. PDE10A is expressed in two types of striatal neurons: D_1 receptor-expressing striatonigral neurons and D_2 receptor-expressing striatopallidal neurons. Pharmacological inhibition of PDE10A by papaverine potentiates the adenosine A2A receptor-induced increase in DARPP-32 phosphorylation and opposes the dopamine D_2 receptor-induced decrease in DARPP-32 phosphorylation in striatopallidal neurons. In striatonigral neurons, papaverine potentiates the dopamine D_1 receptor-induced increase in DARPP-32 phosphorylation, whereas activation of mu-opiate receptors by enkephalin opposes the action of D_1 receptors on cAMP production. PDE4 influences protein phosphorylation predominantly at dopaminergic terminals, as inhibition of PDE4 by rolipram results in an increase in TH phosphorylation and dopamine synthesis. The inhibition of PDE4 also increases phosphorylation of DARPP-32 at Thr34, preferentially in striatopallidal neurons.

Inhibitors of PDE10A Mimic Antipsychotic Effects of D2 Receptor Antagonists

A defining feature of effective antipsychotic medications is the ability to block activity of dopamine D_2-type receptors in the striatum [5,161]. The efficacy of D_2 receptor blockade in the treatment of psychosis has led to the hypothesis that overactivity within dopamine D_2 pathways is a major contributing factor to the disease [162]. This view is supported by brain imaging studies that confirm elevated striatal dopamine release in nonmedicated schizophrenic patients [163,164]. Current hypotheses on the basis of psychoses also emphasize the role of aberrant glutamatergic neurotransmission in psychosis [165]. The blockade of NMDA-type glutamate receptors by phencyclidine (PCP), for example, results in psychotomimetic behavior in humans [166]. The D_2 antagonist-like biochemistry of PDE10A inhibitors may explain the "antipsychotic-like" behavioral activity of enzyme inhibitors and genetic knockout models. PDE10A inhibitors, including papaverine and the more selective MP-10, and mice bearing a knockout of the gene for PDE10A display similar behavioral phenotypes that resemble the effects of antipsychotic medication, including the reversal of sensorimotor gating deficits caused by the NMDA receptor antagonist, MK-801 [152,155,167]. Locomotor behavior is also affected by PDE10A inhibition or knockout in a manner consistent with blockade of D_2 receptor activity. PDE10A inhibitors block hyperlocomotion induced by psychostimulant drugs, such as amphetamine. Importantly, both PDE10A inhibitors and PDE10A gene knockout reduce spontaneous locomotor activity in mice and can result in catalepsy reminiscent of the motor side effects, which is a common feature of typical antipsychotic drugs like haloperidol [168]. PDE10A inhibitors potentiate the cataleptic effects of low doses of haloperidol and PDE10A knockout mice display an exaggerated sensitivity to the cataleptic actions of this drug. Thus, interference with PDE10A activity has been used to support a role for enzyme inhibitors as schizophrenia medication [153]. An important question, however, is whether the influence of PDE10A inhibitors on dopamine-regulated motor function, associated with effects in striatum, will complicate their utility in the control of the psychiatric symptoms of schizophrenia, associated with mesolimbic and mesocortical dopamine systems [1,2].

Current antipsychotic medications, like traditional D_2 receptor antagonists, are effective against positive symptoms of psychosis (e.g., hallucinations), but do not address the negative symptoms and cognitive dysfunction that are highly debilitating core features of the disease [169–171]. There is evidence that inhibition of PDE10A can reverse behaviors in rodents that model these features of psychosis. Rodefer et al. [167] have reported that deficits in executive function (i.e., attentional set shifting) induced by subchronic PCP treatment in rats are attenuated by treatment with papaverine. Papaverine treatment also enhanced memory performance in rats, as measured in the novel object recognition (NOR) task [155]. Memory was enhanced by papaverine across a broad range of doses and at dose levels effective in antipsychotic screens. Papaverine also improved memory performance in mice, as assayed in the social odor recognition (SOR) paradigm [155]. Recently, a novel and more potent PDE10A inhibitor, THPP-1, was shown to enhance performance of rats in the NOR paradigm [168]. THPP-1 also proved effective in reversing ketamine-induced memory deficits in an object retrieval detour task in nonhuman primates (NHPs). Interestingly, in contrast to the effects of papaverine and THPP-1, the PDE10A inhibitor, MP-10, tested at a range of doses, did not significantly alter memory performance in NOR and had only modest positive effects in SOR [155]. The basis for the observed differences in effects of PDE10A inhibitors of varying potency and selectivity requires further investigation.

PDE4

PDE4 Regulates Signaling in Dopamine Terminals and in Striatopallidal MSNs

The impact of PDE4 inhibition on dopamine signaling has been studied using inhibitors like rolipram that block activation of all four of the PDE4 subfamilies [172]. Rolipram increases release of dopamine *in vivo* (see above) [85], as well as from isolated striatal tissue and primary cultures of dopamine neurons. Rolipram also increases the turnover of dopamine, consistent with the effects of cAMP on neurotransmitter release and supporting a presynaptic site of action in striatum [84,173]. These observations are consistent with the distribution of PDE4 isoforms, in particular PDE4B and PDE4D [57]. The data are also supported by the results of a recent report, showing that treatment of mice *in vivo* with rolipram increases the phosphorylation state of striatal TH at Ser40 [86]. The regulation of striatal TH by the PDE4 inhibitor is likely mediated via local effects on striatal dopamine terminals because the regulation of TH is recapitulated by rolipram treatment of striatal slices *in vitro*. In contrast to previous studies performed in rats [174,175], rolipram treatment alone was insufficient to increase striatal dopamine turnover in mice *in vivo*, as measured by the ratio of the major dopamine metabolites, 3,4-dihydroxy-phenylacetic acid (DOPAC) and homovanillic acid (HVA), relative to tissue dopamine content. However, rolipram treatment did significantly potentiate the increase in the DOPAC-to-dopamine ratio seen after treatment of mice with haloperidol [86], a drug that increases dopamine metabolism by blocking presynaptic, release-modulating autoreceptors [175]. Interestingly, rolipram treatment selectively potentiated haloperidol-induced increases in DOPAC, the product of dopamine metabolism by monoamine oxidase (MAO) occurring primarily in dopamine terminals. Increases in HVA, the product of dopamine metabolism via catechol-O-methyltransferase (COMT), occurring primarily in extracellular locations, was increased by haloperidol, but not further altered by rolipram treatment. These data suggest that rolipram preferentially enhances the effects of D_2 receptor blockade on dopamine synthesis, rather than dopamine release in striatum. Thus, the presynaptic effect of rolipram on dopamine metabolism indicates a role for PDE4 in setting dopamine tone in striatum.

In contrast to the strong regulation of TH phosphorylation elicited by rolipram treatment, phosphoproteins restricted to or enriched in striatal MSNs are only weakly affected by rolipram treatment *in vitro* or *in vivo* [57]. For example, the phosphorylation state of Thr34 DARPP-32 and Ser845 GluR1 is only modestly affected by treatment of striatal slices with dose levels of rolipram that elevate TH phosphorylation [57]. Furthermore, the functional impact of PDE4 inhibition appears to predominate in striatopallidal MSNs. PDE4 inhibition fails to potentiate cAMP-dependent signaling in striatal slices if stimulated by a D_1 receptor agonist, but does significantly elevate phosphorylation evoked by an adenosine A2A receptor agonist. These data are supported by studies using D_1-flag/D_2-myc DARPP-32 tagged mice in which rolipram treatment increases Thr34 DARPP-32 phosphorylation associated with D_2-myc tag (Figure 12.2b and c) [86]. These data, then, argue for PDE4 as a regulator of dopamine protein phosphorylation predominantly in striatopallidal neurons, similar to that seen for PDE10A, based on the effects of papaverine. Collectively, the data support a role for PDE4, distinct from PDE10A, in the regulation of presynaptic dopamine activity, with a shared role for both PDE4 and PDE10A in the modulation of dopamine signaling in D_2 receptor-enriching MSNs. Although PDE4 immunoreactivity is relatively stronger in striatopallidal neurons, there appears to be no strong anatomical basis for functional segregation of PDE4 effects in this subpopulation of MSNs (Figure 12.2b and c and Figure 12.3).

Inhibitors of PDE4 Suppress Dopamine-Mediated Behaviors Relevant to Psychosis

The striking biochemical effects seen in striatum after PDE4 inhibition with rolipram resemble the effects of dopamine D_2 receptor antagonists, including both an increase in dopamine synthesis in nigrostriatal dopamine terminals and the preferential regulation of dopamine signaling in D_2 receptor-expressing striatopallidal MSNs. Rolipram's effects on striatopallidal neurons, in fact, are similar to those seen with PDE10A inhibitors and mimic the behavioral profile of an antipsychotic drug [176]. For instance, rolipram treatment blocks the hyperlocomotion caused either by dopamine overactivity (i.e., amphetamine) or by NMDA receptor blockade (i.e., PCP) and reverses deficits in sensorimotor processing produced by amphetamine treatment, as measured in the prepulse inhibition paradigm [177]. The PDE4 inhibitor also suppressed responding in the CAR paradigm [176], a model that is particularly sensitive to pharmacological effects of D_2 antagonists [178]. The suppression of hyperlocomotion and avoidance responding in the various antipsychotic behavioral screens can result from drug effects on motor performance. This does not appear to account for the behavioral effects of rolipram, as the drug produces only modest reductions in spontaneous motor activity at dose levels far exceeding those used in CAR and hyperlocomotion assays [176]. These data support the hypothesis that PDE4 inhibition normalizes aberrant dopamine and glutamate neurotransmission to address symptoms of psychosis with minimal effect on movement. As with PDE10A inhibitors, the relative lack of motor interference found with PDE inhibition compared with dopamine receptor blockade, despite profound changes in dopamine signaling pathways, is an attractive, yet poorly understood, effect of PDE inhibitors.

Severely impaired working memory function is a core feature of schizophrenia, which is not effectively treated by current antipsychotic medications, including compounds with potent D_2 receptor antagonist activity. Cortical circuits under control of dopamine D_1-type receptors have been shown to promote normal working memory function [179–182]. PDE4 isoforms, including PDE4B and PDE4D, are expressed in cortical pyramidal cells, such as those that receive dopamine inputs, and in cell populations in hippocampus that are essential for normal spatial memory function [57]. In cortex, recent immunohistochemical studies localize PDE4B to cortical neurons that express DARPP-32 and demonstrate large increases in DARPP-32 Thr34 phosphorylation in response to rolipram. These observations indicate that PDE4 exerts strong biochemical control over D_1/cAMP activity in cortex, supporting a biochemical enhancement of D_1 signaling by PDE4 [183]. The relationship between the level of activity within D_1 receptor/PKA-dependent pathways and working memory performance is complex. Current data suggest an "inverted-U" response where prefrontal memory function is impaired by abnormal increases or decreases in cAMP activity [184–186]. Nonetheless, impaired working memory can be improved by activation of the D_1 receptor/cAMP pathway within prefrontal cortex [187]. Age-dependent deficits in spatial memory function in rats are also attenuated by increases in dopamine D_1 activity in the hippocampus [149]. Inhibition of all PDE4 isoforms with rolipram mimics the effect of a D_1 agonist, repairing hippocampal long-term potentiation (LTP) deficits and spatial memory loss in aged animals [149,187]. These data support the concept that PDE4 enzyme families modulate D_1 receptor-dependent memory circuits in cortex and hippocampus that are impaired by aging and disease. This quality distinguishes the approach of using PDE4 inhibitors from that of using traditional antipsychotic drugs that often further impair cognition [188].

PDE4B Interactions with DISC1

PDE4 inhibitors have biochemical and behavioral effects similar to those of dopamine D_2 receptor antagonists, supporting a potential role for PDE4 in psychosis. The link between the PDE4B isoform and schizophrenia is further supported by its association with DISC1, a 100 kDa adapter protein that is encoded by a major schizophrenia susceptibility gene called *disrupted in schizophrenia 1 (DISC1)*. *DISC1* was first identified as a factor cosegregating with psychiatric illnesses in the pedigree of a large Scottish family [189]. The DISC1 protein was subsequently shown to directly bind PDE4B to regulate cAMP signaling. Available data support the idea that DISC1 functions as a targeting protein for PDEs analogous to the well-known targeting proteins for PKA, called AKAPs [80,190]. DISC1 may serve to sequester PDE4 enzymes, and thus, control their access to cyclic nucleotides within specific cellular compartments to effect the regulation of subcellular actions of cAMP in neurons [79,82,190]. *DISC1*, its interactions with PDE4 enzymes, and its possible role in psychosis are discussed in detail in Chapter 8.

Gene Knockout of PDE4 Isoforms: Biochemical and Behavioral Outcomes

As discussed above, nonselective PDE4 inhibitors, like rolipram, mimic many of the biochemical and behavioral effects of dopamine D_2 antagonists and D_1 agonists. In the absence of pharmacological inhibitors with selectivity for PDE4 subfamilies, gene knockout mice have been used in an effort to attribute specific behavioral roles to individual PDE4 enzymes. Unfortunately, current data from the available gene knockouts fail to recapitulate the pharmacological profile of rolipram. To date, knockout mouse lines for the two prominent striatal PDE4 isoforms, PDE4B and PDE4D, have been generated and characterized [191–193]. Studies with the PDE4B knockout mouse have focused on the anticipated role of this enzyme in dopamine signaling and psychosis, as defined by the pharmacological effects of rolipram.

Surprisingly, PDE4B gene knockout and rolipram elicit opposing phenotypes in a number of behavioral models. For example, knockout mice display impaired PPI responses, and enhanced locomotor responses to amphetamine that are accompanied by reduced tissue levels and turnover of dopamine and 5-HT [191]. These findings suggest that a downregulation of monoamines occurs in PDE4B-deficient mice, compared with increased dopamine turnover after PDE4 inhibition [84]. The surprising "propsychotic" phenotypes seen in PDE4B knockout mice are similar to the reported effect of certain DISC1 missense mutations. For example, the L100P mutation, induced in mice using the technique of *N*-nitroso-*N*-ethylurea (ENU) mutagenesis [194], results in the loss of DISC1 binding to PDE4B [195]. Mice expressing DISC1 bearing the L100P mutation display impaired PPI responses indicative of a "propsychotic" phenotype, which was partially reversed by either atypical (clozapine) or typical (haloperidol) antipsychotic drugs. Impaired PPI in DISC1 mutant mice is fully restored to normal levels by treatment with the PDE4 inhibitor, rolipram [195]. In a separate study, PDE4B knockout mice displayed anxiogenic behaviors [192], not seen in DISC1 mutant mice [195]. PDE4D knockout mice exhibit deficits in LTP, rather than the enhanced LTP seen after rolipram treatment in wild-type mice [193]. Thus, to date, gene knockouts of individual PDE family enzymes do not adequately explain the procognitive and antipsychotic profile of the PDE4 inhibitor, rolipram. Rather, they indicate opposing effects of inactivation of the enzymes using pharmacological tools and genetic approaches that abolish expression of single PDE4 enzyme isoforms. Together with data from DISC1 mutant mice, the results suggest that it is the selective loss of normal PDE4B activity mediated either by PDE4B gene knockout, or loss of its normal PDE4B sequestration by adaptor proteins such as DISC1, which reveals psychosis-like behaviors.

OTHER PDE ISOFORMS WITH EMERGING ROLES IN VOLITIONAL BEHAVIOR

Recent studies have highlighted brain distribution patterns for several PDE isoforms, including 9A, 2A, and 7B, within regions controlling motivated behaviors and movement, including basal ganglia, cortex, and hippocampus [54,56,60,65,96]. Considerably less is known regarding the roles of these PDE isoforms in volitional behavior and their interactions with dopamine signaling cascades. The recent introduction of isoform-selective inhibitors is facilitating the study of these isoforms. Here, we briefly review available data on the role of PDE2A and PDE9A in the regulation of cyclic nucleotide signaling and motivated behaviors.

Role of PDE2A in Cyclic Nucleotide Signaling

In addition to its interesting brain distribution (see Chapter 2), PDE2A has been studied as a drug target for cardiac and inflammatory indications, based on its presence in cardiac myocytes and endothelial cells [88,89,196]. Early studies investigated brain and peripheral roles of this PDE using low-affinity, isoform-selective inhibitors such as EHNA [197]. Biochemical studies in primary cortical neuron cultures show that inhibition of PDE2A activity with EHNA significantly increases bulk cGMP levels. Moreover, comparison with other isoform-preferring inhibitors indicates a significant role for PDE2A in control of cGMP levels, compared to inhibitors of other cGMP PDEs [198]. The use of EHNA as a PDE2A inhibitor is complicated by the activity of this compound as an inhibitor of adenosine deaminase, an enzyme that prevents the breakdown of adenosine. Thus, effects of EHNA on adenosine availability could well influence cyclic nucleotide levels as both a PDE inhibitor and as an adenosine agonist (especially in striatopallidal MSNs), making data interpretation difficult. The introduction of PDE2A inhibitors, like BAY 60-7550, with greater potency and fewer off-target effects, has confirmed a role for PDE2A in regulating cGMP and cAMP levels in primary neurons from cortex and hippocampus [199]. Importantly, BAY 60-7550 has little effect on basal levels of cyclic nucleotides in cultured neurons. Rather, effects of the PDE2A inhibitor were revealed only in the presence of stimulation with a cGMP agonist [199]. Similarly, cAMP elevations are seen only in the presence of stimulation using high levels of the adenylyl cyclase activator, forskolin. These data indicate a preferential role for this PDE in the regulation of cGMP, and, to a lesser degree, cAMP, only under stimulated conditions. This interpretation is consistent with the rather low affinity of PDE2A for cGMP ($10 \mu M$ range) [88], compared with cGMP-selective PDEs (nanomolar range), like PDE9A [64]. The data also are consistent with the observation that cAMP breakdown by PDE2A is cGMP-stimulated [88]. Although they did not report the data, Boess et al. [199] also indicate that BAY 60-7550 increases cGMP immunoreactivity in other brain regions, including striatum, indicating that PDE2A may significantly impact motivated behavior. At the present time it is unclear whether PDE2A, like PDE10A, PDE1B, and PDE4, affects dopamine system function.

Role of PDE9A in the Regulation of cGMP Levels and Memory Performance

PDE9A is a high-affinity, cGMP-selective PDE with abundance in hippocampus, cortex, and striatum [65]. Several agents are recognized as weak and relatively nonselective inhibitors of PDE9A, including zaprinast, sildenafil, SCH 51866, and vardenafil, all with IC_{50} values in the micromolar concentration range. The recent introduction of several brain-permeant, high-affinity, and selective inhibitors of this enzyme has exponentially advanced research into the biochemical and behavioral roles of PDE9A (see Chapters 4 and 5) [200–202]. The first of these recent compounds, BAY 73-6691, possesses nanomolar potency for inhibition of the enzyme and modest (25–75-fold) selectivity against other PDE isoforms. Using a cGMP reporter cell line expressing recombinant PDE9A [200], BAY 73-6691 was found to have no effect on basal cGMP levels but did elicit dose-dependent increases in cGMP concentration when paired with submaximal concentrations of the activator of soluble guanylate cyclase BAY 58-2667. No studies have reported the effect of BAY 73-6691 on cGMP levels in the brain. However, in functional studies, BAY 73-6691 influences cellular correlates of memory, as measured by LTP responses in the

CA1 region of hippocampus evoked by activation of Schaeffer collaterals. PDE9A inhibition with BAY 73-6691 had no effect on basal synaptic transmission in hippocampal slices from young rats of either Wistar or FBNF1 strains, but did preferentially increase both basal synaptic transmission and LTP in hippocampus from aged FBNF1 rats that display a reduced capacity for LTP [203]. The positive effects of BAY 73-6691 in this model are consistent with the reported role for cGMP in LTP and indicate that cGMP and PKG pathways are capable of restoring cellular plasticity that underlies certain forms of learning [204]. Furthermore, the data are consistent with the idea that PDE9A may preferentially support cGMP-dependent signaling under conditions in which neurotransmission is impaired by age or disease.

The effect of BAY 73-6691 on LTP suggests a preferential role of this PDE in addressing the age-related declines in cellular communication. The behavioral effects of this molecule also support an efficacy of selective inhibitors in alleviating cognitive decline in aging or age-related diseases [103]. In parallel with the enhancement of hippocampal LTP, inhibition of PDE9A by BAY 73-6691 partially restores normal memory performance in two pharmacological models. The muscarinic cholinergic antagonist, scopolamine was used to impair memory performance of C57BL/6 mice in the passive avoidance paradigm, a model for cholinergic neuron loss relevant in cognitive decline in aging and in Alzheimer's disease (AD). NMDA antagonist treatment was used to induce memory deficits in the T-maze alternation task. Treatment with BAY 73-6691 significantly repaired performance in both tasks [203].

Recently a new family of PDE9A inhibitors possessing greater isoform selectivity and excellent CNS availability has been reported [201,202]. Studies with these compounds largely confirm and extend the results obtained with the earlier BAY 73-6991 compound. One point of difference, compared with early studies of BAY 73-6991, is the observation of enhanced basal level of cGMP in brain and cerebrospinal fluid (CSF) after systemic administration of the new Pfizer PDE9A inhibitors, namely PF-4181366 and PF-04447943 [202,205,206]. The new data support the idea that PDE9A plays a role in controlling basal levels of cGMP in the brain that may explain the observed behavioral effects of inhibitors. The new inhibitors exert positive effects on cognitive performance in rodents tested in a variety of paradigms. For example, PDE9A inhibitors enhance memory performance in normal animals and reverse scopolamine-induced memory deficits in the NOR paradigm [205,206]. Deficits in Morris water maze performance produced by ketamine treatment are also attenuated by PDE9A inhibition [205]. The ability of PDE9A inhibition to address cognitive impairment induced by disruption of both cholinergic (i.e., scopolamine) and glutamatergic (i.e., ketamine) neurotransmitter systems suggests wide-ranging effects of this PDE isoform on circuits involved in memory and learning [206]. The memory effects of PDE9A inhibition may be related to ability of increased cGMP levels [207] to prevent deficits in LTP produced by amyloid beta protein [204]. In fact, the loss of hippocampal dendritic spine density observed in the amyloid precursor protein (APP)-overexpressing Tg2576 mouse model is blocked by treatment with PDE9A inhibitors [202,208]. The robust effects of PDE9A inhibitors in cognition-impaired animals have been critical for the advancement of PF-04447943 into clinical testing as a cognitive therapy in AD patients.

PDE ISOFORMS, DOPAMINE SIGNALING, AND DISEASE: IMPLICATIONS FOR TREATMENT

There is a high degree of overlap between expression of PDE isoforms and dopamine signaling pathways that mediate motor movement and motivated behaviors. Moreover, the use of new genetic models and novel pharmacological inhibitors has shown that many of the prominent brain PDEs directly impact cyclic nucleotide-dependent dopamine signaling pathways in a region- and cell type-specific manner. Certain PDE isoforms, then, represent targets for novel pharmaceutical approaches to a wide array of neuropsychiatric and neurodegenerative diseases that affect dopamine neurons and their target cells, including schizophrenia, Parkinson's disease, Huntington's disease, and depression. Although the role of PDE isoforms in brain and behavioral disease is the subject of other chapters in this book (see Chapters 7–11), we review briefly, here, the potential utility of PDE inhibitors for the treatment of the dopamine-related neurodegenerative disease, Parkinson's disease.

Parkinson's Disease and Symptomatic Treatment

PD develops as a result of a progressive degeneration of dopamine-containing neurons in the brain and is most often characterized by motor deficits, notably bradykinesia, limb rigidity, and resting tremor. Dopamine depletion—most notable as the degeneration of nigrostriatal dopamine neurons—is considered the primary cause for the loss of volitional movement in Parkinson's disease and may contribute to the associated nonmotor symptoms of the disease, including cognitive dysfunction, loss of affect, and depression [3]. Although mutations in specific genes (e.g., LRRK2, α-synuclein, PINK-1) appear responsible for certain familial forms of the disease [209], in the vast majority (>95%) of patients, the cause of PD is unknown. There is no cure for PD. Rather, the disease is currently addressed with symptomatic treatments that temporarily restore motor function, such as L-DOPA (levodopa), the immediate precursor for dopamine synthesis [210]. Unfortunately, L-DOPA therapy becomes ineffective in treating motor disabilities after chronic use. The period of time for effective maintenance of motor disability, or "on" time, lessens after repeated drug exposure, and uncontrollable, involuntary movements, known as L-DOPA-induced dyskinesias (LIDs), emerge [211,212].

L-DOPA Effects on Disease Progression and Motor Side Effects

The biological basis for the loss of L-DOPA effectiveness is controversial; several factors may contribute, including disease progression (i.e., reduced capacity for L-DOPA conversion to dopamine), abnormal receptor changes resulting from phasic L-DOPA dosing, or toxic effects of L-DOPA or its metabolites that promote the neurodegenerative process. Despite the lack of agreement regarding the precise biological basis for LIDs, there is reason to believe that the use of adjunctive or alternate first-line therapies that delay the introduction of L-DOPA therapy or reduce the required dose of L-DOPA can positively affect the course of the disease and the appearance of motor side effects [213]. For example, the use of dopamine receptor agonists (e.g., pramipexole, ropinirole) is reported to slow disease progression [214]. Although not as effective as L-DOPA in moderate to severe PD, these agents have become standard effective treatments for early-stage disease, and appear associated with fewer drug-induced dyskinesias [215]. Two notable clinical trials (REAL-PET and CALM-PD-CIT) have compared dopamine agonists to L-DOPA treatments for effects on disease progression; patients receiving dopamine agonists showed a reduced rate of biomarker decline, indicating a slower disease progression [216,217]. Although complementary *in vitro* preclinical studies indicate that dopamine agonists may protect by exerting antiapoptotic effects directly on dopamine neurons (e.g., via autoreceptors) [214], dopamine agonists act at postsynaptic sites (downstream of actions on dopamine neurons) to normalize dopamine activity within the basal ganglia motor system and slow disease progression. The strategy of normalizing motor symptoms to slow disease progression by intervening at points distant from the affected dopamine neurons is the basis for several novel therapeutic approaches. For instance, existing PD therapies, including subthalamic nucleus surgery and early-stage pharmaceutical programs (e.g., mGluR receptor ligands; NMDA receptor antagonists), manipulate targets (e.g., glutamate and glutamate receptors) down-stream of/convergent with dopamine terminals to provide symptomatic relief and protective effects in PD.

A second benefit of reducing L-DOPA dosing early in PD is a lower incidence of LIDs. As demonstrated by the ELLDOPA study, motor complications arising after prolonged L-DOPA treatment, including L-DOPA-induced dyskinesias and wearing-off, are dose dependent. Patients who received higher maintenance doses of L-DOPA (e.g., 200 mg t.i.d.) displayed more severe motor side effects than patients receiving a lower dose level (e.g., 50 mg t.i.d). In addition, those patients receiving the highest maintenance doses of L-DOPA developed dyskinesias more rapidly than other patients, sometimes within months after initiating drug treatment [218]. Once established, few therapies are available to treat LIDs. At present, the only medication recognized by the FDA for efficacy in the treatment of LIDs is amantadine [219], a mixed-action drug that produces a modest attenuation of LIDs in some patients via molecular mechanisms that are unclear, but which likely involve NMDA receptor blockade and dopamine agonist activities [210]. Based on these data, a major avenue of PD therapy is the discovery of standalone or adjunctive therapies that will increase "on" time, enable the use of lower maintenance doses of L-DOPA, and prolong the useful lifetime of PD therapy, delaying or preventing the appearance of motor fluctuations and LIDs.

PDEs as Novel Targets for PD Therapy

PDE1B represents a potentially interesting target for addressing the motor symptoms of PD. This is based, in part, upon the high levels of expression and enrichment of the PDE1B isoform in basal ganglia, and the demonstrated actions in enhancing dopamine-mediated cAMP signaling at the biochemical and behavioral level. For instance, PDE1B is an attractive candidate based on its striatal enrichment [49,50]. Furthermore, PDE1B gene knockout amplifies dopamine signaling via D_1 receptor pathways and motor activity stimulated by low-level dopamine agonist administration [146–148]. The data support the idea that PDE1B inhibition enhances dopamine signaling in a stimulus-bound manner, as knockout mice display no significant change in basal protein phosphorylation and negligible changes in basal locomotor activity. This quality is consistent with the unique regulatory properties of the PDE1 family of enzymes. As all three family members are stimulated by calcium/calmodulin, their activity is likely controlled by neuronal activity. This property confers an "on demand" quality to PDE1 activity that would be anticipated to provide phasic amplification of dopamine signaling contingent upon stimulation of MSNs by endogenous factors. The "on demand" activity of PDE1B may be a superior attribute as a drug target compared with dopamine agonists that tonically activate dopamine receptors. The tonic activation of receptors by agonists is one factor that results in dopamine receptor changes that contribute to the development of motor fluctuations [214]. The utility of PDE1B inhibitors as PD therapeutics needs to be evaluated with potent and selective inhibitors of the enzyme. The recent discovery of potent and selective PDE1 inhibitors indicates that these molecules can amplify low-dose L-DOPA effects in animal models of PD, an observation, which is consistent with the hypothesized role of PDE1B in the maintenance of normal striatal dopamine tone (P. Li, G.L. Snyder, and L. Wennogle, unpublished observations).

PDE4 enzymes have also been proposed as drug targets of interest for PD [84]. PDE4 family members, including PDE4B, are expressed in nigrostriatal dopamine terminals and in striatal MSNs, in proximity to presynaptic and postsynaptic dopamine signaling machinery. Pharmacological inhibition of PDE4 with rolipram effectively increases dopamine synthesis in cultured mesencephalic dopamine neurons [84]. This effect is consistent with the ability of rolipram to increase cAMP levels in these neurons, leading to phosphorylation of TH at a site (S40) that catalyzes dopamine synthesis. These data are supported by *in vivo* studies that demonstrate increases in TH phosphorylation state and dopamine turnover in response to rolipram treatment [86]. PDE4 inhibitors also exhibit antidepressant and procognitive effects in animal models [111,112,117,186,187,220] that would address nonmotor symptoms of PD that are largely unresponsive to L-DOPA pharmacotherapy [221]. The ability of PDE4 inhibitors to drive dopamine synthesis and release and provide nonmotor support suggest they might be useful in addressing early- stage PD.

The positive pharmacological effects of PDE4 inhibitors, however, are complicated by postsynaptic effects that mimic the actions of dopamine D_2 receptor antagonists. D_2 receptor blockers, such as the antipsychotic medication haloperidol, produce profound catalepsy in animals and result in extrapyramidal motor symptoms and tardive dyskinesia [222]. Dopamine receptor antagonists can interfere with the restoration of motor activity by L-DOPA in animal models of PD [223,224]. The PDE4 inhibitor, rolipram, preferentially increases Thr34 DARPP-32 phosphorylation in striatopallidal neurons [86], an effect characteristic of D_2 receptor blockers such as haloperidol, and shared with PDE10A inhibitors, such as papaverine [152].

Furthermore, inhibition of either PDE10A or PDE4 modestly reduces motor activity in mice and rats and potentiates the catalepsy produced by neuroleptic drugs [152,153,176,177]. It is unclear whether the motor effects of these PDE inhibitors will be as debilitating in humans.

Despite concerns that D_2 antagonist properties of PDE4 or PDE10A inhibitors could further compromise motor activity, mild D_2 antagonist activity provided by these agents might be of value for the management of LIDs. Dopamine receptor sensitization that results from the loss of striatal dopamine innervation and the effects of dopamine replacement therapy likely contributes significantly to the development of LIDs [225]. Thus, mild D_2 antagonist-like activity that would normalize dopamine receptor responses to L-DOPA might delay the onset or lessen the severity of motor responses to replacement therapy. D_2 receptor antagonists have been found to effectively suppress the appearance of specific behaviors that are analogous to human dyskinesias, including axial, limb and orolingual movements, without significantly comprising L-DOPA effects on spontaneous motor activity [226,227]. Antagonists of specific receptors within the D_2 family, like the D_3-type dopamine receptor, have been shown to suppress LIDs, supporting the idea that molecules with D_2 antagonist-like activity may be useful antidyskinetic agents [228]. Clinically, several studies support the efficacy of atypical antipsychotic drugs, such as clozapine and aripiprazole, for the control of LIDs. The positive effects of these drugs may owe to their complex pharmacology, incorporating several different receptor activities that may contribute to their positive motor effects. For example, both of these antipsychotic medications possess potent activity as 5-HT2A antagonists and 5-HT1A agonists, with mixed actions as modest D_2 antagonists with partial D_2 receptor agonism [229,230]. Selective 5-HT2A antagonists and 5-HT1A agonists can reduce dyskinetic behaviors in rodent models of PD and in PD patients [231–234]. Thus, the interesting dopamine signaling effects of PDE4 and PDE10A inhibitors may warrant their consideration for neurological indications such as LIDs. Abnormal orolingual movements induced by chronic neuroleptic drug treatment, a model for dyskinetic behaviors seen after L-DOPA, are attenuated by rolipram treatment [235].

PDE7B is also being pursued as a potential target for PD therapy development. To date, only a few studies have been published regarding the expression pattern and regulation of this PDE isoform, though available data demonstrate its expression in striatal MSNs and its translational regulation under control of the dopamine D_1 receptor [78]. Recently, double *in situ* hybridization analysis has shown that the PDE7B signal localizes to D_2 receptor-containing striatal neurons [236], further supporting the potential significance as a target for therapy development in PD. A link between familial PD genes and PDE7B is also supported by a recent study in mice overexpressing mutant A53T-α synuclein [237]. Mutant α-synuclein, which was associated with reduced levels of several indices for striatal dopamine signaling, was found to negatively regulate striatal gene expression, reducing transcription of several genes, including the gene encoding PDE7B. High-affinity inhibitors of this PDE are currently being investigated for motor benefit in animal models of PD (http://www.michaeljfox.org/).

Cyclic GMP and Corticostriatal Correlates of Dyskinesia: Possible Role for PDE Inhibitors

The potential significance of PDE inhibitors and, in particular, inhibitors for cGMP-preferring or cGMP-specific PDEs, is highlighted by recent research on the molecular basis of dyskinesia in PD. These studies have identified a dysfunction in striatal cGMP signaling as an electrophysiological correlate of LIDs in animals. Work by Calabresi and coworkers has identified a deficit in long-term depression (LTD) of striatal responses to high-frequency stimulation (HFS) of corticostriatal slices in dopamine-depleted animals displaying dyskinesia after chronic L-DOPA treatment [238]. Rats depleted of striatal dopamine with the neurotoxin 6-OHDA received repeated daily doses of L-DOPA that resulted in a subset of animals that developed and expressed LIDs and a subset of rats that remained nondyskinetic under similar treatment conditions. Corticostriatal slices from dyskinetic rats were distinguishable from those of nondyskinetic animals based on the loss of LTD responses. Thus, the absence of LTD provides an electrophysiological correlate for dyskinesia [238]. Rats with established LIDs expressed lower striatal levels of cAMP and cGMP compared with normal animals. Treatment of these animals with agents such as zaprinast, an inhibitor of cGMP-preferring PDEs, partially restores striatal cyclic nucleotides and attenuates dyskinesias [239]. Zaprinast and other cGMP-elevating treatments further induce LTD responses in corticostriatal slices [240,241]. Together, these data indicate that LIDs are associated with abnormal corticostriatal plasticity that results from deficits in cGMP levels. Thus, PDE inhibitors that elevate striatal cGMP levels and signaling are potentially beneficial for attenuating LIDs. These data support a possible therapeutic role for inhibitors of several striatal-enriched PDEs, including PDE1B and PDE9A, in this indication.

REFERENCES

1. Palmiter, R.D. (2008) Dopamine signaling in the dorsal striatum is essential for motivated behaviors: lessons from dopamine-deficient mice. *Ann. N.Y. Acad. Sci.*, **1129**:35–46.

2. Wise, R.A. (2004) Dopamine, learning and motivation. *Nat. Rev. Neurosci.*, **5**:1–12.

3. Hornykiewicz, O. (1966) Dopamine (3-hydroxytyramine) and brain function. *Pharmacol. Rev.*, **18**:925–964.

4. Bernheimer, H., Birkmayer, W., Hornykiewicz, O., Jellinger, K., and Seitelberger, F. (1973) Brain dopamine and the syndromes of Parkinson and Huntington: clinical, morphological and neurochemical correlations. *J. Neurol. Sci.*, **20**:415–455.

5. Seeman, P., Lee, T., Chau-Wong, M. and Wong, K. (1976) Antipsychotic drug doses and neuroleptic/dopamine receptors. *Nature*, **261**:717–719.

6. Willuhn, I., Wanat, M.J., Clark, J.J., and Phillips, P.E. (2010) Dopamine signaling in the nucleus accumbens of animals self-administering drugs of abuse. *Curr. Top. Behav. Neurosci.*, **3**:29–71.

7. Carlsson, A. (1959) The occurrence, distribution and physiological role of catecholamines in the nervous system. *Pharmacol. Rev.*, **11**:490–493.

8. Anden, N.E., Carlsson, A., Dahlstrom, A., Fuxe, K., Hillarp, N.A., and Larsson, K. (1964) Demonstration and mapping out of nigroneostriatal dopamine neurons. *Life Sci.*, **3**:523–530.

9. Dahlstrom, A. and Fuxe, K. (1964) Evidence for the existence of monoamine-containing neurons in the central nervous system. I. Demonstration of monoamine in the cell bodies of brain stem neurons. *Acta Physiol. Scand. (Suppl.)*, **232**:1–55.

10. Ungerstedt, U. (1971) Adipsia and aphagia after 6-hydroxdopamine induced degeneration of the nigrostriatal dopamine system. *Acta Physiol. Scand. (Suppl.)*, **367**:95–122.

11. Kelly, P.H. and Iversen, S.D. (1975) Selective 6OHDA-induced destruction of mesolimbic dopamine neurons: abolition of psychostimulant-induced locomotor activity in rats. *Eur. J. Pharmacol.*, **40**:45–56.

12. Zhou, Q.Y. and Palmiter, R.D. (1995) Dopamine-deficient mice are severely hypoactive, adipsic, and aphagic. *Cell*, **83**:1197–1209.

13. Kemp, J.M. and Powell, T.P.S. (1971) The structure of the caudate nucleus of the cat: light and electron microscopy. *Phil. Trans. R. Soc. Lond.*, **262**:383–401.

14. Kita, H. and Kitai, S.T. (1988) Glutamate decarboxylase immunoreactive neurons in rat neostriatum: their morphological types and populations. *Brain Res.*, **447**:346–352.

15. Gerfen, C.R. (1990) The neostriatal mosaic: multiple levels of compartmental organization. *Trends Neurosci.*, **15**:133–139.

16. Graybiel, A.M. (1990) Neurotransmitters and neuromodulators in the basal ganglia. *Trends Neurosci.*, **13**:244–254.

17. Gerfen, C.R., Enger, T.M., Mahan, L.C., Susel, Z., Chase, T.N., Monsma, F.J., and Sibley, D.R. (1990) D_1 and D_2 dopamine receptor-regulated gene expression of striatonigral and striatopallidal neurons. *Science*, **250**:1429–1432.

18. Heiman, M., Schaefer, A., Gong, S., Peterson, J.D., Day, M., Ramsey, K.E., Suarez-Farinas, M., Schwarz, C., Stephan, D.A., Surmeier, D.J., Greengard, P., and Heintz, N. (2008) A translational profiling approach for the molecular characterization of CNS cell types. *Cell*, **135**:738–748.

19. Kebabian, J.W. and Calne, D.B. (1979) Multiple receptors for dopamine. *Nature*, **277**:93–96.

20. Missale, C., Nash, S.R., Robinson, S.W., Jaber, M., and Caron, M.G. (1998) Dopamine receptors: from structure to function. *Physiol. Rev.*, **78**:189–225.

21. Valjent, E., Bertran-Gonzalez, J., Herve, D., Fisone, G., and Girault, J.A. (2009) Looking BAC at striatal signaling: cell-specific analysis in new transgenic mice. *Trends Neurosci.*, **32**:538–547.

22. Kebabian, J.W. and Greengard, P. (1971) Dopamine-sensitive adenylyl cyclase: possible role in synaptic transmission. *Science*, **174**:1346–1349.

23. Stoof, J.C. and Kebabian, J.W. (1981) Opposing roles for D-1 and D-2 dopamine receptors in efflux of cyclic AMP from rat neostriatum. *Nature*, **294**:366–368.

24. Herve, D., Levi-Strauss, M., Marey-Semper, I., Verney, C., Tassin, J.P., Glowinski, J., and Girault, J.A. (1993) G(olf) and Gs in rat basal ganglia: possible involvement of G(olf) in the coupling of dopamine D_1 receptor with adenylyl cyclase. *J. Neurosci.*, **13**:2237–2248.

25. Lindskog, M., Svenningsson, P., Fredholm, B.B., Greengard, P., and Fisone, G. (1999) Activation of dopamine D_2 receptors decreases DARPP-32 phosphorylation in striatonigral and striatopallidal projection neurons via different mechanisms. *Neuroscience*, **88**:1005–1008.

26. Schiffmann, S.N., Jacobs, O., and Vanderhaeghen, J.J. (1991) Striatal restricted adenosine A2A receptor (RCD8) is expressed by enkephalin but not by substance P neurons: an *in situ* hybridization histochemistry study. *J. Neurochem.*, **57**:1062–1067.

27. Schiffmann, S.N., Fisone, G., Moresco, R., Cunha, R.A., and Ferre, S. (2007) Adenosine A2A receptors and basal ganglia physiology. *Prog. Neurobiol.*, **83**:277–292.

28. Svenningsson, P., Nishi, A., Fisone, G., Girault, J.A., Nairn, A.C., and Greengard, P. (2004) DARPP-32: an integrator of neurotransmission. *Annu. Rev. Pharmacol. Toxicol.*, **44**:269–296.

29. Surmeier, D.J., Ding, J., Day, M., Wong, Z., and Shen, W. (2007) D_1 and D_2 dopamine-receptor modulation of striatal glutamatergic signaling in striatal medium spiny neurons. *Trends Neurosci.*, **30**:228–235.

30. Hofmann, T., Spano, P.F., Trabucchi, M., and Kumakura, K. (1977) Guanylate cyclase activity in various rat brain areas. *J. Neurochem.*, **29**:395–396.

31. Ariano, M.A. (1983) Distribution of components of the guanosine 3′,5′-phosphate system in rat caudate-putamen. *Neuroscience*, **10**:707–723.

32. Furuyama, T., Inagaki, S., and Takagi, H. (1993) Localizations of alpha 1 and beta 1 subunits of soluble guanylate cyclase in the rat brain. *Brain Res. Mol. Brain Res.*, **20**:335–344.

33. Ding, J.D., Burette, A., Nedvetsky, P.I., Schmidt, H.H., and Weinberg, R.J. (2004) Distribution of soluble guanylyl cyclase in the rat brain. *J. Comp. Neurol.*, **472**:437–448.

34. Walaas, S.I. (1981) The effects of kainic acid injections on guanylate cyclase activity in the rat caudatoputamen, nucleus accumbens, and septum. *J. Neurochem.*, **36**:233–241.

35. Walaas, S.I., Girault, J.A., and Greengard, P. (1989) Localization of cGMP-dependent protein kinase in rat basal ganglia neurons. *J. Mol. Neurosci.*, **1**:243–250.

36. Snyder, S.H. and Bredt, D.S. (1991) Nitric oxide as a neuronal messenger. *Trends Pharmacol.*, **12**:125–128.

37. Bredt, D.S. and Snyder, S.H. (1992) Nitric oxide, a novel neuronal messenger. *Neuron*, **8**:2–11.

38. Dawson, T.M., Dawson, V.L., and Snyder, S.H. (1992) A novel neuronal messenger molecule in brain: the free radical, nitric oxide. *Ann. Neurol.*, **32**:297–311.

39. Vincent, S.R. and Kimura, H. (1992) Histochemical mapping of nitric oxide synthase in the rat brain. *Neuroscience*, **46**:755–784.

40. Garthwaite, J. (1991) Glutamate, nitric oxide, and cell–cell signaling in the central nervous system. *Trends Neurosci.*, **14**:60–67.

41. West, A.R. and Tseng, K.Y. (2011) Nitric oxide-soluble guanylate cyclase-cyclic GMP signaling in the striatum: new targets for the treatment of Parkinson's disease? *Front. Syst. Neurosci.*, **5**:55.

42. Beavo, J. (1995) Cyclic nucleotide phosphodiesterases: functional implications of multiple isoforms. *Physiol. Rev.*, **75**:725–748.

43. Conti, M. and Jin, S.L. (1999) The molecular biology of cyclic nucleotide phosphodiesterases. *Prog. Nucl. Acid Res. Mol. Biol.*, **63**:1–38.

44. Soderling, S.H. and Beavo, J.A. (2000) Regulation of cAMP and cGMP signaling: new phosphodiesterases and new functions. *Curr. Opin. Cell Biol.*, **12**:174–179.

45. Menniti, F.S., Chappie, T.A., Humphrey, J.M., and Schmidt, C.J. (2007) Phosphodiesterase 10A inhibitors: a novel approach to the treatment of the symptoms of schizophrenia. *Curr. Opin. Investig. Drugs*, **8**:54–59.

46. Seeger, T.F., Bartlett, B., Coskran, T.M., Culp, J.S., James, L.C., Krull, D.L., Lanfear, J., Ryan, A.M., Schmidt, C.J., Strick, C.A., Varghese, A.H., Williams, R.D., Wylie, P.G., and Menniti, F.S. (2003) Immunohistochemical localization of PDE10A in the rat brain. *Brain Res.*, **985**:113–126.

47. Coskran, T.M., Morton, D., Menniti, F.S., Adamowicz, W.O., Kleiman, R.J., Ryan, A.M., Strick, C.A., Schmidt, C.J., and Stephenson, D.T. (2006) Immunohistochemical localization of phosphodiesterase 10A in multiple mammalian species. *J. Histochem. Cytochem.*, **54**:1205–1213.

48. Xie, Z., Adamowicz, W.O., Eldred, W.D., Jakowski, A.B., Kleiman, R.J., Morton, D.G., Stephenson, D.T., Strick, C.A., Williams, R.D., and Menniti, F.S. (2006) Cellular and subcellular localization of PDE10A, a striatum-enriched phosphodiesterase. *Neuroscience*, **139**:597–607.

49. Polli, J.W. and Kincaid, R.L. (1994) Expression of a calmodulin-dependent phosphodiesterase isoform (PDE1B1) correlates with brain regions having extensive dopaminergic innervation. *J. Neurosci.*, **14**:1251–1261.

50. Yan, C., Bentley, J.K., Sonnenburg, W.K., and Beavo, J.A. (1994) Differential expression of the 61 kDa and 63 kDa calmodulin-dependent phosphodiesterases in the mouse brain. *J. Neurosci.*, **46**:755–784.

51. Lakics, V., Karran, E.H., and Boess, F.G. (2010) Quantitative comparison of phosphodiesterase mRNA distribution in human brain and peripheral tissues. *Neuropharmacology*, **59**:367–374.

52. Hetman, J.M., Soderling, S.H., Glavas, N.A., and Beavo, J.A. (2000) Cloning and characterization of PDE7B, a cAMP-specific phosphodiesterase. *Proc. Natl. Acad. Sci. USA*, **97**:472–476.

53. Sasaki, T., Kotera, J., Yuasa, K., and Omori, K. (2000) Identification of human PDE7B, a cAMP-specific phosphodiesterase. *Biochem. Biophys. Res. Commun.*, **271**:575–583.

54. Bloom, T.J. and Beavo, J.A. (1996) Identification and tissue-specific expression of PDE7 phosphodiesterase splice variants. *Proc. Natl. Acad. Sci. USA*, **91**:4188–4192.

55. Smith, S.J., Brookes-Fazakerley, S., Donnelly, L.E., Barnes, P.J., Barnette, M.S., and Giembycz, M.A. (2003) Ubiquitous expression of phosphodiesterase 7A in human proinflammatory and immune cells. *Am. J. Physiol. Lung Cell. Mol. Physiol.*, **284**:L279–289.

56. Sasaki, T., Kotera, J., and Omori, K. (2002) Novel alternative splice variants of rat phosphodiesterase 7B showing unique tissue-specific expression and phosphorylation. *Biochem. J.*, **361**:211–220.

57. Cherry, J.A. and Davis, R.L. (1999) Cyclic AMP phosphodiesterases are localized in regions of the mouse brain associated with reinforcement, movement, and affect. *J. Comp. Neurol.*, **407**:476–509.

58. Borisy, F., Ronnett, G., Cunningham, A., Juilfs, D., Beavo, J., and Snyder, S. (1992) Calcium/calmodulin activated phosphodiesterase expressed in olfactory receptor neurons. *J. Neurosci.*, **12**:915–923.

59. Cherry, J.A. and Davis, R.L. (1995) A mouse homolog of dunce, a gene important for learning and memory in *Drosophila*, is preferentially expressed in olfactory receptor neurons. *J. Neurobiol.*, **28**:102–113.

60. Stephenson, D.T., Coskran, T.M., Wilhelms, M.B., Adamowicz, W.O., O'Donnell, M.M., Muravnick, K.B., Menniti, F.S., Kleiman, R.J., and Morton, D. (2009)

Immunohistochemical localization of phosphodiesterase 2A in multiple mammalian species. *J. Histochem. Cytochem.*, **57**:933–949.

61. Reinhardt, R.R. and Bondy, C.A. (1996) Differential cellular pattern of gene expression for two distinct cGMP-inhibited cyclic nucleotide phosphodiesterases in developing and mature rat brain. *Neuroscience*, **72**:567–578.

62. Giordano, D., De Stefano, M.E., Citro, G., Modica, A., and Giorgi, M. (2001) Expression of cGMP-binding cGMP-specific phosphodiesterase (PDE5) in mouse tissues and cell lines using an antibody against the enzyme amino-terminal domain. *Biochim. Biophys. Acta*, **1539**:16–27.

63. Van Staveren, W.C.G., Steinbusch, H.W.M., Markerink-van Ittersum, M., Repaske, D.R., Goy, M.F., Kotera, J., Omori, K., Beavo, J.A., and De Vente, J. (2003) mRNA expression patterns of the cGMP-hydrolyzing phosphodiesterases types 2, 5, and 9 during development of the rat brain. *J. Comp. Neurol.*, **467**:566–580.

64. Fischer, D.A., Smith, J.F., Pillar, J.S., St. Denis, S.H., and Cheng, J.B. (1998) Isolation and characterization of PDE9A, a novel human cGMP-specific phosphodiesterase. *J. Biol. Chem.*, **273**:15559–15564.

65. Van Staveren, W.C.G., Glick, J., Markerink-van Ittersum, M., Shimizu, M., Beavo, J.A., Steinbusch, H.W.M., and DeVente, J. (2002) Cloning and localization of the cGMP-specific phosphodiesterase 9 in the rat brain. *J. Neurocytol.*, **31**:729–741.

66. Fawcett, L., Baxendale, R., Stacey, P., McGrouther, C., Harrow, I., Soderling, S., Hetman, J., Beavo, J.A., and Phillips, S.C. (2000) Molecular cloning and characterization of a distinct human phosphodiesterase gene family: PDE11A. *Proc. Natl. Acad. Sci. USA*, **97**:3702–3707.

67. Yuasa, K., Ohgaru, T., Asahina, M., and Omori, K. (2001) Identification of rat cyclic nucleotide phosphodiesterase 11A (PDE11A): comparison of rat and human PDE11A splicing variants. *Eur. J. Biochem.*, **268**:4440–4448.

68. Kelly, M.P., Logue, S.F., Brennan, J., Lakkaraju, S., Jiang, L., Presman, E., Tam, M., Rizzo, S., Platt, B.J., Dwyer, J.M., Neal, S., Pulito, V.L., Grauer, S., Navarra, R.L., Kelley, C., Kramer, A., Comery, T.A., Marquis, K., Murrills, R.J., and Brandon, N.J. (2009) PDE11A expression in the brain is enriched in ventral hippocampus and deletion results in schizophrenia-related phenotypes. *Soc. Neurosci. Abstr.*, **39**:248.4.

69. Bentley, J.K., Kadlecek, A., Sherbert, C.H., Seger, D., Sonnenburg, W.K., Charbonneau, H., Novack, J.P., and Beavo, J.A. (1992) Molecular cloning of cDNA encoding a "63"-kDa calmodulin-stimulated phosphodiesterase from bovine brain. *J. Biol. Chem.*, **267**:18676–18682.

70. Sonnenburg, W.K., Seger, D., and Beavo, J.A. (1993) Molecular cloning of a cDNA encoding the "61-kDA" calmodulin-stimulated cyclic nucleotide phosphodiesterase: tissue-specific expression of structurally related isoforms. *J. Biol. Chem.*, **268**:645–652.

71. Rybalkin, S.D., Yan, C., Bornfeldt, K.E., and Beavo, J.A. (2003) Cyclic GMP phosphodiesterases and regulation of smooth muscle function. *Circ. Res.*, **93**:280–291.

72. Vandeput, F., Wolda, S.L., Krall, J., Hambleton, R., Uher, L., McCaw, K.N., Radwanski, P.B., Florio, V., and Movsesian, M.A. (2007) Cyclic nucleotide phosphodiesterase PDE1C1 in human cardiac myocytes. *J. Biol. Chem.*, **282**:32749–32757.

73. Fujishige, K., Kotera, J., Michibata, H., Yuasa, K., Takebayashi, S., Okumura, K., and Omori, K. (1999) Cloning and characterization of a novel human phosphodiesterase that hydrolyzes both cAMP and cGMP (PDE10A). *J. Biol. Chem.*, **274**:18438–18445.

74. Loughney, K., Snyder, P.B., Uher, L., Rosman, G.J., Ferguson, K., and Florio, V.A. (1999) Isolation and characterization of PDE10A, a novel human 3′,5′-cyclic nucleotide phosphodiesterase. *Gene*, **234**:109–117.

75. Fujishige, K., Kotera, J., and Omori, K. (1999) Striatum- and testis-specific phosphodiesterase PDE10A isolation and characterization of a rat PDE10A. *Eur. J. Biochem.*, **266**:1118–1127.

76. Kotera, J., Sasaki, T., Kobayashi, T., Fujishige, K., Yamashita, Y., and Omori, K. (2004) Subcellular localization of cyclic nucleotide phosphodiesterase type 10A variants, and alteration of the localization by cAMP-dependent protein kinase-dependent phosphorylation. *J. Biol. Chem.*, **279**:4366–4375.

77. Charych, E., Jiang, L.X., Lo, F., Sullivan, K., and Brandon, N.J. (2010) Interplay of palmitoylation and phosphorylation in the trafficking and localization of phosphodiesterase 10A: implications for the treatment of schizophrenia. *J. Neurosci.*, **30**:9027–9037.

78. Sasaki, T., Kotera, J., and Omori, K. (2004) Transcriptional activation of phosphodiesterase 7B1 by dopamine D_1 receptor stimulation through the cyclic AMP/cyclic AMP-dependent protein kinase/cyclic AMP-response element binding protein pathway in primary striatal neurons. *J. Neurochem.*, **89**:474–483.

79. Conti, M., Richter, W., Mehats, C., Livera, G., Park, J.Y., and Jin, C. (2003) Cyclic AMP-specific PDE4 phosphodiesterases as critical components of cyclic AMP signaling. *J. Biol. Chem.*, **278**:5493–5496.

80. Houslay, M.D. and Adams, D.R. (2003) PDE4 cAMP phosphodiesterases: modular enzymes that orchestrate signaling cross-talk, desensitization and compartmentation. *Biochem. J.*, **370**:1–18.

81. Colledge, M. and Scott, J.D. (1999) AKAPs: from structure to function. *Trends Cell Biol.*, **9**:216–221.

82. Houslay, M.D., Schafer, P., and Zhang, Y.J. (2005) Phosphodiesterase-4 as a therapeutic target. *Drug Discov. Today*, **22**:1503–1519.

83. Brandon, N.J., Handford, E.J., Schurov, I., Rain, J.C., Pelling, M., Duran-Jimeniz, B., Camargo, L.M., Oliver, K.R., Beher, D., Shearman, M.S., and Whiting, P.J. (2004) Disrupted in Schizophrenia 1 and Nudel form a neurodevelopmentally regulated protein complex: implications for schizophrenia and other major neurological disorders. *Mol. Cell. Neurosci.*, **25**:42–55.

84. Yamashita, N., Miyashiro, M., Baba, J., and Sawa, A. (1997) Rolipram, a selective inhibitor of phosphodiesterase type 4, pronouncedly enhanced the forskolin-induced promotion of dopamine biosynthesis in primary cultured rat mesencephalic neurons. *Jpn. J. Pharmacol.*, **75**:91–95.

85. West, A.R. and Galloway, M.P. (1996) Regulation of serotonin-facilitated dopamine release *in vivo*: the role of protein kinase A activating transduction mechanisms. *Synapse*, **23**:20–27.

86. Nishi, A., Kuroiwa, M., Miller, D.B., O'Callaghan, J.P., Bateup, H.S., Shuto, T., Sotogaku, N., Fukuda, T., Heintz, N., Greengard, P., and Snyder, G.L. (2008) Distinct roles of PDE4 and PDE10A in the regulation of cAMP/PKA signaling in the striatum. *J. Neurosci.*, **28**:10460–10471.

87. Engels, P., Abdel'Al, S., Hulley, P., and Lübbert, H. (1995) Brain distribution of four rat homologues of the Drosophila dunce cAMP phosphodiesterase. *J. Neurosci. Res.*, **41**:169–178.

88. Martins, T.J., Mumy, M.C., and Beavo, J.A. (1982) Purification and characterization of a cyclic GMP-stimulated cyclic nucleotide phosphodiesterase from bovine tissues. *J. Biol. Chem.*, **257**:1972–1979.

89. Rosman, G.J., Martins, T.J., Sonnenburg, W.K., Beavo, J.A., Ferguson, K., and Loughney, K. (1997) Isolation and characterization of human cDNAs encoding a cGMP- stimulated 3′5′-cyclic nucleotide phosphodiesterase. *Gene*, **191**:89–95.

90. MacFarland, R.T., Zelus, B.D., and Beavo, J.A. (1991) High concentrations of a cGMP-stimulated phosphodiesterase mediate ANP-induced decreases in cAMP and steroidogenesis in adrenal glomerulosa cells. *J. Biol. Chem.*, **266**:136–142.

91. Repaske, D.R., Corbin, J.G., Conti, M., and Goy, M.F. (1993) A cyclic GMP stimulated cyclic nucleotide phosphodiesterase gene is highly expressed in the limbic system of the rat brain. *Neuroscience*, **56**:673–686.

92. Meacci, E., Taira, M., Moos, J.M., Smith, C.J., Movsesian, M.A., Degerman, E., Belfrage, P., and Manganiello, V.C. (1992) Molecular cloning and expression of human myocardial cGMP-inhibited cAMP phosphodiesterase. *Proc. Natl. Acad. Sci. USA*, **89**:3721–3725.

93. Taira, M., Hockman, S., Calvo, J.C., Belfrage, P., and Manganiello, V.C. (1993) Molecular cloning of the adipocyte hormone-sensitive cyclic GMP-inhibited cyclic nucleotide phosphodiesterase. *J. Biol. Chem.*, **268**:18573–18579.

94. Pisani, A., Bernardi, G., Ding, J., and Surmeier, D.J. (2007) Re-emergence of striatal cholinergic interneurons in movement disorders. *Trends Neurosci.*, **30**:545–553.

95. Guipponi, M., Scott, H.S., Kudoh, J., Kawasaki, K., Shibuya, K., Shintani, A., Asakawa, S., Chen, H., Lalioti, M.D., Rossier, C., Minoshima, S., Shimizu, N., and Antonarakis, S.E. (1998) Identification and characterization of a novel cyclic nucleotide phosphodiesterase gene (PDE9A) that maps to 21q22.3: alternative splicing of mRNA transcripts, genomic structure and sequence. *Hum. Genet.*, **103**:386–392.

96. Andreeva, S.G., Dikkes, P., Epstein, P.M., and Rosenberg, P.A. (2001) Expression of cGMP-specific phosphodiesterase 9A mRNA in the rat brain. *J. Neurosci.*, **21**:9068–9076.

97. Loughney, K., Hill, T.R., Florio, V.A., Uher, L., Rosman, G.J., Wolda, S.L., Jones, B.A., Howard, M.L., McAllister-Lucas, L.M., Sonnenburg, W.K., Francis, S.H., Corbin, J.D., Beavo, J.A., and Ferguson, K. (1998) Isolation and characterization of cDNAs encoding PDE5A, a human cGMP-binding, cGMP-specific 3′,5′-cyclic nucleotide phosphodiesterase. *Gene*, **216**:139–147.

98. Wyatt, T.A., Naftilan, A.J., Francis, S.H., and Corbin, J.D. (1998) ANF elicits phosphorylation of the cGMP phosphodiesterase in vascular smooth muscle cells. *Am. J. Physiol.*, **274**:H448–H455.

99. Boolell, M., Allen, M.J., Ballard, S.A., Gepi-Attee, S., Muirhead, G.J., Naylor, A.M., Osterloh, I.H., and Gingell, C. (1996) Sildenafil: an orally active type 5 cyclic GMP-specific phosphodiesterase inhibitor for the treatment of penile erectile dysfunction. *Int. J. Impot. Res.*, **8**:47–52.

100. Prickaerts, J. (2004) Phosphodiesterase type 5 inhibition improves early memory consolidation of object information. *Neurochem. Int.*, **45**:915–928.

101. Prickaerts, J., Van Staveren, W.C.G., Sik, A., Markerink-van Ittersum, M., Niewöhner, U., Van der Staay, F.J., Blokland, A., and De Vente, J. (2002) Effects of two selective phosphodiesterase type 5 inhibitors, sildenafil and vardenafil, on object recognition memory and hippocampal cyclic GMP levels in the rat. *Neuroscience*, **72**:567–578.

102. Puzzo, D., Staniszewski, A., Deng, S.X., Privitera, L., Leznik, E., Liu, S., Zhang, H., Feng, Y., Palmeri, A., Landry, D.W., and Arancio, O. (2009) Phosphodiesterase 5 inhibition improves synaptic function, memory and amyloid-B load in an Alzheimer's disease mouse model. *J. Neurosci.*, **29**:8075–8086.

103. Reyes-Irisarri, E., Markerink-van Ittersum, M., Mengod, G., and de Vente, J. (2007) Expression of the cGMP-specific phosphodiesterases 2 and 9 in normal and Alzheimer's disease human brains. *Eur. J. Neurosci.*, **25**:3332–3338.

104. Goff, D.C., Cather, C., Freudenreich, O., Henderson, D.C., Evins, A.E., Culhane, M.A., and Walsh, J.P. (2009) A placebo-controlled study of sildenafil effects on cognition in schizophrenia. *Psychopharmacology*, **202**:411–417.

105. Hebb, A.L.O., Robertson, H.A., and Denovan-Wright, E.M. (2004) Striatal phosphodiesterase mRNA and protein levels are reduced in Huntington's disease transgenic mice prior to the onset of motor symptoms. *Neuroscience*, **123**:967–981.

106. Hu, H., McCaw, E.A., Hebb, A.L.O., Gomez, G.T., and Denovan-Wright, E.M. (2004) Mutant huntingtin affects the rate of transcription of striatum-specific isoforms of phosphodiesterase 10A. *Eur. J. Neurosci.*, **20**:3351–3363.

107. Sancesario, G., Giorgi, M., D'Angelo, V., Modica, A., Martorana, A., Morello, M., Bengtson, C.P., and Bernardi, G. (2004) Down-regulation of nitrergic transmission in the rat striatum after chronic nigrostriatal deafferentation. *Eur. J. Neurosci.*, **20**:989–1000.

108. Dlaboga, D., Hajjhussein, H., and O'Donnell, J.M. (2008) Chronic haloperidol and clozapine produce different patterns of effects on phosphodiesterase-1B, -4B, and -10A expression in rat striatum. *Neuropharmacology*, **54**:745–754.

109. Duman, R.S. and Vaidya, V.A. (1998) Molecular and cellular actions of chronic electroconvulsive seizures. *J. ECT*, **14**:181–183.

110. Suda, S., Nibuya, M., Ishiguro, T., and Suda, H. (1998) Transcriptional and translational regulation of phosphodiesterase type IV isozymes in rat brain by electroconvulsive seizure and antidepressant drug treatment. *J. Neurochem.*, **71**:1554–1563.

111. Takahashi, M., Terwilliger, R., Lane, C., Mezes, P.S., Conti, M., and Duman, R.S. (1999) Chronic antidepressant administration increases the expression of cAMP-specific phosphodiesterase 4A and 4B isoforms. *J. Neurosci.*, **19**:610–618.

112. D'Sa, C., Eisch, A.J., Bolger, G.B., and Duman, R.S. (2005) Differential expression and regulation of the cAMP-selective phosphodiesterase type 4A splice variants in rat brain by chronic antidepressant administration. *Eur. J. Neurosci.*, **22**:1463–1475.

113. Ye, Y., Conti, M., Houslay, M.D., Faroqui, S.M., Chen, M., and O'Donnell, J.M. (1997) Noradrenergic activity differentially regulates the expression of rolipram-sensitive, high-affinity cyclic AMP phosphodiesterase (PDE4) in rat brain. *J. Neurochem.*, **69**:2397–2404.

114. Ye, Y., Jackson, K., and O'Donnell, J.M. (2000) Effects of repeated antidepressant treatment of type 4 A phosphodiesterase (PDE4A) in the rat brain. *J. Neurochem.*, **74**:1257–1262.

115. Lourenco, C.M., Kenk, M., Beanlands, R.S., and DaSilva, J.N. (2006) Increasing synaptic noradrenaline, serotonin and histamine enhances *in vivo* binding of phosphodiesterase-4 inhibitor-[^{11}C]rolipram in rat brain, lung, and heart. *Life Sci.*, **79**:356–364.

116. Shakur, Y., Wilson, M., Pooley, L., Lobban, M., Griffiths, S.L., Campbell, A.M., Beattie, J., Daly, C., and Houslay, M.D. (1995) Identification and characterization of the type-IVA cyclic AMP-specific phosphodiesterase RD1 as a membrane-bound protein expressed in cerebellum. *Biochem. J.*, **306**:801–809.

117. Bolger, G.B., Rodgers, L., and Riggs, M. (1994) Differential CNS expression of alternative mRNA isoforms of the mammalian genes encoding cAMP-specific phosphodiesterases. *Gene*, **149**:237–244.

118. Rena, G., Begg, F., Ross, A., MacKenzie, C., McPhee, I., Campbell, L., Huston, E., Sullivan, M., and Houslay, M.D. (2001) Molecular cloning, genomic positioning, promoter identification, and characterization of the novel cyclic AMP-specific phosphodiesterase PDE4A10. *Mol. Pharmacol.*, **59**:996–1011.

119. Reierson, G.W., Mastronardi, C.A., Licinio, J., and Wong, M.-L. (2009) Repeated antidepressant therapy increases cyclic GMP signaling in hat hippocampus. *Neurosci. Lett.*, **466**:149–153.

120. Greengard, P., Allen, P.B., and Nairn, A.C. (1999) Beyond the dopamine receptor: the DARPP-32/protein phosphatase-1 cascade. *Neuron*, **23**:435–447.

121. Walaas, S.I., Aswad, D.W., and Greengard, P. (1983) A dopamine- and cyclic AMP-regulated phosphoprotein enriched in dopamine-innervated brain regions. *Nature*, **301**:69–71.

122. Hemmings, Jr., H.C., Williams, K.R., Konigsberg, W.H., and Greengard, P. (1984) DARPP-32, a dopamine- and adenosine 3′:5′- monophosphate -regulated neuronal phosphoprotein. I. Amino acid sequence around the phosphorylated threonine. *J. Biol. Chem.*, **259**:14486–14490.

123. Hemmings, Jr., H.C. and Greengard, P. (1986) DARPP-32, a dopamine- and adenosine 3′:5′-monophosphate-regulated phosphoprotein: regional, tissue, and phylogenetic distribution. *J. Neurosci.*, **6**:1469–1481.

124. Ouimet, C.C., Miller, P.E., Hemmings, Jr., H.C., Walaas, S.I., and Greengard, P. (1984) DARPP-32, a dopamine- and adenosine 3′:5′- monophosphate -regulated phosphoprotein enriched in dopamine-innervated brain regions III. Immunocytochemical localization. *J. Neurosci.*, **4**:111–124.

125. Gustafson, E.L., Ouimet, C.C., and Greengard, P. (1989) Spatial relationship of the striatonigral and mesostriatal pathways: double-label immunocytochemistry for DARPP-32 and tyrosine hydroxylase. *Brain Res.*, **491**:297–306.

126. Ouimet, C.C., LaMantia, A.S., Goldman-Rakic, P., Rakic, P., and Greengard, P. (1992) Immunocytochemical localization of DARPP-32, a dopamine and cyclic-AMP-regulated phosphoprotein, in the primate brain. *J. Comp. Neurol.*, **323**:209–218.

127. Ouimet, C.C., Langley-Guillon, K.C., and Greengard, P. (1998) Quantitative immunocytochemistry of DARPP-32-expressing neurons in the rat caudatoputamen. *Brain Res.*, **808**:8–12.

128. Ouimet, C.C. and Greengard, P. (1990) Distribution of DARPP-32 in the basal ganglia: an electron microscopic study. *J. Neurocytol.*, **19**:39–52.

129. Hemmings, Jr., H.C., Greengard, P., Tung, H.Y.L., and Cohen, P. (1984) DARPP-32, a dopamine-regulated neuronal phosphoprotein, is a potent inhibitor of protein phosphatase-1. *Nature*, **310**:503–508.

130. Huang, F.L. and Glinsmann, W.H. (1976) Preparation and characterization of two phosphorylase phosphatase inhibitors from rabbit skeletal muscle. *Eur. J. Biochem.*, **70**:419–426.

131. Nimmo, G.A. and Cohen, P. (1978) The regulation of glycogen metabolism: purification and characterization of protein phosphatase inhibitor-1 from rabbit skeletal muscle. *Eur. J. Biochem.*, **87**:341–351.

132. Aitken, A. and Cohen, P. (1982) Isolation and characterization of active fragments of protein phosphatase inhibitor-1 from rabbit skeletal muscle. *FEBS Lett.*, **147**:54–58.

133. Williams, K.R., Hemmings, Jr., H.C., LoPresti, M.B., Konigsberg, W.H., and Greengard, P. (1986) DARPP-32, a dopamine- and cyclic AMP-regulated neuronal phosphoprotein. Primary structure and homology with protein phosphatase inhibitor-1. *J. Biol. Chem.*, **261**:1890–1903.

134. Ariano, M.A., Lewicki, J.A., Brandwein, H.J., and Murad, F. (1983) Immuno-histochemical localization of guanylate cyclase within neurons of rat brain. *Proc. Natl. Acad. Sci. USA*, **79**:1316–1320.

135. Tsou, K., Snyder, G.L., and Greengard, P. (1993) Nitric oxide/cGMP pathway stimulates phosphorylation of DARPP-32, a dopamine- and cAMP-regulated phosphoprotein, in the substantia nigra. *Proc. Natl. Acad. Sci. USA*, **90**:3462–3465.

136. Nishi, A., Watanabe, Y., Higashi, H., Tanaka, M., Nairn, A.C., and Greengard, P. (2005) Glutamate regulation of DARPP-32 phosphorylation in neostriatal neurons involves activation of multiple signaling cascades. *Proc. Natl. Acad. Sci. USA*, **102**:1199–1204.

137. Goto, S., Matsukado, Y., Miyamoto, E., and Yamada, M. (1987) Morphological characterization of striatal neurons expressing calcineurin immunoreactivity. *Neuroscience*, **22**:189–201.

138. King, M.M., Huang, C.Y., Chock, P.B., Nairn, A.C., Hemmings, Jr., H.C., Chan, K.F.J., and Greengard, P. (1984) Mammalian brain phosphoproteins as substrates for calcineurin. *J. Biol. Chem.*, **259**:8080–8083.

139. Halpain, S., Girault, J.A., and Greengard, P. (1990) Activation of NMDA receptors induces dephosphorylation of DARPP-32 in rat striatal slices. *Nature*, **343**:369–371.

140. Nishi, A., Snyder, G.L., Nairn, A.C., and Greengard, P. (1999) Role of calcineurin and protein phosphatase-2A in the regulation of DARPP-32 dephosphorylation in neostriatal neurons. *J. Neurochem.*, **72**:2015–2021.

141. Nishi, A., Snyder, G.L., and Greengard, P. (1997) Bidirectional regulation of DARPP-32 phosphorylation by dopamine. *J. Neurosci.*, **17**:8147–8155.

142. Sharma, R.K., Das, S.B., Lakshmikuttyamma, A., Selvakumar, P., and Shrivastav, A. (2006) Regulation of calmodulin-stimulated phosphodiesterase (PDE1): review. *Int. J. Mol. Med.*, **18**:95–105.

143. Kakkar, R., Raju, R.V., and Sharma, R.K. (1999) Calmodulin-dependent cyclic nucleotide phosphodiesterase (PDE1). *Cell. Mol. Life Sci.*, **55**:1164–1186.

144. Hagiwara, M., Endo, T., and Hidaka, H. (1984) Effects of vinpocetine on cyclic nucleotide metabolism in vascular smooth muscle. *Biochem. Pharmacol.*, **33**:453–457.

145. Zhou, X., Dong, X.W., Crona, J., Maguire, M., and Priestley, T. (2003) Vinpocetine is a potent blocker of rat NaV1.8 tetrodotoxin-resistant sodium channels. *J. Pharmacol. Exp. Ther.*, **306**:498–504.

146. Reed, T.M., Repaske, D.R., Snyder, G.L., Greengard, P., and Vorhees, C.V. (2002) Phosphodiesterase 1B knock-out mice exhibit exaggerated locomotor hyperactivity and DARPP-32 phosphorylation in response to dopamine agonists and display impaired spatial learning. *J. Neurosci.*, **22**:5188–5197.

147. Siuciak, J.A., McCarthy, S.A., Chapin, D.S., Teed, T.M., Vorhees, C.V., and Repaske, D. R. (2007) Behavioral and neurochemical characterization of mice deficient in the phosphodiesterase-1B (PDE1B) enzyme. *Neuropharmacology*, **53**:113–124.

148. Ehrman, L.A., Williams, M.T., Schaefer, T.L., Gudelsky, G.A., Reed, T.M., Fienberg, A. A., Greengard, P., and Vorhees, C.V. (2006) Phosphodiesterase 1B differentially modulates the effects of methamphetamine on locomotor activity and spatial learning through DARPP-32-dependent pathways: evidence from PDE1B-DARPP-32 double knockout mice. *Genes Brain Behav.*, **5**:540–551.

149. Bach, M.E., Barad, M., Son, H., Zhou, M., Lu, Y.F., Shih, R., Mansuy, I., Hawkins, R.D., and Kandel, E.R. (1999) Age-related defects in spatial memory are correlated with defects in the late phase of hippocampal long-term potentiation *in vitro* and are attenuated by drugs that enhance the cAMP signaling pathway. *Proc. Natl. Acad. Sci. USA*, **96**:5280–5285.

150. Devan, B.D., Goad, E.H., and Petri, H.L. (1996) Dissociation of hippocampal and striatal contributions to spatial navigation in the water maze. *Neurobiol. Learn. Mem.*, **66**:305–322.

151. Devan, B.D. and White, N.M. (1999) Parallel information processing in the dorsal striatum: relation to hippocampal function. *J. Neurosci.*, **19**:2789–2798.

152. Siuciak, J.A., Chapin, D.S., Harms, J.F., Lebel, L.A., McCarthy, S.A., Chambers, L., Shrikhande, A., Wong, S., Menniti, F.S., and Schmidt, C.J. (2006) Inhibition of the striatum-enriched phosphodiesterase PDE10A: a novel approach to the treatment of psychosis. *Neuropharmacology*, **51**:386–396.

153. Schmidt, C.J., Chapin, D.S., Cianfrogna, J., Corman, M.L., Hajos, M., Harms, J.F., Hoffman, W.E., Lebel, L.A., McCarthy, S.A., Nelson, F.R., Proulx-LaFrance, C., Majchrzak, M.J., Ramirez, A.D., Schmidt, K., Seymour, P.A., Siuciak, J.A., Tingley, 3rd, F.D., Williams, R.D., Verhoest, P.R., and Menniti, F.S. (2008) Preclinical characterization of selective phosphodiesterase 10A inhibitors: a new therapeutic approach to the treatment of schizophrenia. *J. Pharmacol. Exp. Ther.*, **325**:681–690.

154. O'Callaghan, J.P. and Sriram, K. (2004) Focused microwave irradiation of the brain preserves *in vivo* protein phosphorylation: comparison with other methods of sacrifice and analysis of multiple phosphoproteins. *J. Neurosci. Methods*, **135**:159–168.

155. Grauer, S.M., Pulito, V.L., Navarra, R.L., Kelly, M.P., Graf, R., Langen, B., Logue, S., Brennan, J., Jiang, L., Charych, E., Egerland, U., Liu, F., Marquis, K.L., Malamas, M., Hage, T., Comery, T.A., and Brandon, N.J. (2009) Phosphodiesterase 10A inhibitor activity in preclinical models of the positive, cognitive, and negative symptoms of schizophrenia. *J. Pharmacol. Exp. Ther.*, **331**:574–590.

156. Konradi, C., Cole, R.L., Heckers, S., and Hyman, S.E. (1994) Amphetamine regulates gene expression in rat striatum via transcription factor CREB. *J. Neurosci.*, **14**:5623–5634.

157. Snyder, G.L., Allen, P.B., Fienberg, A.A., Nairn, A.C., Huganir, R.L., and Greengard, P. (2000) Regulation of phosphorylation of the AMPA receptors subunit GluR1 in the neostriatum by dopamine and psychostimulants *in vivo*. *J. Neurosci.*, **20**:4480–4488.

158. Doyle, J.P., Dougherty, J.D., Heiman, M., Schmidt, E.F., Stevens, T.R., Ma, G., Bupp, S., Shrestha, P., Shah, R.D., Doughty, M.L., Gong, S., Greengard, P., and Heintz, N. (2008) Application of a translational profiling approach for the comparative analysis of CNS cell types. *Cell*, **135**:749–762.

159. Bateup, H.S., Svenningsson, P., Kuroiwa, M., Gong, S., Nishi, A., Heintz, N., and Greengard, P. (2008) Cell type-specific regulation of DARPP-32 phosphorylation by psychostimulant and antipsychotic drugs. *Nat. Neurosci.*, **11**:932–939.

160. Threlfell, S., Sammut, S., Menniti, F.S., Schmidt, C.J., and West, A.R. (2009) Inhibition of phosphodiesterase 10A increases the responsiveness of striatal projections neurons to cortical stimulation. *J. Pharmacol. Exp. Ther.*, **328**:785–795.

161. Creese, I., Burt, D.R., and Snyder, S.H. (1976) Dopamine receptors and average clinical doses. *Science*, **194**:546.

162. Davis, K.L., Kahn, R.S., Ko, G., and Davidson, M. (1991) Dopamine in schizophrenia: a review and reconceptualization. *Am. J. Psychiatry*, **148**:1474–1486.

163. Laruelle, M., Abi-Dargham, A., van Dyck, C.H., Gil, R., DeSouza, C.D., and Endos, I. (1996) Single photon emission computerized tomography imaging of amphetamine-induced dopamine release in drug-free schizophrenic subjects. *Proc. Natl. Acad. Sci. USA*, **93**:9235–9240.

164. Breier, A., Su, T.P., Saunders, R., Carson, R.E., Kolachana, B.S., de Bartolomeis, A., Weinberger, D.R., Weisenfeld, N., Malhotra, A.K., Eckelman, W.C., and Pickar, D. (1997) Schizophrenia is associated with elevated amphetamine-induced synaptic dopamine concentrations: evidence from a novel positron emission tomography method. *Proc. Natl. Acad. Sci. USA*, **94**:2569–2574.

165. Jentsch, J.D. and Roth, R.H. (1999) The neuropsychopharmacology of phencyclidine: from NMDA receptor hypofunction to the dopamine hypothesis of schizophrenia. *Neuropsychopharmacology*, **20**:201–225.

166. Krystal, J.H., Karper, L.P., Seibyl, J.P., Freeman, G.K., Delaney, R., Bremner, J.D., Heninger, G.R., Bowers, Jr., M.B., and Charney, D.S. (1994) Subanesthetic effects of the noncompetitive NMDA antagonist, ketamine, in humans: psychotomimetic, perceptual, cognitive, and neuroendocrine responses. *Arch. Gen. Psychiatry*, **51**:199–214.

167. Rodefer, J.S., Murphy, E.R., and Baxter, M.G. (2005) PDE10A inhibition reverses subchronic PCP-induced deficits in attentional set-shifting in rats. *Eur. J. Neurosci.*, **21**:1070–1076.

168. Smith, S.M., Uslaner, J.M., Cox, C.D., Huszar, S.L., Cannon, C.E., Vardigan, J.D., Eddins, D., Toolan, D.M., Dandebo, M., Yao, L., Raheem, I.T., Schreier, J.D., Breslin, M. J., Coleman, P.J., and Renger, J.J. (2013) The novel phosphodiesterase 10A inhibitor THPP-1 has antipsychotic-like effects in rat and improves cognition in rat and rhesus monkey. *Neuropharmacology*, **64**:215–223.

169. Arnt, J., Christensen, A.V., Hyttel, J., Larsen, J.J., and Svendsen, O. (1982) Effects of putative dopamine autoreceptor agonists in pharmacological models related to dopaminergic and neuroleptic activity. *Eur. J. Pharmacol.*, **86**:185–198.

170. Jentsch, J., Redmond, D., Elsworth, J., Taylor, J., Youngren, K., and Roth, R. (1997) Enduring cognitive deficits and cortical dopamine dysfunction in monkeys after long-term administration of phencyclidine. *Science*, **277**:953–955.

171. Jentsch, J.D., Taylor, J.R., Elsworth, J.D., Redmond, Jr., D.E., and Roth, R.H. (1999) Altered frontal cortical dopaminergic transmission in monkeys after subchronic phencyclidine exposure: involvement in frontostriatal cognitive deficits. *Neuroscience*, **90**:823–832.

172. Wachtel, H. (1983) Potential antidepressant activity of rolipram and other selective cyclic adenosine 3':5'-monophosphate phosphodiesterase inhibitors. *Neuropharmacology*, **22**:267–272.

173. Zhu, G., Okada, M., Yoshida, S., Hirose, S., and Kaneko, S. (2004) Pharmacological discrimination of protein kinase associated exocytosis mechanisms between dopamine

and 3,4-dihydroxyphenylalanine in rat striatum using *in vivo* microdialysis. *Neurosci. Lett.*, **363**:120–124.

174. Kehr, W., Debus, G., and Neumeister, R. (1985) Effects of rolipram, a novel antidepressant, on monoamine metabolism in rat brain. *J. Neural Transm.*, **63**:1–12.

175. Kehr, W., Carlsson, A., Lindquist, M., Magnusson, T., and Atack, C. (1972) Evidence for a receptor-mediated feedback control of striatal tyrosine hydroxylase activity. *J. Pharm. Pharmacol.*, **24**:744–747.

176. Siuciak, J.A., Chapin, D.S., McCarthy, S.A., and Martin, A.N. (2007) Antipsychotic profile of rolipram: efficacy in rats and reduced sensitivity in mice deficient in the phosphodiesterase-4B (PDE4B) enzyme. *Psychopharmacology (Berl.)*, **192**:415–424.

177. Kanes, S.J., Tokarczyk, J., Siegel, S.J., Bilker, W., Abel, T., and Kelly, M.P. (2006) Rolipram: a specific phosphodiesterase 4 inhibitor with potential antipsychotic activity. *Neuroscience*, **144**:239–246.

178. Wadenberg, M.L. and Hicks, P.B. (1999) The conditioned avoidance response test re-evaluated: is it a sensitive test for the detection of potentially atypical antipsychotics? *Neurosci. Biobehav. Rev.*, **23**:851–862.

179. Sawaguchi, T. and Goldman-Rakic-PS. (1991) D_1 dopamine receptors in prefrontal cortex: involvement in working memory. *Science*, **251**:947–950.

180. Chudasama, Y. and Robbins, T.W. (2004) Dopaminergic modulation of visual attention and working memory in the rodent prefrontal cortex. *Neuropsychopharmacology*, **29**:1628–1636.

181. Fletcher, P.J., Tenn, C.C., Rizos, Z., Lovic, V., and Kapur, S. (2005) Sensitization to amphetamine, but not PCP, impairs attentional set shifting: reversal by a D_1 receptor agonist injected into the medial prefrontal cortex. *Psychopharmacology*, **183**:190–200.

182. Goldman-Rakic, P.S., Castner, S.A., Svensson, T.H., Siever, L.J., and Williams, G.V. (2004) Targeting the dopamine D_1 receptor in schizophrenia: insights for cognitive dysfunction. *Psychopharmacology*, **174**:3–16.

183. Kuroiwa, M., Snyder, G.L., Shuto, T., Fukuda, A., Yanagawa, Y., Benavides, D.R., Nairn, A.C., Bibb, J.A., Greengard, P., and Nishi, A. (2012) A PDE4 inhibitor, rolipram, enhances dopamine D1 receptor/PKA/DARPP-32 signaling in cortical neurons. *Psychopharmacology*, **219**:1065–1079.

184. Williams, G.V. and Castner, S.A. (2006) Under the curve: critical issues for elucidating D_1 receptor function in working memory. *Neuroscience*, **139**:263–276.

185. Taylor, J.R., Birnbaum, S., Ubriani, R., and Arnsten, A.F. (1999) Activation of cAMP-dependent protein kinase A in prefrontal cortex impairs working memory performance. *J. Neurosci.*, **19**:RC23.

186. Bourtchouladze, R., Abel, T., Berman, N., Gordon, R., Lapidus, K., and Tully, T. (1998) Different training procedures recruit either one of two critical periods for contextual memory consolidation, each of which requires protein synthesis and PKA. *Learn. Mem.*, **5**:365–374.

187. Barad, M., Bourtchouladze, R., Winder, D.G., Golan, H., and Kandel, E. (1998) Rolipram, a type IV-specific phosphodiesterase inhibitor, facilitates the establishment of long-lasting long-term potentiation and improves memory. *Proc. Natl. Acad. Sci. USA*, **95**:15020–15025.

188. Nagai, T., Murai, R., Matsui, K., Kamei, H., Noda, Y., Furukawa, H., and Nabeshima, T. (2009) Aripiprazole ameliorates phencyclidine-induced impairment of recognition

memory through dopamine D_1 and serotonin 5-HT1A receptors. *Psychopharmacology*, **202**:315–328.

189. Millar, J.K., Wilson-Annan, J.C., Anderson, S., Christie, S., Taylor, M.S., and Semple, C. A. et al. (2000) Disruption of two novel genes by a translocation co-segregating with schizophrenia. *Hum. Mol. Genet.*, **9**:1415–1423.

190. Houslay, M.D., Schaefer, P., and Zhang, K.Y. (2005) Keynote review: phosphodiesterase-4 as a therapeutic target. *Drug Discov. Today*, **10**:1503–1519.

191. Siuciak, J.A., McCarthy, S.A., Chapin, D.S., and Martin, A.N. (2008) Behavioral and neurochemical characterization of mice deficient in the phosphodiesterase-4B (PDE4B) enzyme. *Psychopharmacology*, **197**:115–126.

192. Zhang, H.T., Huang, Y., Masood, A., Stolinski, L.R., Li, Y., Zhang, L., Dlaboga, D., Jin, S.L.C., Conti, M., and O'Donnell, J.M. (2008) Anxiogenic-like behavioral phenotype of mice deficient in phosphodiesterase 4B (PDE4B). *Neuropsychopharmacology*, **33**:1611–1623.

193. Rutten, K., Misner, D.L., Works, M., Blokland, A., Novak, T.J., Santarelli, L., and Wallace, T.L. (2008) Enhanced long-term potentiation and impaired learning in phosphodiesterase 4D-knockout (PDE4D−/−) mice. *Eur. J. Neurosci.*, **28**:625–632.

194. Coghill, E.L., Hugill, A., Parkinson, N., Davison, C., Glenister, P., Clements, S., Hunter, J., Cox, R.D., and Brown, S.D. (2002) A gene-driven approach to the identification of ENU mutants in the mouse. *Nat. Genet.*, **30**:255–256.

195. Clapcote, S.J., Lipina, T.V., Millar, J.K., Mackie, S., Christie, S., Ogawa, F., Lerch, J.P., Trimble, K., Uchiyama, M., Sakuraba, Y., Kaneda, H., Shiroishi, T., Houslay, M.D., Henkelman, R.M., Sled, J.G., Gondo, Y., Porteous, D.J., and Roder, J.C. (2007) Behavioral phenotypes of DISC1 missense mutations in mice. *Neuron*, **54**:387–402.

196. Seybold, J., Thomas, D., Witzenrath, M., Boral, S., Hocke, A.C., Bürger, A., Hatzelmann, A., Tenor, H., Schudt, C., Krüll, M., Schütte, H., Hippenstiel, S., and Suttorp, N. (2005) Tumor necrosis factor-alpha-dependent expression of phosphodiesterase 2: role in endothelial hyperpermeability. *Blood*, **105**:3569–3576.

197. Podzuweit, T., Nennstiel, P., and Müller, A. (1995) Isozyme selective inhibition of cGMP-stimulated cyclic nucleotide phosphodiesterase by erythro-9-(2-hydroxy-3-nonyl) adenine. *Cell. Signal.*, **7**:733–738.

198. Suvarna, N.U. and O'Donnell, J.M. (2992) Hydrolysis of *N*-methyl-D-aspartate receptor-stimulated cAMP and cGMP by PDE4 and PDE2 phosphodiesterases in primary neuronal cultures of rat cerebral cortex and hippocampus. *J. Pharmacol. Exp. Ther.*, **302**:249–256.

199. Boess, F.G., Hendrix, M., van der Staay, F.J., Erb, C., Schreiber, R., van Staveren, W., de Vente, J., Prickaerts, J., Blokland, A., and Koenig, G. (2004) Inhibition of phosphodiesterase 2 increases neuronal cGMP, synaptic plasticity and memory performance. *Neuropharmacology*, **47**:1081–1092.

200. Wunder, F., Tersteegen, A., Rebmann, A., Erb, C., Fahrig, T., and Hendrix, M. (2005) Characterization of the first potent and selective PDE9 inhibitor using a cGMP reporter cell line. *J. Pharmacol. Exp. Ther.*, **68**:1775–1781.

201. Verhoest, P.R., Proulx-Lafrance, C., Corman, M., Chenard, L., Helal, C.J., Hou, X., Kleiman, R., Liu, S., Marr, E., Menniti, F.S., Schmidt, C.J., Williams, R.D., Nelson, F.R., Fonesca, K.R., and Liras, S. (2009) Identification of a brain penetrant PDE9A inhibitor utilizing prospective design and chemical enablement as a rapid lead optimization strategy. *J. Med. Chem.*, **24**:7946–7949.

202. Verhoest, P.R., Fonseca, K.R., Hou, X., Proulx-LaFrance, C., Corman, M., Helal, C.J., Claffey, M.M., Tuttle, J.B., Coffman, K.J., Liu, S., Nelson, F.R., Kleiman, R.J., Menniti, F.S., Schmidt, C.J., Vanase-Frawley, M.A., and Liras, S. (2012) Design and discovery of 6-[(3S,4S)-4-methyl-2-ylmethyl)purrolidin-3-yl]-1-(tetrahydro-2H-pyran-4-yl)-1,5-dihydro-4H-pyrazolo[3,4-d]purimidin-4-one (PF-04447943), a selective brain penetrant PDE9A inhibitor for the treatment of cognitive disorders. *J. Med. Chem.*, **55**:9045–9054.

203. Reneerkens, O.A.H., Rutten, K., Steinbusch, H.W.M., Blokland, A., and Prickaerts, J. (2009) Selective phosphodiesterase inhibitors: a promising target for cognition enhancement. *Psychopharmacology*, **202**:419–443.

204. Puzzo, D., Vitolo, O., Fabrizio, T., Jacob, J.P., Palmeri, A., and Arancio, O. (2005) Amyloid-beta peptide inhibits activation of the nitric oxide/cGMP/cAMP responsive element-binding protein pathways during hippocampal synaptic plasticity. *J. Neurosci.*, **25**:6887–6897.

205. Hutson, P.H., Finger, E.N., Magliaro, B.C., Smith, S.M., Converso, A., Sanderson, P.E., Mullins, D., Hyde, L.A., Eschle, B.K., Turnbull, Z., Sloan, H., Guzzi, M., Zhang, X., Wang, A., Rindgen, D., Mazzola, R., Vivian, J.A., Eddins, D., Uslaner, J.M., Bednar, R., Gambone, C., Le-Mair, W., Marino, M.J., Sachs, N., Xu, G., and Parmentier-Batteur, S. (2011) The selective phosphodiesterase 9 (PDE9) inhibitor PF-04447943 (6[3S,4S)-4-methyl-1-(pyrimidin-2-ylmethyl)pyrrolidin-3-yl]-1-(tetrahydro-2H-pyran-4-yl)-1,5-dihydro-4H-pyrazolo[3,4-d]pyrimidin-4-one) enhances plasticity and cognitive function in rodents. *Neuropharmacology*, **61**:665–676.

206. Kleiman, R.J., Chapin, D.S., Christoffersen, C., Freeman, J., Fonseca, K.R., Geoghegan, K.F., Grimwood, S., Guanowsky, V., Hajos, M., Harms, J.F., Helal, C.J., Hoffmann, W. E., Kocan, G.P., Majchrzak, M.J., McGinnis, D., McLean, S., Menniti, F.S., Nelson, F., Roof, R., Schmidt, A.W., Seymour, P.A., Stephenson, D.T., Tingley, F.D., Vanase-Frawley, M., Verhoest, P.R., and Schmidt, C.J. (2012) Phosphodiesterase 9A regulates central cGMP and modulates responses to cholinergic and monoaminergic perturbation *in vivo. J. Pharmacol. Exp. Ther.*, **341**:396–409.

207. Zhuo, M., Hu, Y., Schultz, C., Kandel, E.R., and Hawkins, R.D. (1994) Role of guanylyl cyclase and cGMP-dependent protein kinase in long-term potentiation. *Nature*, **368**:635–639.

208. Kleiman, R.J., Lanz, T.A., Finley, J.E., Bove, S.E., Majchrzak, M.J., Becker, S.L., Carvajal-Gonzales, S., Kuhn, A.M., Wood, K.M., Mariaga, A., Nelson, F.R., Verhoest, P. R., Seymour, P.A., and Stephenson, D.T. (2010) Dendritic spine density deficits in the hippocampal CA1 region of young Tg2576 mice are ameliorated with the PDE9A inhibitor PF-04447943. *Alzheimers Dement.*, **6**:S563–S574.

209. Biskup, S., Gerlach, M., Kupsch, A., Reichmann, H., Riederer, P., Vieregge, P., Wüllner, U., and Gasser, T. (2009) Genes associated with Parkinson syndrome. *J. Neurol.*, **255**(Suppl. 5):8–17.

210. Jankovic, J. and Aguilar, L.G. (2008) Current approaches to the treatment of Parkinson's disease. *Neuropsychiatr. Dis. Treat.*, **4**:743–757.

211. Fahn, S. (1974) "On–off" phenomenon with levodopa therapy in Parkinsonism. Clinical and pharmacologic correlations and the effect of intramuscular pyridoxine. *Neurology*, **24**:431–441.

212. Fahn, S. (2000) The spectrum of levodopa-induced dyskinesias. *Ann. Neurol.*, **47**:2–11.

213. Schrag, A.S. and Quinn, N. (2000) Dyskinesias and motor fluctuations in Parkinson's disease. A community-based study. *Brain*, **123**:2297–2305.

214. Olanow, C.W. (2009) Can we achieve neuroprotection with currently available anti-parkinsonian interventions? *Neurology*, **72**:S59–S64.

215. Rascol, O., Brooks, D.F., Korczyn, A.D., et al. (2000) A five-year study of the incidence of dyskinesia in patients with early Parkinson's disease who were treated with ropinirole or levodopa. 056 Study Group. *N. Engl. J. Med.*, **18**:1484–1491.

216. Whone, A., Watts, R., Stoessl, J. et al. (2003) Slower progression of Parkinson's disease with ropinirole versus levodopa: the REAL-PET study. *Ann. Neurol.*, **54**:93–101.

217. Parkinson Study Group. (2002) Dopamine transporter brain imaging to assess the effects of pramipexole vs levodopa on Parkinson disease progression. *JAMA*, **287**:1653–1661.

218. Fahn, S. and Parkinson Study Group. (2005) Does levodopa slow or hasten the rate of progression of Parkinson's disease? *J. Neurol.*, **252**(Suppl. 4):IV37–IV42.

219. Metman, V.L., Del Dotto, P., LePoole, K., Konitsiotis, S., Fang, J., and Chase, T.N. (1999) Amantadine for levodopa-induced dyskinesias: a 1-year follow up study. *Arch. Neurol.*, **56**:1383–1386.

220. Zhang, H.T. (2009) Cyclic AMP-specific phosphodiesterase-4 as a target for the development of antidepressant drugs. *Curr. Pharm. Des.*, **15**:1688–1698.

221. Chaudhuri, K.R. and Schapira, A.H. (2009) Non-motor symptoms of Parkinson's disease: dopaminergic pathophysiology and treatment. *Lancet Neurol.*, **8**:464–474.

222. Klawans, Jr., H.L. and Weiner, W.J. (1974) Attempted use of haloperidol in the treatment of L-DOPA induced dyskinesias. *J. Neurol. Neurosurg. Psychiatry*, **37**:427–430.

223. Boyce, S., Rupniak, N.M., Steventon, M.J., and Iversen, S.D. (1990) Differential effects of D$_1$ and D$_2$ agonists in MPTP-treated primates: functional implications for Parkinson's disease. *Neurology*, **40**:927–933.

224. Grondin, R., Doan, V.D., Gregoire, L., and Bedard, P.J. (1999) D$_1$ receptor blockade improves L-DOPA-induced dyskinesia but worsens parkinsonism in MPTP monkeys. *Neurology*, **52**:771–776.

225. Nutt, J.G. (1990) Levodopa-induced dyskinesia: review, observations, and speculations. *Neurology*, **40**:340–345.

226. Monville, C., Torres, E.M., and Dunnett, S.B. (2005) Validation of the L-DOPA-induced dyskinesia in the 6-OHDA model and evaluation of the effects of selective dopamine receptor agonists and antagonists. *Brain Res. Bull.*, **68**:16–23.

227. Taylor, J.L., Bishop, C., and Walker, P.D. (2005) Dopamine D$_1$ and D$_2$ receptor contributions to L-DOPA-induced dyskinesia in the dopamine-depleted rat. *Pharmacol. Biochem. Behav.*, **81**:887–893.

228. Visanji, N.P., Fox, S.H., Johnston, T., Reyes, G., Millan, M.J., and Brotchie, J.M. (2009) Dopamine D$_2$ receptor stimulation underlies the development of L-DOPA-induced dyskinesia in animal models of Parkinson's disease. *Neurobiol. Dis.*, **35**:184–192.

229. Nakai, S., Hirose, T., Uwahodo, Y., Imaoka, T., Okazaki, H., Miwa, T., Nakai, M., Yamada, S., Dunn, B., Burris, K.D., Molinoff, P.B., Tottori, K., Altar, C.A., and Kikuchi, T. (2003) Diminished catalepsy and dopamine metabolism distinguish aripiprazole from haloperidol or risperidone. *Eur. J. Pharmacol.*, **472**:89–97.

230. Meco, G., Stirpe, P., Edito, F., Purcaro, C., Valente, M., Bernardi, S., and Vanacore, N. (2009) Aripiprazole in L-dopa-induced dyskinesias: a one-year open-label pilot study. *J. Neural Transm.*, **116**:881–889.

231. Oh, J.D., Bibbiani, F., and Chase, T.N. (2002) Quetiapine attenuates levodopa-induced motor complications in rodent and primate parkinsonian models. *Exp. Neurol.*, **177**: 557–564.

232. Durif, F., Debilly, B., Galitzky, M., Morand, D., Viallet, F., Borg, M., Thoboid, S., Broussolle, C., and Rascol, O. (2004) Clozapine improves dyskinesias in Parkinson disease: a double-blink, placebo-controlled study. *Neurology*, **62**:381–388.

233. Vanover, K.E., Betz, A.J., Weber, S.M., Bibbiani, F., Kielaite, A., Weiner, D.M., Davis, R.E., Chase, T.N., and Salamone, J.S. (2008) A 5-HT2A receptor inverse agonist, ACP-103, reduces tremor in a rat model and levodopa-induced dyskinesias in a monkey model. *Pharmacol. Biochem. Behav.*, **90**:540–544.

234. Muñoz, A., Carlsson, T., Tronci, E., Kirik, D., Björklund, A., and Carta, M. (2009) Serotonin neuron-dependent and -independent reduction of dyskinesia by 5-HT1A and 5-HT1B receptor agonists in the rat Parkinson model. *Exp. Neurol.*, **219**:298–307.

235. Sasaki, H., Hashimoto, K., Inada, T., Fukui, S., and Iyo, M. (1995) Suppression of oro-facial movements by rolipram, a cAMP phosphodiesterase inhibitor, in rats chronically treated with haloperidol. *Eur. J. Pharmacol.*, **282**:71–76.

236. de Gortari, P. and Mengod, G. (2010) Dopamine D_1, D_2 and mu-opioid receptors are co-expressed with adenylyl cyclase 5 and phosphodiesterase 7B mRNAs in striatal rat cells. *Brain Res.*, **1310**:37–45.

237. Kurz, A., Double, K.L., Lastres-Becker, I., Tozzi, A., Tantucci, M., Bockhart, V., Bonin, M., Garcia-Arencibia, M., Nuber, S., Schlaudraff, F., Liss, B., Fernandez-Ruiz, J., Gerlach, M., Wüllner, U., Lüddens, H., Calabresi, P., Auburger, G., and Gispert, S. (2010) A53T-alpha-synuclein overexpression impairs dopamine signaling and striatal synaptic plasticity in old mice. *PLoS One*, **5**:1–15.

238. Picconi, B., Centonze, D., Håkansson, K., Bernardi, G., Greengard, P., Fisone, G., Cenci, M.A., and Calabresi, P. (2003) Loss of bidirectional striatal synaptic plasticity in L-DOPA-induced dyskinesia. *Nat. Neurosci.*, **6**:501–506.

239. Giorgi, M., D'Angelo, V., Esposito, Z., Nuccetelli, V., Sorge, R., Martorana, A., Stefani, A., Bernardi, G., and Sancesario, G. (2008) Lowered cAMP and cGMP signaling in the brain during levodopa-induced dyskinesias in hemiparkinsonian rats: new aspects in the pathogenetic mechanisms. *Eur. J. Neurosci.*, **28**:941–950.

240. Calabresi, P., Gubellini, P., Centonze, D., Sancesario, G., Morello, M., Giorgi, M., Pisani, A., and Bernardi, G. (1999) A critical role of the nitric oxide–cGMP pathway in corticostriatal long-term depression. *J. Neurosci.*, **19**:2489–2499.

241. Picconi, B., Bagetta, V., Ghiglieri, V., Paillè, V., Di Filippo, M., Pendolino, M., Tozzi, A., Giampà, C., Fusco, F.R., Sgobio, C., and Calabresi, P. (2011) Inhibition of phospho-diesterases rescues striatal long-term depression and reduces levodopa-induced dyskine-sia. *Brain*, **134**:375–387.

CHAPTER 13

INHIBITION OF PHOSPHODIESTERASES AS A STRATEGY FOR TREATMENT OF SPINAL CORD INJURY

ELENA NIKULINA
The Feinstein Institute for Medical Research, Manhasset, NY, USA

MARIE T. FILBIN
Department of Biological Sciences, Hunter College, City University of New York, New York, NY, USA

SPINAL CORD INJURY: OBSTACLES TO REGENERATION

The central nervous system (CNS) is enormously complex. The brain and spinal cord are the most protected organs in the body. They are shielded from extrinsic insults by the cranium and spine and from metabolic fluctuations in ion concentration and infections by the blood–brain barrier. Injury to the adult CNS often causes irreversible damage to neurons and glia, and the subsequent failure of axons to reconnect to their targets translates into loss of neurological function or, in the case of spinal cord injury (SCI), paralysis. The physical and psychological consequences of SCI can be highly debilitating for patients and their families and costly for society.

The neurological consequences of SCI were previously considered to be untreatable; however, this paradigm has changed. The work of many research groups and clinicians has transformed the field of spinal cord regeneration and turned the development of restorative treatments into an achievable goal. It has been demonstrated that adult CNS neurons are capable of regenerating severed axons in the permissive environment of a peripheral nerve graft [1]. This implies that it is the injured CNS environment that inhibits axonal regeneration and extensive progress has since been made in identifying the inhibitory molecules involved. Based on these findings, a number of treatments have been proposed [2,3]. Additionally, there is

Cyclic-Nucleotide Phosphodiesterases in the Central Nervous System: From Biology to Drug Discovery,
First Edition. Edited by Nicholas J. Brandon and Anthony R. West.
© 2014 John Wiley & Sons, Inc. Published 2014 by John Wiley & Sons, Inc.

growing realization in the field that long-distance regeneration requires changing adult CNS regenerative capacity to become more growth competent [4–6]. Inhibition of phosphodiesterases is a relatively recent and promising method of improving regeneration after CNS injury; it modulates the extrinsic neuronal environment after injury, as well as the neuronal response to inhibitory factors, while boosting the regenerative competence of neurons.

SCI causes initial damage to axonal tracts, oligodendrocytes, and astrocytes [7,8] in the immediate vicinity of the impact site. The damaged cells, ruptured blood vessels, and a compromised blood–brain barrier alter strictly maintained extracellular concentrations of various ions and molecules, most notably glutamate and calcium [9]. After injury, the levels of glutamate quickly rise to a concentration that is toxic to both neurons and glia [10]. Vascular damage leads to edema, inflammatory leukocyte infiltration, and hypoxia [11,12]. This initiates a wave of secondary damage that causes neuronal and glial apoptotic cell death that lasts for days after the initial impact (Figure 13.1) [8,13,14]. Secondary progressive neurodegeneration expands the area of the original injury and significantly worsens the prospect of recovery [15]. Microglia, the resident macrophages of the CNS, become activated, proliferate, and accumulate at the injury site [16]. Activated microglia upregulate a plethora of proinflammatory cytokines, such as Il-1β and TNFα, that have established neurotoxic traits and produce nitric oxide and superoxide anions. Importantly, TNFα increases neuronal surface glutamate receptors [17,18] and therefore exacerbates excitotoxic neuronal death [19]. Conversely however, microglia activation can also be beneficial for CNS repair by removing necrotic debris and secreting anti-inflammatory cytokines and neurotrophins [20]. Recently, it was proposed that macrophages can be differentially activated into two distinct subsets: either proinflammatory or anti-inflammatory [21]. However, the pathophysiological environment of the injured spinal cord pushes macrophage differentiation toward a proinflammatory fate [22]. Numerous studies have demonstrated that pharmacological inhibition of microglia and macrophages or, alternatively, neutralization of the cytokines that they produce, limits secondary injury and improves functional recovery [23].

Reactive astrogliosis is also an integral part of the CNS injury response. Upon injury, astrocytes proliferate, change their morphology, become hypertrophic, and alter their gene expression. Most notably, they upregulate the levels of glial fibrillary acidic protein (GFAP) and chondroitin sulfate proteoglycans (CSPGs) [24]. Reactive astrocytes and the CSPGs that they secrete are major components of the glial scar that surrounds the injury site [25], and are potent physical and biochemical inhibitors of axonal regeneration [26]. At the same time, the glial scar seals off the damaged area, prevents further spread of inflammatory cells, restores the blood–brain barrier, and diminishes the size of the injury [27].

Unlike the CNS, the peripheral nervous system (PNS) regenerates after injury. In the course of the injury response, PNS myelin debris is removed by the orchestrated action of Schwann cells and macrophages. Prior clearance of myelin is critical for regeneration of the injured PNS axons [28]. However, in the injured CNS, myelin debris remains for months after the lesion [29,30]. In the past two decades, tremendous progress has been achieved in identifying proteins associated with myelin that inhibit regeneration and the signaling mechanisms they initiate to restrict axonal

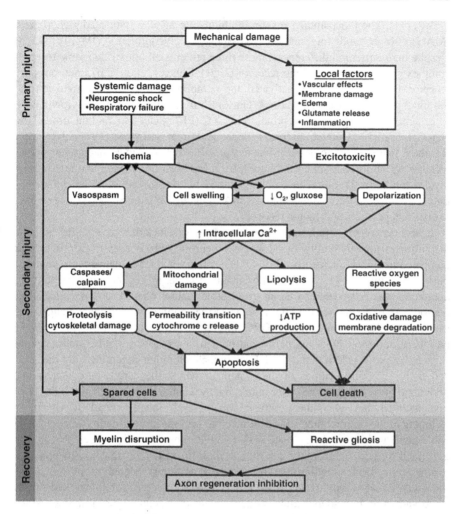

FIGURE 13.1   Schematic of pathophysiology of CNS injury, demonstrating various pathological events during primary spinal cord injury, secondary spinal cord injury, and recovery phases. Following mechanical trauma, there is often hemorrhage, followed by a disruption in blood flow, combined with membrane damage, edema, glutamate release, and severe inflammation. All of these changes may lead to local infarction caused by ischemia and hypoxia. Secondary injury is characterized by glutamate-mediated cytotoxicity, calcium-mediated secondary injury, lipid peroxidation, electrolyte imbalance, and apoptosis. In recovery phase, dysfunction resulting from spinal cord or other CNS injuries occurs because of loss of neurons and the production of an environment that is refractory for axon regeneration (adapted from Ref. [14] with permission from Elsevier, Copyright 2012).

growth [31]. The most studied myelin inhibitors are myelin-associated glycoprotein (MAG), Nogo, and oligodendrocyte-associated glycoprotein (OMgp) [31,32]. Despite their structural differences, these three proteins all bind to the same receptor complex consisting of the Nogo receptor (NgR), Lingo 1, and the p75 neurotrophin receptor (p75^{NTR}) [33]. When activated, this complex signals downstream to activate the small guanosine triphosphatase (GTPase) RhoA and to block axonal growth [34]. This signaling is dependent on protein kinase C (PKC), calcium, and the proteolytic cleavage of p75^{NTR}, which, in turn, is dependent on α- and γ-secretase activity [35]. Recently, paired immunoglobulin-like receptor-B (PirB) was identified as yet another receptor for all three myelin inhibitors [36]. It has been established that interrupting myelin inhibitor signaling by blocking ligand binding to the NgR receptor complex [37] or by inhibiting downstream RhoA signaling [38] improves, to some degree, axonal regeneration and motor recovery.

It has been long recognized that another reason for a lack of recovery after injury is the intrinsic growth ability of adult CNS neurons, which appears to be limited compared to embryonic neurons [39,40]. The embryonic CNS regenerates spontaneously after injury [41,42]. The subsequent decline of regenerative capacity in adulthood was considered to be an irreversible feature of fully differentiated CNS neurons. An experimental paradigm known as the conditioning lesion challenged this assumption. Dorsal root ganglion (DRG) neurons route sensory information from the periphery into the CNS via a peripheral branch that is a part of PNS and a central branch that projects into to the CNS. Each branch, although stemming from the same cell body, exhibits qualities of the system of which they are a part; that is, the peripheral branch regenerates after injury, but the central branch does not. It has been demonstrated that when the peripheral branch is lesioned several days prior to lesioning of the central branch, the central axons become more growth competent [43] and regenerate beyond the SCI site [44]. These findings strongly suggest that signals generated by peripheral injury can initiate molecular changes that facilitate central regeneration. Understanding the mechanisms of injury signaling [45] and the molecular changes they evoke [6,46] points to new strategies for spinal cord repair.

Advances in the understanding of mechanisms of secondary death in SCI and the molecular mechanisms precluding axonal regeneration will help to develop restorative therapies. These treatments should aim to prevent the progression of secondary injury, promote axonal regeneration, and replace lost cells [47]. Both intrinsic and extrinsic factors confine axonal regrowth. Fully differentiated adult neurons have a decreased intrinsic ability to grow compared to embryonic neurons, and the extrinsic factors are the inhibitory molecules associated with both myelin debris and the glial scar. Given the apparent redundancy of myelin inhibitors and their receptors, an increasingly appealing approach to improving regeneration is to alter the intrinsic state of the neuron so that it no longer responds to the inhibitory environment. Elevating cyclic adenosine monophosphate (cAMP) levels in neurons stimulates axonal growth in the presence of myelin inhibitors *in vitro* and promotes regeneration in the animal models of SCI *in vivo* [48–50]. Pharmacological inhibition of phosphodiesterases (PDEs) to increase cAMP levels after SCI has already been shown to improve axonal regeneration as well as decrease glial cell death, thus preventing the spread of secondary damage.

cAMP Levels Dictate Neuronal Regenerative Ability

Signaling through the intracellular second messengers cAMP and cyclic guanosine monophosphate (cGMP) regulates crucial and diverse neuronal functions such as survival [51], axonal guidance [52], synaptic plasticity [53,54], and regeneration [55]. It was demonstrated that elevation of cAMP in neurons with a cAMP analog, dibutyryl-cAMP (db-cAMP), overcomes MAG and myelin-induced inhibition of neurite outgrowth *in vitro* [48] and allows postnatal CNS neurons to grow on inhibitory substrates at a rate similar to untreated embryonic neurons [40,56]. Indeed, embryonic neurons have high levels of cAMP that precipitously decline after birth [56]. In addition, cAMP also mediates the effects of a conditioning lesion. Levels of cAMP in DRG neurons were elevated more than twofold 1 day after peripheral lesion and returned to basal levels after 1 week [50]. In a different study, cAMP was reported to be still elevated 1 week after peripheral lesion [57]. Importantly, injection of db-cAMP directly into the DRG mimics the effects of a peripheral lesion [49,50], and neurite outgrowth of neurons from both conditioning lesioned and db-cAMP-treated DRGs was significantly improved on permissive and inhibitory substrates. When a lesion of the central branch was performed 2 days after db-cAMP injection and regeneration was assessed 6 weeks postlesion, numerous regenerating fibers penetrated into the lesion site, which rarely happened in control untreated animals [49]. When the central branch was lesioned 1 week after db-cAMP treatment, in addition to the robust regeneration into the lesion site, some growth beyond the lesion site was detected [50]. Overall, these studies established that cAMP elevation is an integral part of the conditioning lesion effect and that pharmacological elevation of cAMP levels can reproduce critical molecular steps leading to an enhanced neuronal growth state.

Intracellular levels of cAMP and cGMP are tightly regulated at the level of synthesis by adenylyl cyclases [58] and guanylyl cyclases [59,60] and at the level of degradation by a diverse family of PDEs [61,62]. Members of all 11 families of PDEs are present in the CNS [63,64]. Notably, different types of neurons and glial cells express different PDE isoforms [65]. Moreover, this pattern is subject to change with development [66] and under pathological conditions. Different PDEs are associated with different cellular compartments, enabling cells to control cyclic nucleotide levels spatially [67]. It is of note that the potential maximal PDE activity in various cell types surpasses the cAMP synthesizing capacity [61], suggesting that under normal conditions activity of PDEs is restricted.

There is a good indication that levels of PDEs may be increased after SCI [68,69]. Gene expression profiling performed 1 h after a spinal cord contusion showed expression of PDE2 mRNA, undetectable before injury, and levels of PDE4 mRNA were increased at least fourfold compared to control [69]. In a separate contusion injury model study, the area immediately adjacent to the impact site (and presumably not directly affected by the impact) was analyzed for changes in gene expression. Increased mRNAs for two isoforms of PDE4, PDE4B2A and PDE4B1A, were detectable at 6 h after injury (1.2- and 4.5-fold changes, respectively) and reached the maximum levels (4.4- and 5.4-fold increases) at 12 h after the impact [68].

Further studies are required to confirm that increased PDE mRNA expression leads to increased protein levels and to determine in which cells of the injured spinal cord they are upregulated.

Postinjury changes in protein expression and, subsequently, in protein–protein interaction may have an impact on PDE activity. The association of the p75[NTR] with PDE4A4 and PDE4A5 studied by Akassoglou and coworkers is an example. They showed that the p75[NTR] receptor interacts with PDE4A4 and PDE4A5 to decrease cAMP levels and that the effect is reversible by the specific PDE4 inhibitor rolipram [70]. Importantly, p75[NTR], barely detectable in adult spinal cord, is considerably upregulated in neurons after injury [71], which may result in the recruitment of PDE4A4 and PDE4A5 to the plasma membrane and a decline in cAMP levels. cAMP levels after peripheral nerve crush are elevated in p75[NTR] knockout mice, an effect largely attributed to an increase in cAMP in Schwann cells [70]. Whether the p75[NTR]–PDE4 interaction plays a role in the cAMP decrease in neurons after CNS injury remains to be investigated.

The best evidence that cAMP levels decrease after SCI comes from studies conducted in Mary Bunge's laboratory. It was established that 1 day after a contusion injury, cAMP levels decreased by 64.3% in the spinal cord, by 68.1% in the brain stem, and by 69.7% in the sensorimotor cortex [72]. Levels of cAMP remained low for at least 2 weeks after the contusion. In addition, changes in cAMP levels have also been observed in other CNS injury models. In a rat model of traumatic brain injury, a decrease in cAMP levels was observed within 15 min of the injury, and levels remained low in the ipsilateral parietal cortex for 2 days [73]. It should be noted that in a rat ischemic model of stroke, a short-term spike in cAMP levels immediately after the onset of ischemia was reported. This coincides with an increased release of glutamate, an established excitotoxic factor [74,75]. It has been recognized that cAMP can enhance the glutamate response [76] and, in the pathological context of ischemia, cAMP may exacerbate the toxic effect of glutamate [77]. It is well documented that the tissue immediately surrounding the impact in SCI becomes ischemic [12]; consequently, although considered generally neuroprotective, pharmacological elevation of cAMP immediately after injury with cAMP-specific PDE inhibitors, such as rolipram, could theoretically harm ischemic cells. This could be one reason why low doses of rolipram (0.3 mg/(kg day)) administered 30 min before traumatic brain injury decreased cortical contusion size more effectively than high doses of rolipram (3 mg/(kg day)) [73]. Blocking non–N-methyl-D-aspartate (NMDA) glutamate receptors with the AMPA-kainate receptor antagonist NBQX reduces tissue loss and functional impairment after contusive SCI [78] and is neuroprotective in ischemia [79]. One would predict that addition of NBQX to early rolipram treatment would protect ischemic neurons and oligodendrocytes against the potential enhancement of glutamate toxicity caused by cAMP and improve the therapeutic effect of rolipram. Further studies are necessary to resolve this issue. Despite this concern, there is no doubt that cAMP levels decrease after traumatic CNS injury and that elevation of cAMP is a promising approach to improve the survival of neurons and glia and enhance axonal regeneration.

Less is known about changes in the levels of cGMP after SCI. One day after spinal cord compression injury in the rabbit, cGMP levels were decreased in the ventral white matter of the spinal cord rostral to the lesion and increased in the lateral white matter caudal to the lesion [80]. These results support previous findings reporting similar changes in the activity of nitric oxide synthase (NOS), a major activator of soluble guanylyl cyclase, in corresponding areas [81]. It has also been reported that 3 h after spinal cord contusion, constitutive NOS (cNOS) was decreased while the inducible form of NOS (iNOS) was upregulated [82]. The major source of iNOS appeared to be invading polymorphonuclear leukocytes, and the upregulation of iNOS is linked to development of the inflammatory response [83]. Acute treatment with the iNOS inhibitor, aminoguanidine, decreased lesion volume and improved locomotor recovery [82]. In another study, Suzuki and coworkers showed that within the first 50 min after compression injury, the levels of cNOS were decreased. Importantly, within 2 h of the injury, iNOS levels rose concurrently with the activity of nuclear factor-κB (NF-κB), a transcription factor regulating many pro-inflammatory cytokines. Moreover, immunostaining analysis demonstrated that both iNOS and active NF-κB were present in neurons 2 h after the injury. Suzuki's group proposed that the decrease in cNOS triggers NF-κB activation and iNOS expression in neurons [84]. This complex interaction of different isoforms of NOS signaling to cGMP with apparently different functional outcomes makes it very difficult to predict the consequences of pharmacological interventions [85]. There is a possibility that elevation of cGMP levels with cGMP specific phosphodiesterases within the first hour after the lesion will compensate for decreased cNOS activity and alleviate the inflammatory response.

ROLIPRAM IN SPINAL CORD REGENERATION RESEARCH

Rolipram Is the Inhibitor of Choice in Regeneration Studies

The complex pathology of SCI and the existence of extrinsic and intrinsic barriers to axonal regeneration would appear to make the development of a single, all-encompassing treatment an almost unattainable task. It has been demonstrated in a rat SCI model that only a combination of different treatments resulted in significant regeneration of fibers beyond the lesion site [86]. Importantly, cAMP elevation was an integral part of the combinatorial treatment. Therefore, addition of PDE inhibitors to the list of potential treatments could advance the field of CNS regeneration. Rolipram, an inhibitor of PDE4, has become the PDE inhibitor of choice in regeneration studies since PDE4 is the major cAMP PDE in the brain, representing ∼70% of total PDE activity [87]. PDE4 isoforms are expressed widely in the CNS and are present in areas where descending axonal tracts controlling movement and coordination originate. For example, the corticospinal tract, which is involved in sensory and motor control, originates in layer 5 of rat cerebral cortex [88], where PDE4A, PDE4B, and PDE4D mRNA are expressed [89]. The rubrospinal tract is involved in motor control, and originates in the red nucleus within the midbrain that expresses PDE4D mRNA. In the spinal cord, oligodendrocytes, motor neurons, and microglia express PDE4 [90]. Rolipram increases cAMP in motor neurons *in vitro* [91] and *in vivo* [92]. The difficulties associated with isolating clean preparations of different neuronal subpopulations and the existence of multiple PDE isoforms often precludes assessment of expression profiles for different PDEs in individual cell types. When this information becomes available, it will be possible to specifically target different types of cells. However, based on the existing evidence, rolipram elevates cAMP in the majority of neurons and glial cells, which makes it an excellent research tool. Originally, rolipram, which lacks PDE4 subtype specificity, was developed as an antidepressant. However, clinical trials were unsuccessful because of side effects, including severe nausea [93,94]. The more recently developed PDE4 inhibitors, such as roflumilast, demonstrate a more acceptable therapeutic window [95], although similar to rolipram, they are not subtype specific and produce similar side effects. The long-standing general assumption that subtype-specific and brain- or periphery-specific PDE4 inhibitors will evoke fewer adverse effects was recently confirmed by the surprising discovery of a brain-directed PDE4D-specific inhibitor with low emetic potential [96]. The new small-molecule allosteric modulators of PDE4D are as effective as roflumilast *in vitro* and improve mice performance in cognitive tests similar to rolipram. Importantly, this demonstrated that the development of blood–brain barrier-permeable PDE4 inhibitors with low emetic potential is achievable, which opens a venue for the clinical application of the accumulated scientific data. The results of studies on the ability of rolipram to improve the survival of neurons and glia after SCI and to promote regeneration should be regarded as a proof of principle that inhibition of PDE4 is a potent tool in therapeutic interventions following SCI. When the best conditions and combination of treatments are elucidated, it is possible that other PDE4-specific inhibitors with fewer side effects may be substituted for rolipram (see Chapter 5 for more details on PDE4 subtype-specific inhibitor development).

Inhibitors of cGMP-specific PDE5, and of the cAMP- and cGMP-specific PDE3, have been shown to be neuroprotective and improve outcome in ischemic models of stroke [97–99]. In rat spinal cord contusion model, PDE5 inhibitors provided modest protective effects, although comparable with the effects of standard SCI treatment with methylprednisolone [100]. In mouse contusive injury model, treatment with the inhibitor of PDE5 improved lesion epicenter vascular perfusion, but did improve functional recovery [101]. More recently, resveratrol, a polyphenol found in grapes and red wine and widely studied because of its antiaging properties, was demonstrated to be an inhibitor of PDE1, PDE3, and PDE4 [102]. Resveratrol was neuroprotective in the model of ischemic injury [103], and also improved the behavioral outcome and reduced contusion volume in the rat traumatic brain injury model [104]. Notably, Park et al. reported that rolipram mimicked the metabolic benefits of resveratrol, adding anti-aging to the list of rolipram effects. Further studies will demonstrate whether PDE5 and PDE3 inhibitors, alone or in combination with other treatments, will benefit SCI outcome. Therefore, the main focus of this chapter is the ability of PDE4 inhibition by rolipram to increase cAMP levels in neurons and glia and promote recovery after SCI.

Rolipram Improves Neurite Growth in the Presence of MAG

Rolipram significantly improves the axonal growth of neurons in the presence of MAG *in vitro*, although the effect of rolipram is not as robust as the effect of db-cAMP [105]. This suggests that MAG signaling by as yet unknown mechanisms prevents accumulation of cAMP by rolipram. We reported a similar effect of MAG on neurotrophin signaling [48]. Neurotrophins elevate cAMP in neurons; however, in the presence of MAG, this elevation is blocked [48]. Notably, cAMP accumulation in the course of neurotrophin signaling is mediated by downregulation of PDE4 activity by extracellular signal-regulated kinase (ERK) [106]. Rolipram is less potent in overcoming inhibition by MAG than db-cAMP, but more effective than neurotrophins.

Rolipram Delivery In Vivo Increases the Intrinsic Growth Ability of Neurons

Because it was reported that db-cAMP injections into DRGs mimic critical aspects of the conditioning lesion effect, we tested the effects of prolonged rolipram delivery *in vivo* on the ability of DRG neurons to extend neurites *in vitro*. Prolonged rolipram treatment strongly promoted neurite outgrowth both on permissive and inhibitory substrates [105]. This growth promoting effect of rolipram resembles the effect of a conditioning lesion. Despite similarities between the effects of the conditioning lesion and db-cAMP treatment, there are some important distinctions. Unlike a conditioning lesion, db-cAMP treatment does not elevate GAP-43, a well-characterized growth-promoting gene [107]. The profiles of mRNA upregulated by db-cAMP treatment and a conditioning lesion, while sharing several common mRNAs, appear to be quite different (J. Carmel and M.T. Filbin, unpublished data). It was also demonstrated that a single injection of db-cAMP, unlike a conditioning lesion, does not increase growth into peripheral nerve graft transplanted into the lesioned dorsal column [108]. Notably, it has been shown that rolipram treatment improved the growth of lesioned CNS axons into permissive grafts [72,105]. These results suggest the effects of prolonged rolipram treatment might more faithfully replicate the conditioning lesion effect then a single db-cAMP injection. Although one would expect that prolonged treatment with db-cAMP would be as effective as treatment with rolipram, it was demonstrated that continuous delivery of db-cAMP following a spinal lesion has no beneficial effects at the injury site and has a detrimental effect on locomotor recovery [109]. The discovery of the extraordinary proregenerative effect of the conditioning lesion supported the notion that adult CNS neurons are indeed capable of long-distance regeneration, initiating a multitude of studies aimed to replicate this effect. However, it was never intended to become a therapeutic tool, rather an instrument of discovery, while inhibition of PDE4 provides a starting point to develop treatments that could be translated into the clinic.

Treatment with Rolipram Promotes Regeneration in Spinal Cord Injury Models

Rolipram's effects were tested in several animal models of SCI. In collaboration with Barbara Bregman's laboratory, we assessed the effect of rolipram treatment in a model of cervical spinal cord hemisection [105]. In these experiments, embryonic spinal cord transplants were grafted at the lesion site immediately after the hemisection and rolipram was delivered at a dose of 0.4 μmol/(kg h) (2.6 mg/(kg day)) and 0.8 μmol/(kg h) (5.2 mg/(kg day)) via miniosmotic pump for 10 days, starting 2 weeks after the injury [105]. Rolipram treatment at 0.4 μmol/(kg h) significantly improved the regeneration of serotonergic fibers into the transplant. Higher doses of rolipram did not have an effect. Interestingly, rolipram treatment beginning 2 weeks after injury attenuated reactive astrogliosis. It is not clear at this point whether rolipram caused astrocytes to revert to the less-reactive phenotype or whether it affected maturation of the glial scar. Animals that received rolipram displayed functional recovery with better forelimb control as measured by the paw placement test.

In another study, the effect of acute and delayed rolipram treatment in combination with Schwann cell grafts and db-cAMP injections, both delivered 1 week after injury, was examined in a model of moderate thoracic contusion in rat spinal cord [72]. Rolipram treatment prevented a decrease in cAMP levels after the injury and a substantial increase in the sparing of oligodendrocyte-myelinated axons in the lateral spinal cord (i.e., the portion of spinal cord not immediately affected by the injury) was reported in all the experimental groups that received rolipram acutely. In addition, treatment enhanced the regeneration of serotonergic fibers into the Schwann cell grafts. The best results were achieved with a combined treatment of acute rolipram with delayed Schwann cell graft and db-cAMP injection. In the triple-treated animals, a significant number of regenerating fibers were found to exit the graft and enter the caudal spinal cord. It should be noted that only in triple-treated animals were cAMP levels elevated above levels seen in uninjured rats, and thus it appears that regeneration into the injured CNS environment is supported when cAMP is elevated to levels higher than the normal preinjury range. Increased cAMP levels after the treatment correlated with increased axonal regeneration and improved motor test scores. When assessed using the Basso, Beattie, Bresnahan (BBB) locomotor rating scale [110], functional recovery was significant for the triple-treated animals. Remarkably, in some tests, the acute rolipram group performed almost as well as the triple-treated animals. It also seems to be the case that the beneficial effect of rolipram treatment appears quite early. One week after transplantation, the best-scoring groups were the groups that were treated with rolipram acutely.

The early effect of rolipram treatment on motor recovery was also noted in a study published by Steven Strittmatter's laboratory [111]. In their experiments, rolipram delivery was delayed for 3 days after moderate thoracic contusion injury. Although the difference in BBB scores did not reach statistical significance, there was a trend toward a better performance by rolipram treated rats 2 weeks after injury, which is considered to be early in the testing. This early effect could be explained by the sparing of oligodendrocyte-myelinated axons in the spinal white matter, which are not immediately affected by injury. Results from Stephan Onifer's laboratory support this suggestion. It was shown that besides motor neurons, oligodendrocytes also express PDE4A, PDE4B, and PDE4D. In addition, microglia, the source of the inflammatory cytokine TNFα express PDE4B 3 days after injury [90]. When rolipram (at 0.5 mg/(kg day)) was delivered to the rats after a contusive cervical injury, there was a significant increase at 24 and 72 h postcontusion in the number of spared oligodendrocytes in the ventral lateral funiculus (VLF), which is the area that was not directly injured [90]. A substantial protective effect on oligodendrocytes in the secondary death area was detectable at 5 weeks after injury [112]. The increased number of the spared oligodendrocytes coincided with an increase in axonal conductivity in the VLF. Although the BBB score was not significantly improved, the rolipram-treated animals performed much better in one particular test, namely, hindlimb footfall errors. A recent investigation by Varejão's group has showed that in the rat moderate spinal cord contusion model, the continuous delivery of rolipram via osmotic minipumps significantly improved the white matter sparing and functional recovery [113].

The detrimental impact of SCI on initially spared myelinated fibers has been unequivocally demonstrated by Arvanian et al. [114]. One week after T8 spinal cord hemisection, transmission through the initially spared contralateral myelinated axons is compromised. The decay in conductivity began 2 days after the lesion, was significant after 1 week and reached a maximum at 2 weeks after injury. At the structural level, the loss of myelinated axons has been shown to be significant, peaking at 2 weeks. This work and that of other's [115,116] provide evidence that any treatment that protects oligodendrocyte-myelinated axons from secondary damage could improve behavioral outcomes. The reported data [72,112] strongly suggests that PDE4 inhibitors have potent oligodendrocyte-preserving effects.

Recently, it was reported that rolipram and the anti-inflammatory drug thalidomide work synergistically to preserve tissue and improve motor recovery in a moderate contusion model of SCI [117]. Remarkably, a single dose of thalidomide (at 100 mg/kg) and rolipram (at 3 mg/kg) injected immediately after the injury significantly decreased the levels of IL-1β measured 1 h after the injury. In addition, a trend toward reduction of TNFα was observed. The decrease in proinflammatory cytokines correlated with a significant white matter sparing and better locomotor performance at 42 days after injury. The effects of rolipram or thalidomide when injected separately were not as dramatic as the effects of both drugs applied together. The synergistic effect of rolipram and thalidomide may be accounted for by the difference in their mechanisms as anti-inflammatory agents. Rolipram decreases the levels of TNFα after SCI [72,90], and this effect is cAMP dependent, as suggested by *in vitro* studies [118]. The anti-inflammatory effect of thalidomide is largely a result of the enhancement of degradation of TNF-α mRNA, but not limited to it [119]. The data presented by Koopmans et al. do not allow for assessment of the relative contribution of each treatment to decreased levels of TNFα and IL-1, or the respective kinetics of their separate and combine effects. Nevertheless, the excellent results of this pilot study certainly merit more detailed investigation.

The strategy of combining rolipram with immunomodulatory treatments to promote recovery after SCI was successful in a collaborative study presented by the Steinmetz and Silver laboratories [120]. They showed that a combination of continuous delivery of rolipram with depletion of macrophages by liposomal-encapsulated clodronate resulted in a reduction in lesion volume, sparing of myelinated tissue, and significant improvement of locomotor recovery as measured with BBB score. Taken together, these data clearly demonstrate that the tissue sparing effect of rolipram can be further enhanced when used in combination with anti-inflammatory drugs.

Rolipram can also restore neurological function by reawakening existing synaptic connections, thereby leading to rapid changes in signaling of uninjured axonal pathways and increasing plasticity. The plasticity-inducing effect of rolipram was studied in detail by the Goshgarian laboratory using a model of cervical spinal cord hemisection in rats that interrupts the descending major respiratory pathway to the phrenic motor neurons on the ipsilateral side. This leads to paralysis of the ipsilateral hemidiaphragm. Recovery in rats occurs spontaneously and takes about 6 weeks [121]. Functional improvement was attributed to the activation of the crossed phrenic pathway. The crossed phrenic pathway traverses the midline below the injury site and connects uninjured contralateral, bulbospinal projections to the ipsilateral hemidiaphragm [122]. Although nonfunctional in adult animals, it becomes activated upon injury. It was demonstrated that the crossed phrenic pathway can be activated pharmacologically with rolipram. The injection of rolipram (2 mg/kg) 1 week after C2 hemisection induced respiratory-related activity in the ipsilateral phrenic nerve 1 h after injection [123]. When animals were treated with rolipram for 3 days (2 mg/kg, twice a day intraperitoneally) starting immediately after the lesion, the activity in the phrenic nerve was partially restored and the effect persisted for 10 days [124]. This means that neurological function that would have taken 6 weeks to recover was restored almost immediately. It is unlikely that the immediate plasticity effect of rolipram is limited to the activation of dormant crossed phrenic pathway only. It was also demonstrated in an animal model of stroke that inhibition of PDE4 facilitates rehabilitation-dependent locomotor recovery [125]. It is conceivable that some early effects of rolipram treatment on functional improvement [72,111] could be explained by increased plasticity.

Two mechanisms may contribute to the improved early behavioral recovery of rolipram treated animals, namely, the sparing of oligodendrocyte-myelinated axons and increased plasticity. These mechanisms might not have just an additive effect; they may be, in fact, mutually dependent. Indeed, sparing of oligodendrocyte-myelinated axons could be the result of a direct survival effect of rolipram on oligodendrocytes and an indirect survival effect wherein rolipram stimulates neuronal activity that leads to expression of oligodendrocyte survival factors such as neuregulin [126,127].

Rolipram Promotes Peripheral Nerve Regeneration

Although axons within the peripheral nervous system regenerate spontaneously [128], the rate of regeneration is slow [129]. This presents a serious setback in the recovery of patients, because prolonged muscle denervation leads to muscle atrophy. Therefore, acceleration of axon regeneration is of the utmost importance. The protracted process of peripheral nerve regeneration is attributed to the slow clearance of myelin debris by macrophages and Schwann cells [130] and to the presence of inhibitory proteoglycans [131]. Importantly, denervated Schwann cells cease expression of growth promoting genes over time [132]. If the window of opportunity for regeneration is missed, the regenerating axons do not receive enough support from the Schwann cells and regeneration is stalled [133]. Therefore, it is crucial to speed the rate at which PNS axons regrow through the injury site in order to achieve full recovery. Gordon and coworkers showed that elevation of cAMP by rolipram promotes peripheral nerve regeneration [92]. In the rat peripheral nerve regeneration model, the common peroneal nerve was transected and surgically repaired. Rolipram was delivered with minipumps for the duration of the experiment, that is, for 7, 14, and 21 days. Rolipram treatment accelerated both sensory and motor nerve regeneration across the suture site. After 14 days of treatment with rolipram, the number of regenerating motor and sensory axons that grew 10 mm distal to the injury site doubled when compared to control animals. The rate of Wallerian degeneration was not affected by rolipram. In addition, in rolipram-treated rats, an eightfold increase in the number of myelinated fibers was observed at the distal stump 14 days after injury, which is also indicative of the accelerating effect of rolipram on peripheral nerve repair. These findings suggest that PDE4 inhibitors could benefit patients with peripheral nerve injuries, which include compressive injuries such as carpal tunnel syndrome [133].

Rolipram Improves Survival of Embryonic Stem Cell-Derived Motor Neurons

Stem cell-based therapy holds enormous promise to restore neurons and glial cells lost as a consequence of cell death at the site of SCI and as a result of secondary damage [134,135]. It is also hoped that transplanted cells will bridge the damaged site in the spinal cord, restoring conductivity and thus neurological function. The original idea to repopulate damaged tissue with stem cell-derived neurons, oligo-dendrocytes, and astrocytes in the appropriate proportions was proven to be technically unattainable. Neural stem cells differentiate predominantly into astro-cytes when seeded into the spinal cord [136]. Therefore, the more promising strategy to repair damaged CNS is to transplant embryonic stem (ES) cells predifferentiated into neurons [137].

When ES cell-derived motor neurons (MNs) are transplanted into the injured spinal cord, they encounter an environment that is both inhibitory for neurite extension as a result of inhibitory proteins in myelin and neurotoxic as a result of increased expression of the proinflammatory cytokines TNFα, IL-1, and IL-6 [138]. It has been shown that neurite outgrowth from ES cell-derived MNs is inhibited by myelin *in vitro* and that db-cAMP blocks this inhibition [139]. Furthermore, it was established that rolipram can attenuate both inhibition by myelin and expression of TNFα. Thus, this approach was tested as a method to increase the survival of ES cell-derived MNs [140] in a rat model of paralysis after virally mediated destruction of ventral motor neuron pools [141]. MNs were transplanted 28 days after viral inoculation, with db-cAMP added to the suspension (1 μM). Two days prior to transplantation, different experimental groups were treated with rolipram (0.5 mg/(kg day) continuing for 30 days). In addition, a cell line expressing glial cell-derived neurotrophic factor (GDNF) was injected into the sciatic nerve in order to promote growth of the integrated motor neurons out of the spinal cord into the peripheral nerve. The results of this study clearly show that both db-cAMP infusion and rolipram treatment are critical for long-term ES cell-derived MN survival. This dual treatment combined with the axon-guiding factor GDNF allowed new motor neurons to extend axons that reached their target muscles and formed functional neuromuscular junctions. Only rats that received a combination of all three components of the treatment partially recovered from paralysis. Rolipram was an integral part of this treatment. In the experimental group where rolipram was omitted, the ES cell-derived motor neurons survived poorly, and there was no behavioral recovery.

CONCLUSIONS

Despite the convincing evidence showing that axonal regeneration and functional restoration are attainable, the development of restorative therapies has proven to be enormously challenging. Rolipram has clear advantages as a therapeutic tool to treat SCI; its effects as an axonal growth-promoting, anti-inflammatory, and plasticity-inducing agent have been conclusively demonstrated. Rolipram is an integral part of combinatorial treatments that promote axonal regeneration and recovery in animal models of SCI [72,105,117,120]. These results are consistent with recent developments in the field of CNS regeneration, specifically with the notion that only combination of several treatments will achieve long-distance regeneration [4,5].

Enhancement of plasticity-dependent recovery by PDE4 inhibitors merits further study. The effects of treatments that aim to accelerate the rehabilitation process are hard to assess because of some degree of spontaneous neurological recovery [5]. However, this does not imply that development of these treatments should be abandoned. Specifically, this concerns the older cohort of patients whose prognosis of spontaneously regaining neurological function is much worse than that of young patients [47].

Further studies of the effects of rolipram alone or in combination with anti-inflammatory drugs on secondary neurodegeneration after SCI are required. Most likely, rolipram protects oligodendrocytes by attenuating the inflammatory response mediated by microglia and infiltrating leukocytes; however, a direct effect on oligodendrocyte survival cannot be excluded. It is important to note that, although prevention of secondary damage is one of the central goals of SCI research, none of the suggested treatments demonstrated sufficient efficacy.

It is very likely that the first evidence of regenerative potential of PDE4 inhibitors in humans will come from the field of PNS regeneration. Researchers led by Tessa Gordon and Thomas Brushart have pioneered work on the promotion of peripheral nerve regeneration by rolipram in animal models [142,143]. Their collaborative efforts to incorporate their results into medical practice bring hope that we will soon observe similar restorative effects of rolipram on peripheral nerve regeneration in human patients.

It must be noted, however, that more research is needed before developing PDE4 inhibitors as an effective therapy for SCI. As a result of the differences in experimental settings and regimens of rolipram treatment between the studies discussed, it is difficult at this point to predict the optimal dose or time point of rolipram delivery.

Elucidation of the most effective combination of PDE4 inhibitors with other treatments is another promising avenue of research.

The ability of rolipram to cross the blood–brain barrier makes it a valuable candidate drug to treat CNS traumas. It can be administered systemically, without interference with the injury site, specifically when the extent of the tissue loss is difficult to assess and any interference might potentially cause more damage. There are good reasons to believe that PDE4 inhibitors alone or in combination with other treatments could be developed into efficient therapy for SCI. The benefits of such a treatment could potentially be extended to the treatment of traumatic brain injuries, strokes, and neurodegenerative diseases.

ACKNOWLEDGMENTS

We thank Dr. Christine R. Cain, Dr. Sari S. Hannila, Dr. Ona E. Bloom, and Dr. Peter Bradley for critical reading of the manuscript and their invaluable editorial suggestions. This work was supported by National Institutes of Health (NIH) Grant NS37060, and the New York State Spinal Cord Injury Research Board.

Elena Nikulina expresses her deepest gratitude to Marie T. Filbin, a mentor and a friend.

REFERENCES

1. David, S. and Aguayo, A.J. (1981) Axonal elongation into peripheral nervous system "bridges" after central nervous system injury in adult rats. *Science*, **214**(4523):931–933.

2. Giger, R.J., Hollis, E.R., 2nd, and Tuszynski, M.H. (2010) Guidance molecules in axon regeneration. *Cold Spring Harb. Perspect. Biol.*, **2**(7):a001867.

3. Zorner, B. and Schwab, M.E. (2010) Anti-Nogo on the go: from animal models to a clinical trial. *Ann. N. Y. Acad. Sci.*, **1198**(Suppl. 1):E22–E34.

4. Benowitz, L.I. and Yin, Y. (2007) Combinatorial treatments for promoting axon regeneration in the CNS: strategies for overcoming inhibitory signals and activating neurons' intrinsic growth state. *Dev. Neurobiol.*, **67**(9):1148–1165.

5. Blesch, A. and Tuszynski, M.H. (2009) Spinal cord injury: plasticity, regeneration and the challenge of translational drug development. *Trends Neurosci.*, **32**(1):41–47.

6. Sun, F. and He, Z. (2010) Neuronal intrinsic barriers for axon regeneration in the adult CNS. *Curr. Opin. Neurobiol.*, **20**(4):510–518.

7. Grossman, S.D., Rosenberg, L.J., and Wrathall, J.R. (2001) Temporal-spatial pattern of acute neuronal and glial loss after spinal cord contusion. *Exp. Neurol.*, **168**(2):273–282.

8. Liu, X.Z., et al. (1997) Neuronal and glial apoptosis after traumatic spinal cord injury. *J. Neurosci.*, **17**(14):5395–5406.

9. Park, E., Velumian, A.A., and Fehlings, M.G. (2004) The role of excitotoxicity in secondary mechanisms of spinal cord injury: a review with an emphasis on the implications for white matter degeneration. *J. Neurotrauma*, **21**(6):754–774.

10. McAdoo, D.J., et al. (1999) Changes in amino acid concentrations over time and space around an impact injury and their diffusion through the rat spinal cord. *Exp. Neurol.*, **159**(2):538–544.

11. Donnelly, D.J. and Popovich, P.G. (2008) Inflammation and its role in neuroprotection, axonal regeneration and functional recovery after spinal cord injury. *Exp. Neurol.*, **209**(2):378–388.

12. Tator, C.H. and Fehlings, M.G. (1991) Review of the secondary injury theory of acute spinal cord trauma with emphasis on vascular mechanisms. *J. Neurosurg.*, **75**(1):15–26.

13. Beattie, M.S., Farooqui, A.A., and Bresnahan, J.C. (2000) Review of current evidence for apoptosis after spinal cord injury. *J. Neurotrauma*, **17**(10):915–925.

14. GhoshMitra, S., et al. (2012) Role of engineered nanocarriers for axon regeneration and guidance: current status and future trends. *Adv. Drug Deliv. Rev.*, **64**(1):110–125.

15. Hausmann, O.N. (2003) Post-traumatic inflammation following spinal cord injury. *Spinal Cord*, **41**(7):369–378.

16. Beck, K.D., et al. (2010) Quantitative analysis of cellular inflammation after traumatic spinal cord injury: evidence for a multiphasic inflammatory response in the acute to chronic environment. *Brain*, **133**(Part 2):433–437.

17. Beattie, E.C., et al. (2002) Control of synaptic strength by glial TNFalpha. *Science*, **295**(5563):2282–2285.

18. Stellwagen, D., et al. (2005) Differential regulation of AMPA receptor and GABA receptor trafficking by tumor necrosis factor-alpha. *J. Neurosci.*, **25**(12):3219–3228.

19. Ferguson, A.R., et al. (2008) Cell death after spinal cord injury is exacerbated by rapid TNF alpha-induced trafficking of GluR2-lacking AMPARs to the plasma membrane. *J. Neurosci.*, **28**(44):11391–11400.

20. Streit, W.J. (2005) Microglia and neuroprotection: implications for Alzheimer's disease. *Brain Res. Brain Res. Rev.*, **48**(2):234–239.

21. Mosser, D.M. and Edwards, J.P. (2008) Exploring the full spectrum of macrophage activation. *Nat. Rev. Immunol.*, **8**(12):958–969.

22. Kigerl, K.A., et al. (2009) Identification of two distinct macrophage subsets with divergent effects causing either neurotoxicity or regeneration in the injured mouse spinal cord. *J. Neurosci.*, **29**(43):13435–13444.

23. Popovich, P.G. and Longbrake, E.E. (2008) Can the immune system be harnessed to repair the CNS? *Nat. Rev. Neurosci.*, **9**(6):481–493.

24. Sofroniew, M.V. and Vinters, H.V. (2010) Astrocytes: biology and pathology. *Acta Neuropathol.*, **119**(1):7–35.

25. Fitch, M.T. and Silver, J. (2008) CNS injury, glial scars, and inflammation: inhibitory extracellular matrices and regeneration failure. *Exp. Neurol.*, **209**(2):294–301.

26. McKeon, R.J., Hoke, A., and Silver, J. (1995) Injury-induced proteoglycans inhibit the potential for laminin-mediated axon growth on astrocytic scars. *Exp. Neurol.*, **136**(1):32–43.

27. Faulkner, J.R., et al. (2004) Reactive astrocytes protect tissue and preserve function after spinal cord injury. *J. Neurosci.*, **24**(9):2143–2155.

28. Vargas, M.E. and Barres, B.A. (2007) Why is Wallerian degeneration in the CNS so slow? *Annu. Rev. Neurosci.*, **30**:153–179.

29. George, R. and Griffin, J.W. (1994) Delayed macrophage responses and myelin clearance during Wallerian degeneration in the central nervous system: the dorsal radiculotomy model. *Exp. Neurol.*, **129**(2):225–236.

30. Perry, V.H., Brown, M.C., and Gordon, S. (1987) The macrophage response to central and peripheral nerve injury. A possible role for macrophages in regeneration. *J. Exp. Med.*, **165**(4):1218–1223.

31. Filbin, M.T. (2003) Myelin-associated inhibitors of axonal regeneration in the adult mammalian CNS. *Nat. Rev. Neurosci.*, **4**(9):703–713.

32. Liu, B.P., et al. (2006) Extracellular regulators of axonal growth in the adult central nervous system. *Philos. Trans. R. Soc. Lond. B Biol. Sci.*, **361**(1473):1593–1610.

33. Cao, Z., et al. (2010) Receptors for myelin inhibitors: structures and therapeutic opportunities. *Mol. Cell. Neurosci.*, **43**(1):1–14.

34. Yamashita, T., et al. (2005) Multiple signals regulate axon regeneration through the Nogo receptor complex. *Mol. Neurobiol.*, **32**(2):105–111.

35. Domeniconi, M., et al. (2005) MAG induces regulated intramembrane proteolysis of the p75 neurotrophin receptor to inhibit neurite outgrowth. *Neuron*, **46**(6):849–855.

36. Atwal, J.K., et al. (2008) PirB is a functional receptor for myelin inhibitors of axonal regeneration. *Science*, **322**(5903):967–970.

37. Gonzenbach, R.R. and Schwab, M.E. (2008) Disinhibition of neurite growth to repair the injured adult CNS: focusing on Nogo. *Cell. Mol. Life Sci.*, **65**(1):161–176.

38. Dergham, P., et al. (2002) Rho signaling pathway targeted to promote spinal cord repair. *J. Neurosci.*, **22**(15):6570–6577.

39. Dusart, I., Airaksinen, M.S., and Sotelo, C. (1997) Purkinje cell survival and axonal regeneration are age dependent: an *in vitro* study. *J. Neurosci.*, **17**(10):3710–3726.

40. Goldberg, J.L., et al. (2002) Amacrine-signaled loss of intrinsic axon growth ability by retinal ganglion cells. *Science*, **296**(5574):1860–1864.

41. Bates, C.A. and Stelzner, D.J. (1993) Extension and regeneration of corticospinal axons after early spinal injury and the maintenance of corticospinal topography. *Exp. Neurol.*, **123**(1):106–117.

42. Hasan, S.J., et al. (1993) Axonal regeneration contributes to repair of injured brainstem-spinal neurons in embryonic chick. *J. Neurosci.*, **13**(2):492–507.

43. Richardson, P.M. and Issa, V.M. (1984) Peripheral injury enhances central regeneration of primary sensory neurones. *Nature*, **309**(5971):791–793.

44. Neumann, S. and Woolf, C.J. (1999) Regeneration of dorsal column fibers into and beyond the lesion site following adult spinal cord injury. *Neuron*, **23**(1):83–91.

45. Rishal, I. and Fainzilber, M. (2010) Retrograde signaling in axonal regeneration. *Exp. Neurol.*, **223**(1):5–10.

46. Abe, N. and Cavalli, V. (2008) Nerve injury signaling. *Curr. Opin. Neurobiol.*, **18**(3):276–283.

47. McDonald, J.W. and Sadowsky, C. (2002) Spinal-cord injury. *Lancet*, **359**(9304):417–425.

48. Cai, D., et al. (1999) Prior exposure to neurotrophins blocks inhibition of axonal regeneration by MAG and myelin via a cAMP-dependent mechanism. *Neuron*, **22**(1):89–101.

49. Neumann, S., et al. (2002) Regeneration of sensory axons within the injured spinal cord induced by intraganglionic cAMP elevation. *Neuron*, **34**(6):885–893.

50. Qiu, J., et al. (2002) Spinal axon regeneration induced by elevation of cyclic AMP. *Neuron*, **34**(6):895–903.

51. Cui, Q. and So, K.F. (2004) Involvement of cAMP in neuronal survival and axonal regeneration. *Anat. Sci. Int.*, **79**(4):209–212.

52. Piper, M., van Horck, F., and Holt, C. (2007) The role of cyclic nucleotides in axon guidance. *Adv. Exp. Med. Biol.*, **621**:134–143.

53. Bailey, C.H., Bartsch, D., and Kandel, E.R. (1996) Toward a molecular definition of long-term memory storage. *Proc. Natl. Acad. Sci. USA*, **93**(24):13445–13452.

54. Kleppisch, T. and Feil, R. (2009) cGMP signalling in the mammalian brain: role in synaptic plasticity and behaviour. *Handb. Exp. Pharmacol.*, (191):549–579.

55. Hannila, S.S. and Filbin, M.T. (2008) The role of cyclic AMP signaling in promoting axonal regeneration after spinal cord injury. *Exp. Neurol.*, **209**(2):321–332.

56. Cai, D., et al. (2001) Neuronal cyclic AMP controls the developmental loss in ability of axons to regenerate. *J. Neurosci.*, **21**(13):4731–4739.

57. Kadoya, K., et al. (2009) Combined intrinsic and extrinsic neuronal mechanisms facilitate bridging axonal regeneration one year after spinal cord injury. *Neuron*, **64**(2):165–172.

58. Hanoune, J. and Defer, N. (2001) Regulation and role of adenylyl cyclase isoforms. *Annu. Rev. Pharmacol. Toxicol.*, **41**:145–174.

59. Baltrons, M.A., et al. (2003) Regulation of NO-dependent cyclic GMP formation by inflammatory agents in neural cells. *Toxicol. Lett.*, **139**(2–3):191–198.

60. Kots, A.Y., et al. (2009) A short history of cGMP, guanylyl cyclases, and cGMP-dependent protein kinases. *Handb. Exp. Pharmacol.*, (191):1–14.

61. Bender, A.T. and Beavo, J.A. (2006) Cyclic nucleotide phosphodiesterases: molecular regulation to clinical use. *Pharmacol. Rev.*, **58**(3):488–520.

62. Houslay, M.D. (2001) PDE4 cAMP-specific phosphodiesterases. *Prog. Nucleic Acid Res. Mol. Biol.*, **69**:249–315.

63. Kleppisch, T. (2009) Phosphodiesterases in the central nervous system. *Handb. Exp. Pharmacol.*, (191):71–92.

64. Wong, M.L., et al. (2006) Phosphodiesterase genes are associated with susceptibility to major depression and antidepressant treatment response. *Proc. Natl. Acad. Sci. USA*, **103**(41):15124–15129.

65. Menniti, F.S., Faraci, W.S., and Schmidt, C.J. (2006) Phosphodiesterases in the CNS: targets for drug development. *Nat. Rev. Drug Discov.*, **5**(8):660–670.

66. Zhang, K., Farooqui, S.M., and O'Donnell, J.M. (1999) Ontogeny of rolipram-sensitive, low-K(m), cyclic AMP-specific phosphodiesterase in rat brain. *Brain Res. Dev. Brain Res.*, **112**(1):11–19.

67. McCahill, A.C., et al. (2008) PDE4 associates with different scaffolding proteins: modulating interactions as treatment for certain diseases. *Handb. Exp. Pharmacol.*, (186):125–166.

68. Carmel, J.B., et al. (2001) Gene expression profiling of acute spinal cord injury reveals spreading inflammatory signals and neuron loss. *Physiol. Genomics*, **7**(2):201–213.

69. Nesic, O., et al. (2002) DNA microarray analysis of the contused spinal cord: effect of NMDA receptor inhibition. *J. Neurosci. Res.*, **68**(4):406–423.

70. Sachs, B.D., et al. (2007) p75 neurotrophin receptor regulates tissue fibrosis through inhibition of plasminogen activation via a PDE4/cAMP/PKA pathway. *J. Cell Biol.*, **177**(6):1119–1132.

71. Widenfalk, J., et al. (2001) Neurotrophic factors and receptors in the immature and adult spinal cord after mechanical injury or kainic acid. *J. Neurosci.*, **21**(10):3457–3475.

72. Pearse, D.D., et al. (2004) cAMP and Schwann cells promote axonal growth and functional recovery after spinal cord injury. *Nat. Med.*, **10**(6):610–616.

73. Atkins, C.M., et al. (2007) Modulation of the cAMP signaling pathway after traumatic brain injury. *Exp. Neurol.*, **208**(1):145–158.

74. Prado, R., Busto, R., and Globus, M.Y. (1992) Ischemia-induced changes in extracellular levels of striatal cyclic AMP: role of dopamine neurotransmission. *J. Neurochem.*, **59**(4):1581–1584.

75. Wahl, F., et al. (1994) Extracellular glutamate during focal cerebral ischaemia in rats: time course and calcium dependency. *J. Neurochem.*, **63**(3):1003–1011.

76. Greengard, P., et al. (1991) Enhancement of the glutamate response by cAMP-dependent protein kinase in hippocampal neurons. *Science*, **253**(5024):1135–1138.

77. Tsukada, H., et al. (2004) Transient focal ischemia affects the cAMP second messenger system and coupled dopamine D_1 and 5-HT1A receptors in the living monkey brain: a positron emission tomography study using microdialysis. *J. Cereb. Blood Flow Metab.*, **24**(8):898–906.

78. Wrathall, J.R., Choiniere, D., and Teng, Y.D. (1994) Dose-dependent reduction of tissue loss and functional impairment after spinal cord trauma with the AMPA/kainate antagonist NBQX. *J. Neurosci.*, **14**(11 Part 1):6598–6607.

79. Sheardown, M.J., et al. (1990) 2,3-Dihydroxy-6-nitro-7-sulfamoyl-benzo(F)quinoxaline: a neuroprotectant for cerebral ischemia. *Science*, **247**(4942):571–574.

80. Lukacova, N., et al. (2001) Effect of spinal cord compression on cyclic 3′,5′-guanosine monophosphate in the white matter columns of rabbit. *Neurochem. Int.*, **39**(4):275–282.

81. Lukacova, N., et al. (2000) Effect of midthoracic spinal cord constriction on catalytic nitric oxide synthase activity in the white matter columns of rabbit. *Neurochem. Res.*, **25**(8):1139–1148.

82. Chatzipanteli, K., et al. (2002) Temporal and segmental distribution of constitutive and inducible nitric oxide synthases after traumatic spinal cord injury: effect of amino-guanidine treatment. *J. Neurotrauma*, **19**(5):639–651.

83. Lopez-Figueroa, M.O., et al. (2000) Temporal and anatomical distribution of nitric oxide synthase mRNA expression and nitric oxide production during central nervous system inflammation. *Brain Res.*, **852**(1):239–246.

84. Miscusi, M., et al. (2006) Early nuclear factor-kappaB activation and inducible nitric oxide synthase expression in injured spinal cord neurons correlating with a diffuse reduction of constitutive nitric oxide synthase activity. *J. Neurosurg. Spine*, **4**(6):485–493.

85. Bishop, A. and Anderson, J.E. (2005) NO signaling in the CNS: from the physiological to the pathological. *Toxicology*, **208**(2):193–205.

86. Lu, P., et al. (2004) Combinatorial therapy with neurotrophins and cAMP promotes axonal regeneration beyond sites of spinal cord injury. *J. Neurosci.*, **24**(28):6402–6409.

87. Jin, S.L., et al. (1999) Impaired growth and fertility of cAMP-specific phosphodiesterase PDE4D- deficient mice. *Proc. Natl. Acad. Sci. USA*, **96**(21):11998–12003.

88. Deumens, R., Koopmans, G.C., and Joosten, E.A. (2005) Regeneration of descending axon tracts after spinal cord injury. *Prog. Neurobiol.*, **77**(1–2):57–89.

89. Perez-Torres, S., et al. (2000) Phosphodiesterase type 4 isozymes expression in human brain examined by *in situ* hybridization histochemistry and [^{3}H]rolipram binding autoradiography. Comparison with monkey and rat brain. *J. Chem. Neuroanat.*, **20**(3–4): 349–374.

90. Whitaker, C.M., et al. (2008) Rolipram attenuates acute oligodendrocyte death in the adult rat ventrolateral funiculus following contusive cervical spinal cord injury. *Neurosci. Lett.*, **438**(2):200–204.

91. Aglah, C., Gordon, T., and Posse de Chaves, E.I. (2008) cAMP promotes neurite outgrowth and extension through protein kinase A but independently of Erk activation in cultured rat motoneurons. *Neuropharmacology*, **55**(1):8–17.

92. Udina, E., et al. (2010) Rolipram-induced elevation of cAMP or chondroitinase ABC breakdown of inhibitory proteoglycans in the extracellular matrix promotes peripheral nerve regeneration. *Exp. Neurol.*, **223**(1):143–152.

93. Hebenstreit, G.F., et al. (1989) Rolipram in major depressive disorder: results of a double-blind comparative study with imipramine. *Pharmacopsychiatry*, **22**(4):156–160.

94. Zhu, J., Mix, E., and Winblad, B. (2001) The antidepressant and antiinflammatory effects of rolipram in the central nervous system. *CNS Drug Rev.*, **7**(4):387–398.

95. Hatzelmann, A., et al. (2010) The preclinical pharmacology of roflumilast: a selective, oral phosphodiesterase 4 inhibitor in development for chronic obstructive pulmonary disease. *Pulm. Pharmacol. Ther.*, **23**(4):235–256.

96. Burgin, A.B., et al. (2010) Design of phosphodiesterase 4D (PDE4D) allosteric modulators for enhancing cognition with improved safety. *Nat. Biotechnol.*, **28**(1):63–70.

97. Lee, J.Y., et al. (2012) Ibudilast, a phosphodiesterase inhibitor with anti-inflammatory activity, protects against ischemic brain injury in rats. *Brain Res.*, **1431**:97–106.

98. Menniti, F.S., et al. (2009) Phosphodiesterase 5A inhibitors improve functional recovery after stroke in rats: optimized dosing regimen with implications for mechanism. *J. Pharmacol. Exp. Ther.*, **331**(3):842–850.

99. Wakita, H., et al. (2003) Ibudilast, a phosphodiesterase inhibitor, protects against white matter damage under chronic cerebral hypoperfusion in the rat. *Brain Res.*, **992**(1):53–59.

100. Serarslan, Y., et al. (2010) Protective effects of tadalafil on experimental spinal cord injury in rats. *J. Clin. Neurosci.*, **17**(3):349–352.

101. Myers, S.A., et al. (2012) Sildenafil improves epicenter vascular perfusion but not hindlimb functional recovery after contusive spinal cord injury in mice. *J. Neurotrauma*, **29**(3):528–538.

102. Park, S.J., et al. (2012) Resveratrol ameliorates aging-related metabolic phenotypes by inhibiting cAMP phosphodiesterases. *Cell*, **148**(3):421–433.

103. Simão, F., et al. (2012) Resveratrol prevents CA1 neurons against ischemic injury by parallel modulation of both GSK-3beta and CREB through PI3-K/Akt pathways. *Eur. J. Neurosci.*, **36**(7):2899–2905.

104. Singleton, R.H., et al. (2010) Resveratrol attenuates behavioral impairments and reduces cortical and hippocampal loss in a rat controlled cortical impact model of traumatic brain injury. *J. Neurotrauma*, **27**(6):1091–1099.

105. Nikulina, E., et al. (2004) The phosphodiesterase inhibitor rolipram delivered after a spinal cord lesion promotes axonal regeneration and functional recovery. *Proc. Natl. Acad. Sci. USA*, **101**(23):8786–8790.

106. Gao, Y., et al. (2003) Neurotrophins elevate cAMP to reach a threshold required to overcome inhibition by MAG through extracellular signal-regulated kinase-dependent inhibition of phosphodiesterase. *J. Neurosci.*, **23**(37):11770–11777.

107. Andersen, P.L., et al. (2000) Cyclic AMP prevents an increase in GAP-43 but promotes neurite growth in cultured adult rat dorsal root ganglion neurons. *Exp. Neurol.*, **166**(1): 153–165.

108. Han, P.J., et al. (2004) Cyclic AMP elevates tubulin expression without increasing intrinsic axon growth capacity. *Exp. Neurol.*, **189**(2):293–302.

109. Fouad, K., et al. (2009) Dose and chemical modification considerations for continuous cyclic AMP analog delivery to the injured CNS. *J. Neurotrauma*, **26**(5):733–740.

110. Basso, D.M., Beattie, M.S., and Bresnahan, J.C. (1995) A sensitive and reliable locomotor rating scale for open field testing in rats. *J. Neurotrauma*, **12**(1):1–21.

111. Wang, X., et al. (2006) Delayed Nogo receptor therapy improves recovery from spinal cord contusion. *Ann. Neurol.*, **60**(5):540–549.

112. Beaumont, E., et al. (2009) Effects of rolipram on adult rat oligodendrocytes and functional recovery after contusive cervical spinal cord injury. *Neuroscience*, **163**(4):985–990.

113. Costa, L.M., et al. (2013) Rolipram promotes functional recovery after contusive thoracic spinal cord injury in rats. *Behav. Brain Res.*, **243**:66–73.

114. Arvanian, V.L., et al. (2009) Chronic spinal hemisection in rats induces a progressive decline in transmission in uninjured fibers to motoneurons. *Exp. Neurol.*, **216**(2):471–480.

115. Nashmi, R. and Fehlings, M.G. (2001) Changes in axonal physiology and morphology after chronic compressive injury of the rat thoracic spinal cord. *Neuroscience*, **104**(1): 235–251.

116. Shi, R., Kelly, T.M., and Blight, A.R. (1997) Conduction block in acute and chronic spinal cord injury: different dose-response characteristics for reversal by 4-aminopyridine. *Exp. Neurol.*, **148**(2):495–501.

117. Koopmans, G.C., et al. (2009) Acute rolipram/thalidomide treatment improves tissue sparing and locomotion after experimental spinal cord injury. *Exp. Neurol.*, **216**(2): 490–498.

118. Prabhakar, U., et al. (1994) Characterization of cAMP-dependent inhibition of LPS-induced TNF alpha production by rolipram, a specific phosphodiesterase IV (PDE IV) inhibitor. *Int. J. Immunopharmacol.*, **16**(10):805–816.

119. Franks, M.E., Macpherson, G.R., and Figg, W.D. (2004) Thalidomide. *Lancet*, **363**(9423):1802–1811.

120. Iannotti, C.A., et al. (2011) A combination immunomodulatory treatment promotes neuroprotection and locomotor recovery after contusion SCI. *Exp. Neurol.*, **230**(1):3–15.

121. Goshgarian, H.G. (2009) The crossed phrenic phenomenon and recovery of function following spinal cord injury. *Respir. Physiol. Neurobiol.*, **169**(2):85–93.

122. Moreno, D.E., Yu, X.J., and Goshgarian, H.G. (1992) Identification of the axon pathways which mediate functional recovery of a paralyzed hemidiaphragm following spinal cord hemisection in the adult rat. *Exp. Neurol.*, **116**(3):219–228.

123. Kajana, S. and Goshgarian, H.G. (2009) Systemic administration of rolipram increases medullary and spinal cAMP and activates a latent respiratory motor pathway after high cervical spinal cord injury. *J. Spinal Cord. Med.*, **32**(2):175–182.

124. Kajana, S. and Goshgarian, H.G. (2008) Administration of phosphodiesterase inhibitors and an adenosine A1 receptor antagonist induces phrenic nerve recovery in high cervical spinal cord injured rats. *Exp. Neurol.*, **210**(2):671–680.

125. MacDonald, E., et al. (2007) A novel phosphodiesterase type 4 inhibitor, HT-0712, enhances rehabilitation-dependent motor recovery and cortical reorganization after focal cortical ischemia. *Neurorehabil. Neural. Repair.*, **21**(6):486–496.

126. Eilam, R., et al. (1998) Activity-dependent regulation of Neu differentiation factor/neuregulin expression in rat brain. *Proc. Natl. Acad. Sci. USA*, **95**(4):1888–1893.

127. Fernandez, P.A., et al. (2000) Evidence that axon-derived neuregulin promotes oligodendrocyte survival in the developing rat optic nerve. *Neuron*, **28**(1):81–90.

128. Radtke, C. and Vogt, P.M. (2009) Peripheral nerve regeneration: a current perspective. *Eplasty*, **9**:e47.

129. Gordon, T., Sulaiman, O., and Boyd, J.G. (2003) Experimental strategies to promote functional recovery after peripheral nerve injuries. *J. Peripher. Nerv. Syst.*, **8**(4):236–250.

130. Hirata, K. and Kawabuchi, M. (2002) Myelin phagocytosis by macrophages and non-macrophages during Wallerian degeneration. *Microsc. Res. Tech.*, **57**(6):541–547.

131. Zuo, J., et al. (2002) Regeneration of axons after nerve transection repair is enhanced by degradation of chondroitin sulfate proteoglycan. *Exp. Neurol.*, **176**(1):221–228.

132. Hoke, A., et al. (2002) A decline in glial cell-line-derived neurotrophic factor expression is associated with impaired regeneration after long-term Schwann cell denervation. *Exp. Neurol.*, **173**(1):77–85.

133. Gordon, T., et al. (2007) The potential of electrical stimulation to promote functional recovery after peripheral nerve injury: comparisons between rats and humans. *Acta Neurochir. Suppl.*, **100**:3–11.

134. Goldman, S. (2005) Stem and progenitor cell-based therapy of the human central nervous system. *Nat. Biotechnol.*, **23**(7):862–871.

135. Jain, K.K. (2009) Cell therapy for CNS trauma. *Mol. Biotechnol.*, **42**(3):367–376.

136. Cao, Q.L., et al. (2001) Pluripotent stem cells engrafted into the normal or lesioned adult rat spinal cord are restricted to a glial lineage. *Exp. Neurol.*, **167**(1):48–58.

137. Wichterle, H., et al. (2002) Directed differentiation of embryonic stem cells into motor neurons. *Cell*, **110**(3):385–397.

138. Okano, H., et al. (2003) Transplantation of neural stem cells into the spinal cord after injury. *Semin. Cell. Dev. Biol.*, **14**(3):191–198.

139. Harper, J.M., et al. (2004) Axonal growth of embryonic stem cell-derived motoneurons *in vitro* and in motoneuron-injured adult rats. *Proc. Natl. Acad. Sci. USA*, **101**(18): 7123–7128.

140. Deshpande, D.M., et al. (2006) Recovery from paralysis in adult rats using embryonic stem cells. *Ann. Neurol.*, **60**(1):32–44.

141. Kerr, D.A., et al. (2003) Human embryonic germ cell derivatives facilitate motor recovery of rats with diffuse motor neuron injury. *J. Neurosci.*, **23**(12):5131–5140.

142. Gordon, T., et al. (2009) Accelerating axon growth to overcome limitations in functional recovery after peripheral nerve injury. *Neurosurgery*, **65**(4 Suppl.):132–144.

143. Udina, E., et al. (2010) Rolipram-induced elevation of cAMP or chondroitinase ABC breakdown of inhibitory proteoglycans in the extracellular matrix promotes peripheral nerve regeneration. *Exp. Neurol.*, **223**(1):143–152.

INDEX

Adaptor protein, 69
Adenosine, 306
  EHNA , effects, 329
Adenosine triphosphate (ATP), 269
Adenylyl cyclases (ACs), 7, 59, 306
  AKAP-mediated clustering, 79
  Ca^{2+} sensitive/insensitive isoforms, 68
  membrane-bound, coupled to AKAP
    complexes, 68
A-kinase anchoring proteins (AKAPs), 20
  feature, 67
  local signal integration, 67
  nucleate signaling domains, 68
Allosteric inhibitors, 20
Allosteric modulators, 78, 100, 102
  interaction with UCR2 helix and active site
    of PDE4D, 163
  structures of large PDE fragments and
    implication on design of, 162–163
Allosteric regulation, 3, 84, 90
  of PDE activities, 3
Alzheimer's disease (AD), 18, 59, 85, 172,
  330

APP/PS1 double-transgenic mouse model,
  89, 278
brains examined by *in situ* hybridization,
  14
cholinergic neuron loss relevant in
  cognitive decline in, 330
development of PDE4 inhibitors, 289
effects of PDE5A inhibition and, 88
mouse model, 18
mRNA levels for PDE8B, 18
PDE5A inhibition, therapeutic utility in,
  88
PDE9A inhibitors for, 103
PDE9 target for treatment of symptoms of,
  190
PF-04447943 into clinical testing as
  cognitive therapy in, 330
rolipram reverses memory deficits
  produced by pharmacological
  agents, 179
γ-Aminobutyric acid (GABA) receptor, 59,
  191, 273, 304, 305
AMPA receptor, 68

Cyclic-Nucleotide Phosphodiesterases in the Central Nervous System: From Biology to Drug Discovery,
First Edition. Edited by Nicholas J. Brandon and Anthony R. West.
© 2014 John Wiley & Sons, Inc. Published 2014 by John Wiley & Sons, Inc.

Wiley Series in Drug Discovery and Development

Binghe Wang, Series Editor

Drug Delivery: Principles and Applications
Edited by Binghe Wang, Teruna Siahaan, and Richard A. Soltero

Computer Applications in Pharmaceutical Research and Development
Edited by Sean Ekins

Glycogen Synthase Kinase-3 (GSK-3) and Its Inhibitors: Drug Discovery and Development
Edited by Ana Martinez, Ana Castro, and Miguel Medina

Drug Transporters: Molecular Characterization and Role in Drug Disposition
Edited by Guofeng You and Marilyn E. Morris

Aminoglycoside Antibiotics: From Chemical Biology to Drug Discovery
Edited by Dev P. Arya

Drug-Drug Interactions in Pharmaceutical Development
Edited by Albert P. Li

Dopamine Transporters: Chemistry, Biology, and Pharmacology
Edited by Mark L. Trudell and Sari Izenwasser

Drug Design of Zinc-Enzyme Inhibitors: Functional, Structural, and Disease Applications
Edited by Claudiu T. Supuran and Jean-Yves Winum

ABC Transporters and Multidrug Resistance
Edited by Ahcene Boumendjel, Jean Boutonnat, and Jacques Robert

Kinase Inhibitor Drugs
Edited by Rongshi Li and Jeffrey A. Stafford

Evaluation of Drug Candidates for Preclinical Development: Pharmacokinetics, Metabolism, Pharmaceutics, and Toxicology
Edited by Chao Han, Charles B. Davis, and Binghe Wang

HIV-1 Integrase: Mechanism and Inhibitor Design
Edited by Nouri Neamati

Carbohydrate Recognition: Biological Problems, Methods, and Applications
Edited by Binghe Wang and Geert-Jan Boons

Chemosensors: Principles, Strategies, and Applications
Edited by Binghe Wang and Eric V. Anslyn

Medicinal Chemistry of Nucleic Acids
Edited by Li He Zhang, Zhen Xi, and Jyoti Chattopadhyaya

Oral Bioavailability: Basic Principles, Advanced Concepts, and Applications
Edited by Ming Hu and Xiaoling Li

Dendrimer-Based Drug Delivery Systems: From Theory to Practice
Edited by Yiyun Cheng

Plant Bioactives and Drug Discovery: Principles, Practice, and Perspectives
Edited by Valdir Cechinel-Filho

Cyclic-Nucleotide Phosphodiesterases in the Central Nervous System: From Biology to Drug Discovery
Edited by Nicholas J. Brandon and Anthony R. West